Lecture Notes in Computer Science   10585

Commenced Publication in 1973
Founding and Former Series Editors:
Gerhard Goos, Juris Hartmanis, and Jan van Leeuwen

More information about this series at http://www.springer.com/series/7409

Hujun Yin · Yang Gao
Songcan Chen · Yimin Wen
Guoyong Cai · Tianlong Gu
Junping Du · Antonio J. Tallón-Ballesteros
Minling Zhang (Eds.)

Intelligent Data Engineering and Automated Learning – IDEAL 2017

18th International Conference
Guilin, China, October 30 – November 1, 2017
Proceedings

 Springer

Editors
Hujun Yin
University of Manchester
Manchester
UK

Yang Gao
School of Electronic and Electrical
 Engineering
Nanjing University
Nanjiing, China

Songcan Chen
Nanjing University of Aeronautics
 and Astronautics
Nanjing, China

Yimin Wen
Guilin University of Electronic Technology
Guilin, China

Guoyong Cai
Guilin University of Electronic Technology
Guilin, China

Tianlong Gu
Guilin University of Electronic Technology
Guilin, China

Junping Du
Beijing University of Posts
 and Telecommunications
Beijing, China

Antonio J. Tallón-Ballesteros
University of Seville
Seville
Spain

Minling Zhang
Southeast University
Nanjing
China

ISSN 0302-9743   ISSN 1611-3349   (electronic)
Lecture Notes in Computer Science
ISBN 978-3-319-68934-0   ISBN 978-3-319-68935-7   (eBook)
https://doi.org/10.1007/978-3-319-68935-7

Library of Congress Control Number: 2017956071

LNCS Sublibrary: SL3 – Information Systems and Applications, incl. Internet/Web, and HCI

Printed on acid-free paper

This Springer imprint is published by Springer Nature
The registered company is Springer International Publishing AG
The registered company address is: Gewerbestrasse 11, 6330 Cham, Switzerland

Preface

In the era of big data and deep learning, the IDEAL conference has been playing an important role as an established forum for active, new, or leading researchers in the world to exchange the latest results and report new findings. The IDEAL conference has continued to devotedly serve the community over the last 19 years and has witnessed the ever fast-changing world of data science and machine learning. It has become one of the leading platforms for data-driven technology and learning algorithms with an emphasis on real-world problems and turning data into information, knowledge, and solutions. The IDEAL conference attracts international experts, new researchers, leading academics, practitioners, and industrialists from the communities of machine learning, computational intelligence, novel computing paradigms, data mining, knowledge management, biology, neuroscience, bio-inspired systems and agents, distributed systems, and robotics. It continues to evolve to embrace emerging topics and exciting trends.

This year IDEAL was held in one of most beautiful cities in mainland China, Guilin. The conference received 110 submissions, which were rigorously peer-reviewed by the Program Committee members and other experts. Only the papers judged to be of highest quality were accepted and included in the proceedings. This volume contains 65 papers accepted and presented at the 18th International Conference on Intelligent Data Engineering and Automated Learning (IDEAL 2017), held from October 30 to November 1, 2017, in Guilin, China. These papers provided a valuable and timely sample of the latest research outcomes in data engineering and automated learning, from methodologies, frameworks, and techniques to applications. They cover various topics such as evolutionary algorithms, deep learning neural networks, probabilistic modeling, particle swarm intelligence, big data analytics, and applications in image recognition, regression, classification, clustering, medical and biological modeling and prediction, text processing, and social media analysis. IDEAL 2017 also enjoyed stimulating keynotes from leaders in the field – Hojjat Adeli, Xiaoyang (Sean) Wang, and Xizhao Wang.

We would like to thank all the people who devoted so much time and effort to the successful running of the conference, in particular the members of the Program Committee and reviewers, as well as the authors who contributed to the conference. We are also very grateful for the hard work of the local organizing team at Guilin University of Electronic Technology, especially Prof. Yimin Wen, in local arrangements, as well as the help from Miss Yao Peng at the University of Manchester in checking through all the camera-ready files. The continued support and collaboration from Springer's LNCS team are also greatly appreciated.

August 2017

Hujun Yin
Yang Gao
Songcan Chen
Yimin Wen
Guoyong Cai
Tianlong Gu
Junping Du
Antonio J. Tallón-Ballesteros
Minling Zhang

Organization

Honorary Chair

Hojjat Adeli   Ohio State University, USA

General Chairs

Hujun Yin   University of Manchester, UK
Tianlong Guo   Guilin University of Electronic Technology, China
Yang Gao   Nanjing University, China

Program Co-chairs

Guoyong Cai   Guilin University of Electronic Technology, China
Songcan Chen   Nanjing University of Aeronautics and Astronautics, China
Junping Du   Beijing University of Posts and Telecommunications, China
Antonio   University of Seville, Spain
    J. Tallón-Ballesteros

International Advisory Committee

Lei Xu (Chair)   Chinese University of Hong Kong, Hong Kong, SAR China
Yaser Abu-Mostafa   CALTECH, USA
Shun-ichi Amari   RIKEN, Japan
Michael Dempster   University of Cambridge, UK
José R. Dorronsoro   Autonomous University of Madrid, Spain
Nick Jennings   University of Southampton, UK
Soo-Young Lee   KAIST, South Korea
Erkki Oja   Helsinki University of Technology, Finland
Latit M. Patnaik   Indian Institute of Science, India
Burkhard Rost   Columbia University, USA
Xin Yao   University of Birmingham, UK

Steering Committee

Hujun Yin (Chair)   University of Manchester, UK
Laiwan Chan (Chair)   Chinese University of Hong Kong, Hong Kong, SAR China
Guilherme Barreto   Federal University of Ceará, Brazil
Yiu-ming Cheung   Hong Kong Baptist University, Hong Kong, SAR China

Emilio Corchado	University of Burgos, Spain
Jose A. Costa	Federal University of Rio Grande do Norte, Brazil
Colin Fyfe	University of the West of Scotland, UK
Marc van Hulle	K.U. Leuven, Belgium
Samuel Kaski	Helsinki University of Technology, Finland
John Keane	University of Manchester, UK
Jimmy Lee	Chinese University of Hong Kong, Hong Kong, SAR China
Malik Magdon-Ismail	Rensselaer Polytechnic Institute, USA
Vic Rayward-Smith	University of East Anglia, UK
Peter Tino	University of Birmingham, UK
Zheng Rong Yang	University of Exeter, UK
Ning Zhong	Maebashi Institute of Technology, Japan

Publicity Co-chairs

Emilio Corchado	University of Salamanca, Spain
Jose A. Costa	Federal University of Rio Grande do Norte, Brazil

International Liaisons

Xiangbin Li	Guilin University of Electronic Technology, China
David Camacho	Universidad Autónoma de Madrid, Spain
Guilherme Barreto	Federal University of Ceará, Brazil
Brijesh Verma	Central Queensland University, Australia

Local Organizing Committee

Yimin Wen (Chair)	Guilin University of Electronic Technology, China
Minling Zhang (Chair)	Southeast University, China

Program Committee

Ajith Abraham	Vicent Botti	Luís Cavique
Paulo Adeodata	Juan A. Botía	Darryl Charles
Jesus Alcala-Fdez	Antonio Braga	Richard Chbeir
Davide Anguita	Fernando Buarque	Songcan Chen
Francisco Assis	Robert Burduk	Xiaohong Chen
Ángel Arcos-Vargas	Luiz Pereira Caloba	Sung-Bae Cho
Romis Attux	José Luis Calvo Rolle	Andrzej Cichocki
Javier Bajo Pérez	David Camacho	Jacek Cichosz
Bruno Baruque	Heloisa Camargo	Stelvio Cimato
Carmelo Bastos Filho	Anne Canuto	André Coelho
Lordes Borrajo	Andre Carvalho	Leandro Coelho

Rafael Corchuelo
Juan Cordero
Francesco Corona
Luís Correia
Paulo Cortez
Jose Alfredo F. Costa
Marcelo A. Costa
Raúl Cruz-Barbosa
Ernesto Cuadros-Vargas
Alfredo Cuzzocrea
Bogusław Cyganek
Ireneusz Czarnowski
Leandro Augusto Da Silva
Ernesto Damiani
Ajalmar Rêgo Darocha
    Neto
Bernard De Baets
Fernando Díaz
Weishan Dong
Jose Dorronsoro
Gérard Dreyfus
Adrião Duarte
Jochen Einbeck
Florentino Fdez-Riverola
Francisco Ferrer
Joaquim Filipe
Juan J. Flores
Gary Fogel
Pawel Forczmanski
Felipe M.G. França
Dariusz Frejlichowski
Hamido Fujita
Bogdan Gabrys
Marcus Gallagher
Matiaz Gams
Salvador Garcia
Ana Belén Gil
María José
    Ginzo-Villamayor
Fernando Gomide
Petro Gopych
Marcin Gorawski
Juan Manuel Górriz
Lars Graening
Manuel Graña
Maciej Grzenda
Jerzy Grzymala-Busse

Alberto Guillen
Juan Manuel Górriz
Barbara Hammer
Ioannis Hatzilygeroudis
Francisco Herrera
Álvaro Herrero
J. Michael Herrmann
James Hogan
Jaakko Hollmén
Vasant Honavar
Wei-Chiang Samuelson
    Hong
Anne Håkansson
Iñaki Inza
Konrad Jackowski
Vahid Jalali
Dariusz Jankowski
Vicente Julian
Ata Kaban
Miroslav Karny
Rheeman Kil
Sung-Ho Kim
Mario Koeppen
Joao E. Kogler Jr.
Andreas König
Rudolf Kruse
Lenka Lhotska
Bin Li
Clodoaldo A.M. Lima
Fernoando B. Lima Neto
Paulo Lisboa
Honghai Liu
Wenjian Luo
José Everardo B. Maia
José Manuel Benitez
Urszula Markowska
    Kaczmar
José F. Martínez
Giancarlo Mauri
José M. Molina
Susana Nascimento
Tim Nattkemper
Antonio Neme
Yusuke Nojima
Fernando Nuñez
Eva Onaindia
Chung-Ming Ou

Vasile Palade
Stephan Pareigis
Juan Pavón
Carlos Pedreira
Sarajane M. Peres
Javier Bajo Pérez
Jorge Posada
Paulo Quaresma
Izabela Rejer
Bernardete Ribeiro
José Riquelme
Ignacio Rojas
Fabrice Rossi
Regivan Santiago
Jose Santos
Javier Sedano
Ivan Silva
Dragan Simic
Anabela Simões
Michael Small
Ying Tan
Ke Tang
Ricardo Tanscheit
Dante Tapia
Peter Tino
Renato Tinós
Stefania Tomasiello
Pawel Trajdos
Carlos M.
    Travieso-González
Alicia Troncoso
Eiji Uchino
José Valente de Oliveira
Marley Vellasco
Alfredo Vellido
José R. Villar
Lipo Wang
Tzai-Der Wang
Wenjia Wang
Dongqing Wei
Michal Wozniak
Wu Ying
Du Zhang
Huiyu Zhou
Andrzej Zolnierek
Rodolfo Zunino

Additional Reviewers

Peter Boyd	Masaharu Hirota	Leandro Pasa
Diego de Siqueira Braga	Bangli Liu	Juan Rada-Vilela
Gaspare Bruno	Faouzi Mhamdi	Luis Rus-Pegalajar
Christoph Doell	Usue Mori	Daniel Sadoc Menasché
Karla Figueiredo	Paulo Oliveira	Fekade Getahun Taddesse

Special Session on Learning from Big Data, Streaming Data and Heterogeneous Multi-source Data: Algorithms, Models and Applications

Organizers

Ming Yang	Nanjing Normal University, China
Yang Gao	Nanjing University, China
Wensheng Zhang	Institute of Automation of Chinese Academy of Sciences, China
Wanqi Yang	Nanjing Normal University, China

Special Session on Finance and Data Mining

Organizers

Peter Mitic	Banco Santander, UK, and University College London, UK
Ángel Arcos-Vargas	University of Seville, Spain
Fernando Núñez Hernández	University of Seville, Spain
Antonio J. Tallón-Ballesteros	University of Seville, Spain

Special Session on Metaheuristics for Data Engineering

Organizers

Milan Tuba	John Naisbitt University, Serbia and State University of Novi Pazar, Serbia
Antonio J. Tallón-Ballesteros	University of Seville, Spain

Special Session on Crisp and Fuzzy Intelligent Systems

Organizers

Antonio J. Tallón-Ballesteros	University of Seville, Spain
Luís Correia	University of Lisbon, Portugal
Juan Rada-Vilela	FuzzyLite Limited, Wellington, New Zealand

Contents

XII   Contents

Learning Convolutional Ranking-Score Function by Query Preference Regularization

Guohui Zhang[1], Gaoyuan Liang[2], Weizhi Li[3], Jian Fang[4], Jingbin Wang[5], Yanyan Geng[6], and Jing-Yan Wang[7(✉)]

[1] Huawei Technologies Co., Ltd., Shanghai, China
[2] Jiangsu University of Technology, Jiangsu 213001, China
[3] Reflektion, Inc., San Mateo 94404, USA
[4] Auburn University, Auburn 36849, USA
[5] Information Technology Service Center,
Intermediate People's Court of Linyi City, Linyi, China
[6] Provincial Key Laboratory for Computer Information Processing Technology,
Soochow University, Suzhou 215006, China
[7] Jiangsu Key Laboratory of Big Data Analysis Technology/B-DAT,
Collaborative Innovation Center of Atmospheric Environment and Equipment
Technology, Nanjing University of Information Science and Technology,
Nanjing, China
jimjywang@gmail.com

Abstract. Ranking score plays an important role in the system of content-based retrieval. Given a query, the database items are ranked according to the ranking scores in a descending order, and the top-ranked items are returned as retrieval results. In this paper, we propose a new ranking scoring function based on the convolutional neural network (CNN). The ranking scoring function has a structure of CNN, and its parameters are adjusted to both queries and query preferences. The learning process guarantees that the ranking score of the query itself is large, and also the ranking scores of the positives (database items which the query wants to link) are larger than those of the negatives (database items which the query wants to avoid). Moreover, we also impose that the neighboring database items have similar ranking scores. An optimization problem is formulated and solved by Estimation-Maximization method. Experiments over the benchmark data sets show the advantage over the existing learning-to-rank methods.

Keywords: Convolutional neural network · Content-based retrieval · Ranking score · Estimation-maximization method

1 Introduction

In content-based retrieval problem, calculating the ranking scores is a critical process [17,21,26,27,33]. Recently, learning-to-rank has been proposed to learn

---

The study was supported by the open research program of the Jiangsu Key Laboratory of Big Data Analysis Technology/B-DAT, Nanjing University of Information Science and Technology, Nanjing, China, (Grant No. KBDat1602).

H. Yin et al. (Eds.): IDEAL 2017, LNCS 10585, pp. 1–8, 2017.
https://doi.org/10.1007/978-3-319-68935-7_1

machine learning models to calculate the ranking scores from the data [25]. Up to now, many learning-to-rank algorithms have been proposed. These algorithms used the manifold regularisation to learn the ranking scores directly, or use a linear model to estimate the ranking scores from the feature vectors from the data items of the database. For example, He et al. [32] constructed a neighbourhood graph, and use it to regularise the ranking scores. Yang et al. [31] adopted the local linear regression method to learn the ranking score. Burges et al. [1] proposed to use gradient descent methods to learn the ranking score estimation functions. Cao et al. [2] proposed to use the lists of objects as the training instances in the learning process of the ranking function, and a probabilistic framework is proposed for this purpose.

Recently, convolutional neural network (CNN) has been a popular data representation method for sequence data. It represents the sequence of instances by a four-layer neural network, including convolution layer, non-linear activation layer, max-pooling layer, and the full-connection layer. It has been approved to be successful in representation and classification of image, text, voice, and video data [5,19,24,28]. However, it is still unknown how the performance of CNN over the problem of learning-to-rank. In this paper, we study to use the CNN as the model of ranking score estimation function. Moreover, in the traditional learning-to-rank problem settings, the only input of the query is the database index of the query, indicating which data point in the database is the query. However, in real-world applications, more information from the query can be provided and utilised to learn the ranking scores. For example, Liu et al. [13] propose to use both the query itself and the query preferences over the database items to learn a optimal similarity matrix to measure the similarities between each pair of data points of the database. The query preferences are defined as a set of positive data items which the query wants to connect, and a set of negative data items which the query wants to avoid. Yang et al. [31] proposed to use a query to retrieval some initial top ranked data items from the database, and then ask the query to provide some feedback to identify some positive items from the initial top rated data items. The identified positive items are then used as the new set of queries to perform the ranking again.

In this paper, we propose a novel learning-to-rank method based on CNN and the query-specific preference information. Firstly, we proposed to build the ranking score estimation function (ranker) from the CNN model. Secondly, we proposed to used the query-specific preference information to regularise the learning of the CNN parameters. Given the data items of the positive set and the negative set specified by the query, we argue that the output of the CNN of a positive data point should be larger than that of a negative data point, and construct a hinge loss function to panellise the validations. Moreover, we also argue that the ranking scores of the CNN outputs should be smooth over the neighbourhoods of the data items of the database. To this end, we impose the ranking score of a data point is imposed to be close to that of its nearest neighbour. The nearest neighbour estimation is also based on the output of the max-pooling of the CNN, and probabilities of neighbours are used to regularise the learning of the ranking scores. We build a unified objective function to impose the query regularise, query-specific preference regularisation, and the neighbourhood regularisation. The optimisation problem is solved by a gradient decent algorithm to learn the parameters of the CNN model.

2   Proposed Method

We assume we have a database of n data objects, denoted as $\{X_1, \cdots, X_n\}$, and the i-th data object is a sequence of instances, $X_i = \{\mathbf{x}_{i1}, \cdots, \mathbf{x}_{im_i}\}$. To represent each data object, we use the CNN model. Our CNN model has four layers.

- **Sliding window layer:** We first use a sliding window of size w to split the sequences to a set of overlapping frames, $\{\mathbf{z}_{i1}, \cdots, \mathbf{z}_{i(m_i-w+1)}\}$, where $\mathbf{z}_{ij} = [\mathbf{x}_{ij}^\top, \cdots, \mathbf{x}_{i(j+w-1)}^\top]^\top$ is the j-th frame.
- **Convolutional layer:** Secondly we use a set of filters, $\mathbf{w}_1, \mathbf{w}_q$, to filter the frames. The outputs of the filtering regarding the k-th filter are

$$f_k(X_i) = \mathbf{w}_k^\top \mathbf{z}_{i1}, \cdots, \mathbf{w}_k^\top \mathbf{z}_{i(m_i-w+1)}. \tag{1}$$

- **Activation layer:** Thirdly, we apply an activation function to the outputs of filtering, $g(f_k(X_i)) = g(\mathbf{w}_k^\top \mathbf{z}_{i1}), \cdots, g(\mathbf{w}_k^\top \mathbf{z}_{i(m_i-w+1)})$, where $g(x) = 1/(1 + \exp(-x))$ is a non-linear activation function.
- **Max-pooling layer:** Fourthly, we perform the max-pooling to the activation outputs to select the maximum response of each filter, and the output regarding the k-th filter is given as

$$\max(g(f_k(X_i))) = \max_{j=1}^{m_i-w+1} g(\mathbf{w}_k^\top \mathbf{z}_{ij}). \tag{2}$$

- **Full-connection layer:** Finally, we apply a full-connection layer to the output of the max-pooling layer to obtain the ranking score of X_i, $s(X_i)$,

$$s(X_i) = \sum_{k=1}^{q} \omega_k \max(g(f_k(X_i))), \tag{3}$$

where $\omega_k, k = 1, \cdots, q$ are the weights of the full-connection layer.

In this CNN based ranker function, there are two groups of parameters, the filter set, $\mathbf{w}_1, \cdots, \mathbf{w}_q$, and the full-connection weights, $\omega_1, \cdots, \omega_q$. To learn the parameters, we use both the query object and the query preference information to regularize the learning of the ranking scores. We also use the outputs of the max-pooling layer to define the neighborhoods and use it to regularize the ranking scores. We propose the following optimization problem for the learning of the convolutional neural ranking scoring function,

$$
\min_{\Phi} \left\{ \frac{1}{2} \|\Phi\|_2^2 + C_1 \sum_{i=1}^{n} \phi_i \|s(X_i) - \lambda\|_2^2 + C_2 \sum_{i,j:X_i \in \mathcal{P}^+, X_j \in \mathcal{P}^-} \max(0, s(X_j) \right.
$$

$$
\left. -s(X_i) + 1) + C_3 \sum_{i=1}^{n} \sum_{j:X_j \in \mathcal{N}_i} P_{ij} \|s(X_i) - s(X_j)\|_2^2 \right\}. \tag{4}
$$

where Φ is the set of parameters of CNN, including the filters, and the full-connection weights. C_1, C_2 and C_3 are the weights of the regularisation terms. The first term $\|\Phi\|_2^2 = \sum_{k=1}^{q} \left(\omega_k^2 + \|\mathbf{w}_k\|_2^2 \right)$ is a ℓ_2 norm regularisation term over the parameters to avoid the over-fitting problem. This term can reduce the complexity of the CNN model. The second term is the query object regularisation. To indicate which object is query in the database, we define a query indicator vector, $\phi = [\phi_1, \cdots, \phi]^\top \in \{1, 0\}^n$, where $\phi_i = 1$ if the i-th object is query, and 0 otherwise. Naturally, we hope the ranking score of the query itself to be large, thus we propose to minimize $\min \sum_{i=1}^{n} \phi_i \|s(X_i) - \lambda\|_2^2$, where λ is large constant. The third term is the query preference regularisation term. We have two sets of reference data objects specified by the query. One set is the positive set, $\mathcal{P}^+$, which contains the data objects the query wants to link. The other set is the negative set, $\mathcal{P}^-$, which contains the data objects the query wants to avoid. Naturally, we hope any data objective in $\mathcal{P}^+$ has a larger score than of any data object in $\mathcal{P}^-$, $s(X_i) > s(X_j), \forall X_i \in \mathcal{P}^+, X_j \in \mathcal{P}^-$. We define a hinge loss function to panellise the validation of this condition,

$$
\ell(X_i, X_j) = \max(0, s(X_j) - s(X_i) + 1), \ \forall X_i \in \mathcal{P}^+, X_j \in \mathcal{P}^-. \tag{5}
$$

The hinge losses are applied over the positive-negative data object pairs. By minimising the losses of the validations over the query specified preference, we further regularise the learning of the CNN ranker to align it to the query preferences. The last term is the neighbourhood regularisation term. We hope the ranking scores of CNN can be smooth over the neighbourhoods of the data objects, thus we impose the ranking score of a data object to be close to its nearest neighbour as possible. The nearest neighbours are defined in the space of the outputs of the max-pooling layer of CNN. The output vector of this layer of X_i is given as $\mathbf{r}(X_i) = [\max(g(f_1(X_i))), \cdots, \max(g(f_q(X_i)))]^\top$. Since this space is also dependent on the CNN models, we proposed to firstly find the K-nearest neighbours of X_i using these representation vectors, $\mathcal{N}_i$, and then estimate the probability of the data objects in $\mathcal{N}_i$ of being the nearest neighbour. To this end, we first calculate the similarity between X_i and a data object in $\mathcal{N}_i$, X_j, $S_{ij} = \exp\left(-\frac{\|\mathbf{r}(X_i) - \mathbf{r}(X_j)\|_2^2}{2\sigma^2}\right)$, then calculate the probability by normalising the similarities,

$$
P_{ij} = \frac{S_{ij}}{\sum_{j':X_j' \in \mathcal{N}_i} S_{ij}}. \tag{6}
$$

We propose to minimise the expected squared ℓ_2 distance between the ranking score of X_i and X_j in $\mathcal{N}_i$, $\sum_{i=1}^{n} \sum_{j:X_j \in \mathcal{N}_i} P_{ij} \| s(X_i) - s(X_j) \|_2^2$.

The direct optimisation of the problem in (4) is difficult, due to the complexity of the objective function $O(\Phi)$. We use the Exception-Maximization method. In an iterative algorithm, we first fix the CNN parameters to estimate the frames which are selected by the max-pooling layer, and the positive-negative pairs which validates the condition of (5), and to find the nearest neighbour set and estimate the probabilities of being the nearest neighbour (E-step). Then we use the mini-batch method to update the parameters by using the gradient descent algorithm (M-step).

3   Experiments

3.1   Data Sets and Experimental Setting

To evaluate the proposed method, we conduct experiments over two benchmark data sets, including Caltech 101 data set [8], and INRIA Holidays Dataset [11]. The Caltech 101 dataset contains images of 101 classes. For each class, the number of images varies from 40 to 800. For most classes, the number of images is 50. The size of each is 300×300. To represent each image as a sequence of instances, we spilt the image to small image patches. The sliding winding is a square window. The INRIA Holidays dataset contains images of personal holiday photos. The images of this dataset belong to 500 different classes. In each class, there are one single query image, and the remaining images are the database images. There are 1,491 images in total.

To perform the experiments, we firstly split a data set to a database set and a query set. To do this, we use the 10-fold cross-validation. To obtain the query preference data, we randomly select some relevant and irrelevant items from the database as the input of our method. For each query, we use our method to learn the query-specific CNN ranker to estimate the ranking scores. The ranking scores are used to retrieve the database items. To measure the retrieval performances of the proposed methods, we use the recall, precision, the F-measure.

3.2   Results

In the experiments, we compared the proposed CNN based query-specific ranker (CNN-QSR) method against six stat-of-the-art learning-to-rank methods. The compared methods are Manifold Ranking (MR) [32], Local Regression and Global Alignment (LRGA) [31], Learning to Rank with Neural Nets (RankNet) [1], ListNet [2], and Query-Specific Optimal Networks (QUINT) [13]. The compression results over the benchmark datasets are shown in Fig. 1. According to the results reported in Fig. 1, the proposed method CNN-QSR outperforms the compared method over all two data sets and three performance measures. This is not surprising because this method uses the advanced data ranking and representation technology, CNN, and it uses the query preference information to improve the performance of the CNN model. Over the recall and precision performances, it seems the recall values are higher than the precisions. This indicates that for the retrieval problems over the three benchmark data sets, it is easier to obtain a good recall, but a good precision remains difficult to achieve. To have good precision, we still need to design a precision-targeted ranker based on CNN [6,9,10,14,17,18].

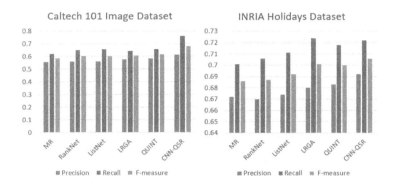

Fig. 1. Comparison results over benchmark data sets.

4 Conclusion

In this paper, we proposed a novel learning-to-rank method. We use the convolutional model to construct the ranking score estimation model. We also use the query-preference information to refine the learning of the convolutional model parameters. The query-preference regularisation is implemented by imposing the ranking scores of any positive item to be higher than that of the negative items. The experiments over three benchmark data sets show the advantage of the proposed method over the existing learning-to-rank methods. In the future, we will consider using the proposed method for problems beyond database retrieval, such as medical imaging [12,15,16,23], computer vision [3,4,20,22], and mobile networks [7,29,30].

References

1. Burges, C., Shaked, T., Renshaw, E., Lazier, A., Deeds, M., Hamilton, N., Hullender, G.: Learning to rank using gradient descent. In: Proceedings of the 22nd International Conference on Machine Learning, pp. 89–96. ACM (2005)
2. Cao, Z., Qin, T., Liu, T.Y., Tsai, M.F., Li, H.: Learning to rank: from pairwise approach to listwise approach. In: Proceedings of the 24th International Conference on Machine Learning, pp. 129–136. ACM (2007)

3. Chen, W., Ma, L., Shen, C.C.: Congestion-aware MAC layer adaptation to improve video telephony over Wi-Fi. ACM Trans. Multimedia Comput. Commun. Appl. (TOMM) **12**(5s), 83 (2016)
4. Chen, W., Ma, L., Sternberg, G., Reznik, Y.A., Shen, C.C.: User-aware dash over Wi-Fi. In: 2015 International Conference on Computing, Networking and Communications (ICNC), pp. 749–753. IEEE (2015)
5. Duan, Y., Liu, F., Jiao, L., Zhao, P., Zhang, L.: Sar image segmentation based on convolutional-wavelet neural network and markov random field. Patt. Recogn. **64**, 255–267 (2017)
6. Fan, J., Liang, R.Z.: Stochastic learning of multi-instance dictionary for earth mover's distance-based histogram comparison. Neural Comput. Appl. 1–11 (2016)
7. Fang, J., Lim, A., Yang, Q.: TOA ranging using real time application interface (RTAI) in IEEE 802.11 networks. In: Rodrigues, J.J.P.C., Zhou, L., Chen, M., Kailas, A. (eds.) GreeNets 2011. LNICSSITE, vol. 51, pp. 88–98. Springer, Heidelberg (2012). doi:10.1007/978-3-642-33368-2_8
8. Fei-Fei, L., Fergus, R., Perona, P.: One-shot learning of object categories. IEEE Trans. Patt. Anal. Mach. Intell. **28**(4), 594–611 (2006)
9. Geng, Y., Liang, R.Z., Li, W., Wang, J., Liang, G., Xu, C., Wang, J.Y.: Learning convolutional neural network to maximize pos@top performance measure. In: ESANN (2017)
10. Geng, Y., Zhang, G., Li, W., Gu, Y., Liang, G., Wang, J., Wu, Y., Patil, N., Wang, J.Y.: A novel image tag completion method based on convolutional neural network. In: International Conference on Artificial Neural Networks. Springer (2017)
11. Jegou, H., Douze, M., Schmid, C.: Hamming embedding and weak geometric consistency for large scale image search. In: Forsyth, D., Torr, P., Zisserman, A. (eds.) ECCV 2008. LNCS, vol. 5302, pp. 304–317. Springer, Heidelberg (2008). doi:10.1007/978-3-540-88682-2_24
12. King, D.R., Li, W., Squiers, J.J., Mohan, R., Sellke, E., Mo, W., Zhang, X., Fan, W., DiMaio, J.M., Thatcher, J.E.: Surgical wound debridement sequentially characterized in a porcine burn model with multispectral imaging. Burns **41**(7), 1478–1487 (2015)
13. Li, L., Yao, Y., Tang, J., Fan, W., Tong, H.: QUINT: on query-specific optimal networks. In: Proceedings of the 22nd ACM SIGKDD International Conference on Knowledge Discovery and Data Mining, pp. 985–994. ACM (2016)
14. Li, Q., Zhou, X., Gu, A., Li, Z., Liang, R.Z.: Nuclear norm regularized convolutional max pos@top machine. Neural Comput. Appl. 1–10 (2016)
15. Li, W., Mo, W., Zhang, X., Lu, Y., Squiers, J.J., Sellke, E.W., Fan, W., DiMaio, J.M., Thatcher, J.E.: Burn injury diagnostic imaging device's accuracy improved by outlier detection and removal. In: Algorithms and Technologies for Multispectral, Hyperspectral, and Ultraspectral Imagery XXI, vol. 9472 (2015)
16. Li, W., Mo, W., Zhang, X., Squiers, J.J., Lu, Y., Sellke, E.W., Fan, W., DiMaio, J.M., Thatcher, J.E.: Outlier detection and removal improves accuracy of machine learning approach to multispectral burn diagnostic imaging. J. Biomed. Opt. **20**(12), 121305 (2015)
17. Liang, R.Z., Shi, L., Wang, H., Meng, J., Wang, J.J.Y., Sun, Q., Gu, Y.: Optimizing top precision performance measure of content-based image retrieval by learning similarity function. In: 2016 23rd International Conference on Pattern Recognition (ICPR). IEEE (2016)

18. Liang, R.Z., Xie, W., Li, W., Wang, H., Wang, J.J.Y., Taylor, L.: A novel transfer learning method based on common space mapping and weighted domain matching. In: 2016 IEEE 28th International Conference on Tools with Artificial Intelligence (ICTAI), pp. 299–303. IEEE (2016)
19. Liu, Y., Chen, X., Peng, H., Wang, Z.: Multi-focus image fusion with a deep convolutional neural network. Inf. Fusion **36**, 191–207 (2017)
20. Ma, L., Chen, W., Veer, D., Sternberg, G., Liu, W., Reznik, Y.: Early packet loss feedback for WebRTC-based mobile video telephony over Wi-Fi. In: 2015 IEEE Global Communications Conference (GLOBECOM), pp. 1–6. IEEE (2015)
21. Ma, L., Liu, X., Gao, Y., Zhao, Y., Zhao, X., Zhou, C.: A new method of content based medical image retrieval and its applications to CT imaging sign retrieval. J. Biomed. Inform. **66**, 148–158 (2017)
22. Ma, L., Veer, D., Chen, W., Sternberg, G., Reznik, Y.A., Neff, R.A.: User adaptive transcoding for video teleconferencing. In: 2015 IEEE International Conference on Image Processing (ICIP), pp. 2209–2213. IEEE (2015)
23. Mo, W., Mohan, R., Li, W., Zhang, X., Sellke, E.W., Fan, W., DiMaio, J.M., Thatcher, J.E.: The importance of illumination in a non-contact photoplethysmography imaging system for burn wound assessment. In: Photonic Therapeutics and Diagnostics XI, vol. 9303 (2015)
24. Ren, X., Chen, K., Yang, X., Zhou, Y., He, J., Sun, J.: A novel scene text detection algorithm based on convolutional neural network. In: VCIpp, 2016–30th Anniversary of Visual Communication and Image Processing, p. 7805444 (2017). doi:10.1109/VCIP.2016.7805444
25. Tian, Q., Li, B.: Weakly hierarchical lasso based learning to rank in best answer prediction. In: Proceedings of the 2016 IEEE/ACM International Conference on Advances in Social Networks Analysis and Mining, ASONAM 2016, pp. 307–314 (2016)
26. Wang, H., Wang, J.: An effective image representation method using kernel classification. In: 2014 IEEE 26th International Conference on Tools with Artificial Intelligence (ICTAI 2014), pp. 853–858 (2014)
27. Xia, Z., Xiong, N., Vasilakos, A., Sun, X.: EPCBIR: an efficient and privacy-preserving content-based image retrieval scheme in cloud computing. Inf. Sci. **387**, 195–204 (2017)
28. Xu, J., Xu, B., Wang, P., Zheng, S., Tian, G., Zhao, J., Xu, B.: Self-taught convolutional neural networks for short text clustering. Neural Netw. **88**, 22–31 (2017)
29. Yang, Q., Lim, A., Li, S., Fang, J., Agrawal, P.: ACAR: adaptive connectivity aware routing protocol for vehicular ad hoc networks. In: Proceedings of 17th International Conference on Computer Communications and Networks, 2008, ICCCN 2008, pp. 1–6. IEEE (2008)
30. Yang, Q., Lim, A., Li, S., Fang, J., Agrawal, P.: ACAR: adaptive connectivity aware routing for vehicular ad hoc networks in city scenarios. Mob. Netw. Appl. **15**(1), 36–60 (2010)
31. Yang, Y., Nie, F., Xu, D., Luo, J., Zhuang, Y., Pan, Y.: A multimedia retrieval framework based on semi-supervised ranking and relevance feedback. IEEE Trans. Patt. Anal. Mach. Intell. **34**(4), 723–742 (2012)
32. Zhou, D., Weston, J., Gretton, A., Bousquet, O., Schölkopf, B.: Ranking on data manifolds. In: NIPS, vol. 3 (2003)
33. Zhu, L., Shen, J., Xie, L., Cheng, Z.: Unsupervised visual hashing with semantic assistant for content-based image retrieval. IEEE Trans. Knowl. Data Eng. **29**(2), 472–486 (2017)

Dynamic Community Detection Algorithm Based on Automatic Parameter Adjustment

Kai Lu[1,2], Xin Wang[1,2(✉)], and Xiaoping Wang[1,2]

[1] College of Computer Science, National University of Defense Technology, Changsha, People's Republic of China
{kailu,xiaopingwang}@nudt.edu.cn, dywangxin@foxmail.com
[2] Science and Technology on Parallel and Distributed Processing Laboratory, National University of Defense Technology, Changsha, People's Republic of China

Abstract. Community detection is widely used in social network analysis. It clusters densely connected vertices into communities. As social networks get larger, scalable algorithms are drawing more attention. Among those methods, the algorithm named Attractor is quite outstanding both in terms of accuracy and scalability. However, it is highly dependent on the parameter, which is abstract for users. The improper parameter value can bring about some problems. There can be a huge community (monster) sometimes; other time the communities are generally too small (fragments). The existing fragments also need eliminating. Such phenomenon greatly deteriorates the performance of Attractor. We modify the algorithm and propose mAttractor, which adjusts the parameter automatically. We introduce two constraints to limit monsters and fragments and to narrow the parameter range. An optional parameter is also introduced. The proposed algorithm can choose to satisfy or ignore the optional parameter by judging whether it is reasonable. Our algorithm also eliminates the existing fragments. A delicate pruning is designed for fast determination. Experiments show that our mAttractor outperforms Attractor by 2%–270%.

Keywords: Community detection · Social network · Data mining

1 Introduction

Community detection is to cluster nodes into communities so that nodes are densely connected inside communities [19]. It can reveal fundamentally related entities in various networks [7], and is widely used in statistics, biology, and computer science [4, 17].

There are various methods of community detection, but most of them are not satisfying either in accuracy or in scalability. The algorithm named Attractor [22] is outstanding in both areas. Nonetheless, Attractor highly depends on the parameter, which is quite abstract for users. The blindly determined parameter is likely to bring about some problems. There can be a huge community sometimes, which is usually called a monster; other time the communities are

H. Yin et al. (Eds.): IDEAL 2017, LNCS 10585, pp. 9–19, 2017.
https://doi.org/10.1007/978-3-319-68935-7_2

generally too small, which are named fragments. Moreover, whatever the parameter value is, the fragments always exist, which impact the performance unless properly eliminated. Such phenomenon makes it difficult for Attractor to achieve its satisfying performance.

We proposed a modified algorithm, mAttractor, which can adjust the parameter and eliminate the fragments. By two constraints it regulates the quality and the parameter range. It also eliminates the fragments and utilizes the requirement of users. Our main contributions are listed as follows:

1. We produce a mechanism to adjust the parameter automatically.
2. Our algorithm can limit the monsters and the fragments.
3. It can eliminate fragments and maintain the scalability.
4. Our algorithm can choose to satisfy or ignore the user-specified parameter by judging whether it will deteriorate result.

The rest of the paper is organised as follows. Section 2 gives the related works. Section 3 discusses the algorithm and Sect. 4 shows the experiments. The paper is concluded in Sect. 5.

2   Related Works

As social networks get larger, community detection is required to be both accurate and scalable. Most algorithms, however, are not always satisfying in either aspect. The Markov Cluster Algorithm (MCL) [24] is quite reasonable, but it suffers from scalability. The Label Propagation Algorithm (LPA) [15,20,26] is fast, but it is neither accurate nor stable. Algorithms based on artificial criteria, such as Modularity [18] and Normalized Cut [23], are influential because of their high performance. But they can lead to the resolution limit [5], in which case the small communities are neglected.

Among those methods, Attractor is outstanding both in terms of accuracy and scalability. It views the graph as a dynamical system, where nodes interact with each other. By the interactions, some nodes are moving nearer while others are further, and communities are detected then. Attractor is scalable, stable and accurate. Because it works naturally, it is free from the resolution limit.

Nonetheless, Attractor highly depends on a parameter λ, which determines the direction of influence of exclusive adjacent nodes. To determine λ, therefore, is abstract for users. Users may determine λ blindly, which can result in poor performance. Sometimes there can be a huge community (monster); other time too many communities become fragments which have only a few nodes. Even though we limit the percentage of fragments, some still remain. In a word, because the parameter is hard to determine and the fragments need eliminating, the algorithm performance is greatly deteriorated.

In our proposed mAttractor, λ is adjusted automatically and the fragments are eliminated while the scalability is maintained.

3   The Detailed Description of the Algorithm

This section firstly studies λ. Then we add constraints on the result. The fragments elimination is given then, followed by the entire algorithm.

3.1   The Influence of Parameter

We use the Lancichinetti-Fortunato-Radicchi (LFR) benchmark [12,13] to generate graphs with 200 nodes. The average degree ranges from 10 to 50. For each network, we repeatedly run Attractor with increasing λ, and study the community number. We also extend the range of λ from $[0, 1]$ to $[0, +\infty)$. In Fig. 1, as λ grows, the community number roughly increases and the average community size decreases. So we can adjust λ by such correlation. If there are only a few communities, we can increase λ; otherwise, we can decrease λ.

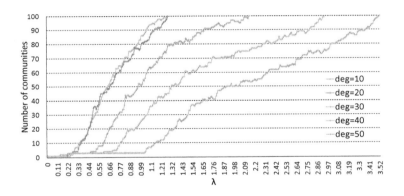

Fig. 1. The influence of λ

The theoretical analysis also demonstrates our study. The parameter λ controls the direction of the influence of exclusive adjacent nodes. The higher λ, the more repulsive forces and the less attractive forces. So the algorithm yields more communities if λ is higher and larger communities if λ is lower.

3.2   The Two Constraints

Because λ impacts the result, we introduce two constraints to regulate the result as well as λ. Firstly we define the undirected graph, which is the algorithm input.

Definition 1 (The Undirected Graph). *Let* $G = (V, E)$ *donate an undirected graph, where* $V = \{v_1, v_2 \ldots v_n\}$ *is the set of nodes,* $E = \{e_1, e_2 \ldots e_m\}$ *is the set of edges and* $e = \{v_i, v_j\} \in E$ *indicates an edge between* v_i *and* v_j.

After community detection, the graph is divided into parts and each node is mapping to its community.

Definition 2 (The Community Affiliation). *Given a graph $G = (V, E)$ and the set of community labels $L = \{l_1, l_2 \ldots l_k\}$, the community affiliation $C : V \rightarrow L$ maps the given node $v \in V$ to its community $C(v) \in L$.*

Then we can analyse the result. In case of fragments, we claim that the smallest community has three nodes. So smaller communities are viewed as fragments, whose nodes are outliers. The percentage of outliers reflects whether the result is generally fragmentary. We also study the largest community in case of a monster.

Definition 3 (The Fragment). *Given a graph $G = (V, E)$ and a community affiliation C, if the community $l_f \in L$ has less than 3 nodes, then it is called a fragment or a fragmentary community, and its nodes are called outliers.*

Definition 4 (The Fragment Rate). *Given a graph $G = (V, E)$ and a community affiliation C, suppose V_f is the set of all the outliers, then the fragment rate $R_f(G, C)$ is the percentage of outliers.*

$$R_f(G, C) = \frac{|V_f|}{|V|} \tag{1}$$

Definition 5 (The Monster Rate). *Given a graph $G = (V, E)$ and a community affiliation C, if the largest community is $l_m \in L$, then the monster rate $R_m(G, C)$ is the ratio of its size to the total number of nodes.*

$$R_m(G, C) = \frac{|\{v \in V | C(v) = l_m\}|}{|V|} \tag{2}$$

We introduce two constraints to limit the fragments and the monster.

1. Given a graph $G = (V, E)$ and a community affiliation C, the fragment rate $R_f(G, C)$ should be less than 25%.
2. Given a graph $G = (V, E)$ and a community affiliation C, the monster rate $R_m(G, C)$ should be less than 70%.

Then we can adjust λ by the constraints. Communities are generally fragmentary if Constraint 1 is violated, so λ is supposed to decrease. If Constraint 2 is violated, there is a monster community and we need to increase λ for more communities. In fact, the constraints regulate the upper and lower bound of λ, respectively. Each time one constraint is violated, mAttractor narrows the range of λ. We keep adjusting λ until both constraints are satisfied, or they are judged never to be satisfied at the same time.

3.3 The Elimination of Fragments

After we limit the fragments, we still need to eliminate the remaining fragments.

The key is to maintain the scalability of the algorithm. Inspired by the Label Propagation Algorithm (LPA) [20] which is linear to the graph size, we eliminate the outliers in descending order by the degree. In Algorithm 1, each outlier is appointed to the majority community in its neighbourhood. The majority community is the one with the most votes after each neighbour node votes for its community by the weight of degree.

We keep running Algorithm 1 until there are only isolated nodes remaining, which cannot be eliminated because they have no edge at all.

Algorithm 1. Elimination

1: **Input:** $G = (V, E)$, $C : V \rightarrow L$, fragment node list $Flist$;
2: Reorder the nodes in $Flist$ in descending order by the degree;
3: **for** each $v \in Flist$ **do**
4:     Find the neighbour set $\Gamma(v) = \{v' \in V | \{v, v'\} \in E\}$;
5:     Initialize the vote function $f(l) = 0, \forall l \in L$;
6:     **for** each $v' \in \Gamma(v)$ **do**
7:         Update the vote function $f(C(v')) = f(C(v')) + deg(C(v'))$;
8:     **end for**
9:     Find $l_{max} \in L$ with the maximum votes, randomly choose one if many;
10:    Update the affiliation $C(v) = l_{max}$;
11: **end for**
12: **Return:** the modified community affiliation C;

3.4   The Algorithm

Algorithm 2 shows the parameter adjustment, whose step halves if the direction toggles or λ steps out of the regulated range. Algorithm 3 shows mAttractor, which is designed to call Attractor and analyse the result repeatedly. After calling Attractor, we run Algorithm 1 and study $R_f(G, C)$ and $R_m(G, C)$. We consider the fragments before elimination when studying $R_f(G, C)$. We adjust λ, call Attractor and restart the analysis until the constraints are satisfied or they are discovered impossible to be met.

Then we adjust λ so that the community number is close to an optional parameter, $target$, which is the targeted number of communities. The algorithm will adjust λ until the number is within 10% more or less than $target$. Also, $target$ should satisfy the two constraints. The constraints regulate the range of the reasonable community number. If $target$ becomes out of the range, it violates the constraints and then becomes illegal. But as long as it is legal, it is influential. We do not encourage users to appoint $target$ unless they know the exact number. We set $target = 0$ if it is not appointed.

After the two constraints are satisfied, the algorithm terminates if the community number is close to the parameter $target$, or $target$ is illegal or not appointed. It also terminates when the adjustment step or the range of λ is smaller than the arbitrarily small positive quantity ϵ.

Algorithm 2. Change

1: **Input:** λ, $step$, dir, dir_{old}, $ubound_\lambda$, $lbound_\lambda$;
2: **if** $dir_{old} \neq dir$ **then**
3:　　 $step = step/2$;
4: **end if**
5: **if** $dir == `up'$ **then**
6:　　 $\lambda = \lambda + step$;
7:　　 **if** $\lambda \geq ubound_\lambda$ **then**
8:　　　　 $\lambda = (\lambda - step + ubound_\lambda)/2$;
9:　　　　 $step = step/2$;
10:　　 **end if**
11: **else**
12:　　 $\lambda = \lambda - step$;
13:　　 **if** $\lambda \leq lbound_\lambda$ **then**
14:　　　　 $\lambda = (\lambda + step + lbound_\lambda)/2$;
15:　　　　 $step = step/2$;
16:　　 **end if**
17: **end if**
18: **Return:** λ, $step$, dir;

4　Experiments

4.1　Experiment Setup

We choose ten networks for experiments, as shown in Table 1. Six networks have ground truth, including the Karate [25], the Dolphins [16], the Lesmis [11], the Mexican [3,6], the Polblogs [1] and the Company [2]. The ground truth can give the correct result and the number $target$. There are also four graphs without ground truth, including the Email [9], the Jazz [8], the Ca-GrQc [14] and the Ca-HepTh [14]. Circles are removed from those graphs.

We use Rand Index (RI) [10,21] and Adjusted Rand Index (ARI) [10,21] to study the accuracy performance on networks with ground truth. Then we use all networks to study Modularity [18], an influential quality measurement of community detection. At last, we study the scalability of the algorithm.

In Attractor, $\lambda \in [0,1]$, so we set $\lambda = 0.3, 0.5, 0.7$ to represent the blind parameter determination and the best performance of the three is viewed as Attractor's performance. Our mAttractor runs in two modes. In the targeted mode, $target$ equals to the real number of communities; in the untargeted mode, $target$ is not appointed and equals to 0. For graphs without ground truth, mAttractor only runs in the untargeted mode. The arbitrarily small positive quantity $\epsilon = 0.01$.

4.2　Accuracy Performance

The accuracy performance is shown in Fig. 2. In the Karate and the Dolphins, Attractor produces a monster when $\lambda = 0.3, 0.5$. ARI is sensitive to such terrible

Algorithm 3. mAttractor

1: **Input:** $G = (V, E)$, $target$, ϵ;
2: $dir_{old} = {}'up'$, $\lambda = 0.5$, $step = 0.25$;
3: $lbound_\lambda = 0$, $ubound_\lambda = \infty$, $lbound_c = 1$, $ubound_c = |V|/3$;
4: **while** $(step \geq \epsilon$ and $(ubound_\lambda - lbound_\lambda) \geq \epsilon)$ **do**
5:     Call $C = Attractor(G, \lambda)$;
6:     Store outliers into $Flist$ and $Flist_{original}$, $Flist_{old} = V$;
7:     **while** $|Flist| < |Flist_{old}|$ **do**
8:         $C = Elimination(G, C, Flist)$;
9:         $Flist_{old} = Flist$, store all outliers into $Flist$;
10:     **end while**
11:     $n = $ the number of communities in C;
12:     **if** $|Flist_{original}| \geq 0.25|V|$ **then**
13:         $ubound_\lambda = min(ubound_\lambda, \lambda)$, $ubound_c = min(ubound_c, n)$;
14:         $\lambda, step, dir_{old} = Change(\lambda, step, {}'down', dir_{old}, ubound_\lambda, lbound_\lambda)$;
15:         **continue**;
16:     **end if**
17:     **if** $R_m(G, C) \geq 0.7$ **then**
18:         $lbound_\lambda = max(lbound_\lambda, \lambda)$, $lbound_c = max(lbound_c, n)$;
19:         $\lambda, step, dir_{old} = Change(\lambda, step, {}'up', dir_{old}, ubound_\lambda, lbound_\lambda)$;
20:         **continue**;
21:     **end if**
22:     **if** (not $(lbound_c < target < ubound_c)$ or $(0.9target < n < 1.1target))$ **then**
23:         **break**;
24:     **end if**
25:     **if** $target > n$ **then**
26:         $\lambda, step, dir_{old} = Change(\lambda, step, {}'up', dir_{old}, ubound_\lambda, lbound_\lambda)$;
27:     **else**
28:         $\lambda, step, dir_{old} = Change(\lambda, step, {}'down', dir_{old}, ubound_\lambda, lbound_\lambda)$;
29:     **end if**
30: **end while**
31: **Output:** the community affiliation C;

---

results, and it is either zero or minus. It works well when $\lambda = 0.7$. But mAttractor outperforms Attractor by 6% in RI and 13% in ARI on the Karate, and 9% in RI and 156% in ARI on the Dolphins. Particularly, mAttractor achieves a perfectly correct result on the Karate. For the Lesmis, Attractor works well when $\lambda = 0.5, 0.7$, and mAttractor is better in the targeted mode. For the Polblogs, Attractor yields a monster when $\lambda = 0.3$ and too many fragments when $\lambda = 0.7$. It is satisfying when $\lambda = 0.5$, but mAttractor still outperforms Attractor by 132% in ARI and 14% in RI. Things are similar for the Mexican and the Company. Attractor always produces a huge monster, while mAttractor performs 25% better in RI on the Mexican and 270% better in RI on the Company.

Our mAttractor runs in two modes, but the result is usually the same. Sometimes the targeted mode can be better (the Mexican and the Lesmis). The algorithm can work well without $target$, the parameter can produce better results.

Table 1. Basic information of datasets

Dataset	#Vertices	#Edges	#Communities
Karate	34	78	2
Dolphins	62	159	2
Lesmis	77	254	11
Mexican	35	117	2
Polblogs	1224	19087	2
Company	77	2326	4
Email	1133	10902	/
Jazz	198	5484	/
Ca-GrQc	5241	28968	/
Ca-Hepth	9875	51946	/

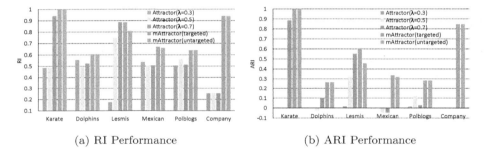

(a) RI Performance   (b) ARI Performance

Fig. 2. The accuracy performance

4.3 Modularity Performance

The Modularity performance is shown in Fig. 3. For the first three graphs, Attractor performs well when $\lambda = 0.7$, but mAttractor outperforms Attractor by 2% on the Karate, 17% on the Dolphins, 7% (targeted) and 3% (untargeted) on the Lesmis. As for the Mexican, mAttractor performs much better than Attractor. Attractor performs well on the Polblogs when $\lambda = 0.5$, while mAttractor performs a little better. In the Company, although Attractor has higher Modularity, such performance is meaningless because of the monster it produces.

For the networks without ground truth, we cannot appoint *target*, so we only consider the untargeted mode. In the Email, mAttractor performs 6% better. In the Jazz, Attractor produces a huge monster when $\lambda = 0.3$ and 0.5 but performs well when $\lambda = 0.7$. Our mAttractor performs 3% better by eliminating the fragments. As for the Ca-GrQc and the Ca-Hepth, Attractor performs well most of the time, but mAttractor still outperforms Attractor by 3%–4%.

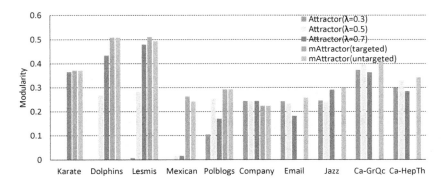

Fig. 3. The modularity performance

4.4 Scalability Study

We use LFR benchmark to generate networks for scalability study. The node size ranges from 2000 to 30000, and the average degree is set to 10. For Attractor, $\lambda = 0.5$; for mAttractor, it works in the untargeted mode.

In Fig. 4, we can see that mAttractor is scalable with its linear time complexity. It calls Attractor repeatedly, so it is slower. However, mAttractor achieves a better result than the three results of Attractor ($\lambda = 0.3, 0.5, 0.7$), so we claim such cost is acceptable.

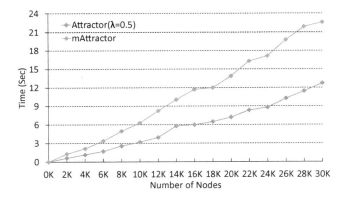

Fig. 4. The running time

5   Conclusion

We propose mAttractor to adjust the parameter based on the current result. Introduced constraints can guarantee the quality and narrow the range of λ. It also eliminates the fragments to optimize the result. It can also choose to utilize the user-specified parameter by judging whether it is reasonable. Experiments demonstrate mAttractor can achieve higher accuracy and Modularity, with a linear time complexity.

Acknowledgement. This work is partially supported by The National Key Research and Development Program of China (2016YFB0200401), by program for New Century Excellent Talents in University, by National Science Foundation (NSF) China 61402492, 61402486, 61379146, by the laboratory pre-research fund (9140C810106150C81001).

References

1. Adamic, L.A., Glance, N.: The political blogosphere and the 2004 US election: divided they blog. In: Proceedings of the 3rd International Workshop on Link Discovery, pp. 36–43. ACM (2005)
2. Cross, R., Parker, A., Christensen, C.M., Anthony, S.D., Roth, E.A.: The hidden power of social networks. J. Appl. Manag. Entrepreneurship **9** (2004)
3. De Nooy, W., Mrvar, A., Batagelj, V.: Exploratory Social Network Analysis with Pajek, vol. 27. Cambridge University Press, Cambridge (2011)
4. Fortunato, S.: Community detection in graphs. Phys. Rep. **486**(3), 75–174 (2010)
5. Fortunato, S., Barthelemy, M.: Resolution limit in community detection. Proc. Nat. Acad. Sci. U.S.A. **104**(1), 36–41 (2007)
6. Gil-Mendieta, J., Schmidt, S.: The political network in Mexico. Soc. Netw. **18**(4), 355–381 (1996)
7. Girvan, M., Newman, M.E.J.: Community structure in social and biological networks. Proc. Nat. Acad. Sci. U.S.A. **99**(12), 7821–7826 (2002)
8. Gleiser, P.M., Danon, L.: Community structure in jazz. Adv. Complex Syst. **6**(04), 565–573 (2003)
9. Guimera, R., Danon, L., Diaz-Guilera, A., Giralt, F., Arenas, A.: Self-similar community structure in a network of human interactions. Phys. Rev. E **68**(6), 065103 (2003)
10. Hubert, L., Arabie, P.: Comparing partitions. J. Classif. **2**(1), 193–218 (1985)
11. Knuth, D.E.: The Stanford GraphBase: A Platform for Combinatorial Computing, vol. 37. Addison-Wesley, Reading (1993)
12. Lancichinetti, A., Fortunato, S.: Benchmarks for testing community detection algorithms on directed and weighted graphs with overlapping communities. Phys. Rev. E Stat. Nonlinear Soft Matter Phys. **80**(1), 016118 (2009)
13. Lancichinetti, A., Fortunato, S., Radicchi, F.: Benchmark graphs for testing community detection algorithms. Phys. Rev. E Stat. Nonlinear Soft Matter Phys. **78**(4), 046110 (2008)
14. Leskovec, J., Kleinberg, J., Faloutsos, C.: Graph evolution: densification and shrinking diameters. ACM Trans. Knowl. Discov. Data (TKDD) **1**(1), 2 (2007)
15. Leung, I.X.Y., Hui, P., Lio, P., Crowcroft, J.: Towards real-time community detection in large networks. Phys. Rev. E Stat. Nonlinear Soft Matter Phys. **79**(6), 066107 (2009)
16. Lusseau, D., Schneider, K., Boisseau, O.J., Haase, P., Slooten, E., Dawson, S.M.: The bottlenose dolphin community of doubtful sound features a large proportion of long-lasting associations. Behav. Ecol. Sociobiol. **54**(4), 396–405 (2003)
17. Mislove, A., Marcon, M., Gummadi, K.P., Druschel, P., Bhattacharjee, B.: Measurement and analysis of online social networks. In: Proceedings of the 7th ACM SIGCOMM Conference on Internet Measurement, pp. 29–42. ACM (2007)

18. Newman, M.E.J.: Modularity and community structure in networks. Proc. Nat. Acad. Sci. U.S.A. **103**(23), 8577–8582 (2006)
19. Porter, M.A., Onnela, J.P., Mucha, P.J.: Communities in networks. Not. Am. Math. Soc. **56**(9), 4294–4303 (2009)
20. Raghavan, U.N., Albert, R., Kumara, S.: Near linear time algorithm to detect community structures in large-scale networks. Phys. Rev. E Stat. Nonlinear Soft Matter Phys. **76**(3), 036106 (2007)
21. Rand, W.M.: Objective criteria for the evaluation of clustering methods. J. Am. Stat. Assoc. **66**(336), 846–850 (1971)
22. Shao, J., Han, Z., Yang, Q., Zhou, T.: Community detection based on distance dynamics. In: Proceedings of the 21st ACM SIGKDD International Conference on Knowledge Discovery and Data Mining, pp. 1075–1084. ACM (2015)
23. Shi, J., Malik, J.: Normalized cuts and image segmentation. IEEE Trans. Pattern Anal. Mach. Intell. **22**(8), 888–905 (2000)
24. Van Dongen, S.M.: Graph clustering by flow simulation. Ph.D. thesis, University of Utrecht (2001)
25. Zachary, W.W.: An information flow model for conflict and fission in small groups. J. Anthropol. Res. **33**(4), 452–473 (1977)
26. Zhang, X.-K., Fei, S., Song, C., Tian, X., Ao, Y.-Y.: Label propagation algorithm based on local cycles for community detection. Int. J. Mod. Phys. B **29**(05), 1550029 (2015)

An Ant Colony Random Walk Algorithm for Overlapping Community Detection

TianRen Ma, Zhengyou Xia[✉], and Fan Yang

College of Computer Science and Technology,
Nanjing University of Aeronautics and Astronautics, Nanjing 210015, China
zhengyou_xia@nuaa.edu.cn

Abstract. Discovery of communities is a very effective way to understand the properties of complex networks. An improved ant colony algorithm based on random walk has been proposed in this paper. Inspired by the framework proposed in AntCBO, firstly, a list of node importance is obtained through calculation. The nodes in the network will be sorted in descending order of importance. Secondly, on the basis of random walk, a matrix is constructed to measure the similarity of nodes and we can use this matrix and pheromone to get the heuristic information. Thirdly, an improved ant's location discovery strategy is proposed. After the movement of ants, every node will keep a list of labels and the proposed post processing will give the result of overlapping community detection. Finally, a test in real-world networks is given. The result shows that this algorithm has better performance than existing methods in finding overlapping community structure.

Keywords: Overlapping community detection · Ant colony · Random walk

1 Introduction

In the real world, there are a variety of complex systems and these systems can often be expressed in the form of the network to facilitate the analysis and management [1,2]. As we know, a potential attribute in networks is community structure.

It can be seen that community division is a hot topic and a lot of methods have been proposed in recent years. Most of them focus on disjoint communities. Nevertheless, communities often overlap in real networks. In 2009, Local Fitness Maximization [3] (LFM) was proposed by Lancichinetti et al., which is a hierarchical overlapping community detection method and can identify a community from any starting node. Gregory proposed the COPRA [4] in 2010, and it was the first time using the label propagation method to solve the problem of overlapping community detection. SLPA [5] was proposed by Jierui Xie et al. in 2011. It is also an overlapping community discovery algorithm. LinkLPA [6] transforms node partition problem into link partition problem and propagate labels on links instead of nodes to detect communities. AntCBO [7] was proposed by X Zhou et al. in 2015. This algorithm extended ACO to solve the overlapping community detection problem and redefined a heuristic formula to measure the similarity between nodes. The ant colony random walk algorithm (ACRWA)

© Springer International Publishing AG 2017
H. Yin et al. (Eds.): IDEAL 2017, LNCS 10585, pp. 20–26, 2017.
https://doi.org/10.1007/978-3-319-68935-7_3

proposed in this article is based on the concept of random walk [8] and the framework proposed in AntCBO.

The remainder of this paper is organized as follows. The improved ant colony initialization process and definition mentioned in this paper are described in Sect. 2. In Sect. 3, an ant colony random walk algorithm (ACRWA) is particularly studied, followed by experiments in Sect. 4. Finally, we conclude the paper in Sect. 5.

2   The Improved Ant Colony Initialization Process and Definition

In this section, we formalize the definition of node importance used in the paper and briefly describe the improved ant colony initialization process.

2.1   Definitions

Inspired by the PageRank algorithm [9], a new method to measure the importance of nodes is designed.

Definition 1 (Node Importance). Given a network $G = \{V, E\}$, for any node $v_i \in V$, the node importance is:

$$\mathrm{NI}(v_i) = \sum_{v_i' \in V_i'} \frac{1}{d(v_i')} \tag{1}$$

V_i' is the neighbor node set of node v_i and $d(v_i')$ is the degree of node v_i'. The importance of a node v_i depends on the number and degree of its neighbor nodes. If the degree of the neighbor node is larger, the contribution to the importance of the node v_i is relatively smaller and vice versa.

2.2   The Description of the Improved Ant Colony Initialization Process

The specific initialization algorithm is described as follows:

1. For every node *node_i* in V which match the attribute *available* of it is True and the degree of it is no less than two, running (2)–(5).
2. Finding the most important neighbors of *node_i*, recording it as *most_impo_neigs*.
3. Finding the first available node *node_j* in *most_impo_neigs*. If no one is available in *most_impo_neigs*, return to (1).
4. Setting the label of *node_i* and *node_j* as the id of *node_i*; setting the attribute *available* of these two nodes as False. Recording *common_neigs* as the common neighbors of these two nodes.
5. If the set *common_neigs* isn't equal to null, finding the first available node *node_com* for every node in *common_neigs*. Setting the label of it as the id of *node_i*; setting the *available* of it as False. Setting the *common_neigs* as intersection of neighbors

of *node_com* and *common_neigs*. Return to the beginning of (5). These nodes with the same label are regarded as ants in the same colony.

3   ACRWA: An Ant Colony Random Walk Algorithm for Overlapping Community Detection

3.1   Concept and Calculation of Random Walk

In literature [10], a local random walk (LRW) is proposed, which is similar to the one with low time complexity. For the convenience of description, we use P_{xy} to represent the probability that the random walker can move to the node y from node x in a step, and use $\pi_{xy}(t)$ to denote the probability that the random walker can move to node y from node x in t steps. Let $\pi_x(t)$ denote the x-th row vector of the matrix $\pi(t)$. P_{xy} and $\pi_x(t)$ are calculated as follows:

$$P_{xy} = \frac{a_{xy}}{k_x} \tag{2}$$

$$\pi_x(t) = \pi_x(t-1)P \tag{3}$$

When there is an edge between node x and y, the value of a_{xy} is 1. Otherwise, the value of a_{xy} is 0. The value of k_x is the degree of node x. The initial value of $\pi_x(t)$ is P. The local random walk is defined as follow:

$$S_{xy}^{OLRW}(t) = k_x \cdot \pi_{xy}(t) + k_y \cdot \pi_{yx}(t) \tag{4}$$

But due to the randomness of random walk, random walker may mistakenly go to another node in other community. To solve this problem, the specific definition is as follow:

$$S_{xy}^{OSRW}(t) = \sum_{i=1}^{t} S_{xy}^{OLRW}(i) \tag{5}$$

In this paper, we choose $S_{xy}^{OSRW}(t)$ as the measure of node similarity and the heuristic information for ants. The parameter t is set to 4.

3.2   Description of ACRWA

ACRWA is described as follows:

(1)   Initializing nodes in the network and appending its ID to its label list; initializing edges and setting its pheromones to 10. Setting the maximum number of iterations T = 200, pheromone evaporation rate ρ = 0.1, pheromone increment a = 0.1, pheromone threshold value b = 15, θ for post-processing is equal to 0.4.

(2) Calculating the importance of each node, calculating the similarity matrix S_{xy}^{OSRW}; Quickly sorting the node set N according to the node importance;
(3) Running the process mentioned in Sect. 2.2.
(4) Spreading labels by ants. In this stage, every ant will spread label from current node i to next node j based on SP(i,j). Recording the pheromone on edge (i,j) as phe(i,j). We can obtain the value of SP(i,j) according to the following formula:

$$SP(i,j) = S_{ij}^{OSRW}(t) * phe(i,j) \qquad (6)$$

If a random number rn < 0.1, then the ant will select a neighbor which has the largest SP as the destination node. Otherwise, the ant selects the destination node randomly. This strategy will let ants traverse around the whole network. When the next node has been determined, ant will choose a label that appears most frequently from its label list to bring to the next node. If there are more than one label having the maximal frequency, ants will select one of them randomly.

After spreading, ants will deposit pheromone on the edge they pass by. As a result, the pheromone on this edge will increased by a. To avoid premature convergence of this process to non-global optimal solution, when the pheromone is larger than b, it will be set to b. This operation can effectively prevent the algorithm from trapping in local optimal. After all ants have moved, for every edge in E, the pheromone of it will evaporate. This means at the end of every spreading stage, for every edge (i,j) in E, the following assignment statement is executed:

$$phe(i,j) = (1 - \rho) * phe(i,j) \qquad (7)$$

(5) Post processing.

The post-processing procedure contains two steps: dealing with over-overlapping nodes and deleting too small communities. The first step deals with the over-overlapping problem and we adopt the solution mentioned in LinkLPA. But after each delete operation, we will normalize the label list of nodes which is different from the solution in LinkLPA. After dealing with over-overlapping problem, some labels whose frequency are smaller than θ will be deleted.

4   Experiments Result

In this section, the performance of ACRWA is tested on standard data sets which are commonly used in the real world.

4.1   Data Source

There are many real-world networks that have been abstracted into community detection fields, such as Zachary's Karate Club, which is constructed by observing an American university karate club. We also choose the Dolphin social network. It is the dolphin social network that D. Lusseau et al. used for seven years to observe the dolphin population of Doubtful Sound, New Zealand. The third network we choose is Polbooks, which was created by V. Krebs. In this section, we use these three networks to perform community partitioning experiments. The detail parameters of these networks are shown in Table 1.

Table 1. Real-world networks

Name	Nodes	Edges	Average degree
Karate	34	78	4.59
Dolphins	62	159	5.13
Polbooks	105	441	8.40

4.2 Experimental Results

We use total seven different algorithms including ACRWA for overlapping community detection. Each algorithm is repeated 100 times for each network to obtain more efficient values of modularity, eliminating the effects of random selection in the algorithm.

The first experiment was test on Zachary network. The overlapping modularity is shown as Fig. 1. ACRWA obtains the best Q_{ov} value of 0.7348. AntCBO, SLPA and LinkLPA which were proposed in recent years only obtain the Q_{ov} value of 0.7243, 0.6541 and 0.5010, respectively. COPRA and LFM performed not very well on this network. The overlapping modularity of these two algorithms does not exceed 0.5.

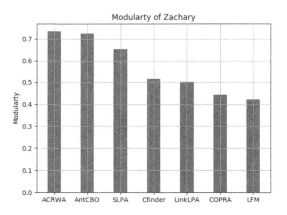

Fig. 1. Modularity comparison for the Zachary's Karate network

The second experiment was run on dolphins network. The overlapping modularity is shown in Fig. 2.

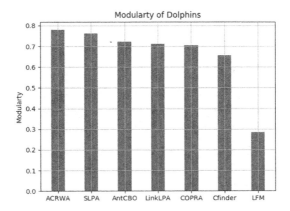

Fig. 2. Modularity comparison for the Dolphins social network

On dolphins social network, ACRWA has obtained the highest modularity value of 0.7795. SLPA algorithm also has a nice performance. The Q_{ov} value of AntCBO is 0.7227. LinkLPA and COPRA also have a not bad performance.

The last experimental network is polbooks network. The overlapping modularity is shown in Fig. 3. We can find that all of these algorithms have a high modularity value. Top four of them are SLPA, ACRWA, COPRA and AntCBO. The value of them that we obtained are 0.8307, 0.8228, 0.8218 and 0.8211, respectively. Based on this result on polbooks network, a conclusion that polbooks network has a obvious community structure is accepted.

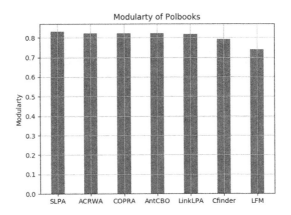

Fig. 3. Modularity comparison for the Polbooks network

5   Summary and Conclusion

In this paper, an ant colony random walk algorithm based on ACO and random walk is proposed to solve the problem of overlapping community partitioning, and we compare

ACRWA with SLPA, COPRA, LFM, Cfinder, LinkLPA and AntCBO by using three real-world networks.

The ACRWA improves the ant colony initialization operation in AntCBO algorithm and can detect overlapping communities structure. In our algorithm, nodes are sorted by computing the importance of them and every ant only belongs to one colony. The initialization process proposed in this paper can avoid the phenomenon of some colonies may overlapping. A new heuristic information was proposed, which we can used to make ants more intelligent and guide ants to move correctly when pheromones have not yet accumulated.

The experiment in this paper is divided into three real-world networks. The value of overlapping modularity is used to evaluate the performance. Experimental results of these three networks show that the algorithm proposed in this paper has achieved better experimental results.

References

1. Barabási, A.L., Crandall, R.E.: The new science of networks. Phys. Today **6**(5), 243–270 (2003)
2. Newman, M.E.J.: The structure and function of complex networks. Siam Rev. **45**(2), 167–256 (2006)
3. Lancichinetti, A., Fortunato, S., Kertész, J.: Detecting the overlapping and hierarchical community structure of complex networks. New J. Phys. **11**(3), 19–44 (2008)
4. Gregory, S.: Finding overlapping communities in networks by label propagation. New J. Phys. **12**(10), 2011–2024 (2009)
5. Xie, J., Szymanski, B.K., Liu, X.: SLPA: uncovering overlapping communities in social networks via a speaker-listener interaction dynamic process. In: IEEE, International Conference on Data Mining Workshops IEEE Computer Society, pp. 344–349 (2011)
6. Sun, H., et al.: LinkLPA: a link-based label propagation algorithm for overlapping community detection in networks. Comput. Intell. (2016)
7. Zhou, X., et al.: An ant colony based algorithm for overlapping community detection in complex networks. Phys. A Stat. Mech. Appl. **427**, 289–301 (2015)
8. Zhang, X.K., et al.: An improved label propagation algorithm based on the similarity matrix using random walk. Int. J. Mod. Phys. B **30**(16), 1650093 (2016)
9. Brin, S., Page, L.: Reprint of: the anatomy of a large-scale hypertextual web search engine. Comput. Netw. **56**(18), 3825–3833 (2012)
10. Liu, D.: Methods of community detection in complex networks and their responses to perturbations. Dissertation. Tianjin University (2013). In Chinese

UK - Means Clustering for Uncertain Time Series Based on ULDTW Distance

Xiaoping Zhu$^{(\boxtimes)}$, Zongmin Ma, and Qijie Tang

College of Computer Science and Technology,
Nanjing University of Aeronautics and Astronautics, Nanjing, China
{xpzhu_ly,15605185109}@163.com, zongminma@nuaa.edu.cn

Abstract. The probability density function represents the uncertainty of time series at each time point. In this paper, based on probability density function, we adopt the ULDTW distance for uncertain time series and apply it to the traditional UK-Means clustering. Combining the property that ULDTW distance has a one-to-many correspondence between time points in the matching process, we propose a 1ToNCenter calculation method replacing the traditional mean cluster-center calculation method to improve the accuracy of clustering results. Experiments show that the Adjusted Rand Index (ARI) of UKMeansULDTW clustering results have an obviously higher accuracy than the existing UK-Means algorithms in the high dimensional uncertain time series cases.

Keywords: Uncertain time series · DTW with limited width · UK-Means clustering · ARI

1 Introduction

A time series is a set of records that are identified by time. It can reflect the change characteristics of a certain attribute value of the record object in the chronological order. Being one of the most common data types in data mining, time series widely exist in financial, medical, voice and image processing and other fields [1,2]. Its unique time-dependent and high-dimensional characteristics makes the trend of the property information it contains more visible.

Clustering is an important technology in data mining, not only because of its powerful exploration ability, but also as the pretreatment steps or subroutines in many other techniques [3]. There are many algorithms for deterministic time series clustering [4–6], the most commonly used is the K-Means clustering algorithm, which puts the samples into clusters of the nearest cluster center repeatedly. Considering that the large dimension of the time series, K-Means algorithm is widely used to deal with it, due to the high efficiency.

In practical applications, data, including time series data, is not always accurate. Due to the influence of the subjective and objective factors, the original data collected by people usually contain uncertainty [7]. Here are a few representative scenarios [1,7,8], such as accuracy of equipment, location tracking system, personal privacy encryption and so on.

© Springer International Publishing AG 2017
H. Yin et al. (Eds.): IDEAL 2017, LNCS 10585, pp. 27–35, 2017.
https://doi.org/10.1007/978-3-319-68935-7_4

Because uncertain data is widely used in practical applications, the study of the representation and processing of uncertain data has become an important research in the field of database. From the data granularity perspective, there are two main forms of data uncertainty [7]. The first one is the uncertainty of the object, mainly in the inability to determine whether the data object itself exists; and latter is the uncertainty of the data point values, mainly in the object attribute values are not accurate. With the increasing focus on uncertain data, the study of uncertain time series also raises more and more interest. In this paper, we mainly study the time series with uncertain time-point values. For the uncertainty of the values at each time point, we use the probability density function to represent the values at each time point, and calculate the similarity between samples through the probability density.

The main contributions of this paper are as follows:

(1) Based on the existing UK-Means clustering algorithm of uncertain data, we propose an UKMeansULDTW algorithm. The algorithm considers the displacement error between time series in time.
(2) We improve the way of seeking the cluster center based on the corresponding relationship between time points.

The rest of this paper is organized as follows. Section 2 compares the research work of static uncertain data clustering and uncertain time series clustering algorithm. In Sect. 3, we briefly introduce the contents of the UK-means clustering algorithm and propose the UKMeansULDTW algorithm and the 1ToCenter cluster-center solution for the uncertain time series based on the dynamic time warping distance with bounded bandwidth. Section 4 shows the results of the experiment. Finally, a summary is made in Sect. 5.

2 Related Work

Time series clustering is an important research in data mining technology. Considering the large dimension of time series, clustering studies for time series are mainly focused on the method of partitioning and density-based methods.

In the literature [5], a hybrid algorithm is proposed based on the advantages of Fuzzy C-Means clustering (FCM) clustering and Fuzzy C-Medoids clustering (FCMdd). And in literature [9,10], HD Menndez et al. proposed a medoid-based ACO clustering algorithm by using ant colony optimization. The K-Shape algorithm, proposed in literature [3], is a time series clustering algorithm based on the extensible iterative refinement process. And HD Menndez et al. used Genetic Algorithms for partitional clustering in literature [11–13] that has competitive performance in comparison with classical clustering methods. For clustering of uncertain data, Michael Chau et al. proposed the UK-Means clustering algorithm in literature [14], which is the first uncertain data-clustering algorithm for the location of devices. It uses the probability density function to represent the next possible position of the object.

However, most researches of uncertain time series clustering are directly use static uncertain data clustering algorithm, and ignore the interdependence of time series at time points, which can cause the clustering results have a certain error. Different from static data, the time series is a set of values in order of time and each point represents the same attributes and the relative position between them can not be exchanged. Therefore, the clustering algorithm of static data can not be applied to the uncertain time series directly. In this paper, we improve the traditional UK-Means clustering algorithm by using the ULDTW distance to calculate the similarity between uncertain time series and cluster center, so that it can more accurately for the uncertain time series clustering.

3   Improved UK-Means Clustering of Uncertain Time Series

3.1   The Disadvantage of UK-Means for Uncertain Time Series

The UK-Means algorithm proposed by Michael Chau et al. [14] is based on the traditional K-Means algorithm. It employs the expected distance to calculate the distance between the samples and the cluster centers. In this algorithm, the expected distance between the uncertain time series X and any cluster center c is calculated as follows:

$$ED(X, c) = E(\|X - c\|^2) = \sum_{i=1}^{m} \int \|x_i - c_i\|^2 f(x_i) dx_i \tag{1}$$

Where $\|\bullet\|^2$ indicates the Euclidean distance between any two time points and $f(x_i)$ represents the probability density function of X at the time point i.

However, this method only considers the distance between the corresponding time points between uncertain time series, rather than considering the time displacement error between them. Figure 1 shows the trend of the four sets of Trace time series in UCR Time Series Classification Archive [15]. In each class set, there is a significant time displacement relationship between the samples, which is called displacement error. As the time series tends to have a large time span, such displacement errors are prevalent with time series. To solve these kind of time series set, the Euclidean distance is obviously lack of accuracy [16].

In order to solve this kind of problem, we introduce the probability error function to represent the uncertain of time series, and the improved uncertain dynamic time warping (ULDTW) distance calculation method is applied to the clustering of uncertain time series, which can solve the problem of time displacement error between time series.

3.2   ULDTW Distance for Uncertain Time Series

Dynamic time Warping distance (DTW) was initially applied in the speech recognition field to find the speech samples with the most similar information.

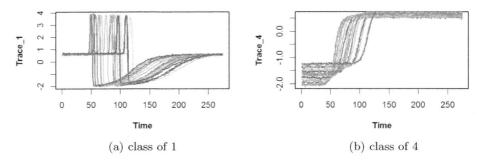

(a) class of 1   (b) class of 4

Fig. 1. Trace data set of UCR

Through adjusting the correspondence of point-to-point, this method can find the optimal matching path. In the uncertain time series, the value of each sample point is uncertain. We use the probability estimation method to predict the error function $\varepsilon(x)$ at each time point. Then in the process of clustering uncertain time series, each time we choose the cluster center c, we can calculate the expected distance between the corresponding time points (i, j) of an uncertain time series X and cluster center c:

$$d(x_i, c_j) = ED(x, c) = \int \|x + r(x_i) - c_j\|^2 \varepsilon(x_i) dx \qquad (2)$$

Where $r(x)$ represents the raw value of X at each time point.

Considering the time complexity of DTW, we apply the method mentioned in literature [17], using a window to limit the width of the matching path in a certain area. $path[i, j]$ must satisfy the limit that $j - r \leq i \leq j + r$, where $path[i, j]$ records the corresponding points (i, j) of the matching path and r represents the limited width of the window. Finally, we apply the Eq. (2) to the traditional DTW distance to get the ULDTW distance. The similarity between the uncertain time series and cluster center should satisfy the following.

(1) If $i = 1, j = 1$, $udtw(X_1, Y_1) = d(x_1, y_1) = ED(x_1, y_1)$
(2) otherwise, in $j - r \leq i \leq j + r$,

$$udtw(X_i, Y_j) = d(i, j) + \min \begin{cases} udtw(X_{i-1}, Y_j) & ,i > 1, j = 1 \\ udtw(X_{i-1}, Y_{j-1}) & ,i > 1, j > 1 \\ udtw(X_i, Y_{j-1}) & ,i = 1, j > 1 \end{cases} \qquad (3)$$

Where X_i represents the subsequence $(x_1, x_2, \cdots, x_i)$ of X. From the Eq. (3), we can see that in the process of matching the optimal path, the matching path must be incremental in time, that is prohibit return during the matching process, which also satisfies the time order of the time series. As a result the ULDTW distance between the uncertain time series X and the deterministic time series Y can be expressed as:

$$ULDTW(X, Y) = \sqrt{udtw(X_p, Y_q)} \qquad (4)$$

3.3   Optimization the Calculation of Cluster Center

Different from the point-to-point calculation method of the Euclidean distance, the time point is no longer a simple one-to-one corresponding relationship when calculating the expected distance between the sample and the cluster center by using ULDTW distance. Due to the existence of the time displacement error, each time point of a time series may match with a number of consecutive time points of the cluster center. Therefore, the traditional methods like mean value cluster center calculation method(hereinafter referred to as MeanCenter) for corresponding time points is no longer applicable to calculate ULDTW distance. In this paper, we employ the one-to-many mean method to calculate the cluster center based on the global matching in the literature [18]. The concrete implementation can be shown as Algorithm 1.

Algorithm 1. 1ToNCenter

1: Input the samples of the cluster $C = \{s_1, s_2, \cdots, s_{|C|}\}$ and the current cluster center c

2: Let vector(list): $assocTab$ records the matched time points of time series

3: **for** each time series X of cluster C **do**

4:     $p \leftarrow$ the length of X **and** $q \leftarrow$ the length of cluster center c

5:     **while** $p \geq 1$ **and** $q \geq 1$ **do**

6:         $asscoTab[q] \leftarrow assocTab[q] \cup X_p$

7:         $(p, q) \leftarrow path[p, q]$

8:     **end while**

9: **end for**

10: **for** i in 1:m **do**

11:     $c^*_i = mean(assocTab(i))$

12: **end for**

13: **return** (c^*)

We improved the UK-Means algorithm based on the ULDTW distance to calculate the expected distance between the uncertain time series and the cluster center. Finally, we get the UKmeansULDTW clustering algorithm that is more suitable for uncertain time series. Specific implementation is described in Algorithm 2.

4   Experimental

4.1   The Construction of Uncertain Time Series

In this paper, we utilize the sample sets in UCR [15] time series database as the testing data, which contains many time series sets in different areas. Since it covers a variety of time series sets of different time dimension, different classes, and different numbers of samples, it can efficiently evaluate the clustering performance of UKMeansDTW clustering algorithm. In order to compare with the

---

Algorithm 2. UKmeansULDTW

1: Input the uncertain time series set D, the number of clusters to be clustered K, maximum number of iterations T, and convergence threshold θ

2: Initialize the cluster, select K cluster centers randomly, denoted as $c = \{c_1, c_2, \cdots, c_K\}$

3: **repeat**

4:     Calculate the $udtw(X_i, c_j)$ distance between X_i and each cluster center c_j

5:     Use formula $j = \arg\min_j udtw(X_i, c_j)$ to find the cluster whose cluster center is closest to X_i, assign X_i to cluster C_j

6:     Update the cluster center c^* of each cluster according to the Algorithm 1

7:     Update $t = t + 1$

8: **until** The cluster centers of all cluster are no longer changed or $t > T$

9: **return** the collection of all cluster C

---

traditional UK-Means algorithm, we use the same method in literature [14] to construct the uncertain time series data sets, using different norm distribution functions with expectation of 0 and variance of $0.1\sigma \sim 0.2\sigma$ as the error function at different time periods, where σ is the standard variance of each raw sample.

Through the uncertain construction method in the above, we randomly selected eight sets of data with different number of classes as the test sets. The characteristic parameters of the original test data sets are shown in Table 1, where K is the number of sample classes, N is the number of samples of each data set, and M is the sample dimension, that is, the time span.

Table 1. Experimental data sets

Data set	K/N/M	Data set	K/N/M
Coffe	2/56/286	Plane	7/105/144
BeetleFly	2/40/512	OliveOil	4/40/570
DistalPhalanxOutlineAgeGroup	3/400/80	Symbols	6/995/398
ProximalPhalanxOutlineAgeGroup	3/400/80	Synthetic_Control	6/300/60

4.2   Experimental Results and Analysis

In order to compare with the traditional UK-Means algorithm, we adopt Adjusted Rand Index (ARI) [19], the same evaluation criterion in literature [14], to evaluate the clustering results of UKMeansDTW algorithm. By comparing the arbitrary pair of all the samples, we calculate the probability of two samples belonging to the same class that are grouped into the same cluster, and the probability of two samples that are not belonging to the same class that are grouped into different clusters. The value of ARI is in $[-1, 1]$, the closer the

ARI is to 1, the more accurate the clustering results are. The experiments in this paper are implemented by R. Table 2 shows the results of the traditional UK-Means algorithm and the UKMeansDTW algorithm.

Table 2. The results of UKMeansDTW and UKMeans

Data set	ARI		% of improvement
	UKMeansDTW	UKMeans	
Plane	1	0.832	20.19
Coffe	0.794	0.669	18.68
Symbols	0.753	0.711	5.91
OliveOil	0.718	0.541	**32.72**
BeetleFly	0.805	0.477	**68.73**
Synthetic_Control	0.77	0.592	**30.07**
DistalPhalanx-OutlineAgeGroup	0.611	0.527	15.94
ProximalPhalanx-OutlineAgeGroup	0.545	0.528	3.22

Compared with the traditional UK-Means algorithm, the ARI of UK-Means algorithm with ULDTW similarity calculation for most of the uncertain time series has a significant improvement. As the ARI of UK-Means for BeetleFly is only 0.477, while that of the proposed algorithm for it is 0.805, which increased by 68.73%. Furthermore, the ARI of the proposed algorithm for Plane is 1, which means the classification is totally correct. On some individual data sets (e.g., ProximalPhalanx-OutlineAgeGroup), with the ULDTW distance as a similarity measure, the improvement of the results is ineffective. The reason is that the time displacement error between the uncertain time series is not obvious, which resulting in little difference between the ULDTW distance and the expected Euclidean distance in samples and cluster center. However, considering the time displacement error is prevalent in the time series, the algorithm proposed in this paper to clustering uncertain time series is more ubiquitous.

With different value of window width r, Fig. 2 shows the comparison of the correctness of clustering results when using the traditional mean method (Mean-Center) and the one-to-many method (1ToNCenter) to update the cluster centers. The straight line represents the correct rate of UK-Means algorithm clustering results. As we can see, the clustering results are improved by using the 1ToNCenter method for the updating of cluster center compared with using the point-to-point MeanCenter method. Because cluster center updated by 1ToN-Center method is closer to the overall trend of uncertain time series, which can improve the quality of clustering results effectively.

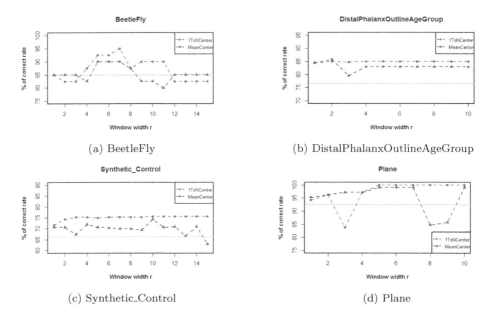

Fig. 2. The correct rate of clustering by MeanCenter and 1ToNCenter

5   Conclusion and Future Work

DTW distance is widely used in the similarity analysis of high dimensional time series. In this paper, we adopt ULDTW distance with limited width to calculate the similarity between uncertain time series and cluster center. Comparing with the traditional UK-Means clustering algorithm, we takes into account the displacement error between the uncertain time series, it has a better performance when reflecting the time series in each time point. During the experiment, eight sets of time series with different number of classes are tested. Compared with the traditional UK-Means clustering algorithm, the clustering results of the UKMeansULDTW clustering algorithm proposed in this paper are obviously improved. Considering the time complexity of DTW distance, in the future work, we will further consider the pruning of the algorithm and the reduced dimension of the time series, so as to improve the efficiency of the algorithm.

Acknowledgment. This work was supported in part by the National Natural Science Foundation of China (61370075).

References

1. Zuo, Y., Liu, G., Yue, X., Wang, W., Wu, H.: Similarity matching over uncertain time series. In: Seventh International Conference on Computational Intelligence and Security, pp. 1357–1361 (2011)

2. Izakian, H., Pedrycz, W.: Anomaly detection and characterization in spatial time series data: a cluster-centric approach. IEEE Trans. Fuzzy Syst. **22**(6), 1612–1624 (2014)
3. Paparrizos, J., Gravano, L.: k-Shape: efficient and accurate clustering of time series. ACM SIGMOD Rec. **45**(1), 69–76 (2016)
4. Aghabozorgi, S., Shirkhorshidi, A.S., Wah, T.Y.: Time-series clustering - a decade review. Inf. Syst. **53**(C), 16–38 (2015)
5. Izakian, H., Pedrycz, W., Jamal, I.: Fuzzy clustering of time series data using dynamic time warping distance. Eng. Appl. Artif. Intell. **39**, 235–244 (2015)
6. Menéndez, H.D., Barrero, D.F., Camacho, D.: A co-evolutionary multi-objective approach for a k-adaptive graph-based clustering algorithm. In: 2014 IEEE Congress on Evolutionary Computation (CEC), pp. 2724–2731. IEEE (2014)
7. Xu, L., Hu, Q., Hung, E., Chen, B., Tan, X., Liao, C.: Large margin clustering on uncertain data by considering probability distribution similarity. Neurocomputing **158**(C), 81–89 (2015)
8. Qin, B., Xia, Y., Wang, S., Xiaoyong, D.: A novel bayesian classification for uncertain data. Knowl.-Based Syst. **24**(8), 1151–1158 (2011)
9. Menéndez, H.D., Otero, F.E.B., Camacho, D.: Medoid-based clustering using ant colony optimization. Swarm Intell. **10**(2), 123–145 (2016)
10. Menéndez, H.D., Otero, F.E.B., Camacho, D.: MACOC: a medoid-based ACO clustering algorithm. In: Dorigo, M., Birattari, M., Garnier, S., Hamann, H., Montes de Oca, M., Solnon, C., Stützle, T. (eds.) ANTS 2014. LNCS, vol. 8667, pp. 122–133. Springer, Cham (2014). doi:10.1007/978-3-319-09952-1_11
11. Bello-Orgaz, G., Menéndez, H.D., Camacho, D.: Adaptive k-means algorithm for overlapped graph clustering. Int. J. Neural Syst. **22**(05), 1250018 (2012)
12. Menéndez, H., Camacho, D.: A genetic graph-based clustering algorithm. In: Yin, H., Costa, J.A.F., Barreto, G. (eds.) IDEAL 2012. LNCS, vol. 7435, pp. 216–225. Springer, Heidelberg (2012). doi:10.1007/978-3-642-32639-4_27
13. Menendez, H.D., Barrero, D.F., Camacho, D.: A genetic graph-based approach for partitional clustering. Int. J. Neural Syst. **24**(03), 1430008 (2014)
14. Chau, M., Cheng, R., Kao, B., Ng, J.: Uncertain data mining: an example in clustering location data. In: Ng, W.-K., Kitsuregawa, M., Li, J., Chang, K. (eds.) PAKDD 2006. LNCS (LNAI), vol. 3918, pp. 199–204. Springer, Heidelberg (2006). doi:10.1007/11731139_24
15. Chen, Y., Keogh, E., Hu, B., Begum, N., Bagnall, A., Mueen, A., Batista, G.: The UCR time series classification archive (2015). www.cs.ucr.edu/~eamonn/time-series_data
16. Qu, J., Shao, Z., Liu, X.: Mixed PSO clustering algorithm using point symmetry distance. J. Comput. Inf. Syst. **20**, 53–65 (2010)
17. Keogh, E., Ratanamahatana, C.A.: Exact indexing of dynamic time warping. Knowl. Inf. Syst. **7**(3), 358–386 (2005)
18. Petitjean, F., Ketterlin, A., Gançarski, P.: A global averaging method for dynamic time warping, with applications to clustering. Patt. Recogn. **44**(3), 678–693 (2011)
19. Zhang, S., Wong, H.-S., Shen, Y.: Generalized adjusted rand indices for cluster ensembles. Patt. Recogn. **45**(6), 2214–2226 (2012)

Predicting Physical Activities from Accelerometer Readings in Spherical Coordinate System

Kittikawin Lehsan and Jakramate Bootkrajang[(✉)]

Department of Computer Science, Chiang Mai University, Chiang Mai, Thailand
{kittikawin.l,jakramate.b}@cmu.ac.th

Abstract. Recent advances in mobile computing devices enable smartphone an ability to sense and collect various possibly useful data from a wide range of its sensors. Combining these data with current data mining and machine learning techniques yields interesting applications which were not conceivable in the past. One of the most interesting applications is user activities recognition accomplished by analysing information from an accelerometer. In this work, we present a novel framework for classifying physical activities namely, walking, jogging, push-up, squatting and sit-up using readings from mobile phone's accelerometer. In contrast to the existing methods, our approach first converts the readings which are originally in Cartesian coordinate system into representations in spherical coordinate system prior to a classification step. Experimental results demonstrate that the activities involving rotational movements can be better differentiated by the spherical coordinate system.

Keywords: Activity recognition · Classification · Spherical coordinate system

1 Introduction

Advances in semi-conductor and sensors technology foster the development of practical wearable devices. Many of such devices, for example a heart rate monitor or a GPS unit, can be easily spot in our daily life activities. People use them either for recreational, health or security purposes. The interesting thing is that raw data recorded by those devices can often be useful in understand the nature of the activities. With the help of data analysis techniques currently available, we can now gain more insights into regularities and patterns in the data.

The mobile phone industry also benefits from the technological advancement. We now see manufacturers packed a number of sensors into its mobile phones. Unlike wearable devices which is quite specific to its task, mobile phone is more ubiquitous. This leads to the idea of alternatively using mobile phone to sense the world instead of using specialised wearable devices.

There are increasing number of applications which make use of sensory data gathered from mobile phone sensors. For example, [10] used mobile phone

© Springer International Publishing AG 2017
H. Yin et al. (Eds.): IDEAL 2017, LNCS 10585, pp. 36–44, 2017.
https://doi.org/10.1007/978-3-319-68935-7_5

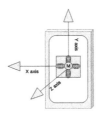

Fig. 1. An illustration of an accelerometer of a typical mobile phone. The sensor measures accelerations in x, y, z axes.

data to detect person's mode of transportation. Readings from mobile phone's accelerometer can be used to signal the falling of elderly people [4]. Similarly, [6] proposed a methodology for efficiently recognising human activities as well as a method for detecting the falling. A work in [5] studied the using of mobile phone's GPS unit together with predefined points of interest for inferring user's activities based on current position of the user. Along the same line, a combination of an accelerometer and a gyroscope was used to recognise daily activities such as walking, standing and sitting [12]. Also, [3] proposed a method to detect walking as well as counting steps using smartphone's sensory data. Apart from sensory data from mobile phone, [8] studied activity classification using pressure sensors attached to five different spots on the body.

In this work, we are interested in inferring physical activities using data readings from mobile phone's accelerometer. Briefly speaking, an accelerometer is an electro-mechanical sensor used for measuring acceleration. The sensor is composed of a tiny mass attached to springs. A change in acceleration causes the springs to compress or extend, and the degree of compression/extension is translated to acceleration accordingly. In a typical mobile phone, accelerations in the x, y, z axes are usually provided. A picture of an accelerometer and its axes is illustrated in Fig. 1.

A notable work that pioneered activities recognition using data from an accelerometer is probably the one in [9]. In that work, a special accelerometer unit is paired with an IPAQ personal digital assistant for data processing purpose. The work considered the recognition of eight different daily activities. The authors found that it is sufficient to use the mean, the standard deviation and the correlation of readings in the x-axis, the y-axis and the z-axis of the accelerometer to correctly recognise most of the activities. From that moment, researchers then started to investigate the topic from different perspectives and in different environments. The work in [12] compared various classification algorithms for activities recognition. They found that k-NN and boosting algorithms are among the top performer for the task. Studies that focus on the effect of smartphone's mounting positions on recognition rate can be found in [1,2].

The work which are most related to our work are probably the work in [7,11]. Both studied the recognition of physical activities using only the accelerometer. The empirical results using different classification algorithms suggested that accurate recognition can be achieved only with the data from the accelerometer. However, and rather interestingly, all of the above work only consider the data represented in the Cartesian coordinate system. Motivated by the fact that a spherical coordinate system is often used to represent data which contains rotational movements, we postulate that the spherical coordinate system might be more suitable in capturing the dynamics of activities which involve rotational movements compared to the typical Cartesian coordinate system. Accordingly, we consider the following points to be our contributions.

- We investigate the using of the spherical coordinate system to represent the data instead of the typical Cartesian coordinate system.
- We empirically study an appropriate time frame[1] for extracting data from a stream of sensor readings.
- We extensively test the proposed representation using five different well-known classifiers.

The rest of the paper is organised as follow. Section 2 presents our approach that uses the spherical coordinates to represent the data readings. Section 3 then presents empirical evaluations while Sect. 4 concludes the study.

2   The Proposed Framework

The movement of smartphone during activity recognition can be of both translational and rotational motions. For example, the movements for sit-up and squating are rather similar except that the movement for sit-up additionally contains rotational motions. Differentiating the two activities can be challenging. Motivated by a capability to capture rotational motions, in this work we will investigate the using of a spherical coordinates to represent our data. The spherical coordinate system is a generalisation of a polar coordinate system to three dimensional vector space. Specifically, a point (x, y, z) in the Cartesian coordinate system can be converted to a point (r, θ, ϕ) in the spherical coordinate system by

$$r = \sqrt{x^2 + y^2 + z^2} \tag{1}$$

$$\theta = \arccos \frac{z}{r} \tag{2}$$

$$\phi = \arctan \frac{y}{x} \tag{3}$$

Essentially, the quantity r in (r, θ, ϕ) tuple represents the length of a vector measured from the origin, θ represents an angle between the vector and the z-axis while ϕ represents an angle that the projection of the vector on the xy plane makes with the x-axis.

One interesting property of the spherical coordinate system is that the distance between any two points is no longer a straight line. Instead the distance

---

[1] Later on, we will use the term 'window' and 'time frame' interchangeably.

Table 1. A list of 9 features extracted from raw data collected from the smartphone.

Feature	Calculation	Description
μ_r	$\frac{\sum_{i=1}^{n} r_i}{n}$	The mean of r
μ_θ	$\frac{\sum_{i=1}^{n} \theta_i}{n}$	The mean of θ
μ_ϕ	$\frac{\sum_{i=1}^{n} \phi_i}{n}$	The mean of ϕ
σ_r	$\sqrt{\frac{\sum_{i=1}^{n}(r_i - \mu_r)^2}{n-1}}$	The standard deviation of r
σ_θ	$\sqrt{\frac{\sum_{i=1}^{n}(\theta_i - \mu_\theta)^2}{n-1}}$	The standard deviation of θ
σ_ϕ	$\sqrt{\frac{\sum_{i=1}^{n}(\phi_i - \mu_\phi)^2}{n-1}}$	The standard deviation of ϕ
$corr(r, \theta)$	$\frac{cov(r,\theta)}{\sigma_r \sigma_\theta}$	The correlation between r and θ
$corr(r, \phi)$	$\frac{cov(r,\phi)}{\sigma_r \sigma_\phi}$	The correlation between r and ϕ
$corr(\theta, \phi)$	$\frac{cov(\theta,\phi)}{\sigma_\theta \sigma_\phi}$	The correlation between θ and ϕ

is the length of the curvature which is known as a geodesic. This non-linearity of the distance measure is beneficial for the case of non-linearly separable data. Therefore, distance-based classifiers such as a nearest-neighbour classifier might be able to take advantage of this transformation. Further, as mentioned earlier, we believe that geodesic distance is also more natural for this task since we are interested in movements which is composed of rotational elements.

Next, we will follow the steps previously used in [9] for summarising the stream of sensor readings. The steps involve dividing the stream of data into equal window of length n seconds. Commonly in the literature including that found in [9], a 10-second window is used. In this work, we will additionally investigate windows of different lengths namely 5, 10, 15 and 20 s. A set of sensor readings that fit in one time frame will undergo feature extraction steps in order to produce a point representation for the readings in that time frame. To extract the features, we calculate the means, the standard deviations of each of the three variables, i.e., r, θ and ϕ, as well as the correlations between all pairs of the variables. This, in total, results in a 9-dimensional point representation for the raw sensor readings in one time frame. Although, other statistical features, i.e., max, min, inter quartile range, energy, can be extracted from the set of sensor readings, we empirically observed that these 9 features can adequately capture the dynamics of the activities under consideration. The finding is also in agreement with [9]. Table 1 summarises the 9 features extracted from one time frame of length n.

In summary, our framework involves first converting the readings into the spherical coordinates. A set of 9 features are then extracted from the converted data producing a set of 9-dimensional input vectors. The set of input vectors together with manually assigned class labels are used to train a classifier.

3   Experiments

3.1   Data Collection

We manually collected data by asking volunteers to perform each of the physical activities namely, walking, jogging, push-up, squating and sit-up for 50 repetitions. In each repetition we asked the volunteers to exercise for 20 s. We mounted the mobile phone on the right shoulder of the participants as shown in Fig. 2.

During data collection, we sampled data from the accelerometer sensor which are accelerations in the x-coordinate, the y-coordinate and the z-coordinate using 1 Hz sampling rate. We converted the raw readings into the spherical coordinates so that the 9 features can later be extracted. In total, there are 5000 s of sensory data available (1000 s for each of the five activities). Now, labelled datasets of different size can be constructed from the sensory data depending on the length of the time frame. For example, using 10-second time frame would result in a dataset containing 100 input instances for each class, while using 20-second time frame would produce 50 input instances for each class.

(a) Walk-ingand Jogging   (b) Squat-ting   (c) Sit-up   (d) Push-up

Fig. 2. Smartphone mounting positions for each of the activities.

3.2   Protocol

In this study, we are interested in whether transforming accelerometer readings into the spherical coordinate representation is beneficial. To evaluate this, we will compare the performances obtained from five well-known classifiers namely, k-nearest neighbour, linear discriminant analysis, naive Bayes, Support Vector Machine with linear kernel and Support Vector Machine with polynomial degree 2 operating in the Cartesian coordinate system and the newly proposed spherical coordinate system. We note that, due to the small sample size nature of the data, we will use leave-one-out cross validation technique to measure the predictive performance of the classifiers. In addition, we are also interested in the effect of the length of the time frame on the classification accuracy. For this purpose, we will additionally study the comparative performance of the classifiers learning from data sets produced by setting the time frame to 5 s, 10 s, 15 s and 20 s, respectively.

3.3   Results

Firstly, we would like to establish a window length that gives the best recognition performance. Table 2 summarises recognition accuracies for 4 different windows length tested. We can see, from the table, that larger time frame is generally better compared to the smaller ones in both coordinate systems. Comparing the two coordinate systems, we observe that for smaller time frame the Cartesian coordinate system has an upper hand but as window length increases to 20 s we see that using the spherical coordinate system is preferable. The accuracy averaged over all the classifiers is 90.33% for the spherical system while the accuracy for the Cartesian coordinate system slightly lags behind. We did not test window lengths beyond 20 s but we speculate that the result might not be worse than the 20 s mark. Moreover, as we shall subsequently see that using 20 s window is already very satisfying as we achieved perfect classification performance. Therefore, increasing the length of the time frame might be overkill.

Table 2. Mean accuracies (%) of the two coordinate systems as a function of window length averaged over all classifiers. Boldface entries indicates the best window length.

Window length (seconds)	Cartesian	Spherical
5	84.35 ± 12.49	74.63 ± 12.33
10	87.72 ± 10.76	85.03 ± 11.74
15	88.69 ± 9.34	88.40 ± 9.98
20	**89.59 ± 8.75**	**90.33 ± 9.08**

Next, we take what we have learnt from the previous experiment that the 20 s time frame yielded the best averaged performance. In this experiment, we fix the window length to 20 s and compare the performances of the two coordinate systems using five classifiers. The classification accuracies of the five classifiers trained using data represented in the Cartesian and the spherical coordinate systems are summarised in Table 3. From the table, it is quite clear that the spherical coordinates representation gives relatively better predictive performance compared to the Cartesian coordinates counterpart. Interestingly, we also see that three out of five classifiers namely, 5-NN, SVM with linear kernel and Linear Discriminant Analysis, all learnt from the data represented in the spherical coordinate system achieved perfect classification results.

To gain further insights regarding why using spherical coordinates is superior to using the Cartesian coordinates, Table 4 presents the confusion matrix of the 5-NN classifier operated in the Cartesian coordinate system where it achieved 97.2% classification accuracy. We noticed that the 5-NN misclassified sit-up as squatting and vice-versa. It is worth noting that the orientation of the smartphone in these two activities are almost identical (see Fig. 2). Further, the movement of the device during the action are also similar, with an exception that there exists rotational movements around the hip when performing sit-up.

Table 3. Predictive accuracy of 5 classifiers. Window length is fixed to 20 s. Boldface highlights the coordinate system in which each of the classifier works best.

Classifier	Cartesian	Spherical
5-NN	97.20	**100.00**
Decision tree	**98.00**	95.60
SVM (Linear Kernel)	96.80	**100.00**
SVM (Polynomial Kernel degree 2)	98.80	**99.60**
Naive Bayes	97.20	**99.20**
Linear Discriminant Analysis	97.60	**100.00**

Table 4. Confusion matrix for the 5-NN under the Cartesian coordinate system.

	Walking	Jogging	Squatting	Push-up	Sit-up
Walking (predicted)	50	0	0	0	0
Jogging (predicted)	0	50	0	0	0
Squatting (predicted)	0	0	47	0	3
Push-up (predicted)	2	0	0	48	0
Sit-up (predicted)	0	0	2	0	48

It seems that the Cartesian representation cannot capture this rotational movements adequately. However, this does not seem to be a problem for the spherical representation. The claim is confirmed by the perfect predictive performance of the 5-NN operates in the spherical coordinate system. The explanation also applies to the case of walking and push-up. For the sake of exposition, let us consider 2-dimensional trajectories of a smartphone in Fig. 3. The blue trajectory contains rotational elements while the red one does not. If we were to extract, for example, the mean of x and y components from the readings (represented by the blue and red dots) from the two trajectories under the Cartesian coordinate system, we would end up with the similar means. As such, differentiating the two trajectories using the means would be problematic. However, converting the reading into the polar coordinates (a special case of the spherical coordinates in 2D) can clearly alleviate the problem since the means of r and θ from the two trajectories are quite distinct. We believe that this is a reason why working in the spherical coordinate system is desirable.

Overall, based on the empirical evidences we can conclude that transforming sensor readings from the Cartesian coordinates to the spherical coordinates is advantageous for physical activities recognition especially when the activities involve rotational motions.

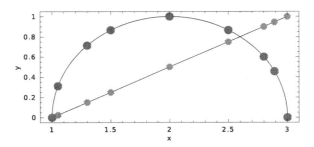

Fig. 3. Two trajectories that produce similar features under the Cartesian coordinate system but not under the polar coordinate systems as well as the proposed spherical coordinate system. (Color figure online)

4   Conclusion and Future Work

In this work, we studied comparative performance of representing accelerometer readings in the Cartesian and the spherical coordinate system for human activities recognition. As part of the experiment we also investigated a suitable window length for summarising a stream of sensory data sampled from the accelerometer. Experimental results based on recognition rates of five well-known classifiers suggest that a 20 s window and the spherical coordinate system is well-suited for the task, and especially so for capturing the dynamics of physical activities that involve rotational motions. Future work includes the relaxation on the mounting position of the smartphone. For that purpose, a rotation-invariant data representation is required, so that the accelerometer readings are relatively the same regardless of the position of the mobile device.

References

1. Antos, S.A., Albert, M.V., Kording, K.P.: Hand, belt, pocket or bag: practical activity tracking with mobile phones. J. Neurosci. Methods **231**, 22–30 (2014)
2. Ayu, M.A., Ismail, S.A., Matin, A.F.A., Mantoro, T.: A comparison study of classifier algorithms for mobile-phone's accelerometer based activity recognition. Procedia Eng. **41**, 224–229 (2012)
3. Brajdic, A., Harle, R.: Walk detection and step counting on unconstrained smartphones. In: Proceedings of the 2013 ACM International Joint Conference on Pervasive and Ubiquitous Computing, pp. 225–234. ACM (2013)
4. Dai, J., Bai, X., Yang, Z., Shen, Z., Xuan, D.: Perfalld: a pervasive fall detection system using mobile phones. In: 2010 8th IEEE International Conference on Pervasive Computing and Communications Workshops (PERCOM Workshops), pp. 292–297. IEEE (2010)
5. Furletti, B., Cintia, P., Renso, C., Spinsanti, L.: Inferring human activities from GPS tracks. In: Proceedings of the 2nd ACM SIGKDD International Workshop on Urban Computing, p. 5. ACM (2013)

6. Kozina, S., Gjoreski, H., Gams, M., Luštrek, M.: Efficient activity recognition and fall detection using accelerometers. In: Botía, J.A., Álvarez-García, J.A., Fujinami, K., Barsocchi, P., Riedel, T. (eds.) EvAAL 2013. CCIS, vol. 386, pp. 13–23. Springer, Heidelberg (2013). doi:10.1007/978-3-642-41043-7_2

7. Kwapisz, J.R., Weiss, G.M., Moore, S.A.: Activity recognition using cell phone accelerometers. ACM SIGKDD Explor. Newsl. **12**(2), 74–82 (2011)

8. Moncada-Torres, A., Leuenberger, K., Gonzenbach, R., Luft, A., Gassert, R.: Activity classification based on inertial and barometric pressure sensors at different anatomical locations. Physiol. Meas. **35**(7), 1245 (2014)

9. Ravi, N., Dandekar, N., Mysore, P., Littman, M.L.: Activity recognition from accelerometer data. In: AAAI, vol. 5, pp. 1541–1546 (2005)

10. Reddy, S., Mun, M., Burke, J., Estrin, D., Hansen, M., Srivastava, M.: Using mobile phones to determine transportation modes. ACM Trans. Sens. Netw. (TOSN) **6**(2), 13 (2010)

11. Weiss, G.M., Lockhart, J.W.: The impact of personalization on smartphone-based activity recognition. In: AAAI Workshop on Activity Context Representation: Techniques and Languages, pp. 98–104 (2012)

12. Wu, W., Dasgupta, S., Ramirez, E.E., Peterson, C., Norman, G.J.: Classification accuracies of physical activities using smartphone motion sensors. J. Med. Internet Res. **14**(5), e130 (2012)

A Community Detection Algorithm Based on Jaccard Similarity Label Propagation

Meng Wang, Xiaodong Cai$^{(\boxtimes)}$, Yan Zeng, and Xiaoxi Liang

School of Information and Communication,
Guilin University of Electronic Technology, Guilin 541004, Guangxi, China
caixiaodong@guet.edu.cn

Abstract. Due to the randomness of label propagation algorithm, the stability is poor and the accuracy is low for community detection results in complex networks. In order to solve the problem, this paper proposes a novel community detection algorithm based on Jaccard similarity label propagation. Firstly, the Jaccard similarity is used to measure nodes importance. Then, the importance of nodes is utilized to reduce the randomness in label selection. Finally, nodes with the highest importance are selected to update labels in iteration, which improves the stability for community detection. Stability and accuracy of data sets of real networks are measured by modularity and normalized mutual information, respectively. Experimental results show that, the proposed algorithm for community detection results is more stable and more accurate than LPA, LPA_SI and KLPA algorithms in the cases of near linear time complexity.

Keywords: Complex network · Community detection · Label propagation · Jaccard similarity

1 Introduction

In nature and social activities, there are many universal connections and interactive relationships. This phenomenon is expressed with complex network models. Network nodes represent specific objects, while edges stand for links between objects, such as biology network [1], citation network [2] and social network [3]. It was found that community structures are a common feature of these networks. The community is composed of nodes with similar characteristics. Furthermore, the internal connection of communities is dense, the external correlation is relatively sparse [4]. Besides, it is important for research not only on the community structure in terms of biology, sociology and e-commerce, but also on applications in network security and criminal organizations recognition.

A large number of algorithms have been proposed with the widespread concern for community detection in complex networks. The classical community detection algorithms proposed graph segmentation, hierarchical clustering and modularity optimization method. In 2007, a fast label community detection algorithm is proposed based on label propagation [5]. The algorithm does not require any input parameters, such as the number and size of the community, and performs a linear time complexity. The efficiency of community detection results is improved extensively. It is suitable for

H. Yin et al. (Eds.): IDEAL 2017, LNCS 10585, pp. 45–52, 2017.
https://doi.org/10.1007/978-3-319-68935-7_6

large-scale complex networks. However, the label propagation algorithm is random in iterative update process, which leads to instability of the result for community detection. In order to solve the problem, the optimization and improvement methods were provided in many papers. In [6], a semi-synchronous label propagation algorithm overcoming the node oscillation problem in complex networks is proposed, but the stability is not improved for community detection. In LPA_SI [7], the stability of community detection is improved by integrating node importance and label influence. However, the computational complexity is large, which makes it difficult for large-scale complex networks to divide. In KLPA [8], the author combined the K value and local influence to achieve community detection, the transmission of label nodes is reduced. Nevertheless, the time complexity is increased, and the efficiency of large-scale networks for community detection is affected.

In this paper, a novel community detection method is proposed to solve the problem of randomness in label propagation algorithm. It is inspired by Jaccard similarity measure [9]. Firstly, the similarity is used to measure nodes importance. Then, the importance of node is utilized to reduce the randomness in label selection. Finally, the nodes with the highest importance are selected to update labels, which improves the stability for community detection.

2   Label Propagation Algorithm for Community Detection Based on Jaccard Similarity

2.1   Label Propagation Algorithm

In the literature [5], a community detection method is proposed based on Label Propagation Algorithm (LPA). Each node chooses the label with the largest number of neighbors as its own label by sending the label information between the node and its neighbors. After numerous iterations, node labels belong to the same community tend to be consistent. The main process of LPA is described as follows:

Input: Network model is $G = (V, E)$, where $V = \{v_i, i = 1, 2, 3 \ldots n\}$ represents the collection of nodes, set $C = \{C_1, C_2, \ldots, C_k\}$ denotes individual; where $E = \{e_i, i = 1, 2, 3, \ldots, n\}$ represents the collection of edges, $e_{ij} = \{(v_i, v_j), v_i \in V, v_j \in V\}$ denotes relationship.

Output: Set $C = \{C_1, C_2, \ldots, C_k\}$, where k represents the number of community.

(1)   Each node v is assigned a unique label as the identity which stands for the community.
(2)   An ordered list is generated by the random ordering of node sets.
(3)   Following the ordered list, the node is updated. The label of each node v is updated by the label with the largest number of neighbors in the network. When there are multiple labels with the same maximum number, a label is chosen randomly as the label of the node.
(4)   After several iterations, the label tends to be stable in each node neighbors.
(5)   Nodes with the same label are classified to the same community.

In LPA, the order of each iteration is random. When a neighbor node has the same number of labels, a label is randomly selected to update, so that the stability of the community detection result is greatly affected. In the process of label propagation, the importance of the node itself is not considered, which leads to some less important nodes affect some of the more important nodes in turn, resulting in "counter current" phenomenon. In addition, the nearby small communities are swallowed by the first formation of community labels. Finally, a very large community is formed and even the labels of nodes in the entire network are the same.

2.2   Label Propagation Algorithm Based on Jaccard Similarity

In order to solve the "counter current" phenomenon in LPA, this paper considers the influence of the node importance and proposes an algorithm based on Jaccard similarity label propagation for community detection.

Related Definitions. The proposed algorithm is based on Jaccard similarity of label propagation. The related definitions are as follows:

Definition 1: Node Neighborhood.

$$T_{(v_i)} = \{v_i\} \cup \{(v_i, v_j) \in V\} \tag{1}$$

$T_{(v_i)}$ denotes the set of neighbors of node i.

Definition 2: Jaccard Similarity between Nodes.

$$S(v_i, v_j) = \frac{\left| T_{(v_i)} \cap T_{(v_j)} \right|}{\left| T_{(v_i)} \cup T_{(v_j)} \right|} \tag{2}$$

The degree of similarity is used to measure the tightness of any two neighboring nodes. In formula (2), $\left| T_{(v_i)} \cap T_{(v_j)} \right|$ is the number of communal neighbors between node v_i and v_j in networks. $\left| T_{(v_i)} \cup T_{(v_j)} \right|$ denotes the number of all neighbors between node v_i and v_j. When node v_i and v_j have more jointed neighbors, while the similarity is bigger, the node is more important to others. The range of $S(v_i, v_j)$ is $[0, 1]$.

Definition 3: Similarity between Nodes and Communities.

$$S(v_i, C_j) = \frac{\left| T_{(v_i)} \cap C_j \right|}{\left| T_{(v_i)} \cup C_j \right|} \tag{3}$$

The similarity between the node and community is used to measure the degree of the connection between nodes and adjacent communities. Where $\left| T_{(v_i)} \cap C_j \right|$ represents the number of common nodes, $|T_{(v_i)} \cup C_j|$ denotes the number of all neighbors between node v_i and community C_j.

Definition 4: Node Importance.

$$\text{important}(v_i) = argmax\left\{S(v_i, C_j)\delta(v_i, C_j)\right\} \qquad (4)$$

Where $\delta(v_i, C_j)$ is the Kronecker function, if v_i is identical to C_j, $\delta(v_i, C_j)$ equals to 1, otherwise equals to 0. When the node importance is bigger, the influence to others is larger. The label of node spreads more easily.

In the process of updating labels, the Jaccard similarity of adjacent nodes is firstly calculated according to formula (2). Based on the similarity descending sort, node labels are updated in order and the randomness is avoided. After several iterations, the node is updated when there are multiple adjacent label collections. The similarity between adjacent label collection is calculated according to formula (3). It represents the tightness of the association between label nodes and adjacent label sets. A higher tightness indicates that the node label is more important for the adjacent label sets. The probability is greater to be the label of the node updated. The most important label set is selected to update the node label by the formula (4) to avoid the "reverse flow" phenomenon in the process of label propagation. The formula of node label updated is as follows:

$$\text{LabelNew}(v_i) = \arg\max_{l}\{important(v_i)\} \qquad (5)$$

Where l is the adjacent label collection of the node.

Algorithm Description. The steps of the algorithm for community detection are as follows:

Input: Complex network $G = (V, E)$, set the maximum number of iterations $t = maxIter$. Node v_i has n neighbors. $Lv_{i(t)}$ is the label of node v_i after the tth iteration.
Output: SetC $= \{C_1, C_2, \ldots, C_k\}$, where k represents the number of communities.

1. The label of each node $v \in V$ is initialized. Each node is assigned a unique label. The iteration time is initialized, $t = 1$.
2. The similarity between the label nodes is calculated by formula (2). According to the similarity from high to low sort, an ordered list $V' = \{v_1, v_2, \ldots, v_n\}$ is generated.
3. According to the ordered list V', the label of node is updated. According to formula (3), the node is updated to the most important label.
4. If iteration $t = maxIter$ and $Lv_{i(t)}$ no longer changes, the nodes with the same label are grouped into same community. If $t = t + 1$, return to step 3.

Compared with LPA for community detection, in the proposed algorithm, the randomness of label propagation is reduced by considering the transmission characteristics of node labels and the tightness of connection among nodes. In addition, the influence of node is integrated, which makes it more accurate in label selection. The stability is improved for community detection.

3   Experimental Results and Analysis

To evaluate the proposed algorithm, real network data sets by Mark Newman [10] and the information of data sets is shown in Table 1 are used. The modularity and normalized mutual information are used as the measurement standard for community detection of these structure known data sets. For unknown structure data sets Hep-th and Internet, the modularity is used to test. The stability and accuracy of the proposed algorithm are analyzed, and compared by using the LPA [5], LPA_SI [7] and KLPA [8]. The time complexity is also analyzed. The experimental platform configuration consists of Intel (R) Core (TM) i5-4460 CPU @ 3.20 GHz processor, 8 GB memory, Microsoft Windows 7 operating system and JDK1.8, and Java programming language. In the testing process, to reduce the random caused by different operations, the average of 50 runs is taken. The maximum number iteration of the algorithm is set to 100.

Table 1. Data sets information.

Datasets	Node (N)	Edge (E)	Description
Karate	N = 34	E = 78	Zarchary's karate club network
Dolphins	N = 62	E = 159	Dolphin social network
Polbooks	N = 105	E = 441	Books about US politics network
Footballs	N = 115	E = 616	NCAA College-Football network
Hep-th	N = 8361	E = 15751	High-energy theory collaborations network
Internet	N = 22963	E = 48436	Internet at the level of autonomous systems network

3.1   Modularity

Modularity (Q) is the measure of quality for community structure proposed by Girvan and Newman [4], which can be used to measure the stability for community detection. It is defined as follows:

$$Q = \sum_1^k \left[\frac{link(C_j)}{m} - \left(\frac{\deg(C_j)}{2m} \right)^2 \right] \qquad (6)$$

Where k is the number of community, $link(C_j)$ is the number of C_j, and $\deg(C_j)$ is the sum of all nodes in the community C_j, and m is the number of edges in the network. The value range of Q is $[-1, 1]$. If Q is larger, the quality is better for community detection and the result is more stable.

The result of Q with the proposed algorithm, LPA [5], LPA_SI [7] and KLPA [8] in different data sets are shown in Table 2. It can be seen that the value of Q is obtained by using the Karate, Dolphins, Polbooks, Footballs, Hep-th and Internet data sets. Since the algorithm considers the importance of the node itself, the value of Q is obviously higher than that of the LPA [5], and the stability of the algorithm is improved significantly. Compared with LPA_SI [7], it considers importance of nodes and influence of labels together, there is a slightly higher Q value of the three small data

Table 2. Comparison the modularity Q of the algorithm.

Datasets	Algorithm			
	LPA [5]	LPA_SI [7]	KLPA [8]	Proposed
Karate	0.338	0.395	0.382	0.397
Dolphins	0.465	0.512	0.506	0.524
Polbooks	0.474	0.521	0.583	0.601
Footballs	0.465	0.612	0.607	0.632
Hep-th	0.604	0.645	0.625	0.634
Internet	0.365	0.497	0.505	0.528

sets: Dolphins, Polbooks and Footballs. In addition, Karate data set of the Q value is similar and Hep-th is lower. However, the Q of the big data set Internet is higher. Compared with KLPA [8], the value of Q is higher than that of the LPA [5].

3.2 Normalized Mutual Information

The accuracy for community detection is evaluated in NMI [11, 12] by comparing the similarity between prediction and the true community structure. It is defined as follows:

$$NMI(p, r) = \frac{2MI(p, r)}{H(p) + H(r)} \tag{7}$$

Where $MI(p, r)$ denotes the mutual information between prediction community p and the real community r, and the formula is:

$$\text{MI(p, r)} = \sum_i \sum_j P\left(C_{p_i} \cap C_{r_j}\right) log \frac{P\left(C_{p_i} \cap C_{r_j}\right)}{P\left(C_{p_i}\right) P\left(C_{r_j}\right)} \tag{8}$$

H(p) represents the entropy of the community p, the formula is:

$$\text{H(p)} = -\sum_j P\left(C_{p_i}\right) \log\left(P\left(C_{p_i}\right)\right) \tag{9}$$

Where $P\left(C_{p_i}\right)$ and $P(C_{r_i})$ respectively represent the probability of a node in the network divided into the community p and in the real community r. The range of NMI is [0, 1]. The NMI is greater and the result is more accurate.

The result of NMI with the proposed algorithm, LPA [5], LPA_SI [7] and KLPA [8] in different data sets are shown in Fig. 1. The values of NMI is obtained by using the Karate, Dolphins and Footballs data sets. The NMI of the proposed are close to 1, which is larger than that of LPA, LPA_SI and KLPA algorithm. For the Dolphins data set, NMI does not reach 1. However, it is also higher than LPA, LPA_SI and KLPA. Experimental results show that, the proposed algorithm improves the accuracy of community detection.

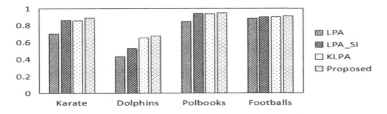

Fig. 1. The normalized mutual information (NMI).

3.3 Computational Complexity

If in a given network, n represents the number of nodes, and m represents the number of edges. All nodes in the network initialize the label, all nodes traverse once, the time complexity is $O(n)$. When calculating the similarity between nodes, all edges traverse once and the time complexity is $O(m)$. Similarity computing node according to importance and sorting node update, the time is required for most $O(n+m)$. The time complexity of each node is divided into $O(n)$, and the time complexity is $O(ms)$ after iteration s times. Thus, the total time complexity is $O(n+ms)$. Compared with the LPA algorithm, the time complexity increases, mainly on the ordering of nodes, but still close to the linearity. Compared with LPA_SI and KLPA, the time complexity is similar (Table 3).

Table 3. Computational complexity.

Algorithm	LPA [5]	LPA_SI [7]	KLPA [8]	Proposed
Computational complexity	$O(n+m)$	$(n+ms)$	$O(n+ms)$	$O(n+ms)$

4 Conclusion

The proposed algorithm integrates label propagation and the importance of nodes, and adopts the new node importance measurement for label iteration process. The improvement is shown by comparing the modularity and normalized mutual information of the LPA, LPA_SI and KLPA algorithms in small and large data sets of real networks. It can be confirmed that the stability and accuracy of the community detection are improved when the importance of node itself and the impact of labels in the label propagation algorithm are considered. With the increase of the number of nodes in networks, the performance of the proposed algorithm is further improved.

Acknowledgment. This work was supported by the 2016 Key Laboratory of Cognitive Radio and Information Processing, Ministry of Education Project No. CRKL160102 and 2016 Guangxi Science and Technology Project No. AB16380264.

References

1. Mering, V.C., Krause, R., Snel, B., et al.: Automated inference of highly conserved protein interaction networks. J. Nat. **417**(6998), 399–403 (2002)
2. Xiao, X., Chen, Y., Deng, Y.: Development of community discovery in citation networks. J. Inf. **35**(4), 125–130 (2016)
3. Yuta, K., Ono, N., Fujiwara, Y.: A gap in the community-size distribution of a large-scale social networking site. Physics (2007)
4. Newman, M.E.J., Girvan, M.: Finding and evaluating community structure in networks. J. Phys. Rev. **E69**(2), 026113 (2004)
5. Raghavan, U.N., Albert, R., Kumara, S.: Near linear time algorithm to detect community structures in large - scale networks. J. Phys. Rev. **E76**(3), 036106 (2007)
6. Cordasco, G., Gargano, I.: Community detection via semi synchronous label propagation algorithms (2011)
7. Huang, J., Guo, K., Guo, H.: Label propagation algorithm for community detection based on vertex significance and label influence. J. Chin. Comput. Syst. **36**(6), 1171–1175 (2015)
8. Deng, G.: Community detection base on mixed k-influence. J. Inf. Commun. **2**, 61–63 (2016)
9. Pan, L., Lei, Y., Wang, C., et al.: Method on entity identification using similarity measure based on weight of Jaccard. J. Beijing Jiaotong Univ. **33**(6), 141–145 (2009)
10. Mark Newman network data. http://www-personal.umich.edu/~mejn/netdata
11. Danon, L., Diaz-Guilera, A., Duch, J., et al.: Comparing community structure identification. J. Stat. Mech. Theor. Exp. **2005**(09), P09008 (2005)
12. Lancichinetti, A., Fortunato, S., Kertész, J.: Detecting the overlapping and hierarchical community structure of complex networks. New J. Phys. **11**(3), 033015 (2009)

A Robust Object Tracking Method Based on CamShift for UAV Videos

Chang Zhao$^{(\boxtimes)}$, Jiabin Yuan, and Huiting Zheng

College of Computer Science and Technology,
Nanjing University of Aeronautics and Astronautics, Nanjing, China
{zhaochang,jbyuan,zhenghuiting}@nuaa.edu.cn

Abstract. Unmanned aerial vehicles (UAVs) equipped with monitoring systems have played an important role in various fields in recent years. An object tracking algorithm is necessary in order to process information in the wide range of UAV videos. CamShift algorithm is outstanding as its efficient pattern matching and fast convergence. This paper presents an excellent method based on CamShift to implement precise target tracking in UAV videos. This method integrates multi-feature fusion (MF), CamShift, and Kalman filter (KF) called the MF-KF-Camshift algorithm. Experimental results show that the method achieves great performance in dealing with different scenes and meets the real-time requirements.

Keywords: Unmanned aerial vehicles · Target tracking · CamShift

1 Introduction

The development of unmanned aerial vehicle (UAV) technology in the recent years has facilitated the wide application of UAVs in military and commercial areas, such as meteorological exploration, disaster monitoring, geological survey, map mapping, traffic control, and video capture [1]. The number of UAV videos increases exponentially, and the processing of these videos is extremely important.

Visual tracking is an active vision-based research topic and can assist UAVs to achieve various tasks, such as air monitoring, aircraft avoidance and aerial pre-warning [2].

Several tracking algorithms are currently used in UAV videos. In [3], Kalman filter (KF) and global motion estimation were combined. In [4], SURF feature was extracted to track small objects. In [5], particle filter was applied to this field. In [6,7], the CamShift algorithm was used in UAV video tracking.

Numerous visual tracking algorithms are recently proposed in the computer vision community, such as tracking with Random Projection and Weighted Least Squares [8], Minimum Output Sum of Squared Error, Spatio-Temporal Context,

© Springer International Publishing AG 2017
H. Yin et al. (Eds.): IDEAL 2017, LNCS 10585, pp. 53–62, 2017.
https://doi.org/10.1007/978-3-319-68935-7_7

Kernel Correlation Filter and NormalHedge [9]. Target tracking in UAV videos remains as a challenging task because of many factors, such as change in lighting, complex background, uncertain perspective, small size of tracking target, rotation, change in size of tracking window target occlusion and so on [10].

In this study, an improved CamShift is proposed to track target in UAV videos. The results show that the proposed algorithm can solve the above-mentioned problems. The main contributions of the study are as follows: KF is integrated to the original CamShift to predict the target position and thus deal with the target occlusion. The original color feature is replaced by multi-feature fusion (MF) to enable good performance in complex environment, target deformation and other scenes.

The rest of the paper is organized as follows. Section 2 describes the principle of the original CamShift. Section 3 introduces the MF-KF-CamShift algorithm and its flow. Section 4 demonstrates the experimental results and analysis. Finally, Sect. 5 elaborates the conclusions of the study.

2   Object Tracking by CamShift

The MeanShift algorithm is a kernel tracking method based on non-parametric probability density estimation. The algorithm was first introduced into the object tracking field by Comaniciu and Ramesh in [11]. MeanShift presents good real-time performance and is easy to be implemented and insensitive to rotation and slight deformation. At present, MeanShift is widely used in clustering, image smoothing, segmentation, and object tracking [12,13]. CamShift as an improvement of MeanShift was first proposed by Gary and was applied to track people's face. CamShift is currently popular in the object tracking field [14,15].

2.1   Principle of MeanShift

To achieve successful object tracking, the object and the candidate object should be described separately first and then compared with a few criteria. In Mean-Shift, the probability density of the object model is:

$$q_u = C \sum_{i=1}^{n} K(\|x_i^*\|^2)\delta(b(x_i) - u) \tag{1}$$

where x_i^* is the normalized pixel position with the object center as the origin, C is the normalized coefficient, K is a kernel function, and $\delta(b(x_i) - u)$ is used to determine whether the pixel belongs to the u interval of the color histogram. If the said condition is true, then the value is 1; otherwise, the value is 0.

In meanshift algorithm, the probability density of the candidate model is:

$$p_u(f) = C \sum_{i=1}^{n} K(\left\|\frac{(f - x_i)}{h}\right\|^2)\delta(b(x_i) - u) \tag{2}$$

where f is the center of the candidate object, h is the bandwidth of kernel function.

The similarity function is used to describe the degree of similarity between the object and the candidate object. In meanshift algorithm, the Bhattacharyya coefficient is used as a similarity function:

$$\rho(p, q) = \sum_{u=1}^{m} \sqrt{p_u(f)q_u} \tag{3}$$

The larger the value of the similarity function, the more similar the candidate model to the object, so the object is positioned at the maximum of the similarity function. With Taylor expansion for Eq. (3):

$$\rho(p, q) \approx \frac{1}{2} \sum_{u=1}^{m} \sqrt{p_u(f_0)q_u} + \frac{C}{2} \sum_{i=1}^{n} w_i K\left(\left\|\frac{f - x_i}{h}\right\|^2\right) \tag{4}$$

$$w_i = \sum_{u=1}^{m} \sqrt{\frac{q_u}{p_u(f)}} \delta(b(x_i) - u) \tag{5}$$

The process of finding the maximization of similar function can be expressed as a process of meanshift iteration from the candidate region to the object region. The object positioning formula of meanshift is:

$$y_1 = \sum_{i=1}^{n} x_i w_i g\left(\left\|\frac{y_0 - x_i}{h}\right\|^2\right) \bigg/ \sum_{i=1}^{n} w_i g\left(\left\|\frac{y_0 - x_i}{h}\right\|^2\right) \tag{6}$$

2.2 Principle of CamShift

CamShift is popularly known as "continuously adaptive MeanShift" which makes a MeanShift operation for each frame and takes the result of the previous frame as the initial value of the next frame. CamShift is fast and easy to implement. However, CamShift can only work under a simple background and lacks robustness to shelter [16].

3 The MK-KF-CamShift Algorithm

3.1 Improvement by KF

KF is simpler and effective in UAV video tracking [7]. When tracking object by UAV, the occlusion is unavoidable and the tracking window cannot remain stable owing to changes in lighting and noise generated by camera shake. In this study, Kalman predictor is used to estimate the position of the object in the next frame and to adjust the tracking window. The use of the predictor improves the accuracy and robustness of the tracking algorithm.

The state variable $X(k)$ and observed value $Z(k)$ are:

$$X(k) = A(k)x(k-1) + B(k)u(k) + w(k) \tag{7}$$

$$Z(k) = H(k)x(k) + v(k) \tag{8}$$

where $A(k)$ is the state transition matrix; $B(k)$ is the input model acting on the control vector $u(k)$; $w(k)$ is the process noise, which is a white noise with a mean of zero; the covariance is $Q(k)$, $w(k) \backsim N(0, Q(k))$. $H(k)$ is the observation matrix, and $v(k)$ is observing noise and $v(k) \backsim N(0.R(k))$.

In this paper, Q and R are defined as the diagonal matrix. $X(k)$ and $Z(k)$ are defined as follows:

$$V(k) = [x_k, y_k, v_x, v_y]^T \tag{9}$$

$$Z(k) = [x_k, y_k]^T \tag{10}$$

where x_k and y_k are the object centers, and v_x and v_y are the speed components. Assuming that the object moves linearly with random acceleration, then $A(k)$ and $H(k)$ can be deduced as follows: $A_k = \begin{bmatrix} 1 & t & 0 & 0 \\ 0 & 1 & 0 & 0 \\ 0 & 0 & 1 & t \\ 0 & 0 & 0 & 1 \end{bmatrix}$, $H(k) = \begin{bmatrix} 1 & 0 & 0 & 0 \\ 0 & 0 & 1 & 0 \end{bmatrix}$. The prediction and correction of KF are:

$$\begin{cases} x_k = Ax_{k-1} + Bu_{k-1} \\ P_k = AP_{k-1}A^T + Q \end{cases} \tag{11}$$

$$\begin{cases} K_k = P_k H^T (HP_k H^T + R)^{-1} \\ x_k = x_k + K_k(Z_k - Hx_k) \\ P_k^+ = (I - K_k H)P_k \end{cases} \tag{12}$$

The cvKalmanPredict function by OpenCV is used to predict and the cvKalmanCorrect function is used to update the position of the object, thereby improving the accuracy of the algorithm and its capability to deal with object occlusion.

3.2   Improvement by MF

The basic CamShift algorithm builds color model using the H component of HSV color space, this method exhibits slight robustness in terms of deformation and light. Given the single feature information, the performance of the algorithm is significantly weakened when the object and background color are similar. Thus, H component histogram and LBP features are fused together to solve the said problem.

The probability density of the H component is

$$q_H = C_H \sum_{i=1}^{n} K(\left\| \frac{x_0 - x_i}{h} \right\|^2) \delta(b(x_i) - H) \tag{13}$$

The LBP feature is an operator used to describe the local texture features of an image, and this feature presents good performance in dealing with rotation and grayscale changes. Figure 1 shows the schematic of the LBP algorithm.

The basic LBP operator is expressed as

$$LBP_{P,R} = \sum_{P=0}^{P-1} S(g_p - g_c)2^p \tag{14}$$

$$S(u) = \begin{cases} 1 & u \geq 0 \\ 0 & u < 0 \end{cases} \tag{15}$$

In order to achieve better tracking effect and reduce the computational complexity, this paper uses the rotation and uniform invariant LBP, the formula is as follows:

$$LBP_{P,R}^{riu2} = \begin{cases} \sum\limits_{p=0}^{P-1} S(g_p - g_c) & U(LBP_{P,R}) \leq 2 \\ P+1 & U(LBP_{P,R}) > 2 \end{cases} \tag{16}$$

$$U(LBP_{P,R}) = |S(g_{P-1} - g_c) - S(g_0 - g_c)| + \sum_{p=0}^{P-1} |S(g_p - g_c) - S(g_{p-1}) - g_c| \tag{17}$$

In this paper, the extracted LBP histogram and H component histogram are merged together as the object model. The probability density of the LBP feature is:

$$q_L = C_L \sum_{i=1}^{n} K\left(\left\|\frac{(x_0 - x_i)}{h}\right\|^2\right)\delta(b(x_i) - L) \tag{18}$$

The probability densities of the candidate model are P_H and P_L.

The similarity measure of the algorithm is:

$$\rho(p,q) = w_H\rho_H + w_L\rho_L = w_H \sum_{i=1}^{n} \sqrt{p_H q_H} + w_L \sum_{i=1}^{m} \sqrt{p_L q_L} \tag{19}$$

where w_H and w_L are the weights of the two features.

In the new algorithm, the prediction value of KF is taken as the initial value of the combined MF and CamShift iterative algorithm. As shown in Eqs. (4), (5), (6), and (19) are used to calculate the optimal value. The value is used by KF to correct the object position.

3.3 Algorithm Implementation

In this study, MF and KF are used to improve CamShift. Figure 1 shows the flow principle of the MF-KF-CamShift algorithm. The new algorithm performs well in UAV video tracking. The pseudocode of the proposed algorithm is shown in Algorithm 1.

Algorithm 1. MF-CM-KF

Initialization:
$vidoe.open$
if $open\ error$ **then**
  $return - 1$
end if
$section \leftarrow select\ the\ target\ by\ mouse$
$video.read(frame)$
if $!paused$ **then**
  $frame \leftarrow frame + 1$
  if $trackobject = 1$ **then**
    $LBP_{value} \leftarrow frame;\ H_{value} \leftarrow frame$
    $get\ the\ Eqs\ 19$
    $trackwindow \leftarrow section$
    $calculate\ the\ backprojection$
    $trackwindow \leftarrow prediction \leftarrow cvKalmanPredict$
    while $N < 10 \cap EPS < 1$ **do**
      $do\ Camshift\ :\ Eqs\ 4,5,6$
      $N \leftarrow N + 1$
    end while
    $measure\ point \leftarrow Camshift$
    $cvKalmanCorrect(kalman, measure\ ment)$
    $cout\ KFPredictCenter, CamshiftCenter, KFcorrect$
  end if
end if

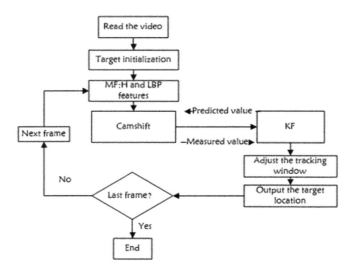

Fig. 1. Principle of the MF-KF-Camshift

4   Experiments and Results Analysis

4.1   Algorithm Performance and Analysis

A total of 21 groups of different clarities and scenes of different objects are tested. The new algorithm achieves great tracking effect and some representative scenes are listed. In Figs. 2, 3 and 4, the green point is the center of the Kalman prediction window, the red point is the center of the combined MF and CamShift prediction window, and the red box is the final tracking window adjusted by KF. The size of the tracking windows is also marked below the pictures.

Figure 2 demonstrates the result of parachute tracking with low visibility and small object. In frame 186, the size of the tracking window is 17 * 9, and the object can barely be distinguished by the human eye. In frame 510, the UAV shakes due to airflow.

Figure 3 demonstrates the result of paragliding tracking with change in lighting and deformation. In frame 192, the size of the object changes. In frame 250, the object begins to turn right. In frame 297, the lighting changes drastically and the shape of the object also changes.

Figure 4 demonstrates the result of red truck tracking. In frames 50 and 147, the object is blocked by a tree with a similar color. In frame 266, the object is seriously blocked; however, KF still predicts the position.

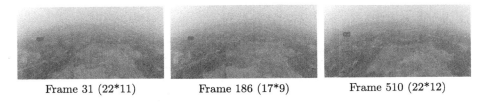

Frame 31 (22*11)   Frame 186 (17*9)   Frame 510 (22*12)

Fig. 2. Tracking of parachute

Frame 192 (69*13)   Frame 250 (51*29)   Frame 297 (54*48)

Fig. 3. Tracking of paragliding

The MK-KF-CamShift algorithm is able to track objects effectively in various scenes of UAV videos such as occlusion, low visibility, small object, complex background, rotation, deformation and adjusts the tracking window sensitively according to the object size.

60   C. Zhao et al.

Frame 50 (35*57)   Frame 147 (29*54)   Frame 266 (47*97)

Fig. 4. Tracking of red truck

4.2 Experimental Comparison and Analysis

Given the improvement effects of KF and MF, the new algorithm achieves good
tracking effect in various scenes of UAV videos. To further verify its performance, several comparative experiments are conducted. The MF-KF-CamShift
algorithm shows great improvement in tracking performance and meets real-time
requirements.

Figure 5 shows a partial comparison of three algorithms including tracking
of people, UAV and cars. Figure 5(a) shows the results of CamShift. Figure 5(b)
shows those for the CamShift only improved by KF. Figure 5(c) shows those for
the MF-KF-CamShift algorithm. Figure 6 shows the time for those algorithm to
process each frame in 8 experiments.

(a)   (b)   (c)

Fig. 5. Comparative experiment

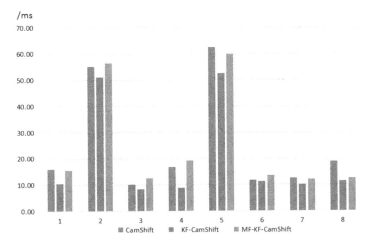

Fig. 6. Comparison of running speed

5 Conclusion and Future Work

A real-time algorithm that combines MF, CamShift, and KF called the MF-KF-Camshift method is proposed to achieve stable object tracking in UAV videos. The experimental results show that the method exhibits more significant improvement effect than the original algorithms and great performance under the conditions of occlusion, low visibility, small object, complex background, rotation, and deformation.

However, there are some small flaws in this algorithm, such as the need to manually select the object. Thus, the proposed algorithm will be optimized in the future to improve its autonomous performance.

Acknowledgments. This work was supported by the "Application platform and Industrialization for efficient cloud computing for Big data" of the Science and Technology Supported Program of Jiangsu Province (BA2015052) and "Research and Industrialization for Intelligent video processing Technology based on GPUs Parallel Computing" of the Science and Technology Supported Program of Jiangsu Province (BY2016003-11).

References

1. Wu, X., Shi, Z., Zhong, Y.: An overview of vision-based UAV navigation. J. Syst. Simul. **A01**, 62–65 (2010)
2. Fang, P., Jianjiang, L., Tian, Y., Miao, Z.: An improved object tracking method in UAV videos. Procedia Eng. **15**, 634–638 (2011)
3. Luzanov, Y., Howlett, T., Robertson, M.: Real-time template based tracking with global motion compensation in UAV video. In: Proceedings of the Second International Conference on Computer Vision Theory and Applications, Visapp 2007, Barcelona, Spain, March, pp. 515–518 (2007)

4. Sun, W., Li, D., Jia, W., Li, P., Zhao, C., Chen, X.: Small moving object tracking in dynamic video. In: 2015 International Conference on Intelligent Information Hiding and Multimedia Signal Processing (IIH-MSP), pp. 239–242. IEEE (2015)
5. Yu, W., Yin, X., Chen, B., Xie, J.: Object tracking with particle filter in UAV video. In: Eighth International Symposium on Multispectral Image Processing and Pattern Recognition, p. 891810. International Society for Optics and Photonics (2013)
6. Zhang, T., Yang, Z., Zhang, X., Shen, Y.: A vision system for multi-rotor aircraft to track moving object. In: 2016 IEEE Chinese Guidance, Navigation and Control Conference (CGNCC), pp. 401–406. IEEE (2016)
7. Qian, Y., Xie, Q.: Camshift and Kalman predicting based on moving target tracking. Comput. Eng. Sci. **8**, 023 (2010)
8. Zhang, S., Zhou, H., Jiang, F., Li, X.: Robust visual tracking using structurally random projection and weighted least squares. IEEE Trans. Circuits Syst. Video Technol. **25**(11), 1749–1760 (2015)
9. Zhang, S., Zhou, H., Yao, H., Zhang, Y., Wang, K., Zhang, J.: Adaptive normal-hedge for robust visual tracking. Sig. Process. **110**, 132–142 (2015)
10. Liu, Y., Li, X.: A survey of target detection and tracking methods in UAV aerial video. Aircr. Navig. Missiles **9**, 53–56 (2016)
11. Comaniciu, D., Ramesh, V., Meer, P.: Real-time tracking of non-rigid objects using mean shift. In: Proceedings of IEEE Conference on Computer Vision and Pattern Recognition, 2000, vol. 2, pp. 142–149. IEEE (2000)
12. Zhou, H., Yuan, Y., Shi, C.: Object tracking using sift features and mean shift. Comput. Vis. Image Underst. **113**(3), 345–352 (2009)
13. Lebourgeois, F., Drira, F., Gaceb, D., Duong, J.: Fast integral meanshift: application to color segmentation of document images. In: 2013 12th International Conference on Document Analysis and Recognition (ICDAR), pp. 52–56. IEEE (2013)
14. Huang, D., Lin, Z., Yao, J., Guo, T.: Monocular gesture tracking based on adaptive extraction and improved CamShift. In: Video Engineering (2016)
15. Doyle, D.D.: Real-time, multiple, pan/tilt/zoom, computer vision tracking, and 3d position estimating system for small unmanned aircraft system metrology. Dissertations & Theses - Gradworks (2013)
16. Xu, K., He, L., Wang, W.: Adaptive color space target tracking algorithm based on CamShift. Comput. Appl. **29**(3), 757–760 (2009)

Multi-output LSSVM-Based Forecasting Model for Mid-Term Interval Load Optimized by SOA and Fresh Degree Function

Huiting Zheng$^{(\boxtimes)}$, Jiabin Yuan, and Chang Zhao

College of Computer Science and Technology, Nanjing University
of Aeronautics and Astronautics, Nanjing, China
zhenghuiting@nuaa.edu.cn

Abstract. Accurate forecasting of mid-term electricity load is an important issue for risk management when making power system planning and operational decisions. In this study we have proposed an interval-valued load forecasting model called SOA-FD-MLSSVM. The proposed model consists of three components, the Human Body Amenity(HBA) indicator is introduced as the input of meteorological factors, Fresh Degree(FD) function is brought into the forecast method based on setting different weight on the historical days and Least Squares Support Vector Machine based on Multi-Output model, called MLSSVM, to make simultaneous interval-valued forecasts. Moreover, the MLSSVM parameters are optimized by a novel seeker optimization algorithm(SOA). Simulations carried out on the electricity markets data from Jiangsu province. Analytical results show that the novel optimized prediction model is superior to others listed algorithms in predicting interval-valued loads with lower U^I, ARV^I and MAPE.

Keywords: Interval-valued load forecasting · Body amenity indicator · Multi-output LSSVM · SOA · Fresh degree function

1 Introduction

Medium-term load forecast has been essential for power security, market operation and power generation expansion. The accuracy of the forecast results is directly related to reasonable power schedule for the smart grid. It would cause enormous loss of economy and resource once there was fail to estimate the loads demand trend in time or give an insufficient power scheduling. In a word, mid-term electricity load plays a vital role in smart grid.

However, one might conclude that the point load time series could be forecasted and thus, there is no need of considering Interval-valued load series. Interval-valued load series, which includes peak loads and valley loads, is an alternative representation to classic load series. Engle and Russell [1] argued that point time series forecasting reveal characteristics that present obstacles for traditional forecasting methods. On the contrary, interval-valued loads have

H. Yin et al. (Eds.): IDEAL 2017, LNCS 10585, pp. 63–72, 2017.
https://doi.org/10.1007/978-3-319-68935-7_8

the advantage of taking into account variability and uncertainty. Garca-Ascanio and Mate [2] pointed out that interval load forecasting plays an instructive role in the electricity industry and can be useful for market operation [3] and scheduling reasonable dispatching plan [4]. Up to now, the issue of point load forecasting has been investigated most intensively while less were focused on the interval-valued load forecasting models [5–7]. Major methods include interval Multi-Layer Perceptron(iMLP) [8], k-nearest neighbors [5] and exponential smoothing [9,10]. Yet most methods focus on the peak load and valley load independently.

Based on the above analysis of interval load forecasting, the one month ahead forecasting capacity methods will be developed to describe the evolution of the daily interval-valued load. MLSSVM(Multi-output Least Squares Support Vector Machine), a generalized multi-dimensional LSSVM [11], with the capabilities of small-batch learning, it has been extensively used in many fields [12]. However, the performance of MLSSVM is affected due to the difficulty in hyper-parameter selection. In this paper, we use the improved Seeker Optimization Algorithm, which is capable of searching the optimal solution, high generalization performance, to optimal the hyper-parameters in MLSSVM. Furthermore, the Fresh Degree function is brought into the MLSSVM forecast method based on setting larger weight on the nearer historical days but smaller weight on the farther days, which can effectively improve the forecasting accuracy.

The main contributions of this study includes the followings:

1. It is found that systematic study on the interval load forecasting is lacking. In this paper, considerating the variability and uncertainty of daily load, we proposed a feasible and integrity interval load forecasting framework, and is capable of the longer-horizon load forecasting(up to one month).
2. The proposed forecasting model choose the human body amenity indicator as the input variables instead of all the other meteorological factors. It is proved by simulation results that replacement has achieved a good performance in decreasing the number of input as well as increasing forecasting accuracy.
3. The seeker optimization algorithm is applied to optimal the hyper-parameters in MLSSVM. Moreover, we improved the SOA to avoid converging to a local optimum.
4. The proposed combination forecasting model using Fresh Degree Function to modify the objective function of MLSSVM, which can deal with forecasting tasks beyond pure time series forecasting and improve the model's veracity.
5. Experimental result proves that the proposed model consider the interrelations between peak points and valley points at a higher forecast accuracy than the other well-established framework.

The rest of this paper is organized as follows: The interval load series and features are analyzed in detail in Sect. 2. In Sect. 3, we construct the SOA-FD-MLSSVM interval load forecasting model. We discuss the experimental results in Sect. 4 and conclude in Sect. 5.

2   Data Description and Analysis

The main purpose of this section is to analyze the characteristics of the daily load curve and to select the necessary exogenous variables for the prediction model. Our analysis in this paper will be based on data from a certain area smart grid of Jiangsu province. Daily interval-valued load ranging from January 1, 2003 to October 12, 2016 are available. The interval-valued load is represented by $[L]_k = \{L_k^V, L_k^P\}$, where L_k^V and L_k^P are the valley load and peak load on day k, respectively.

Figure 1 shows the interval-valued load series. We can see that the valley load had significant correlation with peak load, so the interval prediction can not be separated from the peak and valley forecast independently. In this paper, we use multi-output LSSVM regression to handle with interval load data prediction.

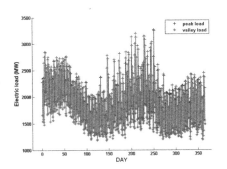

Fig. 1. The daily interval-valued load from January 1, 2015 to December 31, 2015

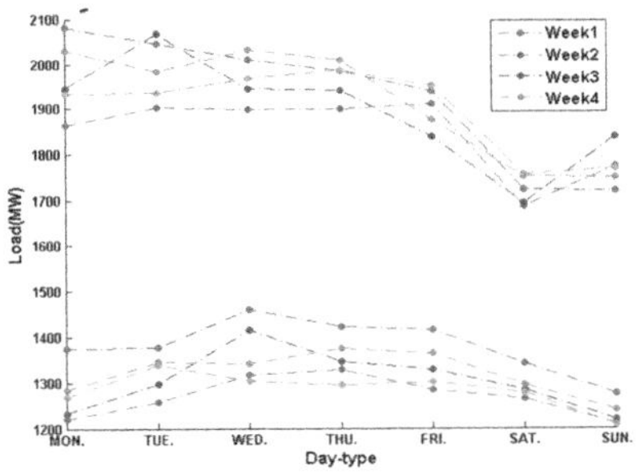

Fig. 2. The daily interval-valued load depending on the Day-type

Meteorological factors, Day-type index etc. have always been important factors in load forecasting, for their strong association with electricity consumption. Therefore the relationship between factors and interval load data are discussed in following.

2.1   Day-Type

According to the statistical data from smart grid, this region's secondary industry electricity consumption accounted for 76.6% of the total electricity consumption. So the working days interval load is higher than the weekend usually due to the decline in industrial load on the non-working day. Figure 2 shows four weeks interval-valued loads from April 4 2016 to May 15 2016. It is can obviously verified that interval-valued loads is very sensitive to the day-type index.

2.2 Body Amenity Indicator

Human activities subject to the combined effects of meteorological factors. Generally speaking, whether the human body feels hot or cold is not only related to temperature, but also has a great relationship with wind speed and relative humidity. Therefore, we should consider the influence of these three meteorological indicators comprehensively. The body comfort index is expressed as:

$$D = 1.8T + 0.55(1 - U) - 3.2\sqrt{V} + 27 \qquad (1)$$

where T represents temperature, U represents humidity and V is wind speed.

It can be seen from the Table 1 that, compared to any single meteorological index, body amenity indicator has a stronger Grey Relational Analysis(GRA) correlation with the daily load. So in interval load forecasting, with body amenity indicator instead of a number of other meteorological indicators as input can more clearly reflect the overall effect of meteorological changes, which is more conducive to improve the accuracy of interval-valued load forecasting.

Table 1. Gray correlative degree between daily load and meteorological factors

Meteorological factors	Temperature	Humidity	Wind speed	Body amenity indicator
GRA	0.834	0.56	0.168	0.887

3   The SOA-FD-MLSSVM Framework

3.1   MLSSVM for Interval Load Forecasting

Giving a training set denoted as $\{x_i, y_i\}, i = 1, 2, ..., n$ where $x_i \in \Re^p$ are the factors that affect the interval-value load, $y_i \in \Re^q$ are the historical daily valley load and peak load. And the multi-output LSSVM model can be expressed by formula as follows:

$$F(x) = W^T \Phi(x) + B \qquad (2)$$

where the $\Phi(x)$ is a nonlinear function maps the input data into a high-dimension space. Compared to the single-output model, all of the variables in this model are in the form of matrix. The W is a high dimensional characteristic space weight matrix, which is an unknown parameter as well as the bias term B.

In multi-output LSSVM, the computation errors come from not only each single output, but also the combined fitting error. Based on the statistical learning theory, the optimization problem can be described as:

$$\min J(w^j, b^j, e^j, E_i) = \frac{1}{2}\sum_{j=1}^{q}(w^j)^T(w^j) + \frac{1}{2}\sum_{j=1}^{q}\sum_{i=1}^{n}\gamma^j(e^j)^2 + \gamma^0\sum_{i=1}^{n}E_i \qquad (3)$$

$$s.t \quad y^j = (w^j)^T\varphi(x) + b^j + e^j \quad j = 1, 2, ..., q$$

$$E_i = \|Y_i - W_i\Phi(x_i) - B\|^2 \quad i = 1, 2, ..., n$$

where W_i is a diagonal matrix denoted as $diag(w_i^1, w_i^2, ..., w_i^q)$, $Y_i = [y_i^1, y_i^2, ..., y_i^q]^T$, $B = [b^1, b^2, ..., b^q]^T$, $\Phi(x_i) = [\varphi(x_i), \varphi(x_i), ..., \varphi(x_i)]^T$. e^j is the single regression error, γ^0 is a penalty parameter and must be greater than zero.

With the introduction of Lagrangian multipliers, the calculation of Eq. (3) can be simplified as follows:

$$L(w^j, b^j, e^j, E_i, \lambda_i, \mu_i^j) = \frac{1}{2}\sum_{j=1}^{q}(w^j)^T(w^j) + \gamma^0\sum_{i=0}^{n}E_i + \frac{1}{2}\sum_{j=1}^{n_y}\sum_{i=1}^{n}\gamma^j(e_i^j)^2$$

$$- \sum_{i=1}^{n}\lambda_i(E_i - Y_i - W_i\Phi(x_i) - B^2)$$

$$- \sum_{j=1}^{q}\sum_{i=1}^{n}\mu_i^j((w_i^j)^T\varphi(x_i) + b^j + e_i^j - y_i^j) \quad (4)$$

where μ_i^j is the Lagrange multiplier. The solution of Eq. (4) is determined by partial derivation of $w^j, b^j, e^j, E_i, \lambda_i, \mu_i^j$. Considering the undetermined expression of the nonlinear mapping and Representer Theorem, we have $w^j = \sum_{i=1}^{n}\alpha_i^j\varphi(x_i) = \varphi^T\alpha^j$, and the kernel function $K^j(x_i, x_k) = \varphi(x_i)\varphi(x_k)^T$. Then the regression function of the interval-value load prediction based on the multi-output LSSVM can be formulated as

$$y^j = \sum_{i=1}^{n}\alpha_i^j K(x, x_i) + b^j, \quad j = 1, 2, ..., q \quad (5)$$

where the Kernel function $K(x, x_i)$ is related to the new input sample x and the modeled sample x_i. j is the number of outputs. In this paper, take the valley and peak load on forecasted day as the outputs(j = 1,2).

3.2 Hyper-parameters Optimized Algorithm: Improved Seeker Optimization Algorithm

Seeker Optimization Algorithm (SOA) [13] has powerful ability of global searching and generalization. In SOA, a search direction $\vec{d_{ij}}(t)$ and a step length $\alpha_{ij}(t)$ are computed separately for each seeker i on each dimension j for each time step t. For each seeker i ($1 \le i \le s$, s is the population size), the position update on each dimension j ($1 \le j \le M$) is given by the following:

$$x_{ij}(t+1) = x_{ij}(t) + \alpha_{ij}(t)d_{ij}(t) \quad (6)$$

Since the subpopulations are searching using their own information, they are easy to converge to a local optimum. In order to avoid this situation, the positions of the worst $K-1$ seekers of each subpopulation are combined with the best one in each of the other $K-1$ subpopulations using the following binomial crossover operator:

$$x_{k_n j,worst} = \{ \begin{matrix} x_{lj,best}, & if\ R_j \le 0.5 \\ x_{k_n j,worst}, & else \end{matrix} \tag{7}$$

where R_j is a uniformly random real number within $[0, 1]$, $x_{k_n j,worst}$ is denoted as the th dimension of the jth worst position in the kth subpopulation, is the jth dimension of the best position in the lth subpopulation with $n, k, l = 1, 2, ..., K-1$ and $k \ne l$. In this way, the good information obtained by each subpopulation is exchanged among the subpopulations and then the diversity of the population is increased.

3.3   The Fresh Degree Function

Due to the practical application, the parameters of the prediction model will change over time. The data near the forecasted day has a greater correlation, and the data away from the forecasted day is not very relevant. In other word, the samples far away from the forecasted day play a less important role in the prediction model. As a result, in this study, we proposed the method to set larger weight on the nearer historical days but smaller weight on the farther days in the MLSSVM regression process

Firstly we modified the objective function of MLSSVM by adding membership to different historical day as demonstrate in Eq. (8).

$$\min J(w^j, b^j, e^j, E_i) = \frac{1}{2} \sum_{j=1}^{q} (w^j)^T (w^j) + \frac{1}{2} \sum_{j=1}^{q} \sum_{i=1}^{n} \gamma^j (e^j)^2 \rho_i + \gamma^0 \sum_{i=1}^{n} E_i \tag{8}$$

Giving a training set $\{x_i, y_i\}$, $i = 1, 2, ..., n$, $\{x_2, y_2\}$ is the farthest sample from the day to be predicted, denote the time distance as $time_1$, $\{x_n, y_n\}$ is the closest sample from the predicted day, the time distance $time_N$ is equal to zero.

Equation (9) was used to construct membership function and confirm ρ_i for different samples.

$$\rho_i = f(i) = (1 - \varepsilon)\sqrt{\frac{time_1 - time_i}{time_1}} + \varepsilon \tag{9}$$

where $\varepsilon \in [0, 1]$, The smaller the ε is, the smaller the weight for remote samples is. It is obviously that $\varepsilon = \rho_1 \le \rho_2 ... \le \rho_N = 1$. In this paper, we set $\varepsilon = 0.6$.

Considering the short-run trend and weekly periodicity characteristics of daily interval-valued load, interval-valued loads of the previous month have been selected as input variables. Consequently, the input vector X and the output vector Y of the SOA-FD-MLSSVM are determined as follows:

X=[Valley body amenity, Peak body amenity, Day-type]

Y=[Valley load, Peak load]

Moreover, Fig. 3 shows the flowchart of the proposed SOA-FD-MLSSVM model for interval load forecasting.

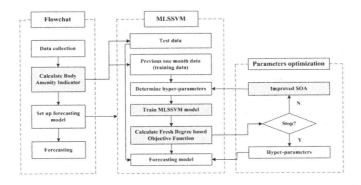

Fig. 3. Flowchart of the prediction model

4 Numerical Experiments

In this paper, we will use three approaches to assess the forecasting performance of the model, including the U^I (interval U of Theil statistics), ARV^I (interval Average Relative Variance) and MAPE (Mean Absolute Percentage Error). They are shown as follows:

$$U^I = \sqrt{\frac{\sum_{j=1}^{m} \left(X_{j+1}^V - \widehat{X}_{j+1}^V\right)^2 + \sum_{j=1}^{m} \left(X_{j+1}^P - \widehat{X}_{j+1}^P\right)^2}{\sum_{j=1}^{m} \left(X_{j+1}^V - X_j^V\right)^2 + \sum_{j=1}^{m} \left(X_{j+1}^P - X_j^P\right)^2}} \tag{10}$$

$$ARV^I = \frac{\sum_{j=1}^{m} \left(X_{j+1}^V - \widehat{X}_{j+1}^V\right)^2 + \sum_{j=1}^{m} \left(X_{j+1}^P - \widehat{X}_{j+1}^P\right)^2}{\sum_{j=1}^{m} \left(X_{j+1}^V - \overline{X}^V\right)^2 + \sum_{j=1}^{m} \left(X_{j+1}^P - \overline{X}^P\right)^2} \tag{11}$$

$$MAPE = \frac{1}{m} \sum_{j=1}^{m} \left| \frac{X_j - \widehat{X}_j}{X_j} \right| \tag{12}$$

where m denotes the number of fitted interval loads, $[X_j^V, X_j^P]$ is the jth actual interval load, $[\widehat{X}_j^V, \widehat{X}_j^P]$ is the jth predicted interval load, $\overline{X}^V$ and $\overline{X}^P$ are the averages of valley load series and peak load series, respectively. X_j and $\widehat{X}_j$ represent the actual and predicted loads (peak or valley). Here, A value of U^I equals to 0 indicates a perfect model, lower ARV^I values lead to better forecasting.

4.1 The Benefit of Body Amenity Indicator

With the human body amenity indicator as the input variables, feature selection can be done more efficient. The forecasting performance of the proposed framework are show in Figs. 4 and 5, where M1 denotes the the forecasting process with human body amenity indicator as the input variables, M2, which denotes the forecasting process with temperature, humidity and wind speed as the input variables.

Through the experiment results shown in Figs. 4 and 5, we can see that compared with the daily interval temperature, humidity and wind speed, the body amenity indicator can show the change of the interval-valued loads more accuracy. In other words, the correlative features have contributed to overfitting learning, whereas the body amenity indicator can largely reduce the curse of dimensionality and the model can outperform the former.

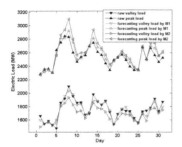

Fig. 4. One-month ahead prediction from January 1, 2015 to January 31, 2015

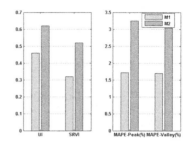

Fig. 5. Comparisons of U^I, ARV^I, MAPE-V, MAPE-P for two different input variables

4.2   The Prediction Model with Different Optimized Methods

To verify the performance of our proposed optimized methods, twelve months in 2016 were selected as the testing periods. What's more, MLSSVM is used with LIBSVM package with penalty coefficients $\gamma_0 = 193$, $\gamma_1 = 398$, $\gamma_2 = 106$, kernel widths $\sigma_1 = 1.03$, $\sigma_2 = 0.99$.

From the results in Figs. 6 and 7, the forecasting curve of the proposed model follows the raw data very well than other alternative models, signfiicant accuracy can be achieved from the proposed approach. The results from the proposed approach are better than from the conventional MLSSVM, SOA-MLSSVM and FD-MLSSVM methods.

4.3   Comparison with Other Forecasting Models

Finally, a simulational experiment is designed to validate the performance between the proposed SOA-FD-MLSSVM forecasting model and well-known interval-valued forecasting models iMLP as well as three classical single-valued forecasting models, i.e. SOA-FD-LSSVM, BPNN and ARIMA. Furthermore, four months in 2015 were selected as the one month ahead interval-valued loads forecasting testing periods.

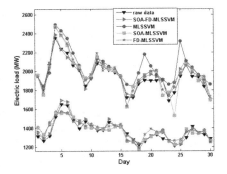

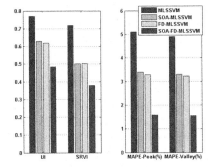

Fig. 6. Predict data of different models

Fig. 7. Results of different optimized methods

Table 2. Comparison of U^I and ARV^I for different forecasting models

	SOA-FD-MLSSVM			SOA-FD-LSSVM			iMLP			BPNN			ARIMA		
	U^I	ARV^I	MAPE(%)	U^I	ARV^I	MAPE(%)	U^I	ARV^I	MAPE(%)	U^I	ARV^I	MAPE(%)	U^I	ARV^I	MAPE(%)
Mar.	0.476	0.299	1.25	0.659	0.649	1.98	0.689	0.685	1.88	0.806	0.916	3.86	0.962	0.964	5.03
Jun.	0.501	0.611	1.35	0.703	0.529	2.19	0.791	0.585	2.28	0.832	0.692	4.69	1.041	1.397	5.86
Sep.	0.652	0.353	1.55	0.758	0.625	2.33	0.776	0.553	2.32	0.886	0.899	4.92	1.256	1.698	6.81
Dec.	0.620	0.569	1.32	0.727	0.799	2.25	0.804	0.524	2.59	0.844	0.988	4.24	1.004	1.754	6.25
Average	0.562	0.458	1.37	0.712	0.650	2.19	0.765	0.587	2.27	0.842	0.874	4.43	1.066	1.453	5.99

Two conclusions can emerge from Table 2. Firstly, The proposed model out-perform the other models in terms of forecasting accuracy. Secondly, multi-output LSSVM do better in interval series than single-output LSSVM.

5   Conclusion

In light of interval load forecasting have the advantage of taking into account variability and uncertainty, so it plays an essential role in power system plan-ning and operational decisions as well as improving the point load forecasting accuracy, in this paper we have proposed an interval load forecasting model based on hyper-parameters selection algorithm SOA and fresh degree function for MLSSVM. The proposed forecasting model use the human body amenity indicator as the input variables instead of all the meteorological factors. It is proved by simulation results that replacement has achieved a good performance in decreasing the number of input as well as increasing forecasting accuracy. Experimental in Sect. 4.2 proves the effectiveness of our optimized methods and Experimental in Sect. 4.3 demonstrate the superior performance of proposed model when comparing with iMLP, BPNN and ARIMA. Accordingly, SOA-FD-MLSSVM can be a alternative model for interval load forecasting.

Acknowledgments. This work was supported by the "Application platform and Industrialization for efficient cloud computing for Big data" of the Science and Technology Supported Program of Jiangsu Province(BA2015052).

References

1. Russell, J.R., Engle, R.F.: Analysis of High-Frequency Data. Elsevier B.V. (2010)
2. Garca-Ascanio, C., Mat, C.: Electric power demand forecasting using interval time series: a comparison between VAR and iMLP. Energy Policy **38**(2), 715–725 (2010)
3. Vaccaro, A., Canizares, C.A., Bhattacharya, K.: A range arithmetic-based optimization model for power flow analysis under interval uncertainty. IEEE Trans. Power Syst. **28**(2), 1179–1186 (2013)
4. Peng, W., Cheng, H., Xing, J.: The interval minimum load cutting problem in the process of transmission network expansion planning considering uncertainty in demand. IEEE Trans. Power Syst. **23**(3), 1497–1506 (2008)
5. Arroyo, J., Nola, R., Maté, C.: Different approaches to forecast interval time series: a comparison in finance. Comput. Econ. **37**(2), 169–191 (2011)
6. Zhongyi, H., Bao, Y., Chiong, R., Xiong, T.: Mid-term interval load forecasting using multi-output support vector regression with a memetic algorithm for feature selection. Energy **84**, 419–431 (2015)
7. Sala, E., Zurita, D., Kampouropoulos, K., Delgado-Prieto, M.: Enhanced load forecasting methodology by means of probabilistic prediction intervals estimation. In: IEEE International Conference on Industrial Technology, pp. 1299–1304 (2015)
8. Roque, A.M., Maté, C., Arroyo, J., Sarabia, A.: iMLP: applying multi-layer perceptrons to interval-valued data. Neural Process. Lett. **25**(2), 157–169 (2007)
9. Maia, A.L.S., De A. T. De Carvalho, F.: Holts exponential smoothing and neural network models for forecasting interval-valued time series. Int. J. Forecast. **27**(3), 740–759 (2011)
10. Gob, R., Lurz, K., Pievatolo, A.: More accurate prediction intervals for exponential smoothing with covariates with applications in electrical load forecasting and sales forecasting. Qual. Reliab. Eng. **31**(4), 669–682 (2015)
11. Pérez-Cruz, F., Camps-Valls, G., Soria-Olivas, E., Pérez-Ruixo, J.J., Figueiras-Vidal, A.R., Artés-Rodríguez, A.: Multi-dimensional function approximation and regression estimation. In: Dorronsoro, J.R. (ed.) ICANN 2002. LNCS, vol. 2415, pp. 757–762. Springer, Heidelberg (2002). doi:10.1007/3-540-46084-5_123
12. Zhang, X., Zhao, J., Wang, W., Cong, L., Feng, W.: An optimal method for prediction and adjustment on byproduct gas holder in steel industry. Expert Syst. Appl. **38**(4), 4588–4599 (2011)
13. Dai, C., Chen, W., Zhu, Y., Zhang, X.: Seeker optimization algorithm for optimal reactive power dispatch. IEEE Trans. Power Syst. **24**(3), 1218–1231 (2009)

A Potential-Based Density Estimation Method for Clustering Using Decision Graph

Huanqian Yan, Yonggang Lu[✉], and Li Li

School of Information Science and Engineering, Lanzhou University, Lanzhou, Gansu, China
ylu@lzu.edu.cn

Abstract. Clustering is an important unsupervised machine learning method which has played an important role in various fields. As suggested by Alex Rodriguez *et al.* in a paper published in Science in 2014, the 2D decision graph of the estimated density value versus the minimum distance from the points with higher density values for all the data points can be used to identify the cluster centroids. However, the traditional kernel density estimation methods may be affected by the setting of the parameters and cannot work well for some complex datasets. In this work, a novel potential-based method is designed to estimate density values, which is not sensitive to the parameters and is more effective than the traditional kernel density estimation methods. Experiments on several synthetic and real-world datasets show the superiority of the proposed method in clustering the datasets with various distributions and dimensionalities.

Keywords: Clustering · Density estimation · Decision graph

1 Introduction

Clustering, as a data mining tool, has its roots in many application areas, such as astronomy [1], biology [2], and anthroposociology [3], *etc*. Clustering algorithms aim to partition a set of data objects into subsets based on their similarities. Similarities are usually assessed based on the attribute values describing the objects using certain distance measures.

Most of the clustering problems are NP-hard [4, 5]. Different clustering methods have been proposed according to different observations. There are two common clustering methods: square-error-based clustering methods, and density-based clustering methods.

For square-error-based clustering methods, such as k-means [6], k-medoids [7], and affinity propagation [8, 9], an objective function, typically the sum of the distance to a set of putative cluster centers, is optimized until the best cluster center candidates are found [9–12]. The main advantage of these methods is that they are easy to implement. However, for k-means and k-medoids, because a data point is always assigned to the nearest center, they cannot be used to detect non-globular clusters [12]. For affinity propagation method, with an improper initial exemplar preference, it may fail to work properly. Besides, many square-error-based methods are greedy algorithms that depend on initial conditions and may converge to suboptimal solutions.

© Springer International Publishing AG 2017
H. Yin et al. (Eds.): IDEAL 2017, LNCS 10585, pp. 73–82, 2017.
https://doi.org/10.1007/978-3-319-68935-7_9

Density-based methods attempt to continue growing a given cluster until the density in some region exceeds some thresholds. The representative methods include DBSCAN [13], mean-shift [14], OPTICS [15], and DENCLUE [15], *etc.* Parameter setting in those algorithms is not a straightforward task. An excellent density-based clustering method published in Science in 2014 is proposed by Alex Rodriguez *et al.* [16]. The method is called Clustering by Fast Search and Find of Density Peaks (CFSFDP). It is based on the simple idea that a cluster centroid has a higher density value than its neighbors and is far away from the other objects with higher density values. It can predict the number of clusters by identifying the cluster centroids in a 2D decision graph. But in our experiments, it is found that the traditional Gaussian kernel-based density estimation method of this algorithm may produce low quality results while dealing with some complex datasets.

To address this issue, we propose a novel density estimation method for clustering called Potential-based Density Estimation (PDE) for clustering using decision graph in this paper. An improved potential-based method is used to estimate density values, which are not sensitive to the parameters. Potential-based clustering is a heuristic method which performs clustering by simulating a natural process: movements under a simulated potential [17]. This method can identify clusters with arbitrary shapes or sizes and can produce more accurate results.

The rest of the paper is organized as follows. The decision graph is introduced in Sect. 2. The proposed PDE method is described in Sect. 3. The experimental results are presented in Sect. 4. And conclusions are drawn in Sect. 5.

2   Decision Graph

CFSFDP generates clusters by assigning data points to the same cluster as its nearest neighbour with higher density after cluster centroids are selected by users. The decision graph is designed to select these centroids. It have two important indicators: local density p_i of each point i, and its minimum distance d_i from points of higher density value. In CFSFDP, the local density p_i is defined through the Gaussian kernel function, as follows:

$$p_i = \sum_j e^{-r_{ij}^2/2r_c^2} \tag{1}$$

where r_{ij} is the distance between point i and point j, and r_c is a cutoff distance which needs to be determined manually. The minimum distance d_i for point i is measured by computing the minimum distance between the point i and any other points of higher density values:

$$d_i = \begin{cases} \min_j (r_{ij}), & if \quad \exists j \, st \, p_j > p_i \\ \max_j (r_{ij}), & otherwise \end{cases} \tag{2}$$

Only the points with relatively high p_i and large d_i are considered as cluster centroids. This observation, which is the core of the algorithm, is illustrated by an example in

Fig. 1. Figure 1A shows 30 points from two normal distributions. Figure 1B is the decision graph which shows the plot of d_i as the function of p_i for each point. From the decision graph, the two points having relatively high local density values and large minimum distances can be easily identified. The two points are identified as cluster centroids, which are shown as filled triangle or square in both Fig. 1A and B.

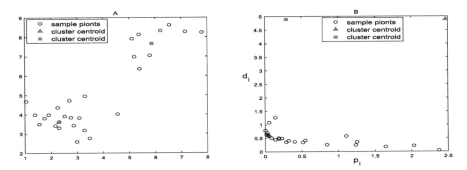

Fig. 1. The algorithm in two dimensions. (A) Points distribution. (B) Decision graph for the data in (A).

Obviously, the density estimation is important for CFSFDP method, and it directly influences the results of clustering. But in our experiments, it is found that the parameter of the kernel-based density estimation is sensitive and is not easy to be selected for different datasets. Besides, for some high dimensional or complex datasets, the produced density values are not accurate and the clustering results are not good.

3   An Novel Potential-Based Density Estimation Method for Clustering Using Decision Graph

Density estimation is crucial for the density-based clustering, such as CFSFDP. In this work, a potential-based method improved using a *Double-KNN (DKNN)* algorithm is used to estimate the density values. It can make the density estimation more effective. The *DKNN* algorithm is used to select neighbors for every point, because it can maintain local information and learn the underlying global geometry of a dataset.

3.1   The *Double-KNN (DKNN)* Algorithm

When the data points are distributed in manifold space, the traditional nearest neighbor algorithm cannot describe the neighbor distribution of each point well. So in our experiments, a *Double-KNN (DKNN)* algorithm is designed to select neighbors for each point.
    The *DKNN* algorithm has mainly two steps:

Step 1:   Getting the *KNN(i)* for each point i, where *KNN(i)* is the K nearest neighbors of point i.

Step 2: Computing the double K nearest neighbors of the each point i, $DKNN(i)$, which is defined as following:

$$DKNN(i) = \bigcup_{j \in KNN(i)} KNN(j) \tag{3}$$

3.2 Potential-Based Density Estimation Method

The potential model is derived from a gravitational force model. For two points i and j, the potential at point i from point j is given by

$$\phi_{ij}(r_{ij}) = \begin{cases} -\dfrac{1}{r_{ij}} & r_{ij} \geq \varepsilon \\ -\dfrac{1}{\varepsilon} & r_{ij} < \varepsilon \end{cases} \tag{4}$$

where ε is a distance parameter which is determined by:

$$\varepsilon = 0.01 * \max_{i=1...N} \left(\min_{j=1...N, j \neq i} (r_{ij}) \right) \tag{5}$$

where N is the total number of points.

In our experiments, the potential value at point i is the sum of the potentials at the point i produced by its nearest neighbors, $NN(i)$:

$$\phi_i = \sum_{j \in NN(i), i \neq j} \phi_{ij}(r_{ij}) \tag{6}$$

To identify nearest neighbors for each data point, $DKNN$ method are used, because different points have different number of neighbors in $DKNN$, the minimum number of $DKNN$ for each point is computed, which is defined as:

$$MinN = \min_j (|DKNN(j)|) \tag{7}$$

Then, only the $MinN$ closest neighbors in the $DKNN(i)$ are selected as $NN(i)$. When computing $DKNN$, the parameter K is determined using the following formula in our experiments:

$$K = \alpha \times \sqrt{N} \tag{8}$$

where α is a parameter which will be discussed in Subsect. 4.2.

In the following, we show that the potential model is closely related to non-parametric probability density estimation: the negative potential value of a data point i computed using (4) and (6) can be viewed as the likelihood of the point i under the probability density function estimated using a non-parametric approach similar to the Parzen window method [18].

First, if the Parzen window is defined as a hypersphere with a radius:

$$r_N = \max_{j \in NN(i)} r_{ij} \qquad (9)$$

The window function can be defined as:

$$\varphi(r) = \begin{cases} 0 & if \quad r > r_N \\ \dfrac{\beta}{r} & if \quad r_N \geq r \geq \varepsilon \\ \dfrac{\beta}{\varepsilon} & if \quad r < \varepsilon \end{cases} \qquad (10)$$

where β is the normalization factor which is used to make sure that the integral of the window function over all the feature space equals to 1.

The probability density at data point i estimated using the above new settings is then

$$\hat{p}_N(i) = \frac{1}{N} \sum_{j=1}^{N} \varphi(r_{ij}) \qquad (11)$$

It follows that $\phi_i = (-N/\beta)\hat{p}_N(i)$, which shows that the total potential value is negatively proportional to the probability density estimated by the non-parametric method. The connection between the potential field and the estimated probability density function indicates that the negative potential value at each point is equivalent to the estimated density value at the point. So the density values are derived from the potential values using $p_i = -\phi_i$ in our experiments.

3.3 Cluster Centroids Identification

After computing two quantities, the density value p_i and the minimum distance d_i, a simple threshold-based method suggested by Alex Rodriguez *et al.* [16] for selecting the cluster centroids is to use the following formula:

$$\gamma_i = p_i \times d_i \geq TH_\gamma \qquad (12)$$

The threshold parameter TH_γ has to be decided by users. In our experiments, let M to be the number of clusters in the dataset, the M_{th} largest value from the sorted set $\{\gamma_i | 1 \leq i \leq N\}$ is used to determine the threshold TH_γ.

4   Experimental Results

In this section, the evaluation of our proposed method on four synthetic and seven real-world datasets is presented. The datasets are all downloaded from the internet. The properties of those datasets are shown in Table 1.

Table 1. Details of datasets in our experiments

Dataset	N[a]	D[b]	M[c]	Source
Aggregation	788	2	7	http://cs.joensuu.fi/sipu/datasets/
Flame	240	2	2	http://cs.joensuu.fi/sipu/datasets/
R15	600	2	15	http://cs.joensuu.fi/sipu/datasets/
Spiral	312	2	3	http://cs.joensuu.fi/sipu/datasets/
Iris	150	4	3	http://archive.ics.uci.edu/ml/datasets/
Seeds	210	7	3	http://archive.ics.uci.edu/ml/datasets/
Ecoli	336	8	8	http://archive.ics.uci.edu/ml/datasets/
Wine quality-red	1599	12	6	http://archive.ics.uci.edu/ml/datasets/
Liver disorders	345	7	2	http://archive.ics.uci.edu/ml/datasets/
Glass identification	214	9	7	http://archive.ics.uci.edu/ml/datasets/
Breast cancer (Wisconsin)	699	9	2	http://archive.ics.uci.edu/ml/datasets/

[a] The number of the data points
[b] The number of features
[c] The actual number of the clusters

4.1 Evaluation Criterion

Because for all the datasets, ground truth cluster labels are available, the Fowlkes-Mallows index (FM-index) is used to evaluate the clustering result [19], which is defined as:

$$FM = \sqrt{\frac{TP}{TP + FP} \times \frac{TP}{TP + FN}} \tag{13}$$

where TP is the number of true positives, FP is the number of false positives, and FN is the number of false negatives. A higher value for the FM-index indicates a greater similarity between the clusters and ground truth.

4.2 Parameter Selection

In our method, the parameter α of *DKNN* method for computing the density values has to be determined. The three datasets including Iris, Seeds and Ecoli are used to cluster under different α values and the results are shown in Fig. 2.

It is found that good clustering results with FM-index > 0.7 can be produced within a relatively wide range of the parameter α values, while $\alpha = 0.8$ is a relatively good choice. So, the parameter $\alpha = 0.8$ is selected in our experiments.

To show the effectiveness of the *DKNN* method used in PDE, the results produced by the traditional *KNN* method using the same number of the nearest neighbors as the *DKNN* method are shown in Fig. 2 for comparison. In addition, the results of the original potential-based method (Original) which doesn't use the nearest neighbors are also shown as dashed line in Fig. 2.

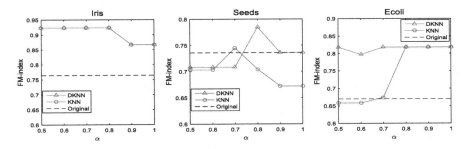

Fig. 2. The results of the clustering with different α values

It can be seen from Fig. 2 that the results of the *KNN* method can be affected more by the number of the nearest neighbors than the results of the *DKNN* method. It can also be seen that the FM-indices produced by the *DKNN* method is better than both the results of the *KNN* and the results of the original potential-based method in most cases.

4.3 Comparison of the Clustering Results

In order to evaluate the proposed PDE method, it is compared with the CFSFDP method which used Gaussian kernel density estimation method to estimate density values. For the Gaussian kernel density estimation method, the parameter r_c can be chose under the condition that the average number of neighbors is around 1% to 2% of the total number of points as suggested by Alex Rodriguez *et al.* In our experiments, the parameter r_c is determined with the condition that the average number of neighbors is around 1.5% of the total number of points. For the two different clustering methods, the FM-indices of the clustering results are recorded in Table 2, where the actual number of clusters is used as input for each dataset.

Table 2. The FM-indices of the clustering results for eleven datasets

	CFSFDP	PDE
Aggregation	0.817	**1**
Flame	**0.776**	**0.776**
R15	**0.993**	**0.993**
Spiral	**1**	**1**
Iris	0.764	**0.923**
Seeds	**0.803**	0.785
Ecoli	0.411	**0.818**
Wine quality-red	0.418	**0.503**
Liver disorders	0.546	**0.712**
Glass identification	0.331	**0.542**
Breast cancer (Wisconsin)	**0.719**	**0.719**

As shown in Table 2, the proposed PDE method has produced the best results for 10 datasets out of the total 11 datasets. Seeds dataset is the only dataset for which PDE does not produce the best results. It can be seen that the gap between the FM-index produced by PDE and the best FM-index is small for Seeds dataset. This shows the superiority of the proposed PDE method to estimate density values when applied to the datasets with various distributions and dimensionalities.

Due to the drawback of threshold-based method for identifying cluster centroids in the decision graph, although the actual number of clusters is given as input, the results of the two methods for some datasets are still bad. Some isolated points or overlapped points may produce large γ values and those points may be identified as cluster centroids. So, in our experiments, the number of clusters are varied by using $M \pm 3$ to determine multiple threshold TH_γ values, where M is the actual number of clusters. The best clustering results from the different threshold values are recorded for the eleven datasets in Table 3.

Table 3. The best clustering results for eleven datasets with the different thresholds

	CFSFDP		PDE	
	FM-index	#cluster	FM-index	#cluster
Aggregation	0.936	6	**1**	7
Flame	**0.865**	3	**0.865**	3
R15	**0.993**	15	**0.993**	15
Spiral	**1**	3	**1**	3
Iris	0.772	2	**0.923**	3
Seeds	**0.803**	3	0.791	4
Ecoli	0.448	5	**0.824**	10
Wine quality-red	0.461	3	**0.51**	3
Liver disorders	0.546	2	**0.712**	2
Glass identification	0.413	4	**0.543**	8
Breast cancer (Wisconsin)	0.719	2	**0.868**	5

As shown in Table 3, when the number of clusters are varied, the results of the two methods are both improved compared to the results in Table 2. The proposed PDE method still has a better performance than the CFSFDP method. The PDE method has produced the best results for 10 datasets out of total 11 datasets, as seen from Table 3. So, the same conclusion can be drawn that the proposed method is better than Gaussian kernel method to estimate density values.

5   Conclusion

In this paper, a novel potential-based density estimation method for clustering using decision graph is proposed. It is shown that the proposed PDE method can deal with datasets of various distributions and dimensionalities, and the improved potential method is better than the original potential method and Gaussian kernel method to estimate density values. The density values produced by improved potential-based density estimation method can describe data distribution more accurately, which leads to the improved identification of the cluster centroids in the decision graph. In future work, we plan to design a novel method to identify cluster centroids in decision graphs automatically and precisely.

Acknowledgments. This work is supported by the National Natural Science Foundation of China (Grants No. 61272213) and the Fundamental Research Funds for the Central Universities (Grants No. lzujbky-2016-k07).

References

1. Satyanarayana, A., Acquaviva, V.: Enhanced cobweb clustering for identifying analog galaxies in astrophysics. In: IEEE Electrical and Computer Engineering, pp. 1–4 (2014)
2. Menéndez, H.D., Plaza, L., Camacho, D.: Combining graph connectivity and genetic clustering to improve biomedical summarization. In: IEEE Congress on Evolutionary Computation, pp. 2740–2747 (2014)
3. Han, E., et al.: Clustering of 770,000 genomes reveals post-colonial population structure of North America. Nat. Commun. **8**, 14238 (2017)
4. Lu, Y., Wan, Y.: Clustering by sorting potential values (CSPV): a novel potential-based clustering method. Pattern Recogn. **45**(9), 3512–3522 (2012)
5. Kleinberg, J.: An impossibility theorem for clustering. In: NIPS, pp. 463–470 (2002)
6. Gan, G., Ng, M.K.-P.: k-means clustering with outlier removal. Pattern Recogn. Lett. 8–14 (2017)
7. Song, H., Yan, H.: Novel K-medoids clustering algorithm based on dynamic search of microparticles under optimized granular computing. Intell. Comput. Appl. (2016)
8. Frey, B.J., Dueck, D.: Clustering by passing messages between data points. Science **315**(5814), 972–976 (2007)
9. Serdah, A.M., Ashour, W.M.: Clustering large-scale data based on modified affinity propagation algorithm. J. Artif. Intell. Soft Comput. Res. **6**(1), 23–33 (2016)
10. Ward Jr., J.H.: Hierarchical grouping to optimize an objective function. J. Am. Stat. Assoc. **58**(301), 236–244 (1963)
11. Höppner, F.: Fuzzy Cluster Analysis: Methods for Classification, Data Analysis and Image Recognition. Wiley, New York (1999)
12. Jain, A.K.: Data clustering: 50 years beyond K-means. Pattern Recogn. Lett. **31**(8), 651–666 (2010)
13. Ester, M., et al.: A density-based algorithm for discovering clusters in large spatial databases with noise. In: KDD, vol. 96(34), pp. 226–231 (1996)
14. Ghassabeh, Y.A., Rudzicz, F.: The mean shift algorithm and its relation to Kernel regression. Inf. Sci. **348**, 198–208 (2016)
15. Campello, R.J.G.B., et al.: Hierarchical density estimates for data clustering, visualization, and outlier detection. ACM Trans. Knowl. Discov. Data (TKDD) **10**(1), 5 (2015)
16. Rodriguez, A., Laio, A.: Clustering by fast search and find of density peaks. Science **344**(6191), 1492–1496 (2014)
17. Lu, Y., Hou, X., Chen, X.: A novel travel-time based similarity measure for hierarchical clustering. Neurocomputing **173**, 3–8 (2016)

>25

18. Parzen, E.: On estimation of a probability density function and mode. Ann. Math. Stat. **33**(3), 1065–1076 (1962)
19. Fowlkes, E.B., Mallows, C.L.: A method for comparing two hierarchical clusterings. J. Am. Stat. Assoc. **78**(383), 553–569 (1983)

Optimization of Grover's Algorithm Simulation Based on Cloud Computing

Xuwei Tang[1], Juan Xu[1,2(✉)], and Ye Zhou[1]

[1] College of Computer Science and Technology,
Nanjing University of Aeronautics and Astronautics, Nanjing, China
xuweitang@nuaa.edu.cn , zhouyenuaa@gmail.com
[2] Key Laboratory of Computer Network and Information Integration,
Ministry of Education, Southeast University, Nanjing, China
juanxu@nuaa.edu.cn

Abstract. At present the scale of true universal quantum computer is still small. The quantum computer has not yet been introduced into practical applications from laboratory. Therefore, quantum simulation has become the main assistant method of verifying quantum algorithms. Grover's quantum search algorithm can speed up many classical algorithms that use search heuristics. In this work, a high performance Grover algorithm simulation is proposed combining the characteristics of Grover's algorithm and the parallelism of cloud computing, which dramatically improves the performance of the load balancing among multi-core, the utilization of memory space and the efficiency of simulation. Moreover, We propose five different computing configurations in the cloud environment suitable for quantum simulation and compare them through experimentation. We also validate the effectiveness of optimization by analysis and experimentation. The experimentation shows the simulation can reach 31 bits depending on the scale of current configuration.

Keywords: Grover's algorithm · High performance computing · Quantum computation · Quantum simulation · Parallelization · Cloud computing

1 Introduction

The jQuantum [1] is one of the quantum simulation softwares, which only supports 15 qubits at most. The simulation software QCCE, developed by G. De Raedt et al. can simulate 42 bits quantum algorithm at most [2] when running on the supercomputer JUGENE at the Ulrich Research Center. However, the supercomputer JUGENE using 262144 CPUs and 64 TB storage costs a lot of time. And the requirement in the aspect of hardware is really high. So the versatility is poor. Xiangwen Lv et al. developed quantum simulation experiment in GPU, which can realize the 25-qubit Grover's algorithm with a speedup of 23 times [3]. Although this method improves the efficiency of simulation to a

H. Yin et al. (Eds.): IDEAL 2017, LNCS 10585, pp. 83–93, 2017.
https://doi.org/10.1007/978-3-319-68935-7_10

certain extent, the problems of high hardware cost, poor versatility still exist. IBM released a free quantum computing cloud service [4] in 2016, simulating a five-qubit quantum computer. Anyone can access the quantum computer with a simple software interface. It reveals that the cloud computing has great advantages in the quantum simulation. Cloud computing now has been widely applied to various fields. It is a kind of distributed computing. Its strong parallelism coincides with the parallel concept of quantum computer. We make full use of the strong computing power of cloud server to support parallel computation. It is possible to simulate multi-bit quantum algorithm.

Grover's algorithm was first proposed by Grover in 1996 to reduce the complexity of unstructured searching problem from $O(N)$ in classical algorithm to $O(\sqrt{N})$ [5]. Grover's algorithm is a search algorithm to find M solutions from $N = 2^n$ unstructured numbers when the number of qubits is n. Due to the wide application of the knowledge in search field, it is foreseeable that Grover's algorithm has great potential.

Our work is to improve the efficiency and the number of qubits of simulation of Grover's algorithm in cloud environment to overcome the problem that the foreign and domestic quantum algorithm simulation can only simulate few qubits and the simulation efficiency is low. Moreover, to save the cost of exponential memory space, the storage of distributed cluster has been built in cloud.

The remainder of the paper is organized as follows: In Sect. 2, the brief overview of Grover's algorithm and cloud computing are given. The memory compression and the optimization of unitary operation are introduced in Sect. 3. Section 4 introduces the design of cluster system in cloud computing platform. Section 5 analyses the results and performance of simulation. Finally, we summarize the whole paper in Sect. 6.

2   Background

2.1   Grover's Algorithm

Grover's quantum search algorithm [5] is aimed to search for the needed data in a class of unstructured data. Grover's algorithm reduces the computational complexity from $O(N)$ to $O(\sqrt{N})$ by using the property of quantum. The quadratic speedup of Grover's algorithm makes it useful for a very wide range of problems. Therefore, Grover search algorithm is of great interest in reality.

Grover's search algorithm executed repeatedly Grover iterations that are written as G. It can be divided into four steps as shown in Fig. 1. The details can refer to [6].

Step I: Applying oracle to recognize the solution, it can be simple expressed as $|x\rangle (-1)^{f(x)} |x\rangle$.

Step II: Apply the Hadamard transform $H^{\otimes n}$.

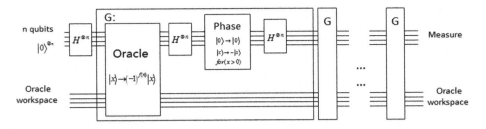

Fig. 1. Circuit frame of Grover algorithm

Step III: Performing a conditional phase shift $2|0\rangle\langle0| - I$, make the probability amplitude of every computational basis state except $|0\rangle$ receive a phase shift of -1.

Step IV: Apply the Hadamard transform $H^{\otimes n}$ again.

Repeating the Grover iteration about $R = O(\sqrt{N/M})$ times we can obtained solutions with a high probability when the first n quantum bits are measured.

2.2 Advantages of Cloud Computing

Cloud computing technology aims to improve resource utilization and resource flexibility by integrating, optimizing resources and sharing virtualized resources. It makes full use of all the hardware resources and reduces the waste of resources. According to the needs, this technique dynamically assigns idle resources in network nodes, which forms a resource pool. It realizes centralized management and output to resources and solves the problem of resource demand of the algorithm. The use of multi-process concurrency strategy, cloud computing's strong computing power and the mainstream parallel configuration technology to build a simulation platform could speed up the computation of exponential level in quantum algorithm simulation greatly.

Microsoft Azure is Microsoft's cloud-based operating system. Azure with flexibility, scalability, stability, versatility, powerful computing power not only allows user to create virtual machines for research conveniently, but also reduces hardware cost of the research. Thus cloud computing platform provides an effective way for quantum algorithm simulation.

3 Optimization

We know the evolution of a closed quantum system is described by a unitary transformation. That is, the state $|\varphi\rangle$ at t_1 of the system is related to the state $|\varphi'\rangle$ at t_2 simply by a unitary operator U: $|\varphi'\rangle = U|\varphi\rangle$.

It can be seen that unitary transformation plays a pivotal role in the quantum algorithms. Thus we have compressed unitary matrix to reduce the memory resources of simulation. At the same time, we have optimized unitary operations to improve the operating efficiency manyfold.

3.1 High Compression of Memory

Grover simulation programs running in stand-alone environment have a serious problem with insufficient memory. It is the primary factor limiting the simulation to continue. In [7], the optimization of quantum circuit simulation is achieved by compressing sparse unitary matrix using a mathematica package. Besides, we found that a large number of zero elements and repetitive elements exist in unitary matrices. In view of this, we proposed a memory compression method with the following data structure. And memory utilization will be greatly improved.

In step I, the Oracle operator is $O = I - 2\,|\beta\rangle\,\langle\beta|$. There are a large number of zero elements in Oracle operator:

$$\begin{cases} i \neq j, O_{ij} = 0; \\ i = j = x, O_{ij} = -1; \\ i = j \neq x, O_{ij} = 1; \end{cases} \tag{1}$$

Combining steps II, III, IV the result is

$$U = H^{\otimes n}(2\,|0\rangle\,\langle 0| - I)H^{\otimes n} = 2\,|\varphi\rangle\,\langle\varphi| - I \tag{2}$$

The elements in matrix U are

$$\begin{cases} i \neq j, U_{ij} = 2^{1-n}; \\ i = j, U_{ij} = 2^{1-n} - 1; \end{cases} \tag{3}$$

Combining with the above two formulas about operator O and U, we get

$$G = UO = (2\,|\varphi\rangle\,\langle\varphi| - I)(I - 2\,|\beta\rangle\,\langle\beta|) \tag{4}$$

And the elements in matrix G are

$$\begin{cases} i \neq j, G_{ij} = 2^{1-n}; \\ i = j \neq x, G_{ij} = 1 - 2^{1-n}; \\ i = j = x, G_{ij} = 2^{1-n} - 1; \end{cases} \tag{5}$$

O_{NN} is a diagonal matrix. We store the N elements in diagonal line of matrix O by a single-dimensional array. After compression, its storage space reduces from $O(N^2)$ to $O(N)$.

For the matrix U, its elements just have two different values. Those are diagonal element value and other value. Considering the structure of matrix U, we construct a single-dimensional array that has two elements: $a[1] = 2^{1-n}$, $a[2] = 2^{1-n} - 1$, making the storage space reduce from $O(N^2)$ to $O(2)$ successfully.

And by the observation about the matrix G, we can find that the effect of combining with U and O is to invert the whole values of the column x in U.

3.2 Speedup of Unitary Operation

We write unitary operation as $Q = M * N$ where M is a $x_1 * y_1$ matrix and N is a $x_2 * y_2$ matrix and $y_1 = x_2$. The tradition algorithm contains three-tier loop that time complexity is $O(x_1 * y_1 * y_2)$, which has low efficiency.

In tradition algorithm, regardless of whether the elements in M or N are zero the operations must be performed. In fact these operations are invalid. Eliminating these ineffective operations the efficiency of simulation can be improved.

Because of temporal and spatial locality, CPU puts not only the data to be accessed into cache but also the adjacent data into cache as a block. In the quantum computation, with the number of qubits increasing, the unitary matrix is quite a large. In the traditional algorithm the size of unitary matrix is stored by rows in memory, but during operation the multiplicand is accessed by column. Because cache can't load with the required data, the computer will keep reading data from memory and constantly flush cache, resulting in cache conflict. It will waste a lot of time. Thus in the operating process we should try to avoid reloading cache.

For this purpose, combining with the formulas of unitary operation and the properties of Grover's algorithm, we propose an optimized unitary matrix multiplication:

$$G_{ij} = \begin{cases} a_1 * b_j, i \neq j; \\ a_2 * b_j, i = j; \end{cases} \tag{6}$$

G act on the equally weighted superposition of states $|\varphi\rangle$, which is single iteration. The effect of $G\,|\varphi\rangle$ is equal to add the whole elements in each row in G and multiply $1/\sqrt{2^n}$ after that.

3.3 Parallel Simulation of Unitary Operation

Hadoop takes HDFS and MapReduce as the core. It provides the software framework for processing massive data in distributed cluster and makes the simulation of multi-qubit unitary operation possible.

The unitary G is the product of the unitary U and unitary O. It is $G_{NN} = U_{NN} * O_{NN}$, where $N = 2^n$ and n is the number of qubits. So we can get expression:

$$G_{ij} = U_{i1} * O_{1j} + U_{i2} * O_{2j} + U_{i3} * O_{3j} + \cdots + U_{iN} * O_{Nj} \tag{7}$$

where $1 \leq i, j \leq N$.

By the calculating process of G_{ij}, we know that the calculation of each element in G is independent. In Map stage, we collect the elements for calculating G_{ij} in the same key. Then in Reduce stage we can parse each element from it to calculate G_{ij}. The algorithm steps are as followed:

Step I: For MapReduce calculating platform in Hadoop, the U and O are structured to be stored in the Hadoop distributed file system HDFS.

Step II: Traverse U and O at the same time, and extract elements of matrix and construct Key/Value. Then these Key/Value are sent to Map().

Step III: Intermediate results produced by Map() are sorted and merged. And all the Values with the same Key are collected to form a new list $<Key, List<Value>>$. Send the new list to Reduce().

Step IV: In Reduce(), the Values are parsed by traversing the list $<Key, List<Value>>$, by which we can get G_{ij}.

We choose a cluster with 8 sub nodes (single-core+3.5 GB memory+500 IPOS each) in cloud environment. We measured the three set of data of qubits $= 11$, 12, 13 and compared with traditional algorithm before optimization.

Table 1. Execution time(ms) of one Grover iteration(OGI) and one Grover optimized iteration(OGOI).

Qubits(n)	11	12	13
OGI	12574	46443	600655
OGOI	1682	5588	22701

It is shown in Table 1 that the time consumption of the traditional algorithm increases sharply with the increase of the number of qubits and the optimized algorithm significantly improves the computing speed. The speed increase is more obvious as the number of qubits increases.

4   Configuration of Simulation on Cloud Computing Platform

At present, there have been some simulations of quantum algorithm in cloud environment [8–10]. However, the simulations of Grover's algorithm are mainly developed in GPU and supercomputer. In view of the advantages of cloud computing, we design cloud computing parallel models on cloud platform to improve the efficiency of simulation.This paper adopts the configuration of 1 host node - n sub nodes on cloud platform. The host node is responsible for sending commands to the sub nodes and coordinating the communication between sub nodes while sub nodes, computing in parallel and sharing memory inside, perform unitary transformations using multi-threaded concurrent operations. Finally, the sub nodes feed back to the host node the results.

We divide the complexity of the original problem into $O(\sqrt{2^a * 2^b})$, where $n = a + b$, by the strategy of divide-and-conquer and parallelism so that the complexity of the original problem reduces from $O(\sqrt{2^n})$ to $O(\sqrt{2^{n-a}})$.

When the virtual machine is single-core, the concurrent model suitable for quantum algorithms is that the whole sub machines execute operations in parallel while inside each sub machine the operations are executed by multi-process, as shown in Fig. 2.

Here we use a crucial loop to control the number of concurrent processes and distribute them averagely to the N sub machines. The sub machines execute sub processes in parallel. In this case, the efficiency is high.

When the virtual machine is multi-core, through some testings we found that when multiple processes work concurrently the CPU is in an overload state so

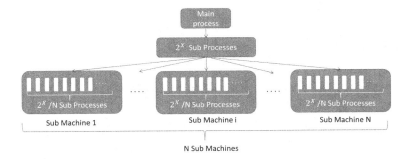

Fig. 2. The concurrent cluster model in single-core virtual machines

that blocking in the execution queues of processes will take place. And the multi-core virtual machines execution queue will conflict. Thus the model suitable for multi-core virtual machines is that sub machines work in parallel and insides sub machines the processes are executed serially.

In addition, during the multi-core computer executing the algorithm, the task allocation of OS is based on the time-sliced circular scheduling. There is a load balancing problem in multi-core CPU. It always happens that the occupancy of one or two CPUs of all is rather high but other CPUs are basically idle. Therefore not all the CPUs work in parallel in fact. In view of this, we use OpenMp to realize parallel programming and to optimize the simulation in multi-core virtual machines. Based on the use of multi-threaded parallel load can be allocated to multiple cores, where the algorithm execution is speeded up and CPUs utilization is improved.

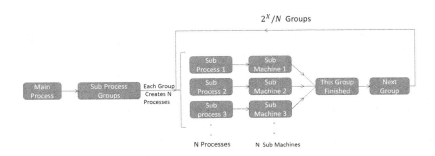

Fig. 3. The concurrent cluster model in multi-core virtual machines

As shown in Fig. 3, the host machine divides 2^X parallel processes into Q groups, where N is the number of sub-machines and Q is $2^X/N$. One Grover's algorithm is implemented by each sub VM in each group. Until all the programs finish the host machine reassign work in the next group.

Through plenty of testings, the parallel computing model in cloud environment is suitable for a variety of simulations of quantum algorithms, which all improves the efficiency of simulation successfully.

5   Experiments and Performance Analysis

The experiments reported in this paper are performed on Microsoft Azure. The simulation of Grover's search algorithm is based on libquantum [11] which is one of the most representative platforms for quantum simulation.

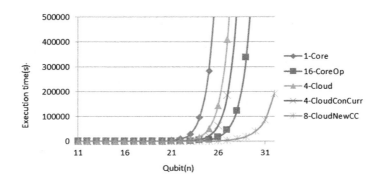

Fig. 4. Execution time of five computing configurations in cloud environment

In order to reduce the impact of network latency and consumption in communication between virtual machines on experimental data, we average three measurements for each experimental data. As shown in Fig. 4, 1-Core is the time curve that single virtual machine (single-core+3.5 GB memory+500 IPOS) implements the quantum search algorithm. It can be seen that as the number of qubits increasing the execution time of simulation increases exponentially. The simulation of 1-Core can reach 23 qubits at most because of insufficient memory. 16-CoreOp is the time curve of simulation in single 16-core virtual machine(16-core+32 GB memory+51200 IOPS) after multi-core optimization. 4-Cloud shows the time curve of simulation in the configuration of 1 host node-4 sub nodes (each is single-core+3.5 GB memory+500 IPOS. The same below.). 4-CloudConCurr replaces the whole single-core machines in 4-Cloud with dual-core machines (dual-core+7 GB memory+6400 IPOS). And 8-CloudNewCC is based on the configuration of 4-CloudConCurr extending to 8 sub machines.

Taking $n = 20$ as an instance, execution time in 1-Core is 1908.463 s. Time in 16-CoreOP is 25.161 s. So through multi-core optimization the performance of 16-Core is about 76 times higher that of single-core virtual machine where the main restraining factors are the number of cores and IOPS parameter. It takes 310.515 s in 4-Cloud and the time reduces further to 71.502 s in 4-CloudConCurr. The experiments finally adopt the configuration of 8-CloudNewCC. Only in 23.210 s the quantum search algorithm simulation can be completed correctly. With the speedup up to 82 times, the efficiency of simulation has been improved dramatically. And the speedup effect is more and more visibly as the number of qubits increasing. But when we add the number of machines to a certain extent, the host's cost in the aspect of communication and load will be further increased. Once the host machine's CPU is nearly full load, the optimization effect will be affected. So the cost-performance ratio of system will be reduced.

In the latest computing configuration 8-CloudNewCC, the efficiency of Grover simulation algorithm is improved up to several times or even thousands of times. As the number of qubit is n and the solution is M, the number of Grover iteration is $O(\sqrt{2^n/M})$. Consider the condition that the solution is 1. And the number of Grover iteration is $O(\sqrt{2^n})$. Thus the complexity of classical Grover simulation process is $O(2^{5.5n})$. The complexity of unitary operation is $O((m+r)2^{3n})$ according to m Map tasks and r Reduce tasks. $2^a/k$ processes (where 2^a is the total number of concurrent processes of the host machine and k is the number of sub nodes) in each sub node execute Grover iterations of $O(\sqrt{2^{n-a}})$ times in parallel. The communication consumption between the host node and the sub nodes is $k\omega$. The network delay is δ. Thus the complexity of simulation process of Grover's algorithm is $O(\sqrt{2^{7n-a}})$ and as the number of qubits is large enough the extra-cost $\Delta_{extra} = k\omega + \delta$ can be approximated by a constant.

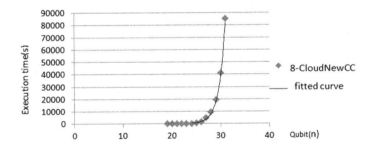

Fig. 5. The fitted curve of 8-CloudNewCC

13 sets of data among simulation results are validated as shown in Fig. 5. We obtain fitted curve which is $t = 4*10^{-6}2^{1.108855n} + 23(s)$. The results and performance analysis are in excellent agreement and the effectiveness of the proposed approach is validated. The expected simulation time according to analysis is $\lambda 2^{\gamma n}$ where the values of both λ and γ are reduced. The decrease of γ is especially obvious.

In respect of memory compression, we use the strategy of divide-and-conquer to divide the cost of memory into 2^a components and compress the memory of unitary matrix, which improves the memory utilization dramatically. The simulation running in 8-CloudNewCC with the virtual machine of 7 GB memory can reach 31 qubits within one day and it takes about 23 h. And with each increasing in the qubit, the memory also needs to be extended severalfold.

In summary, due to the determination of the algorithm framework which has scalability and extendibility increasing the number of virtual machines and enhancing the performance of virtual machines to some extent can reduce the both values of λ and γ so that the efficiency of simulation algorithm can be improved again.

6   Conclusion

Grover's algorithm is one of the two most representative quantum algorithms (Grover's algorithm and Shor's algorithm [12]). In this paper, we propose an optimized method for simulation of Grover's algorithm in cloud environment. For the optimization of unitary operation, an optimized algorithm is proposed which greatly saves storage space and improves the efficiency of unitary operations. We have established multi-qubit parallel computing model by taking advantages of the cloud computing platform and improved the efficiency of Grover algorithm simulation successfully. Finally, the high performance and the expansibility of the simulation system are evaluated by the comparison of experimental results and performance analysis, which provides effective platform supporting for further theoretical research of quantum simulation and quantum algorithm.

Acknowledgments. This work is supported by Natural Science Foundation of Jiangsu Province, China (Grant No. BK20140823), Fundamental Research Funds for the Central Universities (Grant No. NS2014096).

References

1. Petruccione, F., Senekane, M., Mohapi, L.: Simulating quantum circuits with gnu octave and python. AFRICON **2153**(25), 1–4 (2013)
2. EurekAlert: World record: Julich supercomputer simulates quantum computer (2015)
3. Lu, X., Yuan, J., Zhang, W.: Workflow of the Grover algorithm simulation incorporating CUDA and GPGPU. Comput. Phys. Commun. **184**(9), 2035C2041 (2013)
4. Ibm quantum experience (2016)
5. Grover, L.K.: A fast quantum mechanical algorithm for database search. In: Annual ACM Symposium on Theory of Computing, pp. 212–219 (1996)
6. Nielsen, M.A., Chuang, I.L.: Quantum Computation and Quantum Information, 10th Anniversary Edition, vol. 21(1), pp. 1–59 (2010)
7. Gerdt, V.P., Kragler, R., Prokopenya, A.N.: A mathematica package for simulation of quantum computation. In: Gerdt, V.P., Mayr, E.W., Vorozhtsov, E.V. (eds.) CASC 2009. LNCS, vol. 5743, pp. 106–117. Springer, Heidelberg (2009). doi:10. 1007/978-3-642-04103-7_11
8. Huang, H.-L., et al.: Homomorphic encryption experiments on IBMs cloud quantum computing platform. Front. Phys. **12**(1), 120305 (2016)

 9. Zhao, L.: Cloud computing resource scheduling based on improved quantum particle swarm optimization algorithm. J. Nanjing Univ. Sci. Technol. **40**(2), 223–228 (2016)
10. Ding, W., Guan, Z., Shi, Q., Wang, J.: A more efficient attribute self-adaptive co-evolutionary reduction algorithm by combining quantum elitist frogs and cloud model operators. Inf. Sci. **293**(2), 214–234 (2015)
11. Weimer, H., Butscher, B.: Libquantum library. http://jquantum.sourceforge.net/
12. Shor, P.W.: Polynomial-time algorithms for prime factorization and discrete logarithms on a quantum computer. In: Quantum Entanglement and Quantum Information-Proceedings of Ccast, pp. 1484–1509 (1999)

Cross-Media Retrieval of Tourism Big Data Based on Deep Features and Topic Semantics

Yang Li, Junping Du$^{(\boxtimes)}$, Zijian Lin, and Lingfei Ye

Beijing Key Laboratory of Intelligent Telecommunication Software
and Multimedia, School of Computer Science, Beijing University
of Posts and Telecommunications, Beijing 100876, China
junpingdu@126.com

Abstract. As an Internet application, smart tourism has greatly enriched the tourism information. In this paper, we propose a unified modeling and expression method of attraction texts and images based on the text deep representation model and convolution neural network. According to the cross-media characteristics of tourism big data, we propose a semantic learning and analysis method for cross-media data, and correlate tourism texts with images based on deep features and topic semantics. Experimental results show that the proposed method can achieve better results for semantic analysis and cross-media retrieval of tourism big data.

Keywords: Text deep representation model · Convolution neural network · Semantic learning · Topic semantics · Cross-media retrieval

1 Introduction

With the rapid development of cross-media technologies, the tourism information we get contains not only texts, but also images, video and other types of data, which greatly enriched the tourism information sources. However, the content of tourism image is complex, but easy to understand, making it the primary problem of tourism information.

Text mining technologies based on natural language have made great achievements, but we still face the problem of only obtaining limited semantic features. The biggest problem in semantic learning and image recognition is how to cross the semantic gap. Traditional methods only analyze data of one modality and ignore the relations of data of different modalities. It is difficult to establish the relation between text and image, while tourism image has ambiguity and uncertainty, which brings great obstacles to the semantic analysis and recognition of the attractions.

2 Related Work

There have been a lot of methods of cross-media retrieval and image semantic analysis. In view of cross-media retrieval, Rasiwasia [9] learned text features by LDA, extracted image features by SIFT, and the relation was analyzed by canonical correlation

© Springer International Publishing AG 2017
H. Yin et al. (Eds.): IDEAL 2017, LNCS 10585, pp. 94–102, 2017.
https://doi.org/10.1007/978-3-319-68935-7_11

analysis. Feng [10] referred to the shared layer to establish the correlation of two modalities, or used a two-stage framework for two modal semantic learning and established low-dimensional public representation space. As for image semantics, Zhao [3] proposed a deep semantic sorting method based on learning hash function and preserving multi-level semantic similarities between multi-marker images. Ren [8] used region generation method to generate the candidate region. The trained model was used to extract the features on these regions.

All the methods above only obtain basic information from tourism images. Traditional method is not able to analyze the rich semantics of tourism images. Therefore, we propose a learning method based on deep features and topic semantics to improve cross-media retrieval of tourism data.

3   Cross-Media Retrieval Method of Tourism Big Data

3.1   Deep Representation of Tourism Texts and Images

The first part is semantic representation of tourism texts based on text deep representation model. The optimized training model expresses a word into a vector and the tf-idf value is calculated by text deep representation model. According to the tf-idf value and kmeans clustering, we can calculate the text category distribution, and a vectored representation of tourism text data is obtained, as shown in Fig. 1.

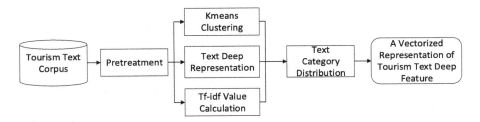

Fig. 1. Vectored representation of tourism text data

The second part is tourism image semantic representation based on convolution neural network. The network consists of 8 layers. The 6th to the 8th layer are fully connected, each layer uses the convolution kernel of size 3*3, and the deep features are normalized by cosine distance measure, as shown in Fig. 2.

3.2   Semantic Learning and Modeling Based on Deep Features and Topic Semantics

We model the text and image deep features by tourism topic semantics, and map into a common potential tourism semantic space. The two types are learned to get the tourism topic semantic distribution, according to the weight of image and text which are determined by the contribution to the semantic content. When the distribution of image

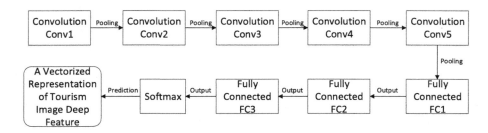

Fig. 2. Vectored representation of tourism image data

or text appear peaks or each dimension is sparse, tourism image or text learning is more reliable, so we give a bigger weight for the related image or text. Assuming that the number of topics of tourism images and texts is m and n, respectively, the number of the relation is k = m + n, the corresponding semantic models of images and texts are represented by i and t, the corresponding topic semantic distributions are expressed as P (i|d) and P(t|d). The correlation semantic distribution is P(z|d), shown as Formula (1).

$$P(z_k|d_i) = \begin{cases} \alpha_{ii}P_i(i_k|d_i), k=1,2,\ldots,m \\ \alpha_{ti}P_t(t_{k-m}|d_i), k=m+1,m+2,\ldots,m+n \end{cases} \tag{1}$$

The α_{ii} and α_{ti} are the weights of images and texts, calculated by Formulas (2) and (3):

$$\alpha_{ii} = \begin{cases} 1, H(v(d_i)) \leq 3 \\ \exp(3-H(v(d_i))), H(v(d_i)) > 3 \end{cases} \tag{2}$$

$$\alpha_{ti} = 1 - \alpha_{ii} \tag{3}$$

The $H(v(di))$ is the entropy of image feature vector distribution $v(di)$.

Our method is shown in Fig. 3. The first part is topic semantic analysis and modeling of tourism cross-media big data. The main steps are as follows:

1. Train the text deep representation model by tourism text corpus, and calculate tourism text category distribution by tf-idf value and kmeans clustering. The text representation is vectored, and the corresponding text is denoted as t(di)
2. Extract deep features of tourism images by convolution neural network, and cluster and normalize the image representation denoted as i(di)
3. Use tourism topic semantic model to model i(di) and t(di) to get the model parameters of text and image: P(i|z), P(Zi|di) and P(t|z), P(Zt|di)
4. Calculate the weights α_{ii} and α_{ti} of texts and images by Formulas (2) and (3)
5. Calculate the correlation semantic distribution P(z|di) by Formula (1)

The second part is cross-media retrieval by tourism topic semantic correlation. Input tourism text and image, and output retrieval results for test tourism texts and images. The main steps are as follows:

1. Get topic semantic distribution P(z|d) from the first part. The model parameters P(i|z) and P(t|z) are estimated by expectation maximization algorithm.

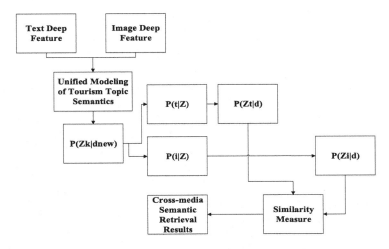

Fig. 3. Tourism cross-media retrieval based on deep features and topic semantics

2. Extract deep features of the new tourism text or image dnew, quantify them by clustering to obtain the vector representation v(dnew)
3. Topic semantic distribution P(Zk|dnew) is calculated by expected maximization algorithm, and get the vector distribution of the image or text by Formulas (4) and (5)

$$P(Zt|d_{new}) = \sum_{k=1}^{k} P(t|z_k)P(z_k|d_{new}) \qquad (4)$$

$$P(Zi|d_{new}) = \sum_{k=1}^{k} P(i|z_k)P(z_k|d_{new}) \qquad (5)$$

4. Measure similarity in the test dataset, and obtain the retrieval results

4   Experimental Results and Analysis

The experimental data is obtained from Baidu tourism, donkey mother, ant honeycomb and other websites, located in Beijing/Tianjin/Hebei regions. The total set is 17239 pairs, divided into training set (12239) and test set (5000). We established the topic semantic model of training set, and get the semantic distributions of texts and images. The initial representations are vectors of 200 dimensions and 4096 dimensions.

The result of semantic correlation is a two-dimensional matrix of 12239 × 200 shown in Table 1. The rows represent each pair of text and image. The columns represent 200 deep topic semantics. Each value represents the probability for each image-text pair belongs to each topic semantic.

Table 1. Semantic correlation of tourism texts and images

Data	Topics				
	1	2	...	199	200
1	0.0059397359	3.9081932e-07	...	6.23917532e-04	5.301861e-06
2	4.8991838e-04	0.0067303681	...	8.13594833e-03	4.155508e-07
...	...	...	...	...	...
12238	8.00598621e-03	6.11219763e-08	...	4.30989933e-06	7.68732428e-05
12239	1.50198018e-04	0.0166889381	...	5.39892578e-03	6.00783767e-08

These topic semantics contain the text and image information in the training set, and reveal the implicit mapping between them. Our method can better express potential semantic relation between tourism data of different modalities. We divided tourism images into different categories: food, transportation, park, architecture, and so on, and are classified as category serial number. As shown in Fig. 4, the retrieval results of Shichahai are: Shichahai, Houhai, Ming Tombs Reservoir, Zizhuyuan, Yuanmingyuan Kyushu area, Yanqi Lake. The retrieval precision is shown in Fig. 5 and Table 2.

Fig. 4. Examples of retrieval results of tourism images

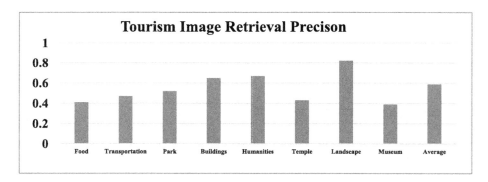

Fig. 5. Tourism image retrieval based on deep features and classification of attractions

As can be seen from the results above, Landscape, Buildings and Humanities have a higher image retrieval precision. It indicates that the characteristics and semantics of these categories are more focused and distinguishable. We need to obtain more representative tourism images, increase the amount of training data, and improve the

Table 2. Toursim attractions image retrieval precision of different categories

Tourism attractions	AP	Tourism attractions	AP
Landscape_Yuanmingyuan Kyushu Scenic Area	0.828824	Park_Taoranting Park	0.291667
Landscape_The Ming Tombs Reservoir	0.950000	Park_Yuyuantan Park	0.500000
Landscape_Shichahai	0.737213	Buildings_People's Heroes Monument	0.525974
Landscape_Jinshanling Great Wall	0.881818	Buildings_Water Cube	0.657778
Museum_Capital Museum	0.712778	Museum_Chinese Art Museum	0.600000
Museum_National Palace Museum	0.798366	Museum_National Museum of China	0.683333
Temple_Dingling underground palace	0.590608	Humanities_Lao She teahouse	0.676505
Temple_Fayuan Temple	0.511905	Humanities_ Tobacco Slant Street	0.517316

classification categories. Based on deep features of convolution neural network and principal component analysis, we reduce the dimension to get better results.

We compared with different methods of tourism cross-media retrieval on our tourism texts and images dataset. The results of different methods of the same attractions, as well as the image-to-text results and text-to-image results from the same attraction with the same method are presented.

According to the deep topic semantic distribution of text and image of the training set, compare similarity measure to test set, and evaluate by the retrieval precision = B/A, where A is the number of all retrieval results, B is the number of times the retrieval result is correct. WCC is Word2vec + CNN + CCA, WSTS is Word2vec + SIFT + Topic Semantics, LCTS is LDA + CNN + Topic Semantics, WCTS is Word2vec + CNN + Topic Semantics, TtoI is the text-to-image retrieval, and ItoT is the image-to-text retrieval. Results are shown in Table 3 and Fig. 6.

Results show that modeling of deep features and topic semantics can improve the accuracy of tourism cross-media retrieval. As the text deep topic semantic is expressed better than the image deep topic semantic, the text-to-image retrieval precision is higher than the image-to-text retrieval precision. Among these representative tourism attractions, the cross-media retrieval of the Great Wall gets the best precision results. It shows that the Great Wall has rich and personalized semantics.

Table 3. Comparison of different methods of tourism cross-media retrieval results

Sight	Forbidden City		Summer Palace		Great Wall		Yuanmingyuan		Average		
Method	TtoI	ItoT	TtoI	ItoT	TtoI	ItoT	TtoI	ItoT	TtoI	ItoT	All
WCC	0.39	0.42	0.48	0.45	0.49	0.54	0.41	0.45	0.41	0.45	0.43
WSTS	0.47	0.51	0.54	0.51	0.58	0.53	0.55	0.52	0.52	0.49	0.51
LCTS	0.53	0.52	0.58	0.53	0.61	0.56	0.54	0.52	0.55	0.52	0.54
WCTS	0.64	0.59	0.57	0.57	0.71	0.69	0.66	0.67	0.62	0.55	0.59

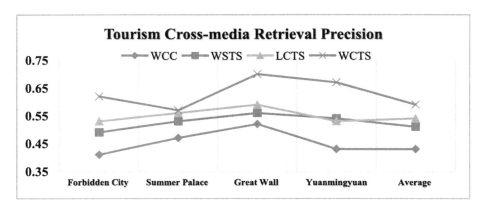

Fig. 6. Comparison of different attractions of tourism retrieval precision

Table 4. Retrieval results of different numbers of tourism topic semantics

Method	Topic		
	20 + 20	50 + 50	100 + 100
WSTS	0.29	0.41	0.51
LCTS	0.35	0.42	0.54
WCTS	0.38	0.46	0.59

We also did experiments of different numbers of tourism topic semantics. We set the number of tourism topic semantics of text and image as 20, 50 and 100.

As shown in Table 4 and Fig. 7, as the number of topic semantics is increasing, the capacity of the model is increasing, the relation between text and image can be fully expressed, and the retrieval result has been improved. With the increase of the number of

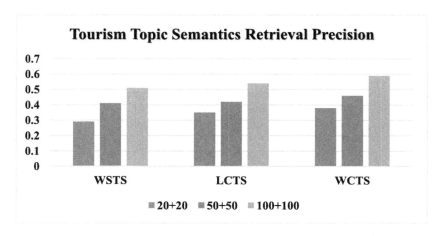

Fig. 7. Comparison of different tourism topic semantics retrieval precision

topic semantics, the retrieval results of the experimental results are flattened, and the increase of the dimension will also affect the computational efficiency. Therefore, we set 100 topic semantics to achieve good tourism cross-media retrieval results, and our method still get better results with different number of tourism topic semantics.

5 Conclusions

We propose a new method of cross-media retrieval of tourism big data. We combined the structural characteristics and information characteristics of tourism-related web pages on the Internet, to obtain valuable tourism data and get the corresponding texts and images of attractions. The correlation of deep feature between texts and images is learned, and the analysis and modeling of tourism topic semantics for attractions are more accurate and specific in semantic level. Experimental results show that the method we proposed can achieve better results of semantic analysis and retrieval of tourism cross-media data.

Acknowledgment. This work is supported by the National Natural Science Foundation of China (No. 61320106006, No. 61532006, No. 61502042).

References

1. Gao, J., Yang, J., Zhang, J., et al.: Natural scene recognition based on convolutional neural networks and deep boltzmann machines. In: IEEE International Conference on Mechatronics and Automation. IEEE (2015)
2. Jiang, Q.Y., Li, W.J.: Deep Cross-Modal Hashing (2016)
3. Zhao, F., Huang, Y., Wang, L., et al.: Deep semantic ranking based hashing for multi-label image retrieval. In: Computer Vision and Pattern Recognition, pp. 1556–1564. IEEE (2015)
4. Donahue, J., Jia, Y., Vinyals, O., et al.: DeCAF: a deep convolutional activation feature for generic visual recognition. Comput. Sci. **50**(1), 815–830 (2013)
5. Lin, G., Shen, C., Wu, J.: Optimizing ranking measures for compact binary code learning. In: Fleet, D., Pajdla, T., Schiele, B., Tuytelaars, T. (eds.) ECCV 2014. LNCS, vol. 8691, pp. 613–627. Springer, Cham (2014). doi:10.1007/978-3-319-10578-9_40
6. Wang, J., Song, Y., Leung, T., et al.: Learning fine-grained image similarity with deep ranking. Comput. Sci., 1386–1393 (2014)
7. Xia, R.: Supervised hashing for image retrieval via image representation learning. In: AAAI Conference on Artificial Intelligence (2014)
8. Ren, S., He, K., Girshick, R., et al.: Faster R-CNN: towards real-time object detection with region proposal networks. IEEE Trans. Pattern Anal. Mach. Intell., 1–11 (2016)
9. Rasiwasia, N., Costa Pereira, J., Coviello, E., et al.: A new approach to cross-modal multimedia retrieval. In: International Conference on Multimedia, pp. 251–260. ACM (2010)
10. Feng, F., Wang, X., Li, R.: Cross-modal retrieval with correspondence autoencoder. In: ACM International Conference on Multimedia, pp. 7–16. ACM (2014)

11. Pereira, J.C., Vasconcelos, N.: On the regularization of image semantics by modal expansion. **157**(10), 3093–3099 (2012)
12. Liu, J.: semantic analysis and classification of food safety emergencies cross media information. Beijing University of Posts and Telecommunications (2013)
13. Wu, F., Lu, X., Song, J., et al.: Learning of multimodal representations with random walks on the click graph. IEEE Trans. Image Process. **25**(2), 1–11 (2015)

Information Retrieval with Implicitly Temporal Queries

Jingjing Wang and Shengli Wu[✉]

School of Computer Science, Jiangsu University, Zhenjiang 212013, China
s.wu1@ulster.ac.uk

Abstract. Time is an important factor that needs to be considered in information retrieval. Often queries raised to search engines may have temporal intention, or the search results that the user interests are time-dependent. In this paper, we deal with one type of temporal queries – implicitly temporal queries. Those queries do not contain explicitly temporal expressions, but the user's information need is related to time. For a given query, we analyze its temporal intention and set up two linear combination models. One of them is based on the language modeling, and the other is based on the metric space model. Both of them consider both aspects of content and time to rank all the documents involved. Experiment results with the dataset that was used in the Temporal Information Access task of NTCIR-11 show that our approaches are effective.

Keywords: Temporal IR · Implicitly temporal query · Query intent · Ranking algorithm

1 Introduction

With the development of the Web, users rely more and more on search engines to find useful information. However, in almost all search engines, any individual query is allowed to include a limited number of words. Sometimes such queries cannot express users' retrieval intention completely and accurately, such as their temporal intention. Researchers found that more than 7% of the web queries contained implicitly temporal intention [1]. For example, a user submits a query "Iraq war". Even it does not express time explicitly, this query is time-sensitive because in 2003 the war started and in 2011 the war ended. Therefore, it is a meaningful undertaking to deal with the potentially temporal intention of queries, and the ignorance of the time factor in query processing may decrease retrieval effectiveness significantly.

One typical method of inferring temporal intention for implicitly temporal queries is to examine the meta-data of related web pages [2, 3]. However, web documents do not always have a credible creation date. Another problem is that the creation date of a web page does not necessarily reflect the temporal intention of the document. In such situations meta-data-based methods do not work. Therefore, we explore content-based approaches. For a given query, we obtain a ranked list of documents from the search engine that only considers content relevance. Then the time expressions in the top k documents are extracted and served as temporal intention of the query.

© Springer International Publishing AG 2017
H. Yin et al. (Eds.): IDEAL 2017, LNCS 10585, pp. 103–111, 2017.
https://doi.org/10.1007/978-3-319-68935-7_12

The documents are re-ranked based on a combination of content relevance and temporal relevance. The main contribution of this paper is as follows:

- For implicitly temporal queries, we propose an approach that considers the top k documents in the resultant list to analyze their temporal intention.
- On the basis of the language modeling and the metric space model, we propose two ranking methods for implicitly temporal queries. The ranking methods consider both temporal relevance score and topical relevance.
- Extensive experiments are conducted to evaluate the performance of the methods proposed by us and several existing methods. Experimental result shows that our methods are effective.

2   Related Work

As a branch of information retrieval, Temporal Information Retrieval (TIR) [4] considers both topical relevance and users' temporal intention. Those queries are referred to as time-sensitive queries. They can be divided into two types: explicit and implicit. As its name indicates, explicit queries include temporal expressions explicitly in the query. Some research has been done to deal with such queries, for example, in [5–8].

The other type is implicit queries. Those queries do not contain temporal expressions explicitly themselves, but relevant documents tend to be related to a certain period of time. In [2], queries were dated by using query keywords, contents of the top-k documents, or the timestamps of the top-k documents. Kanhabua et al. applied some machine learning methods to search logs so as to detect event and query intention [9]. Also using search logs, Shokouhi and Radinsky proposed a method for automatic query completion that considers both time-sensitivity and personalization through time-series analysis [10].

As for ranking methods, [11] incorporated time into language models and assigned each time period an estimated relevance value. Lin et al. proposed a TASE (Time-Aware Search Engine) model [12], which can extract explicitly and implicitly temporal expressions, and calculate temporal relevance score between web pages and time expressions, re-ranking documents according to the combination of temporal relevance and text similarity.

Our research work is mainly focused on implicitly temporal queries. The most relevant paper is [5]. One of our document ranking methods is similar to the one proposed in [5]. However, their method works for explicit queries, while ours is for implicit queries.

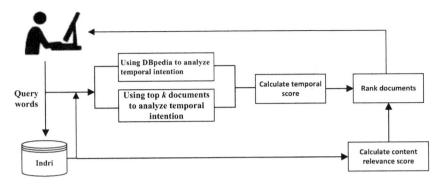

Fig. 1. Modules of the information retrieval system for implicitly temporal queries

3 Metholodogy

The infrastructure of the information retrieval system that supports implicitly temporal queries is shown in Fig. 1. Based on a well-known traditional information search system Indri[1], we add some more modules. For a given query submitted by a user, first we use Indri to make the search and obtain a ranked list of results. At this point relevance is the only concern and time is not considered. Next, the temporal intention of the query is analyzed. This can be done either by DBpedia or examining the top k documents returned from Indri. Then temporal scores and relevance scores are calculated separately and combined to form the final ranking of documents. Next let us detail these modules one by one.

In Fig. 1, there are 4 major steps (lines 1–9, 10–12, 13–19, 20–23). In the following let us discuss them one by one.

3.1 Analyzing Temporal Intention of the Query

If the query is about a major event or celebrities in history, then it is very likely that we may obtain some information from DBpedia; otherwise, we try to find information from the documents themselves in L.

DBpedia[2] is a crowd-sourced community effort that extracts structured information from Wikipedia. In recent years, more and more researchers began to pay attention to the DBpedia knowledge base and set data-level links to DBpedia resources. Gradually DBpedia has been developed into one of the major data repositories on the Web [13]. If the query is about a major event in history, a celebrity, etc., we use the SPARQL language to query the date that is associated with the person or event, so as to determine the temporal intention of the query. For example, the birthday of the English singer "John Lennon" can be obtained by issuing a statement through DBpedia's user interface[3].

---

[1] http://www.lemurproject.org/indri.php.

[2] http://wiki.dbpedia.org/.

[3] http://dbpedia.org/sparql.

Then we can know that his birthday is on the 9th of October, 1940. It is also possible to know that his death date is on the 8th of December, 1980.

For those queries that we cannot get information from DBpedia, we try to obtain their temporal intention using the top k documents returned from Indri, because time expressions frequently appeared in those documents are likely to include some useful information. The procedure of this method is as follows: for a given query, we make a search and obtain a ranked list of results. Then the top k results are examined and the time expressions occurring in them are counted. Those time expressions that occurring over b ($b > 0$, as a given threshold) times are treated as time points of interest to the query, and the whole collection of these time points is regarded as the temporal intention of the query.

3.2   Ranking Methods with Temporal Intentions

First we introduce the terminology used in this paper. C is a corpus and it contains a set of documents $C = \{d_1, d_2, d_3, \ldots, d_n\}$. Any document $d \in c$ is represented by $d = \{d_{term}, d_{time}\}$, in which d_{term} is a set of terms that doesn't have temporal intention and d_{time} is a set of time points. $d_{term} = \{t_1, t_2, \ldots, t_m\}$, and $t_1 \cap t_2 \cap \ldots \cap t_m = \emptyset$; $d_{time} = \{t'_1, t'_2, \ldots, t'_n\}$, and $t'_1 \cap t'_2 \cap \ldots \cap t'_n = \emptyset$. Any query q is also treated as a short document. Initially, q only includes a set of terms $q_{term} = \{u_1, u_2, \ldots, u_o\}$. But after analyzing its temporal intention, it has the second component $q_{time} = \{u'_1, u'_2, \ldots, u'_p\}$.

The score of document d is computed by the following equation

$$S(q, d) = \alpha \cdot S'(q_{word}, d_{word}) + (1 - \alpha) \cdot S''(q_{time}, d_{time}) \tag{1}$$

where α is an adjustable parameter that needs to be set for different policies, $S'(q_{word}, d_{word})$ reflects d's topical relevance to the query, while $S''(q_{time}, d_{time})$ reflects d's temporal relevance to the query. $S'(q_{word}, d_{word})$ can be calculated from a traditional information retrieval system such as Indri, but $S''(q_{time}, d_{time})$ requires special treatment from time sensitive components. Next let us discuss how to calculate $S''(q_{time}, d_{time})$.

3.2.1   The Language Model

Using the language model [5], temporal relevance score $S''(q_{time}, d_{time})$ is defined as follows

$$S''(q_{time}, d_{time}) = P(q_{time}|d_{time}) = P\left(u'_1, u'_2, \ldots, u'_p|d_{time}\right) \tag{2}$$

Assuming that all the time points in q_{time} are independent, then we have

$$P\left(u'_1, u'_2, \ldots, u'_p|d_{time}\right) = \prod_{Q \in q_{time}} P(Q|d_{time}) \tag{3}$$

Next, $P(Q|d_{time})$ can be calculated by

$$P(Q|d_{time}) = (1 - \lambda)\frac{1}{|C_{time}|}\sum_{T \in C_{time}} P(Q|T) + \lambda\frac{1}{|d_{time}|}\sum_{T \in d_{time}} P(Q|T) \qquad (4)$$

where $\lambda \in [0, 1]$ is a smoothing parameter. C_{time} denotes all the time points in all the documents of corpus $C \cdot |C_{time}|$ denotes the number of time points in C_{time}. Similarly, $|d_{time}|$ denotes the number of time points in d_{time}. Time variable T goes through all possible time points in C_{time}. When calculating probabilities for the match of those time points, we allow approximate matches. That is to say, if the two time points are not exactly the same but close to each other, then we claim they have the same temporal intention. $P(Q|T)$ can be calculated as

$$P(Q|T) = \begin{cases} 1, & (|Q - T| \leq \beta \text{ time unit}), \beta \geq 0; \\ 0, & otherwise \end{cases} \qquad (5)$$

An alternative is to enlarge Q and T to include $2\beta + 1$ time units, with β units before and β units afterwards. $P(Q|T)$ can be calculated as

$$P(Q|T) = \frac{|T \cap Q|}{|T| * |Q|} \qquad (6)$$

where $T \cap Q$ denotes the overlapping units between T and Q, and $|T \cap Q|$ is the number of units in $T \cap Q$.

3.2.2   The Metric Space Model

Another method to calculate temporal score is to define distance in the metric space [14]. For any of the time points Q in q_{time}, we try to find the closest point T in d_{time}, then the distance between Q and T is calculated as

$$dist(Q, T) = \begin{cases} 0, & (|Q - T| \leq \eta \text{ time unit}), \eta \geq 0; \\ 2 * |Q - T|, & otherwise \end{cases} \qquad (7)$$

The distance between q_{time} and d_{time} $dist(q_{time}, d_{time})$ is the sum of all such distance values. We may rank those documents according to the values of $dist(q_{time}, d_{time})$ in ascending order.

Finally, for $S'(q_{word}, d_{word})$ and $S''(q_{time}, d_{time})$ obtained by above-mentioned methods, we need to make them comparable by normalization. After that, it can be combined linearly using Eq. 1. If scores are not available, then we may convert ranking information into scores by using the following equation

$$score = \frac{1}{rank + 60} \qquad (8)$$

One commonly used sore normalization method is

$$Score_{norm} = \frac{score_{raw} - score_{min}}{score_{max} - score_{min}} \tag{9}$$

where $score_{raw}$ is the raw score to be normalized, $score_{min}$ and $score_{max}$ are the minimal and maximal scores in the list, and $Score_{norm}$ is the normalized score.

4   Experiments

In order to evaluate the methods proposed in this paper, we use the dataset that was used in the Temporal Information Access task (Temporalia)[4] of NTCIR-11 for experiment.

The data set contains around 3.8 M documents collected from about 1500 different blogs and news sources from May 2011 to March 2013 [15]. There are 50 queries. Indri is used to make the search to produce initial results of content relevance. The results are also served as a component by all the time-sensitive methods. Six methods involved are:

1. DIF uses Eq. 5 to calculate $P(Q|T)$ and Eq. 9 to normalize scores;
2. DIF-rank uses Eq. 5 to calculate $P(Q|T)$ and Eq. 8 to normalize scores;
3. IS uses Eq. 6 to calculate $dist(Q, T)$ and Eq. 9 to normalize scores;
4. IS-rank uses Eq. 6 to calculate $dist(Q, T)$ and Eq. 8 to normalize scores;
5. Metric uses Eq. 7 to calculate $dist(Q, T)$ and Eq. 9 to normalize scores;
6. Metric-rank uses Eq. 7 to calculate $dist(Q, T)$ and Eq. 8 to normalize scores.

We also include two methods proposed by [2] for comparison. They are referred to as QW and NLM. QW uses keywords and NLM takes creation dates of the top k documents as the temporal intention of the query. Similarly, QW-rank and NLM-rank are also involved. Instead of using Eq. 8 as QW and NLM do, QW-rank and NLM-rank use Eq. 9 to normalize scores.

Table 1 shows the performances of all the methods involved and each method is tuned with the optimal setting. Six metrics including MAP (Mean Average Precision), RP (Recall-level Precision), P@5 (Precision at 5 document level), P@10 (Precision at 10 document level), NDCG@20 (Normalized Discounted Cumulative Gain at 20 document level), and ERR@20 (Expected Reciprocal Rank at 20 document level) are used for performance evaluation. All of them are commonly used metrics. The Baseline (Indri) is the ranking results obtained from the Indri system, which only considers content relevance. In general, except a few cases in which Metric and Metric-rank are slightly worse than the baseline in a few metrics, all the time-sensitive methods are slightly better than the baseline. DIF and DIF-rank seem better because each of them achieves the best performance on two metrics, while IS and NLM-rank achieves the best on one metric. However, the difference between all these methods is small.

---

[4] https://sites.google.com/site/ntcirtemporalia/.

Table 1. The performance of all time-sensitive methods in the experiment (the figures in bold indicate the best performer on a specific metric)

Method	MAP	RP	P@5	P@10	NDCG@20	ERR@20
Indri (Baseline)	0.2409	0.2930	0.5490	0.4980	0.3299	0.1737
DIF ($\alpha = 0.92$)	**0.2466** (+2.37%)	0.2991 (+2.08%)	0.5640 (+2.73%)	0.5100 (+2.41%)	**0.3441** (+4.30%)	0.1829 (+5.30%)
DIF-rank ($\alpha = 0.8$)	0.2463 (+2.24%)	0.2960 (+1.02%)	**0.5760** (+4.92%)	**0.5180** (+4.02%)	0.3423 (+3.76%)	0.1846 (+6.28%)
IS ($\alpha = 0.92$)	0.2463 (+2.24%)	**0.3011** (+2.76%)	0.5640 (+2.73%)	0.5040 (+1.20%)	0.3418 (+3.61%)	0.1817 (+4.61%)
IS-rank ($\alpha = 0.8$)	0.2452 (+1.78%)	0.2978 (+1.64%)	0.5640 (+2.73%)	0.5060 (+1.61%)	0.3377 (+2.36%)	0.1787 (+2.88%)
Metric ($\alpha = 0.9$)	0.2415 (+0.25%)	0.2918 (−0.41%)	0.5520 (+0.55%)	0.4800 (−3.61%)	0.3314 (+0.45%)	0.1801 (+3.68%)
Metric-rank ($\alpha = 0.9$)	0.2448 (+1.62%)	0.2964 (+1.16%)	0.5680 (+3.46%)	0.4900 (−1.61%)	0.3415 (+3.52%)	0.1802 (+3.74%)
QW ($\alpha = 0.9$)	0.2460 (+2.12%)	0.2989 (+2.01%)	0.5640 (+2.73%)	0.5040 (+1.20%)	0.3409 (+3.33%)	0.1780 (+2.48%)
QW-rank ($\alpha = 0.9$)	0.2460 (+2.12%)	0.2985 (+1.88%)	0.5640 (+2.73%)	0.5020 (+0.80%)	0.3402 (+3.12%)	0.1799 (+3.57%)
NLM ($\alpha = 0.9$)	0.2460 (+2.12%)	0.2997 (+2.29%)	0.5600 (+2.00%)	0.5000 (+0.40%)	0.3401 (+3.09%)	0.1809 (+4.15%)
NLM-rank ($\alpha = 0.9$)	0.2461 (+2.16%)	0.2984 (+1.84%)	0.5640 (+2.73%)	0.4960 (−0.40%)	0.3425 (+3.82%)	**0.1848** (+6.39%)

It's meaningful to take time uncertainty into consideration while calculating $P(Q|T)$. In Eq. 5, m needs to be set properly. Table 2 shows the performance of DIF under different time intervals (0 days, 7 days, 30 days, and 3 months) for m. Table 2 shows the results. From Table 2, we can see that 7 days is likely the best option for m because either DIF or DIFF-rank achieves the best performance on all those metrics. The results are close when m is set to 0 or 30, but worse off when m is 30 days.

Table 2. The performance of all the methods with different time intervals (the figures in bold indicate the best performer on a specific metric)

Method	MAP	RP	P@5	P@10	NDCG@20	ERR@20
DIF ($\alpha = 0.9$)	0.2421	0.2933	0.5560	0.4980	0.3311	0.1782
DIF-rank ($\alpha = 0.9$)	0.2459	0.2969	0.5640	0.4940	0.3422	0.1845
DIF (7 days, $\alpha = 0.92$)	**0.2466**	**0.2991**	0.5640	0.5100	**0.3441**	0.1829
DIF-rank (7 days, $\alpha = 0.8$)	0.2463	0.2960	**0.5760**	**0.5180**	0.3423	**0.1846**
DIF (30 days, $\alpha = 0.9$)	0.2421	0.2944	0.5600	0.5060	0.3305	0.1763
DIF-rank (30 days, $\alpha = 0.9$)	0.2455	0.2978	0.5640	0.5020	0.3391	0.1788
DIF (3 months, $\alpha = 0.9$)	0.2418	0.2937	0.5520	0.4940	0.3286	0.1767
DIF-rank (3 months, $\alpha = 0.9$)	0.2457	0.2986	0.5640	0.4960	0.3384	0.1797

5 Conclusions and Future Work

In this paper, we have investigated the technique for supporting implicitly temporal queries in information retrieval. Novel methods of analyzing temporal intention of implicitly temporal queries have been introduced and temporal relevance has been taken into account in the ranking model. Experimental results with the NTCIR data set show that our methods do well. Retrieval performance is improved when adding the time factor to the ranking model. Compared with other baseline technologies, our methods are at least as good as them.

There are two directions that we can move forward. One is to combine different methods, such as meta-data-based, keywords-based, top-documents-based, external resource-based, and so on, when analyzing the temporal intension of queries. The other is to consider both time and diversity for search results. Because for some queries, their temporal intentions may include multiple time intervals, so it is necessary to satisfy temporal needs of this kind of queries by considering temporal diversity.

Acknowledgement. This research has been partially supported by Natural Science Foundation of Jiangsu Province (number BK20171303).

References

1. Metzler, D., Jones, R., Peng, F., Zhang, R.: Improving search relevance for implicitly temporal queries. In: Proceedings of SIGIR, pp. 700–701 (2009)
2. Kanhabua, N., Nørvåg, K.: Determining time of queries for re-ranking search results. In: Lalmas, M., Jose, J., Rauber, A., Sebastiani, F., Frommholz, I. (eds.) ECDL 2010. LNCS, vol. 6273, pp. 261–272. Springer, Heidelberg (2010). doi:10.1007/978-3-642-15464-5_27
3. Gupta, D., Berberich, K.: Identifying time intervals of interest to queries. In: Proceedings of CIKM, pp. 1835–1838 (2014)
4. Alonso, O., Strötgen, J., Baeza-Yates, R.A., Gertz, M.: Temporal information retrieval: challenges and opportunities. TWAW **11**, 1–8 (2011)
5. Berberich, K., Bedathur, S., Alonso, O., Weikum, G.: A language modeling approach for temporal information needs. In: Gurrin, C., He, Y., Kazai, G., Kruschwitz, U., Little, S., Roelleke, T., Rüger, S., van Rijsbergen, K. (eds.) ECIR 2010. LNCS, vol. 5993, pp. 13–25. Springer, Heidelberg (2010). doi:10.1007/978-3-642-12275-0_5
6. Jones, R., Diaz, F.: Temporal profiles of queries. ACM Trans. Inf. Syst. **25**(3), 14 (2007)
7. Kanhabua, N., Nørvåg, K.: Learning to rank search results for time-sensitive queries. In: Proceedings of CIKM, pp. 2463–2466 (2012)
8. Chang, P.T., Huang, Y.C., Yang, C.L., Lin, S.D., Cheng, P.J.: Learning-based time-sensitive re-ranking for web search. In: Proceedings of SIGIR, pp. 1101–1102 (2012)
9. Kanhabua, N., Ngoc N.T., Nejdl, W., Wolfgang, N.: Learning to detect event-related queries for web search. In: Proceedings of WWW, pp. 1339–1344 (2015)
10. Shokouhi, M., Radinsky, K.: Time-sensitive query auto-completion. In: Proceedings of SIGIR, pp. 601–610 (2012)
11. Dakka, W., Gravano, L., Ipeirotis, P.: Answering general time-sensitive queries. IEEE Trans. Knowl. Data Eng. **24**(2), 220–235 (2012)
12. Lin, S., Jin, P., Zhao, X., Yue, L.: Exploiting temporal information in Web search. Expert Syst. Appl. **41**(2), 331–341 (2014)

13. Bizer, C., Lehmann, J., Kobilarov, G., Auer, S., Becker, C., Cyganiak, R., Hellmann, S.: DBpedia - a crystallization point for the Web of Data. J. Web Sem. **7**(3), 154–165 (2009)
14. Brucato, M., Montesi, D.: Metric spaces for temporal information retrieval. In: de Rijke, M., Kenter, T., de Vries, A.P., Zhai, C., de Jong, F., Radinsky, K., Hofmann, K. (eds.) ECIR 2014. LNCS, vol. 8416, pp. 385–397. Springer, Cham (2014). doi:10.1007/978-3-319-06028-6_32
15. Joho, H., Jatowt, A., Blanco, R.: NTCIR Temporalia: a test collection for temporal information access research. In: Proceedings of WWW (Companion Volume), pp. 845–850 (2014)

On the Relations of Theoretical Foundations of Different Causal Inference Algorithms

Furui Liu$^{(\boxtimes)}$ and Laiwan Chan

Department of Computer Science and Engineering,
The Chinese University of Hong Kong, Hong Kong, China
{frliu,lwchan}@cse.cuhk.edu.hk

Abstract. Telling cause from effect attracts various attentions from researchers recently. Here we study the algorithms proposed under the postulate of independence between cause and mechanisms. Firstly, we conduct a review of the different definitions of independence in different causal inference algorithms, and show how these theories could lead to practical methodologies. Then, we provide justifications about their links, showing how the seemingly different theoretical foundations could be integrated. This provides new insights of the relations between different inference algorithms, and gives readers a comprehensive understanding of the methods under this research topic.

Keywords: Causality · Independence · Theory

1 Introduction

Causal inference between 2 variables, say X and Y, given purely observational data is an important but challenging problem. Traditional Bayesian network (BN) approaches fail in this task because conditional independence based method does not work in two-variate cases. Additive noise models [2] are proposed to deal with this problem. Given the observational data, it performs a regression of Y on X and vice versus, and test the independence of the residuals and regressors. The direction that admits an ANM is then inferred as the causal direction. However, it may fail to make a decision when X and Y are causally related with stochastic process that cannot be models by ANM. Recently, another view arises to address this problem. It takes use of the postulate that the cause and mechanism are generated independently such that the joint distributions p_{XY} show clear cause-effect asymmetry [4], when one examine certain statistics in causal and anti-causal directions. This postulate is conceptual, and the success of the inference algorithms depends on practical definitions of independence. To our knowledge, several definitions of independence are proposed, and leveraged by different causal inference algorithms for different application scenarios. Given this diversification, one question is left to be answered. Is there any relation between these seemingly different definitions? To provide answers to the question, we here make a study of the definitions of independence, as well as how

© Springer International Publishing AG 2017
H. Yin et al. (Eds.): IDEAL 2017, LNCS 10585, pp. 112–119, 2017.
https://doi.org/10.1007/978-3-319-68935-7_13

they are leveraged by some representative algorithms for causal inference. We also discuss the hidden links between these concepts, and show how the theoretic foundation of one method could lead to another one's. This, provides new insight on the integration of the theoretic foundations of all methods. We believe, this paper could provide readers a guidance about the choice of algorithms when they meet the causal inference problems. The contributions of this paper are two-fold.

1. We make a thorough study of the theoretical foundations of "independence" used by the current causal inference algorithms.
2. We discuss the relations of the concepts, and unveil the hidden links of the seemingly different definitions.

The rest of the paper is structured as following. Section 2 is a brief warm up on the symbols and topics. Section 3 discusses algorithmic independence based causal inference, and Sect. 4 discusses the statistical independence based inference. Section 5 introduces the Bayesian inference, and Sect. 6 lists core points of the whole paper. Section 7 concludes.

2   Preliminary

Before we start, we here give a brief overview of the contents of the paper, and clarify some notations that may appear later. We assume that X is the cause and Y is the effect unless specified. We write p_X as the vector recording the probability of the cause variable X. Similarly we can define $p_{Y|X}$, p_Y and $p_{X|Y}$. We are going to study the precise definitions of the "independence" between p_X and $p_{Y|X}$. In fact, later we will study 3 kinds of inference foundation:

1. p_X and $p_{Y|X}$ are algorithmically independent.
2. p_X and $p_{Y|X}$ are statistically independent.
3. p_X and $p_{Y|X}$ follow independent prior distributions in Bayesian model selection.

Notice that in the 3 points, p_X is all treated as a random variable taking realizations of at different x, and this random variable follow some underlying distributions. But the focus is different: In point 1, we can consider $p_X(X = x)$ as a value from a finite alphabet $\mathcal{A}$, thus having Kolmogorov complexity, and the concatenation of all those values becomes p_X. In point 2, we no longer consider its Kolmogorov complexity, while treating it as a variable. It has its distribution and related statistics like entropy. In point 3, we consider p_X as one point in the high dimensional space, which describes distribution of X under some priors in the likelihood estimation (in fact, for each vector generated as samplings, it could be considered as a point in high dimensional space). In the next section, we are going to talk about algorithmic independence and the related causal inference algorithms.

3   Algorithmic Independence and Kolmogorov Complexity

We first describe the algorithmic independence, which is one face of the independence of cause and mechanism in terms of Kolmogorov complexity. Let s, t denote the two string valued variables that describe observed objects and write $K(s)$ the algorithmic information (or Kolmogorov complexity). It is defined as the length of a shortest program that generates s on a universal Turing machine [1]. Formally, the algorithmic mutual information between two strings s and t is defined as

$$I(s : t) = K(t) - K(t|s^*) \stackrel{\pm}{=} K(s) + K(t) - K(s,t).$$

where s^* is the shortest program that computes s and $K(s,t)$ is the Kolmogorov complexity of the concatenation s and t. The symbol $\stackrel{\pm}{=}$ means this equal condition is up to a constant error, which depends on the Turing machine. One can show that this concept is rather in analogy to the Shannon entropy and mutual information, which writes as

$$MI(s;t) = H(s) - H(t|s) = H(s) + H(t) - H(s,t).$$

where s and t are both random variables. Note that we require computable functions and distributions. A function $f : R^k \to R$ is computable if for every finite length input (x, ϵ), $f(x)$ can be computed up to error $\epsilon > 0$. This is assumed across our analysis. Then we describe the spirit of independence of cause and mechanism. It has been postulated that the shortest description length of p_{YX} is given by separate descriptions of p_X and $p_{Y|X}$ [3–5,7], which is formalized by the following postulate.

Postulate 1. p_X and $p_{Y|X}$ are algorithmically independent.

This leads to the core theorem that is used for causal inference.

Theorem 1 (Causal inference under algorithmic independence). [7] Let p_{XY} be a computable probability function defined over a finite domain, and X causes Y. Then we have

$$I(p_X : p_{Y|X}) \stackrel{\pm}{=} 0.$$

One can interpret the rule as: the Kolmogorov complexity of the joint distribution can be computed as a sum up of complexities of the distribution of cause and effect given cause, since

$$I(p_X : p_{Y|X}) \stackrel{\pm}{=} K(p_X) + K(p_{Y|X}) - K(p_X, p_{Y|X}) = 0,$$

$$I(p_Y : p_{X|Y}) \stackrel{\pm}{=} K(p_Y) + K(p_{X|Y}) - K(p_Y, p_{X|Y}) \geq 0,$$

Bearing in mind that the Kolmogorov complexity of the joint distribution is

$$K(p_X, p_{Y|X}) \stackrel{\pm}{=} K(p_{YX}) \stackrel{\pm}{=} K(p_Y, p_{X|Y}),$$

Thus we have the inference rule

$$K(p_X) + K(p_{Y|X}) \overset{+}{\leq} K(p_Y) + K(p_{X|Y}). \tag{1}$$

Equation (1) is an important formula that lays the foundation of causal inference under this view. The Kolmogorov complexity is not computable, since a Turing machine remains conceptual. Practical causal inference relies on other notions under this concept. By different choices of approximations to Kolmogorov complexity, it leads to different methods. Suppose this complexity is approximated using $C(\cdot)$. Equation (1) becomes

$$C(p_X) + C(p_{Y|X}) \leq C(p_Y) + C(p_{X|Y}).$$

If we choose $C(\cdot)$ as the kernel seminorm complexity measure, described by [8], then this leads to a kernel method for causal inference. If one choose the relative entropy distance to some manifolds as complexity measure, then this equation leads to information geometric causal inference (IGCI) [3].

The core idea of using algorithmic information theory is clear: the factorization of the joint distribution in causal direction is simpler. In fact, there is a theorem [4] describing that the algorithmic information is approximately given by the entropy of the elements following some probability laws, as showed below.

Theorem 2 (statistical and algorithmic information). *Let p_X be constructed by taking i.i.d. symbols at each point x_i as $p_X(X = x_i)$ from a finite alphabet $\mathcal{A}$, which can be described as a random variable. Denote the n as the length of p_X. Then with probability 1*

$$\lim_{n \to \infty} \frac{1}{n} K(p_X) = H(p_X),$$

where $H(\cdot)$ is the Shannon entropy.

The theorem clearly shows that the average Kolmogorov complexity of the vector p_X, equals to the statistical entropy of the underlying distribution that the p_X follows, when the length of p_X goes to infinity. This means, if the variable realizations inside p_X and $p_{Y|X}$ are recordable, say when data is on discrete domain, we can directly take use of some statistical independence test method. This avoids an approximation step when using the algorithmic independence, which may involve metric dependent errors. This is leveraged by the algorithm called DC [6]. We would discuss this in the next section.

4   Statistical Independence and Distance Correlation

Theorem 2, in fact, already suggests a relation of algorithmic and statistic information. If we replace the Kolmogorov complexity with Shannon entropy given the length of the probability distribution vector p_X and $p_{Y|X}$ becomes infinity, we have

$$\lim_{n \to \infty} \frac{1}{n} I(p_X : p_{Y|X}) \overset{+}{=} MI(p_X; p_{Y|X}).$$

Notice that we here consider the average algorithmic mutual information across every x. This means that if we treat the distribution function as random variables which follow some probability distribution, they are statistically independent. This is leveraged by a new algorithm, which uses distance correlation to measure the dependence of the two variables p_X and $p_{Y|X}$. We first begin with the postulate.

Postulate 2. *p_X and $p_{Y|X}$ are both random variables taking realizations at different x. p_X is independent with $p_{Y|X}$.*

This postulate is used by a characterizing measure called distance correlation [9]. Suppose the random variables $(p_X, p_{Y|X})$ are with characteristic functions f_{p_X} and $f_{p_{Y|X}}$ respectively. Their joint characteristic function is $f_{p_X, p_{Y|X}}$. The distance correlation is defined as below.

Definition 1. *The distance covariance $\mathcal{V}(p_X, p_{Y|X})$ between two random variables $(p_X, p_{Y|X})$ is*

$$\mathcal{V}(p_X, p_{Y|X}) = \|f_{p_X, p_{Y|X}} - f_{p_X} f_{p_{Y|X}}\|^2,$$

and the distance correlation is

$$\mathcal{D}(p_X, p_{Y|X}) = \frac{\mathcal{V}(p_X, p_{Y|X})}{\sqrt{\mathcal{V}(p_X, p_X)\mathcal{V}(p_{Y|X}, p_{Y|X})}}.$$

The distance correlation is 0 if the two variables are independent, which is already verified by $MI(p_X; p_{Y|X}) = 0$. Thus, we have the following causal inference rule.

Theorem 3 (Causal inference under statistical independence). *Let postulate 2 be accepted, we have*

$$\mathcal{D}(p_X, p_{Y|X}) \leq \mathcal{D}(p_Y, p_{X|Y}).$$

This can also be derived from the

$$I(p_X : p_{Y|X}) \overset{+}{\leq} I(p_Y : p_{X|Y}),$$

using the link between algorithmic and statistical information. In fact, the test of the distance correlation can only be done on discrete data. For non-discrete data, we can not get the value of p_X at each point, which makes the test of statistical independence impossible. This is why, we have to shift from the direct independence test to complexity measurements, which can be approximated on continuous data, given a cost of possible errors included in the approximation step. Actually, the interpretation of statistical independence could lead to another method, which implicitly uses the statistical independence, called Bayesian causal inference. We would discuss this in the next section.

5   Likelihood Estimation and Bayesian Inference

Another causal inference method could also be analyzed here [7]. In fact, the Bayesian method based causal inference, still uses the fact that p_X and $p_{Y|X}$ are generated independently in a statistical sense. Suppose the sampled data is (X, Y), and the joint distribution p_{YX} is generated under the characteristic function $f_{p_{YX}}$ as described by Sect. 3. Imagine the hierarchical generation model from the distribution to data sample as

$$f_{p_{YX}} \rightarrow p_{YX} \rightarrow (X, Y).$$

They estimate the model likelihood given data to find $X \rightarrow Y$ or $Y \rightarrow X$ in the following steps:

1. Choose some parameters that characterize the distribution of the p_{YX}.
2. For each p_{YX}, calculate the likelihood in two directions. By integration over the prior distribution, we get the model likelihood.
3. Compare the likelihood and choose the one with higher likelihood as possible causal models.

We write $\alpha = p_X$ and $\beta = p_{Y|X}$, and the true causal direction is $X \rightarrow Y$. Then, if we estimate the model likelihood in the true causal direction, we have

$$\mathcal{L}(X, Y | f_{p_{YX}}) = \iint p(X, Y | \beta, \alpha) p(\beta, \alpha | f_{p_{YX}}) d\beta d\alpha,$$

$$= \iint p(X, Y | \beta, \alpha) p(\beta | f_{p_{YX}}) p(\alpha | f_{p_{YX}}) d\beta d\alpha,$$

$$= \iint p(X | \beta, \alpha) p(Y | X, \beta, \alpha) p(\beta | f_{p_{YX}}) p(\alpha | f_{p_{YX}}) d\beta d\alpha,$$

$$= \int p(X | \alpha) p(\alpha | f_{p_{YX}}) d\alpha \int p(Y | X, \beta) p(\beta | f_{p_{YX}}) d\beta,$$

$$= \mathcal{L}_{X \rightarrow Y}(X, Y | f_{p_{YX}}).$$

Notice that we use the independence between p_X and $p_{Y|X}$ in the whole derivation process. In the second line, we divide the equations

$$p(\beta, \alpha | f_{p_{YX}}) = p(\beta | f_{p_{YX}}) p(\alpha | f_{p_{YX}}), \tag{2}$$

because we have

$$\mathcal{D}(\alpha, \beta) = 0, \tag{3}$$

which is exactly the property that distance correlation technique uses to discover the causal direction. The difference is, distance correlation explicitly tests the independence given the vector form the p_X and $p_{Y|X}$, while Bayesian inference implicitly incorporates this into its estimation procedure. In the anti-causal direction, denote $\alpha' = p_Y$ and $\beta' = p_{X|Y}$, we have

$$
\begin{aligned}
\mathcal{L}(X, Y | f_{p_{YX}}) &= \iint p(X, Y | \beta', \alpha') p(\beta', \alpha' | f_{p_{YX}}) d\beta' d\alpha', \\
&= \iint p(X, Y | \beta', \alpha') (p(\beta' | f_{p_{YX}}) p(\alpha' | f_{p_{YX}}) + \Delta p(\alpha', \beta' | f_{p_{YX}})) d\beta' d\alpha', \\
&= \iint p(Y | \beta', \alpha') p(X | Y, \beta', \alpha') p(\beta' | f_{p_{YX}}) p(\alpha' | f_{p_{YX}}) d\beta' d\alpha' + \Delta\mathcal{L}(X, Y | f_{p_{YX}}), \\
&= \int p(Y | \alpha') p(\alpha' | f_{p_{YX}}) d\alpha' \int p(X | Y, \beta') p(\beta' | f_{p_{YX}}) d\beta' + \Delta\mathcal{L}(X, Y | f_{p_{YX}}), \\
&= \mathcal{L}_{Y \to X}(X, Y | f_{p_{YX}}) + \Delta\mathcal{L}(X, Y | f_{p_{YX}}).
\end{aligned}
$$

The $\Delta\mathcal{L}(X, Y | f_{p_{YX}})$ comes out because of the dependence of the two distributions in the anti-causal direction. Another concept of Bayesian cut, discussed in [10], also uses this definition of independence. Then we have

$$
\mathcal{L}(X, Y | f_{p_{YX}}) = \mathcal{L}_{X \to Y}(X, Y | f_{p_{YX}}) = \mathcal{L}_{Y \to X}(X, Y | f_{p_{YX}}) + \Delta\mathcal{L}(X, Y | f_{p_{YX}}).
$$

The identifiability result of Bayesian causal model inference comes from the assumption that $\Delta\mathcal{L}(X, Y | f_{p_{YX}}) > 0$, while till now there is no direct way to show this being always true. However, one thing is clear: only in the true causal direction, choosing an independent prior can lead to accurate estimation of the likelihoods of the model from X and Y given X; when we flip around, choosing an independent prior leads to errors due to the dependence of the model parameters. We formalize this as a conjecture below.

Conjecture 1 (Causal inference under Bayesian view). *Suppose X causes Y, and $f_{p_{YX}}$ is given. Then, in the causal direction, the model likelihood is no less than that in anti-causal direction as*

$$
\mathcal{L}_{X \to Y}(X, Y | f_{p_{YX}}) \geq \mathcal{L}_{Y \to X}(X, Y | f_{p_{YX}}).
$$

6   A Short Summary

To help the readers understand the paper better, we also list some core points that show the logic of the paper below.

1. Algorithmic independence of the distribution of cause p_X and mechanism represented by conditional distribution $p_{Y|X}$, leads to smaller Kolmogorov complexity measures in the causal direction. By choosing different approximations of Kolmogorov complexity, we can get different algorithms.
2. The link between algorithmic and statistical independence can bridge this concept with distance correlation based causal inference method, when p_X and $p_{Y|X}$ are treated as random variables following some underlying probability law. Only on discrete data, this is testable.
3. The statistical independence of p_X and $p_{Y|X}$ leads to the validity of choosing independent priors in the causal direction, when one infers the causal direction by estimating the model likelihood in two directions, and doing a standard Bayesian model selection.

7 Conclusions

In this paper, we conduct a study of the causal inference algorithms proposed under the concept of independence between the cause and mechanisms. We show precise definitions of independence, as well as how it is used by different causal inference methods. Moreover, we bridge these seemingly different theoretical concepts of independence, and unveil the links of these causal inference methods. We believe, this could provide the readers with not only a comprehensive knowledge of the practical methodologies, but also a deeper understanding of the theories behind them, thus building a big picture of the concepts. We hope that this could trigger more people's interest on this topic, and help them to decide which algorithms to choose when they meet practical causal inference problem.

Acknowledgments. The work described in this paper was partially supported by a grant from the Research Grants Council of the Hong Kong Special Administration Region, China.

References

1. Chaitin, G.J.: A theory of program size formally identical to information theory. J. ACM (JACM) **22**(3), 329–340 (1975)
2. Hoyer, P.O., Janzing, D., Mooij, J.M., Peters, J.R., Schölkopf, B.: Nonlinear causal discovery with additive noise models. In: Advances in Neural Information Processing Systems, pp. 689–696 (2009)
3. Janzing, D., Mooij, J., Zhang, K., Lemeire, J., Zscheischler, J., Daniušis, P., Steudel, B., Schölkopf, B.: Information-geometric approach to inferring causal directions. Artif. Intell. **182**, 1–31 (2012)
4. Janzing, D., Scholkopf, B.: Causal inference using the algorithmic markov condition. IEEE Trans. Inf. Theory **56**(10), 5168–5194 (2010)
5. Lemeire, J., Janzing, D.: Replacing causal faithfulness with algorithmic independence of conditionals. Mind. Mach. **23**(2), 227–249 (2013)
6. Liu, F., Chan, L.: Causal inference on discrete data via estimating distance correlations. Neural Comput. **28**(5), 801–814 (2016)
7. Mooij, J.M., Stegle, O., Janzing, D., Zhang, K., Schölkopf, B.: Probabilistic latent variable models for distinguishing between cause and effect. In: Advances in Neural Information Processing Systems, pp. 1687–1695 (2010)
8. Sun, X., Janzing, D., Schölkopf, B.: Causal reasoning by evaluating the complexity of conditional densities with kernel methods. Neurocomputing **71**(7), 1248–1256 (2008)
9. Székely, G.J., Rizzo, M.L., Bakirov, N.K., et al.: Measuring and testing dependence by correlation of distances. Ann. Stat. **35**(6), 2769–2794 (2007)
10. Zhang, K., Zhang, J., Schölkopf, B.: Distinguishing cause from effect based on exogeneity. arXiv preprint (2015). arXiv:1504.05651

SibStCNN and TBCNN + kNN-TED: New Models over Tree Structures for Source Code Classification

Anh Viet Phan[1,2], Minh Le Nguyen[1(✉)], and Lam Thu Bui[2]

[1] Japan Advanced Institute of Information Technology, Nomi, Japan
{anhphanviet,nguyenml}@jaist.ac.jp
[2] Le Quy Don Technical University, Hanoi, Vietnam
lambt@lqdtu.edu.vn

Abstract. This paper aims to solve a software engineering problem by applying several approaches to exploit tree representations of programs. Firstly, we propose a new sibling-subtree convolutional neural network (SibStCNN), and combination models of tree-based neural networks and k-Nearest Neighbors (kNN) for source code classification. Secondly, we present a pruning tree technique to reduce data dimension and strengthen classifiers. The experiments show that the proposed models outperform other methods, and the pruning tree leads to not only a substantial reduction in execution time but also an increase in accuracy.

Keywords: Abstract Syntax Trees (AST) · Convolutional Neural Networks (CNNs) · k-Nearest Neighbors (kNNs)

1 Introduction

Program classification as well as other software engineering problems are hot research topics. In practice, properly organizing software repositories by categorizing their components according to specific criteria is beneficial to programmer cooperation, maintenance and reuse of software. However, manually assigning a software component into the repository compatible with the criteria is impractical because sizes of projects are very large and increase rapidly [11]. Thus, in the software industry, building a tool to automatically categorize source code repositories is an urgent requirement to manage big projects.

Most studies have focused on using software metrics to build classifiers [6,11]. The metrics-based approaches are also widely applied to solve various software engineering problems such as fault prediction, cost, and effort estimation [4,12]. However, existing metrics have not shown good ability to capture the underlying meaning of programs [5]. Hence although various machine learning algorithms have been applied, the predictors have not achieved so high performance.

Recently, many researchers have been successful in applying language processing approaches to address software engineering problems [1,2]. Mou et al.

H. Yin et al. (Eds.): IDEAL 2017, LNCS 10585, pp. 120–128, 2017.
https://doi.org/10.1007/978-3-319-68935-7_14

proposed a convolutional neural network (TBCNN) on abstract syntax trees (ASTs) of source code for classifying programs by functionalities [7].

In this paper, we design a sibling-subtree convolutional neural network (SibStCNN) to learn features of subtrees including descendants and siblings of each AST node. Furthermore, to take advantages of structural and underlying information of trees, we combine SibStCNN/TBCNN and kNN with tree edit distance (TED), where TED is used to measure the similarities of tree structures[1].

When processing ASTs, we usually face a problem of high-dimensional data. In the experimental dataset, the largest AST contains 7027 nodes, while the program has only 343 lines of code. High-dimensional data not only lead to waste of time and memory but also affect the performance of algorithms. Thus, we apply some techniques to prune redundant branches and reconstructing subtrees of ASTs. The main contribution of this paper can be summarized as follows:

1. Applying pruning techniques to refine AST data.
2. Designing a convolutional neural network (SibStCNN) which can learn information both in depth and width dimensions of tree structures.
3. Proposing combination models of SibStCNN/TBCNN and kNN to enhance classification accuracy.

The rest of the paper is organized as follows: Sect. 2 provides backgrounds on abstract syntax trees (ASTs), and TBCNN. Section 3 describes the details of the proposed models. The data, the data pre-processing technique and the experimental setup of case studies are specified in Sect. 4. We evaluate the results of the methods in Sect. 5 and conclude in Sect. 6.

2 Preliminaries

2.1 Abstract Syntax Trees

In computer science, an abstract syntax tree (AST) is a tree representation of the abstract syntactic structure of source code written in a programming language. Figure 1a illustrates the AST of the C statement "printf("The sum of x+y =%d", x+y);".

Each node of the tree represents an abstract component occurring in the source code. An AST is a product of the syntax analysis phase of a compiler. It serves as an intermediate representation before generating code for the program.

2.2 Tree-Based Convolutional Neural Network

Tree-based convolutional neural network (TBCNN) is proposed by Mou [7]. Figure 1b illustrates the architecture of the TBCNN. Like other convolutional neural networks, the TBCNN includes a sequence of layers: embedding, convolutional, fully-connected, and output. In the embedding layer, AST nodes are represented as real-valued vectors which capture the underlying meaning of the symbols.

---

[1] Our models are publicly available at https://github.com/nguyenlab/TBCNN_kNN_SVM.git.

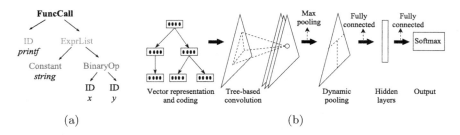

Fig. 1. (a) The AST of the statement "`printf("The sum of x+y =%d", x+y);`".
(b) The architecture of the tree-based convolutional neural network (TBCNN)

Then the author designed a set of fixed-depth subtree detectors sliding over entire AST to extract structural information of the program. The output of the feature detectors is computed by the following equation.

$$y = \tanh(\sum_{i=1}^{n} W_{conv,i} \cdot x_i + b_{conv})\qquad(1)$$

where, $x_1, .., x_n$ are vector representations of nodes inside the sliding window, $y, b_{conv} \in \Re^{N_c}, W_{conv,i} \in \Re^{N_c \times N_f}$. ($N_f$ and N_c are the vector size of symbols and the number of feature detectors).

The highlights of the model are the proposal of "continuous binary trees" notation, which allows computing the output in Eq. 1 with dynamic structures of subtrees. Regarding this, the convolutional layer only uses three weight matrices as parameters including W_{conv}^t, W_{conv}^l, and W_{conv}^r (superscripts t, r, l refer to "top", "left", "right"); the weight matrix for any node x_i is a linear combination of W_{conv}^t, W_{conv}^l, and W_{conv}^r, with coefficients computed as follows:

- $\eta_i^t = \frac{d_i-1}{d-1}$ (d_i: the depth of the node i in the sliding window; d: the depth of the window.)
- $\eta_i^r = (1 - \eta_i^t)\frac{p_i-1}{n-1}$ (p_i: the position of the node; n: the total number of p's siblings.)
- $\eta_i^l = (1 - \eta_i^t)(1 - \eta_i^r)$

The dynamic pooling [9] layer thereafter is stacked to gather the extracted features over parts of the tree. Finally, a fully connected layer and an output layer are added for supervised classification.

3   The Proposed Approaches

3.1   Sibling-Subtree Convolutional Neural Network (SibStCNN)

In the convolutional layer of SibStCNN, feature detectors are applied to exploit information both in depth and width dimensions from different parts of trees. Unlike TBCNN, the local regions for feature extraction of SibStCNN is expanded

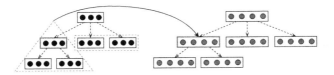

Fig. 2. Sibling-subtree convolution. Nodes on the left are feature vectors of AST nodes. Nodes on the right are feature maps.

to the siblings. In ASTs, sibling nodes have the similar roles, and their information is relevant. For instance, in Fig. 1a, nodes `ID:x` and `ID:y` are two operands of expression `x+y`; `Constant` and `BinaryOp` are two parameters of the function invocation `printf`. Additionally, the output of `printf` depends on both the values of the format string (`Constant`) and the expression (`BinaryOp`). For these reasons, it is more precise when a node is evaluated based on information of itself and the surrounding nodes (descendants and siblings).

In this work, the subtree depth is set to 2; the number and widths of rectangles are varying according to the location of the sliding window and the number of siblings such that the window covers all children, siblings, and the current node (Fig. 2). Formally, in each position, if the current node (x_c^0) has C children and S siblings with the corresponding vector representations $x_c^i, i \in \{1, .., C\}$ and $x_s^j, j \in \{1, .., S\}$, then the feature maps are obtained as follows:

$$y = \tanh(\sum_{i=0}^{C} W_{conv_c^i} \cdot x_c^i + \sum_{j=1}^{S} W_{conv_s^j} \cdot x_s^j + b_{conv}) \tag{2}$$

An obstacle to computing the feature maps (Eq. 2) is that determining the number of weight matrices is unfeasible because AST nodes have different numbers of children, and siblings as well. To solve this problem, we use three weight matrices $(W_{conv}^t, W_{conv}^l, \text{and } W_{conv}^r)$ for the subtree part and one weight matrix (W_{conv}^s) for the rectangle parts. In this scenario, $W_{conv_c^i}$ is a linear combination of $W_{conv}^t, W_{conv}^l, \text{and } W_{conv}^r$, with coefficients $\eta_i^t, \eta_i^l, \text{and } \eta_i^r$ (Sect. 2.2); $W_{conv_s^j}$ is a scale of W_{conv}^s with the ratio $\frac{c_{s_j}}{s_{s_j}}$ (c_{s_j}: the number of s_j's children; s_{s_j}: the number of s_j's siblings).

3.2 Combination Models of kNN-TED and SibStCNN/TBCNN

The tree-based convolution kernels explore the information contained in the AST nodes, while TED measures the similarity between tree structures. Thus, the cooperation between them provides stronger proof for the classifier to determine the label of an unknown instance.

For above reasons, we design two combination models of kNN-TED and SibStCNN/TBCNN, in which the decision values for each unseen instance is estimated by Eq. (3) as follows:

$$DecVal_j^i = (1 - t) * Prob_j^i + t * MF(nnDist_j^i) \tag{3}$$

where $DecVal_j^i$ is the decision value of instance i belonging to class j; $Prob_j^i$ is the prediction probability (produced by TBCNN or SibStCNN) for class j of instance i; $nnDist_j^i$ is the sum of normalized distances between instance i with instances of class j in the set of k neighbors of i; MF is the mapping function, which transforms the value of $nnDist_j^i$ to $[0,1]$; t is the combination factor in the range of $[0,1]$. After that, the label of the instance is determined by Eq. (4):

$$L^i = \begin{cases} L_{n_1} & \text{if } L_{n_1} = L_{n_2} = .. = L_{n_k} \\ l & \text{if } DecVal_l^i = max\{DecVal_j^i\} \end{cases} \tag{4}$$

where L^i is the predicted label of instance i; $L_{n_1}, L_{n_2}, .., L_{n_k}$ are the labels of k neighbors of instance i.

4 Data Preprocessing and Experimental Setup

4.1 Preprocessing AST Data

AST data are high-dimensional and need to be refined. Thus, we present a technique to prune redundant branches and reconstruct sub-tree structures based on programming semantic. The details of this technique are described as follows:

1. Eliminate structures of declaration statements. In a programming language, the properties of an identifier will be revealed in next statements, which manipulate such identifier. Thus, pruning the branches of declaration statements may not lead to a decrease in an amount of information on ASTs.
2. Cut down two branches of For subtrees. For subtrees includes four branches: Init (initialization), Condition (termination expression), Next (increment expression), Statement (body). The main works are usually put in the body section. The termination expression is responsible for controlling For loops. We remove the branches of Init and Next since they contain little information about the tasks of For statements.
3. Cut down the declaration branch of procedure definition subtrees. The Decl subtree depicts the temporary variables used in a procedure. When the procedure is invoked, the actual parameters are passed. We prune Decl branches because the information of temporary and actual parameters is similar.
4. Consider the same roles of For, While, and Do-While statements by rename their root nodes to Loop.

4.2 The Dataset

We verify the proposed approaches by solving program classification problem in which programs performing similar tasks are assigned to the same group. The dataset is shared by Mou [7]. It contains programs for 104 programming problems (considered as target labels) in which each of them includes 500 programs. The dataset was split by 3:1:1 for training, validation, and testing.

Table 1. Statistics of the dataset (ASTs$_{OR}$ are the original ASTs, and ASTs$_{SE}$ are the ASTs after pruning).

Statistics	ASTs$_{OR}$		ASTs$_{SE}$	
	Mean	Std.	Mean	Std.
# of AST nodes	189.6	106.0	134.0	92.1
# of AST leaves	90.5	53.8	66.5	47.7
Avg. leaf nodes' depth in an AST	7.6	1.7	8.3	1.9
# nodes of the smallest AST	29	–	7	–
# nodes of the largest AST	7027	–	6999	–

Table 1 shows statistics on the AST datasets. The statistical figures indicate the challenges of working with AST data due to their shapes and sizes are very large and different. For the original ASTs, the numbers of tree nodes are varying from 29 to 7027; the average number of nodes is 189.6; the standard deviation is 106. The AST nodes reduce notably after pruning, and the average leaf nodes' depth in an AST increases after performing semantic pruning.

4.3 Experimental Setup

To address the program classification problem, we converted programs into ASTs and then surveyed various algorithms, which take the ASTs as the input data. The details and settings of these algorithms are described as follows:

For neural network models, the symbol vectors have size of 30.

TBCNN and SibStCNN. Initial learning rate is 0.3; The convolutional and hidden layers have dimensions of 600; the loop iteration is 60.

The k-Nearest Neighbor (kNN): We use kNN based on tree edit distance (TED) and Levenshtein distance(LD) [8]. k is set to 3.

SibStCNN/TBCNN + kNN-TED. The settings of SibStCNN/TBCNN is kept. The number of nearest neighbors is expanded to 10. The combination factor t is tuned based on training and validation sets.

SVM + Tree Kernel + Bag Of Tree(BOT). For tree kernel SVM [3], we combine both tree kernels and feature vectors of BOT. The decay factor in tree kernels is 0.4. The *one-vs.-all* strategy is used for multi-class classification;

Gated Recurrent Neural Network (GRNN). The GRNN is successful in classifying documents [10]. Each program are converted as a document by traversing the AST using depth-first search algorithm.

5 Results and Discussion

Table 2 shows the performance of classifiers in terms of accuracy and running time. For the case of without pruning, TBCNN yields a remarkable accuracy of 92.63% due to subtree feature detectors which have good ability to capture underlying meanings of AST nodes. Thanks to the expansion of sliding windows, SibStCNN improves the accuracy of TBCNN by 0.63%. The accuracies of the combination models between SibStCNN/TBCNN and kNN indicate that learning trees from many perspectives is beneficial. SibStCNN/TBCNN and kNN outperform other methods, achieving the highest accuracies of 94.13% and 93.48%, respectively.

Table 2. Performance comparison of the pruning approaches in terms of accuracy and execution time (ASTs$_{OR}$ are the original ASTs, and ASTs$_{SE}$ are the ASTs after pruning)

Method	ASTs$_{OR}$		ASTs$_{SE}$	
	Acc. (%)	Avg. time (s)	Acc. (%)	Avg. time (s)
kNN + TED	85.84	259.59	**86.35**	108.8
kNN + LD	83.08	4.78	**85.56**	2.39
Tree kernel SVM	58.39	0.26	**62.65**	0.249
GRNN + LSTM	80.61	454.37	**83.31**	338.2
TBCNN	92.63	1194	**92.88**	810
SibStCNN	**93.26**	1494	93.14	944
TBCNN + kNN-TED	93.48	-	**93.69**	-
SibStCNN + kNN-TED	**94.13**	-	93.98	-

The average computational time of the algorithms is estimated as follows: TBCNN, SibStCNN, and LSTM + GRNN are the running time of each loop iteration; the others are the time to predict an instance. As can be seen, pruning trees not only enhances the accuracies of all classifiers but also decreases the execution time notably. The execution time of kNN-TED and kNN-LD reduces more than two times; the execution time of SibStCNN, TBCNN and GRNN reduces around 1.5 times; the execution time of SVM-Tree kernel reduces slightly. These prove that pruning redundant branches can efficiently eliminate redundant information without loss of useful information.

Figure 3 illustrates the process of tuning the combination factor t of SibStCNN+ kNN-TED model in two scenarios: before and after pruning trees. As can be seen, the two lines representing the change of accuracies on the validation set and the test set according to t have the same trend. The accuracies increase when t is adjusted from 0 to around 0.9; after peaking the top at t around 0.9, the accuracies begin to drop down to the kNN classifier accuracies. We also observed the tuning process of TBCNN+kNN-TED and saw the similar results.

From Eq. (3), t is the trade-off parameter between SibStCNN/TBCNN and kNN of contribution to the models. The down arcs are caused by the bigger contribution of kNN when t increases. Especially, all sub-figures in Fig. 3 show that at $t = 0$, the accuracies of the hybrid models are higher than those of

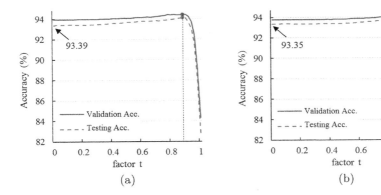

Fig. 3. Tuning the factor t for the combination models of SibStCNN and kNN-TED in two scenarios: before pruning trees (a), after pruning trees (b)

SibStCNN/TBCNN (Table 2). According to the Eq. 4, the label of an unseen instance is predicted based on kNN if ten neighbors are in the same group, otherwise, the label is the output of SibStCNN/TBCNN. These mean that although many structures are similar, SibStCNN/TBCNN could not detect them. Thus, using extra structural information of ASTs is helpful in making the final decision on class labels of instances. In all of the subfigures, the accuracies on the validation sets and the test sets reach maxima approximately at the same values of t. This proves the stability of the combination model TBCNN-kNN on different datasets.

6 Conclusion

In this paper, we present new models including SibStCNN, and combination models of SibStCNN/TBCNN and kNN-TED to solve program classification. The experimental results show that the proposed approaches significantly improve the classification performance. For programming languages with grammars, a source code can be converted to an AST using a parser. Therefore, these models can be applicable to other tasks in the field of software engineering such as software defect prediction, and clone detection.

Acknowledgements. This work was supported partly by JSPS KAKENHI Grant number 15K16048 and the first author would like to thank the scholarship from Ministry of Training and Education (MOET), Vietnam under the project 911.

References

1. Binkley, D., Feild, H., Lawrie, D., Pighin, M.: Software fault prediction using language processing. In: Testing: Academic and Industrial Conference Practice and Research Techniques-MUTATION, 2007. TAICPART-MUTATION 2007, pp. 99–110. IEEE (2007)

2. Huo, X., Li, M., Zhou, Z.-H.: Learning unified features from natural and programming languages for locating buggy source code
3. Joachims, T.: Making large scale SVM learning practical. Technical report, Universität Dortmund (1999)
4. Kaur, J., Singh, S., Kahlon, K.S., Bassi, P.: Neural network-a novel technique for software effort estimation. Int. J. Comput. Theor. Eng. **2**(1), 17 (2010)
5. Menzies, T., Greenwald, J., Frank, A.: Data mining static code attributes to learn defect predictors. IEEE Trans. Softw. Eng. **33**(1), 2–13 (2007)
6. Mo, R., Cai, Y., Kazman, R., Xiao, L., Feng, Q.: Decoupling level: a new metric for architectural maintenance complexity. In: Proceedings of the 38th International Conference on Software Engineering, pp. 499–510. ACM (2016)
7. Mou, L., Li, G., Zhang, L., Wang, T., Jin, Z.: Convolutional neural networks over tree structures for programming language processing. In: Proceedings of the Thirtieth AAAI Conference on Artificial Intelligence (2016)
8. Phan, V.A., Chau, N.P., Nguyen, M.L.: Exploiting tree structures for classifying programs by functionalities. In: 2016 Eighth International Conference on Knowledge and Systems Engineering (KSE), pp. 85–90. IEEE (2016)
9. Socher, R., Huang, E.H., Pennin, J., Manning, C.D., Ng, A.Y.: Dynamic pooling and unfolding recursive autoencoders for paraphrase detection. In: Advances in Neural Information Processing Systems, pp. 801–809 (2011)
10. Tang, D., Qin, B., Liu, T.: Document modeling with gated recurrent neural network for sentiment classification. In: Proceedings of the 2015 Conference on Empirical Methods in Natural Language Processing, pp. 1422–1432 (2015)
11. Ugurel, S., Krovetz, R., Giles, C.L.: What's the code?: automatic classification of source code archives. In: Proceedings of the Eighth ACM SIGKDD International Conference on Knowledge Discovery and Data Mining, pp. 632–638. ACM (2002)
12. Wang, S., Yao, X.: Using class imbalance learning for software defect prediction. IEEE Trans. Reliab. **62**(2), 434–443 (2013)

A Community Detection Algorithm Based on Local Double Rings and Fireworks Algorithm

TianRen Ma and Zhengyou Xia[✉]

College of Computer Science and Technology,
Nanjing University of Aeronautics and Astronautics, Nanjing 210015, China
zhengyou_xia@nuaa.edu.cn

Abstract. In recent years, more and more algorithms have been proposed to detect communities. An improved community detection algorithm based on the concept of local double rings and the framework of fireworks algorithm (LDRFA) has been proposed in this paper. Inspired by the framework of FWA, an improved distinctive fireworks initialization strategy was given. We use this strategy to obtain a more accurate initial solution. Secondly, on the basis of fireworks algorithm, the amplitude of explosion was used to calculate the probability of changing node label. Thirdly, the mutation operator was proposed. Nodes chose labels based on the idea of LPA. Finally, tests on real-world and synthetic networks were given. The experimental results show that the proposed algorithm has better performance than existing methods in finding community structure.

Keywords: Community detection · Fireworks algorithm · Swarm intelligence

1 Introduction

Complex networks have attracted increasing attention of researchers from different fields [1]. Community detection problem is an NP-hard one [2] and has become a research hotspot in recent years.

Label propagation algorithm is one of the most widely used community partitioning algorithms. The most primitive label propagation algorithm [3], proposed by Raghavan et al., has a high accuracy rate and it runs very fast. Swarm intelligence optimization algorithms have great performance on solving lots of real world problems and attract many scholar's attentions. Various algorithms for community detection were proposed by using swarm intelligence optimization algorithms. GA-Net is one of the first approaches to solve detection problem which was proposed by Pizzuti [4]. GA-Net is a single-objective algorithm and optimizes a simple fitness function to identify densely connected groups. CNM was proposed by Clauset et al. in 2004 [5]. This algorithm is a fast greedy modularity optimization algorithm and a fast implementation of the original Girvan-Newman algorithm. Infomap is introduced by Rosvall and Bergstorm [6]. This algorithm used a new information theoretic method to find community structure in a complex network. The Fireworks Algorithm (FWA) [7] is one of the latest swarm intelligence optimization algorithms which was proposed by Tan in 2010. It is a metaheuristic

© Springer International Publishing AG 2017
H. Yin et al. (Eds.): IDEAL 2017, LNCS 10585, pp. 129–135, 2017.
https://doi.org/10.1007/978-3-319-68935-7_15

swarm intelligence algorithm and simulates the explosion of fireworks and the generation of the sparks.

The remainder of this paper is organized as follows. The framework of fireworks algorithm is described in Sect. 2. In Sect. 3, LDRFA is particularly studied, followed by experiments in Sect. 4. Finally, we conclude the paper in Sect. 5.

2   The Framework of Fireworks Algorithm

The fireworks algorithm (FA) was inspired by the behavior of firework. The explosion process of firework can be regarded as the process of searching in a local space around a specific point. When a firework explodes, lots of sparks are generated around it. The number and amplitude of generated sparks will be designed. For each generation of explosion, n locations for n fireworks were selected. After explosion, every firework has generated sparks around it already. The number and amplitude of sparks are designed by the explosion operator. Finally, when the optimal location is found, the algorithm stops. Otherwise, the algorithm keeps the best fireworks or sparks and selects the rest (n-1) fireworks or sparks for the next generation according to their distances. The framework of FWA can be divided into four parts: the explosion operator, the mapping rule, the Gaussian explosion (mutation) operator and the selection strategy. The explosion operator was described as follow.

2.1   Explosion Operator

At each iteration, each firework generates sparks using this explosion operator. This explosion operator plays a key role in the whole algorithm because that it is responsible for calculating the number and amplitude of sparks.

The number of sparks S_i is calculated using the following formula:

$$S_i = \text{m} \times \frac{y_{worst} - f(x_i) + \varepsilon}{\sum_{i=1}^{n} (y_{worst} - f(x_i)) + \varepsilon} \tag{1}$$

Where m is a constant to control the total number of sparks and y_{worst} is the worst value of the objective function among the n fireworks. Function $f(x_i)$ represents the fitness value of individual x_i. The last parameter ε is used to prevent the denominator from becoming 0.

The amplitude of sparks A_i is calculated as follow:

$$A_i = \text{A}' \times \frac{f(x_i) - y_{best} + \varepsilon}{\sum_{i=1}^{n} (f(x_i) - y_{best}) + \varepsilon} \tag{2}$$

Where A' denotes the sum of all amplitudes, while y_{best} is the best value of the objective function among the n fireworks. The meaning of $f(x_i)$ and ε are the same as mentioned in formula (1).

3   LDRFA: A Community Detection Algorithm Based on Local Double Rings and Fireworks Algorithm

In this paper, the number of positions is equal to the number of nodes in the network. An arbitraty fireworks x_i represents the community detection result of the network and x_i^k represents the label of node k.

3.1   Concept of Node Importance

Inspired by the PageRank algorithm [8], a new method to measure the importance of nodes is designed.

Definition 1. *(Node Importance).* Given a network $G = \{V, E\}$, for any node $v_i \in V$, the node importance is:

$$\text{NI}(v_i) = \sum_{v_i' \in V_i'} \frac{1}{d(v_i')} \tag{3}$$

V_i' is the neighbor node set of node v_i and $d(v_i')$ is the degree of node v_i'. The importance of a node v_i depends on the number and degree of its neighbor nodes. If the degree of the neighbor node is larger, the contribution to the importance of the node v_i is relatively smaller and vice versa.

3.2   An Improved Fireworks Initialization Method

In LDRFA, we propose an improved fireworks initialization method. For convenience of description, we denote n as the total number of fireworks, V' as the sorted set of nodes and $V'[i]$ as the ith node in V'.

First of all, the total number of fireworks n is equal to 50. After the whole nodes were sorted in descending order by their importance and initializing the label of node, for each firework x_i, the first local ring is the shortest ring containing $V'[x_i.id \bmod 3]$. The second ring is the shortest ring containing the most important node which is not $V'[x_i.id \bmod 3]$ in the first ring. If there are more than one rings available, then randomly choosing one. After the local double rings were both found, we change the label of nodes in these two rings and the label of the most important neighbor of them to the label of $V'[x_i.id \bmod 3]$. Then the number and amplitude of sparks were calculated according to the formulas (1) and (2). Then the fireworks initialization process is over.

3.3   Detailed Description of LDRFA

LDRFA is described as follows:

(1) Initializing the label of each node in the network; calculating their improtance and sorting them in descending order by importance. Setting m = 50, A′ = 40, the total number of fireworks n = 50. Setting the maximum number of iterations to 100.

(2) Initializing the fireworks according to Sect. 3.2.

(3) Generating sparks and selecting fireworks or sparks for next generation. This part does not stop until the number of iterations reaches its maximum.

In LDRFA, the amplitude of fireworks is used to control the probability of changing node label. And the node label updating stragety updates node's label based on its neighborhood labels. The label with maximum frequency is used as new label. We denote this label as Neig_label$_k$. x_i^k in this part represents the label of node k in firework i.

$$x_i^k = \begin{cases} \text{Neig\_label}_k, & \text{if } \text{random}(0, 1) < \text{sigmoid}(A_i) \\ x_i^k, & \text{else} \end{cases} \tag{4}$$

random(0,1) is a randomly number between 0 and 1. sigmoid(A_i) is the sigmoid function of the amplitude A_i.

After all of sparks were generated, the top n fireworks and sparks which were sorted in desending order by their modularity value will be selected into next iteration. Finally, when the termination condition was reached, the community structure represented by the fireworks or sparks with maximum modularity value was accepted.

4   Experiments Result and Analysis

In this section, the performance of LDRFA is tested on standard data sets which are commonly used in the real world and the synthetic benchmark networks. The former community data is collected from the real world and often measured by modularity [9]. The latter community data is a data set that are constructed from pre-set parameters. The accuracy of this community detection result can be measured by NMI.

4.1   Experimental Result of the Real World

There are many real-world networks that have been abstracted into community detection fields, such as Zachary's Karate Club, which is constructed by observing an American university karate club. In addition to the above network, we also choose the Dolphin social network. It is the dolphin social network that D. Lusseau et al. used for seven years to observe the dolphin population of Doubtful Sound, New Zealand. The third network we choose is Polbooks, which was created by V. Krebs. In this section, we use these three networks to perform community partitioning experiments. The detail parameters of these networks are shown in Table 1.

We use four different algorithms for community detection on these three networks. The other three algorithms are GA-Net, CNM and Infomap which are mentioned in Sect. 1. The values of modularity obtained from the experiment are shown in the following table (Table 2).

Table 1. Real-world networks

Name	Nodes	Edges	Average Degree
Karate	34	78	4.59
Dolphins	62	159	5.13
Polbooks	105	441	8.40

Table 2. Comparison of modularity values

Networks	Q(LDRFA)	Q(GA-Net)	Q(CNM)	Q(Infomap)
Karate	0.3949	0.4019	0.3990	0.3989
Dolphins	0.5264	0.4701	0.4496	0.4310
Polbooks	0.5262	0.5798	0.4193	0.4198

The proposed LDRFA algorithm in the Dolphins network has achieved the highest values of modularity and in the Polbooks network also achieved a nice performance.

4.2 Experimental Result of the Synthetic Benchmark Networks

In this section, we compired the performance on the GN extended benchmark networks. GN extended benchmakr networks proposed by Lancichinetti et al. [10] is an improved version of the classical GN benchmark networks. There is an important mixing parameter μ, which control the tightness of connections between different communities.

We can find that these four algorithms have perfect performance when $\mu \leq 0.15$. With the increase of μ, Infomap algorithm first became unsuccessful and can't discover community structure when μ is greater than 0.25. The detection ability of GA-Net and CNM algorithms began descending when $\mu > 0.3$ and $\mu > 0.2$ respectively. But LDRFA

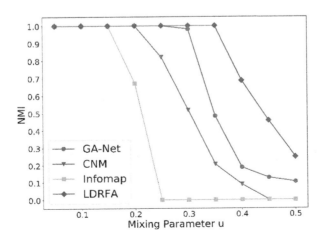

Fig. 1. Average NMI values on GN extended networks

always had a perfect performance when $\mu \leq 0.35$ and a good detection result was obtained when $\mu = 0.4$ (Fig. 1).

The average modularity values obtained by these four algorithms are shown in Fig. 2. From the figure we can see that all algorithms get a consistent modularity value when $\mu \leq 0.15$. As μ increases, the detection task becomes more and more difficult. When $\mu = 0.2$, the Infomap algorithm gave a worst modularity value of 0.32, but others gave a best value of 0.55. When μ is not less than 0.25, the performance of the other three algorithms is significantly reduced. Infomap and CNM algorithms fail to detect the community structure when $\mu = 0.35$. At the same time, the proposed algorithm still detected the community structure successfully with modularity value of 0.4013. When μ continues to increase, the performance of these algorithms began to decline rapidly because the community structure became ambiguous. From these two figures, the advantage of LDRFA against others was shown clearly.

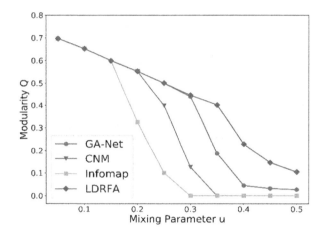

Fig. 2. Average modularity values on GN extended networks

5   Summary

In this paper, a community detection algorithm based on local double rings and fireworks algorithm is proposed to solve the problem of disjoint community partitioning, and we compare LDRFA with GA-Net, Infomap and CNM by using real-world networks and the synthetic networks.

The fireworks algorithm is a new swarm intelligence algorithm which has been widely applied to many fields. The LDRFA improves the initialization operation of fireworks in FA and proposes a new and effective method to discover small groups based on the concept of local double rings. In our algorithm, nodes are sorted by computing the importance of them and the initialization process proposed in this paper can improve the accuracy of detection result. On the basis of fireworks algorithm, the amplitude of explosion was used to calculate the probability of changing node label based on the idea of LPA. Finally, the firework with highest modularity value was accepted and nodes with the same label were considered in the same community.

The experiment in this paper is divided into two parts. The value of modularity and NMI are used to evaluate the performance. Experimental results of these networks show that the algorithm proposed in this paper has achieved better performance.

References

1. John, H., et al.: Natural communities in large linked networks. In: ACM SIGKDD International Conference on Knowledge Discovery and Data Mining, pp. 541–546. ACM (2003)
2. Fortunato, S.: Community detection in graphs. Phys. Rep. **486**(3–5), 75–174 (2010)
3. Raghavan, U.N., Albert, R., Kumara, S.: Near linear time algorithm to detect community structures in large-scale networks. Phys. Rev. E Stat. Nonlin Soft Matter Phys. **76**(2), 036106 (2007)
4. Pizzuti, C.: GA-Net: A Genetic Algorithm for Community Detection in Social Networks. In: Rudolph, G., Jansen, T., Beume, N., Lucas, S., Poloni, C. (eds.) PPSN 2008. LNCS, vol. 5199, pp. 1081–1090. Springer, Heidelberg (2008). doi:10.1007/978-3-540-87700-4_107
5. Clauset, A., Newman, M.E., Moore, C.: Finding community structure in very large networks. Phys. Rev. E Stat. Nonlin. Soft Matter Phys. **70**, 2 (2004).066111
6. Rosvall, M., Bergstrom, C.T.: Maps of random walks on complex networks reveal community structure. Proc. Nat. Acad. Sci. U.S.A **105**(4), 1118–1123 (2007)
7. Tan, Y., Zhu, Y.: Fireworks Algorithm for Optimization. In: Tan, Y., Shi, Y., Tan, K.C. (eds.) ICSI 2010. LNCS, vol. 6145, pp. 355–364. Springer, Heidelberg (2010). doi: 10.1007/978-3-642-13495-1_44
8. Brin, S., Page, L.: Reprint of: The anatomy of a large-scale hypertextual web search engine. Comput. Netw. **56**(18), 3825–3833 (2012)
9. Newman, M.E.J.: Fast algorithm for detecting community structure in networks. Phys. Rev. E: Stat., Nonlin, Soft Matter Phys. **69**(6), 066133 (2004)
10. Lancichinetti, A., Fortunato, S., Radicchi, F.: Benchmark graphs for testing community detection algorithms. Phys. Rev. E Stat. Nonlin. Soft Matter Phys. **78**(2), 046110 (2008)

Cost Sensitive Matrix Factorization
for Face Recognition

Jianwu Wan[1(✉)], Ming Yang[2], and Hongyuan Wang[1]

[1] School of Information Science and Engineering,
Changzhou University, Changzhou, China
`jianwuwan@gmail.com`, `hywang@cczu.edu.cn`
[2] School of Computer Science and Technology,
Nanjing Normal University, Nanjing, China
`m.yang@njnu.edu.cn`

Abstract. In this paper, we propose a cost sensitive matrix factorization (CSMF) for face recognition. To make the face representation cost sensitive, CSMF adopts a more flexible feature embedding strategy. It contains two main steps: (1) matrix factorization for the learning of latent semantic representation and (2) cost sensitive latent semantic regression. In this way, the face images are embedded into their label space with the misclassification loss minimized. The experimental results on Extended Yale B and ORL demonstrate its effectiveness.

Keywords: Cost sensitive learning · Matrix factorization · Face recognition · Latent semantic regression

1 Introduction

Recently, many successful face recognition systems have been developed [1,2], which assume the same loss from all misclassifications and thus aim to attain low recognition error. In many real-world applications such as the door-locker system based on face recognition [3–10], however, this assumption may not hold as different misclassifications could lead to different losses.

According to paper [11], the unequal misclassification losses will change the optimal decision threshold of learner, which makes the conventional methods can not work. To address this problem, cost sensitive learning measures the misclassification losses with different costs and aims to make the optimal decision by minimizing the misclassification loss.

Over the past decades, three kinds of cost sensitive approaches have been proposed. The first is cost sensitive classifier [3,12–14], which aims to make the optimal decision by minimizing the misclassification loss. For example, multi-class cost sensitive kernel logistic regression (McKLR) [3] and multi-class cost sensitive k-nearest neighbor (McKNN) [3] proposed by Zhou et al.

Different from cost sensitive classifier, cost sensitive dimensionality reduction controls the loss in dimensionality reduction phase. To our best knowledge, cost

© Springer International Publishing AG 2017
H. Yin et al. (Eds.): IDEAL 2017, LNCS 10585, pp. 136–145, 2017.
https://doi.org/10.1007/978-3-319-68935-7_16

sensitive principal component analysis, cost sensitive linear discriminant analysis (CSLDA) and cost sensitive locality preserving projections [4] should be the first work. Considering the difficult of obtaining the label information, cost sensitive semi-supervised approaches such as cost sensitive semi-supervised discriminant analysis [5], pairwise costs in semisupervised discriminant analysis (PCSDA) [6] and cost sensitive semi-supervised canonical correlation analysis (CS^3CCA) [8] have been proposed. Recently, researchers also concern the cost sensitive feature selection. For example, cost sensitive laplacian score (CSLS) [15] and discriminative cost sensitive laplacian score (DCSLS) [7].

Inspired by the great success of sparsity learning for face recognition, how to develop a sparse cost sensitive classifier has attracted much attention [9,10]. However, the sparsity assumption may not hold in face recognition [6,8,16]. Based on this, we propose a cost sensitive matrix factorization (CSMF). To make the face representation cost sensitive, CSMF adopts a more flexible feature embedding strategy. It contains two main steps: (1) matrix factorization for the learning of latent semantic representation and (2) cost sensitive latent semantic regression. In this way, the face images are embedded into their label space with the misclassification loss minimized. The experimental results on Extended Yale B [17] and ORL [18] datasets demonstrate its effectiveness.

2   Problem Formulation

In the door-locker system based on face recognition [3–10], there are $c-1$ gallery subjects, denoted by $G_1, G_2, \cdots, G_{c-1}$, the rest are impostors, denoted by I, and the total class number is c. The possible mistakes in predicting a face image include: (1) false rejection, misrecognizing a gallery person as an impostor; (2) false acceptance, misrecognizing an impostor as a gallery person; (3) false identification, misrecognizing a gallery person as another gallery person. To measure the misclassification loss, we let C_{GI}, C_{IG} and C_{GG} be the costs of false rejection, false acceptance and false identification, respectively. A cost matrix C shown in Table 1 therefore can be constructed. Similar with papers [3–10], we assume accepting any impostor will result in the same losses and misclassifying a gallery subject as another gallery subject or an impostor will result in different losses.

Table 1. The cost matrix for cost sensitive face recognition

	G_1	$\cdots$	G_{c-1}	I
G_1	0	$\cdots$	C_{GG}	C_{GI}
$\cdots$	$\cdots$	$\cdots$	$\cdots$	$\cdots$
G_{c-1}	C_{GG}	$\cdots$	0	C_{GI}
I	C_{IG}	$\cdots$	C_{IG}	0

3   Cost Sensitive Matrix Factorization

Let $X_{D \times N} = [x_1, x_2, \cdots, x_N]$ be a training set containing N number of face images from D dimensional space. $Y = [y_1, y_2, \cdots, y_N]_{c \times N}$ is the label matrix, where $y_{ki} = 1$ and $y_{ji} = 0, j = 1, 2, \cdots, c, j \neq k$, if x_i belongs to class k.

Inspired by the great success of matrix factorization for capturing high-level features of data [19,20], we here use it to learn the latent semantic space $B \in \mathbb{R}^{D \times d}$ and the representations $S = [s_1, s_2, \cdots, s_N] \in \mathbb{R}^{d \times N}$ for face images.

$$O_1(B, S) = ||X - BS||_F^2. \tag{1}$$

For Eq. (1), its solution is unstable if existing noise and outliers in datasets [21]. Instead of directly imposing the misclassification costs on the reconstruction coefficients to design cost sensitive classifier, e.g., sparse representation based classification (CS_SRC) [10], we aim to find a $W \in \mathbb{R}^{d \times c}$ that embeds S into label space with misclassification loss minimized, which is more flexible and can be viewed as the regularization for matrix factorization.

$$O_2(W) = \sum_{i=1}^{N} h(l(x_i))||W^T s_i - y_i||_2^2, \tag{2}$$

where $h(l(x_i))$ defines the cost of misclassifying sample x_i from class $l(x_i)$ as the other classes, i.e., the importance of x_i, which is the larger the more important. According to papers [4–8], it can be defined as follows.

$$h(l(x_i)) = \begin{cases} (c-2)C_{GG} + C_{GI}, & if\ l(x_i) = G_1, \cdots, G_{c-1} \\ (c-1)C_{IG}, & if\ l(x_i) = I \end{cases} \tag{3}$$

Eq. (2) minimizes the misclassification loss. To explain this observation, we firstly consider the latent semantic regression with the following formulation.

$$\min_{w_k} : ||S^T w_k - Y^{(k)}||_F^2, \tag{4}$$

where $k = 1, \cdots, c$, $W = [w_1, \cdots, w_c]$ and $Y = [Y^{(1)}, \cdots, Y^{(c)}]^T$. According to Theorem 1, if we substitute X, β and g in Eq. (5) with S, w_k and $Y^{(k)}$ respectively, we have $w_k^T s_i \propto p(k|x_i)$ for any $x_i \in X$.

Theorem 1. *Given a two-class regression problem with the following definition*

$$\min_{\beta} : ||X^T \beta - g||^2, \tag{5}$$

where $X \in \mathbb{R}^{d \times N}$ contains N number of samples, whose first N_i samples belong to class ω_i and the others belong to class ω_j. $g = [g_1, ..., g_N]^T$ is the label vector. $g_k = 1$, if x_k belongs to class ω_i. Otherwise, $g_k = 0$. For any $x \in X$, we have

$$x^T \beta \propto \left(p(\omega_i|x) + \zeta N x^T (XX^T)^{-1} m \right),$$

where m is the mean of X, $p(\omega_i|x)$ denotes the probability of x having label ω_i.

Proof. Suppose each class ω_k, $k = i, j$ satisfies the normal distribution. According to the Bayes theorem, we have

$$p(\omega_i|x) = \frac{p(x|\omega_i)N_i/N}{p(x|\omega_i)N_i/N + p(x|\omega_j)N_j/N} = \frac{1}{1 + \exp(-\epsilon)},$$

$$\epsilon = x^T \Sigma^{-1}(m_i - m_j) - 1/2(m_i + m_j)^T \Sigma^{-1}(m_i - m_j) + \ln N_i/N_j,$$

where m_k and Σ are the mean and covariance matrix of class ω_k, respectively. $p(x|\omega_k)$ is the class-conditional density. As the last two terms in ϵ are constants, the $p(\omega_i|x)$ is proportional to the first term, i.e., $p(\omega_i|x) \propto x^T \Sigma^{-1}(m_i - m_j)$.

Considering the following least square regression problem

$$\min_{\widetilde{\beta}} : ||X^T \widetilde{\beta} + 1\zeta - g||^2, \tag{6}$$

where ζ is an intercept, $1 = [1, ..., 1]^T$. For the optimal coefficients β and $\widetilde{\beta}$ that satisfy Eqs. (5) and (6), we have

$$\begin{cases} X^T \beta = g \\ X^T \widetilde{\beta} = g - 1\zeta \end{cases}$$

By subtracting the both sides of these two equations, we have $X^T(\beta - \widetilde{\beta}) = \zeta 1$. Thus, $\beta = \widetilde{\beta} + \zeta N(XX^T)^{-1}m$. Recall that the optimal regression coefficient $\widetilde{\beta}$ obtained by Eq. (6) is proportional to $\Sigma^{-1}(m_i - m_j)$ [22], i.e., $\widetilde{\beta} \propto \Sigma^{-1}(m_i - m_j)$. For any $x \in X$, we therefore have $x^T \beta \propto (p(\omega_i|x) + \zeta N x^T (XX^T)^{-1}m)$. ∎

Suppose sample x_i is belong to class k, i.e., $y_{ki} = 1$ and $y_{ji} = 0, j = 1, \cdots, c, j \neq k$, we have

$$||W^T s_i - y_i||_2^2 \propto \sum_{j=1, \cdots, c, j \neq k} p(j|x_i)^2 + (1 - p(k|x_i))^2. \tag{7}$$

If the value of $||W^T s_i - y_i||_2^2$ is small, it indicates that x_i is correctly regressed to class k, i.e., $p(k|x_i)$ must be near to 1 with the others $p(j|x_i), j = 1, \cdots, c, j \neq k$ near to 0. Otherwise, the value of $p(k|x_i)$ is small and the others are large, which means x_i is regressed to the other classes. Therefore, $||W^T s_i - y_i||_2^2$ can measure whether x_i is correctly regressed to class k, which can be roughly viewed as the probability of classifying x_i as the other classes. Furthermore, by multiplying the corresponding cost $h(l(x_i))$, the term $h(l(x_i))||W^T s_i - y_i||_2^2$ therefore measures the misclassification loss of misclassifying x_i as the other classes. Considering all of the training samples together, we have the model defined by Eq. (2) that finds the projection W by minimizing the misclassification loss.

3.1 The Overall Objective Function

The overall objective function of CSMF is defined as follows

$$\min_{B,S,W} O = O_1 + \mu O_2 + \lambda R(B, S, W), \tag{8}$$

where the regularization term is defined as $R(B, S, W) = ||B||_F^2 + ||S||_F^2 + ||W||_F^2$ to resist overfitting. μ and λ are two tradeoff parameters.

3.2   Optimization

We adopt the alternate iteration method to solve the optimization problem, which can converge around 20 iterations according to the results in Sect. 4.2.

(1) *Optimal B*: Take the derivative of Eq. (8) with respect to B, we have

$$B = XS^T(SS^T + \lambda I)^{-1}, \tag{9}$$

where $I \in \mathbb{R}^{d \times d}$ is an identity matrix.

(2) *Optimal S*: We optimize N number of $s_i, i = 1, 2, \cdots N$ independently. Specifically, we take the derivative of Eq. (8) with respect to each s_i and get

$$s_i = (B^T B + \mu h(l(x_i))WW^T + \lambda I)^{-1}(B^T x_i + \mu h(l(x_i))W y_i). \tag{10}$$

(3) *Optimal W*: Take the derivative of Eq. (8) with respect to W, we have

$$W = (\mu SHS^T + \lambda I)^{-1}(\mu SHY^T), \tag{11}$$

where $H = diag(h(l(x_1)), \cdots, h(l(x_N)))$.

3.3   Classification Scenario

For a testing image x_t, we first capture its latent semantic representation s_{x_t} with the B learned in training process. And then, s_{x_t} is projected into the c-dimensional label space, i.e., $W^T s_{x_t}$. As W is proven to be cost sensitive, we assign x_t to class k, if the k-th element of $W^T s_{x_t}$ achieves the maximal value.

4   Experimental Results

4.1   Face Datasets and Experimental Settings

Extended Yale B [17]: it contains 2414 front-view face images of 38 individuals under different illuminations. In our experiments, the image size is 32×32 pixels.

ORL [18]: it contains 400 images of 40 people, which were taken at different times, lighting, and facial expressions. The images size is 32×32 pixels.

Similar with papers [3–10], we let $C_{IG} = 20$, $C_{GI} = 2$, $C_{GG} = 1$. M subjects are randomly selected as the gallery person, and the rest are impostors. To form the training set, we randomly select N_G^{tr} images from each gallery person. N_I^{tr} images are selected as the impostors in the training set. The rest N_G^{te} gallery images and N_I^{te} impostors are for testing. The settings are specified in Table 2.

We empirically set $\mu = \lambda = 0.01$ and provide the parameter sensitivity analysis in Sect. 4.3. We repeat 10 times for each dataset and compute the average performance as the final results. Similar with papers [3–10], total cost (TC), total error rate (E), error rate of false acceptance (E_{IG}), false rejection (E_{GI}) and false identification (E_{GG}) are adopted here for performance evaluation.

Table 2. The experimental settings

Datasets	M	N_G^{tr}	N_I^{tr}	N_G^{te}	N_I^{te}
Extended Yale B	30	3	24	1812	488
ORL	20	3	60	140	140

4.2 Comparing with State-of-the-art Cost Sensitive Approaches

In this section, we first investigate the convergence analysis of CSMF. In Fig. 1, we plot its objective function value achieved in each iteration. The result demonstrates that CSMF can converge around 30 iterations.

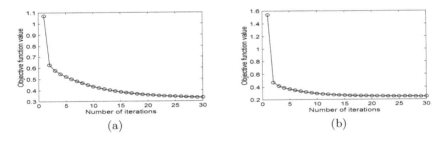

Fig. 1. The convergence analysis of CSMF. (a) Extended Yale B; (b) ORL.

Subsequently, we compare the performance with: (a) cost sensitive classifier: McKLR [3], McKNN [3] and CS_SRC [10]; (b) cost sensitive feature selection: CSLS [15] and DCSLS [7]; (c) cost sensitive dimensionality reduction: CSLDA [4] and CS³CCA [8]. All the methods are compared with their best performance. The results in Table 3 demonstrate that the proposed CSMF can achieve the best performance in TC and E_{IG}, which is resulted by the cost sensitive face representation obtained by CSMF.

4.3 The Influential Factors of CSMF

Firstly, we display the performance of CSMF with different misclassification costs. Let $C_{GI} = 2$, $C_{GG} = 1$, we vary the values of C_{IG} from 5 to 30 with the step length 5. Evaluation criteria TC and E_{IG} are adopted here. The experimental results shown in Fig. 2 demonstrate that the proposed CSMF consistently outperforms state-of-the-art cost sensitive methods in TC with different misclassification costs. Observing the E_{IG} values, we find nearly all of the approaches can control the misclassification of false acceptance by varying the corresponding misclassification cost C_{IG}. This reflects the main strategy of cost sensitive approaches that controls the TC via reducing the high-cost misclassification.

Table 3. The performance of CSMF (mean ± std. The best performance is bolded.)

		McKLR	McKNN	CS_SRC	CSLS	DCSLS	CSLDA	CS³CCA	CSMF
Extended Yale B	TC	3216 ± 476	4924 ± 468	7253 ± 506	6983 ± 346	5816 ± 297	2587 ± 288	3174 ± 133	**2088 ± 252**
	$E_{IG}(\%)$	2.66 ± 26	4.17 ± 30	11.23 ± 27	11.63 ± 18	9.06 ± 15	2.76 ± 15	3.79 ± 7	**1.29 ± 12**
	$E_{GI}(\%)$	36.33 ± 77	56.28 ± 137	22.51 ± 125	**13.66 ± 49**	17.16 ± 61	23.33 ± 65	24.83 ± 56	28.5 ± 34
	$E_{GG}(\%)$	14.02 ± 59	18.12 ± 124	45.6 ± 113	43.69 ± 58	37.27 ± 68	10.58 ± 44	12.52 ± 36	**7.85 ± 19**
	$E(\%)$	53 ± 48	78.57 ± 30	79.38 ± 58	68.98 ± 47	63.5 ± 54	**36.67 ± 39**	41.15 ± 26	37.65 ± 35
ORL	TC	187 ± 18	240 ± 45	370 ± 102	443 ± 90	355 ± 78	147 ± 37	149 ± 49	**83 ± 12**
	$E_{IG}(\%)$	0.214 ± 0.8	1.39 ± 2	5.6 ± 5	6.67 ± 4	5 ± 4	1.11 ± 2	1 ± 2	**0.11 ± 0.6**
	$E_{GI}(\%)$	31.07 ± 6	26.36 ± 8	**6.71 ± 5**	7.5 ± 6	9.86 ± 7	14.46 ± 3	16.32 ± 4	13.18 ± 4
	$E_{GG}(\%)$	0.428 ± 0.7	5.14 ± 5	6.68 ± 5	9.67 ± 5	7.1 ± 5	1.57 ± 2	0.678 ± 1	**1.11 ± 2**
	$E(\%)$	31.71 ± 6	32.89 ± 7	19 ± 9	23.86 ± 9	21.96 ± 7	17.14 ± 3	18 ± 4	**14.39 ± 4**

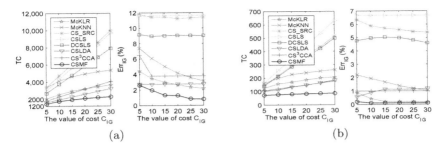

Fig. 2. The influence of misclassification costs. (a) Extended Yale B; (b) ORL.

In the following, we investigate the influence of labeled samples by varying its value from 2 to 6. Evaluation criterion TC is adopted for evaluating the performance. The experimental results shown in Fig. 3 demonstrates that the proposed CSMF consistently outperforms the compared methods under the different settings of labeled samples per class.

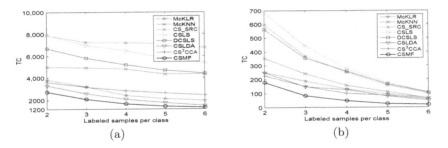

Fig. 3. The influence of labeled samples per class. (a) Extended Yale B; (b) ORL.

μ and λ are two trade-off parameters in CSMF. Here, their values are selected from $\{10^{-3}, 10^{-2}, 10^{-1}, 10^0, 10^1\}$ and $\{4^{-6}, 4^{-5}, 4^{-4}, 4^{-3}, 4^{-2}, 4^{-1}, 4^0\}$, respectively. In Fig. 4, we 3D plot the performance of CSMF versus the varying of μ and λ. We find that CSMF can achieve the best performance in TC with a wide range, i.e., $\mu \in [10^{-2}, 10^1]$ and $\lambda \in [4^{-5}, 4^{-2}]$.

At last, we concern the time complexity of CSMF, including the updating of B, S and W, respectively. According to Eqs. (9), (10) and (11), the corresponding time complexity is $O(N(Dd+d^2)+Dd+d^3)$, $O(N(d(D+c)+d^2(D+c+1)+d^3))$ and $O(N(2d+d^2+dc)+d^2c+d^3)$ respectively. Thus, the overall training time complexity is $O(N)$. As introduced in Sect. 3.3, the testing time complexity is $O(d(D+c)+d^2(D+1)+d^3)$.

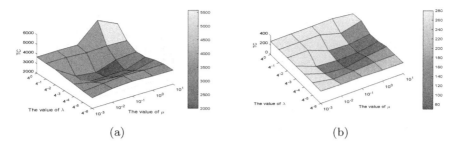

        (a)   (b)

Fig. 4. The influence of parameters μ and λ. (a) Extended Yale B; (b) ORL.

5   Conclusion and Future Works

In this paper, we propose a cost sensitive matrix factorization for face recognition. It combines the parts of matrix factorization and cost sensitive latent semantic regression in a unified model for learning cost sensitive face representation. The experimental results demonstrate its effectiveness. In the future, we will concern the semi-supervised scenario in cost sensitive learning, e.g., designing a unified framework for label propagation and cost sensitive matrix factorization.

Acknowledgements. This work was supported by National Natural Science Foundation of China (61502058, 61572085 and 61432008), Natural Science Foundation of Educational Committee of Jiangsu Province (15KJB520002).

References

1. He, X.F., Yan, S.C., Hu, Y.X., Niyogi, P., Zhang, H.J.: Face recognition using laplacianfaces. IEEE Trans. Patt. Anal. Mach. Intell. **27**(3), 328–340 (2005)
2. Wright, J., Yang, A., Sastry, S., Ma, Y.: Robust face recognition via sparse representation. IEEE Trans. Patt. Anal. Mach. Intell. **31**(2), 210–227 (2009)
3. Zhang, Y., Zhou, Z.H.: Cost-sensitive face recognition. IEEE Trans. Patt. Anal. Mach. Intell. **32**(10), 1758–1769 (2010)
4. Lu, J.W., Tan, Y.P.: Cost-sensitive subspace learning for face recognition. In: Proceedings of the IEEE International Conference on Computer Vision and Pattern Recognition, San Francisco, pp. 2661–2666 (2010)
5. Lu, J.W., Zhou, X.Z., Tan, Y.P., Shang, Y.Y., Zhou, J.: Cost-sensitive semi-supervised discriminant analysis for face recognition. IEEE Trans. Inf. Forensics Secur. **7**(3), 944–953 (2012)
6. Wan, J.W., Yang, M., Gao, Y., Chen, Y.J.: Pairwise costs in semisupervised discriminant analysis for face recognition. IEEE Trans. Inf. Forensics Secur. **9**(10), 1569–1580 (2014)
7. Wan, J.W., Yang, M., Chen, Y.J.: Discriminative cost sensitive Laplacian score for face recognition. Neurocomputing **152**, 333–344 (2015)
8. Wan, J.W., Wang, H.Y., Yang, M.: Cost sensitive semi-supervised canonical correlation analysis for multi-view dimensionality reduction. Neural Process. Lett. **45**(2), 411–430 (2017)

9. Man, J.Y., Jing, X.Y., Zhang, D., Lan, C.: Sparse cost-sensitive classifier with application to face recognition. In: Proceedings of the 18th International Conference on Image Processing, Brussels, pp. 1773–1776 (2011)
10. Zhang, G.Q., Sun, H.J., Ji, Z.X., Yuan, Y.H., Sun, Q.S.: Cost-sensitive dictionary learning for face recognition. Patt. Recogn. **60**, 613–629 (2016)
11. Elkan, C.: The foundations of cost-sensitive learning. In: Proceedings of International Joint Conference on Artificial Intelligence, San Francisco, pp. 973–978 (2001)
12. Lee, Y., Lin, Y., Wahba, G.: Multicategory support vector machines: theory and application to the classification of microarray data and satellite radiance data. J. Am. Stat. Assoc. **99**(465), 67–81 (2004)
13. Lo, H.Y., Wang, J.C., Lin, S.D.: Cost-Sensitive multi-label learning for audio tag annotation and retrieval. IEEE Trans. Multimedia **13**(3), 518–529 (2011)
14. Zhang, G.Q., Sun, H.J., Xia, G.Y., Sun, Q.S.: Multiple kernel sparse representation-based orthogonal discriminative projection and its cost-sensitive extension. IEEE Trans. Image Process. **25**(9), 4271–4285 (2016)
15. Miao, L.S., Liu, M.X., Zhang, D.Q.: Cost-sensitive feature selection with application in software defect prediction. In: Proceedings of the IEEE 21th International Conference on Pattern Recognition, Tsukuba, pp. 976–970 (2012)
16. Shi, Q.F., Eriksson, A., Shen, C.H.: Is face recognition really a Compressive Sensing problem? In: Proceedings of the IEEE International Conference on Computer Vision and Pattern Recognition, Colorado Springs, pp. 553–560 (2011)
17. Georghiades, S., Belhumeur, P.N., Kriegman, D.J.: From few to many: illumination cone models for face recognition under variable lighting and pose. IEEE Trans. Patt. Anal. Mach. Intell. **23**(6), 643–660 (2001)
18. Samaria, F., Harter, A.: Parameterisation of a stochastic model for human face identification. In: Proceedings of IEEE Workshop Applications Computer Vision, pp. 138–142 (1994)
19. Tang, J., Wang, K., Shao, L.: Supervised matrix factorization hashing for cross-modal retrieval. IEEE Trans. Image Process. **25**(7), 3157–3166 (2016)
20. Deerwester, S., Dumais, S.T., Furnas, G.W., Landauer, T.K., Harshman, R.: Indexing by latent semantic analysis. J. Am. Soc. Inf. Sci. **41**(6), 391–407 (1990)
21. Kong, D.G., Ding, C., Huang, H.: Robust nonnegative matrix factorization using L21-norm. In: Proceedings of the 20th ACM International Conference on Information and Knowledge Management, Glasgow, pp. 673–682 (2011)
22. Hastie, T., Tibshirani, P., Friedman, J.: The Elements of Statistical Learning, 2nd edn. Springer, New York (2001)

Research of Dengue Fever Prediction in San Juan, Puerto Rico Based on a KNN Regression Model

Ying Jiang[1(✉)], Guohun Zhu[1,2], and Ling Lin[3]

[1] School of Electronic and Electrical Engineering,
Guilin University of Electronic Technology, Guilin 541004, China
jying1102@163.com
[2] School of ITEE, The University of Queensland, Brisbane QLD 4072, Australia
[3] Jiujiang Power Supply Branch, State Grid Electric Power Company,
Jiangxi, China

Abstract. Existed dengue prediction model associated with temperature data are always based on Poisson regression methods or linear models. However, these models are difficult to be applied to non-stationary climate data, such as rainfall or precipitation. A novel k-nearest neighbor (KNN) regression method was proposed to improve the prediction accuracy of dengue fever regression model in this paper. The dengue cases and the climatic factors (average minimum temperature, average maximum temperature, average temperature, average dew point temperature, temperature difference, relative humidity, absolute humidity, Precipitation) in San Juan, Puerto Rico during the period 1990–2013 were regressed by the KNN algorithm. The performances of KNN regression were studied by compared with correlation analysis and Poisson regression method. Results showed that the KNN model fitted real dengue outbreak better than Poisson regression method while the root mean square error was 6.88.

Keywords: Dengue prediction · Poisson regression · KNN

1 Introduction

Dengue fever is one of the most important mosquito borne infectious diseases in the world. Recent studies showed that about 390 million people were infected with dengue fever each year, in which 96 million people appear obvious symptoms [1]. Dengue fever is mainly transmitted to human by four closely related dengue viruses (DENV) which usually carried by Aedes albopictus [2]. A series of different nonspecific febrile syndromes, such as dengue fever and hemorrhagic shock syndrome, would be occurred if a person was infected by the dengue virus. At present, there is no specific way to treat dengue fever. The prevention strategy of dengue fever is limited to vector control measures [3]. Therefore, a high precision dengue forecast is necessary to prevent dengue outbreaks.

Existed researches have studied the relationship between dengue transmission and climate factors [4–8]. Regarding to model building issues, the influence factors and prediction effects were quite difference [9, 10]. In French Guiana, a literature [11]

© Springer International Publishing AG 2017
H. Yin et al. (Eds.): IDEAL 2017, LNCS 10585, pp. 146–153, 2017.
https://doi.org/10.1007/978-3-319-68935-7_17

identified a positive correlation between climate data and the incidence of dengue fever, where the climatic variables included temperature, rainfall and the Southern Oscillation Index (ENSOI). Banu et al. [12] investigated dengue transmission and verified that the temperature and humidity were positively related to dengue fever incidence in Dhaka, Bangladesh. Dengue incidence increased when temperature became high in Thailand [13], Indonesia [14, 15], Singapore [16], Mexico [17], NewCaledonia [19], Guadeloupe [20], and Sri Lanka [21]. However, it was shown that humidity is a significant factor for dengue fever incidence by affected mosquito density in Taiwan [22]. Banu et al. [12] used lag correlation analysis to analyze the relationship between climate factors and dengue fever, and they established a Poisson model to show that climate factors had 4 months delay in the effects of dengue fever. Antoine et al. [23] established a logical regression model by training climate data for 1993–2013 years in French Guiana. Barrera et al. [24] established a complex linear model based on precipitation and temperature factors to show that the incidence of dengue fever is greatly affected by precipitation in San Juan region.

This paper presented a k-nearest neighbor (KNN) regression method to predict the dengue break in San Juan, Puerto Rico. Eight local weather factor: average minimum temperature, average maximum temperature, average temperature, average dew point temperature, temperature difference, relative humidity, absolute humidity, Precipitation, were analysed from 1993–2013 years. The correlation analysis and Poisson regression model were also chosen to verify temperature, Precipitation and dengue instances in San Juan region.

2 Data

2.1 Research Area

The latitude and longitude of the Puerto Rico region is 18. 28, N, 066. 07, W# where is on the west side of the Atlantic. It is tropical rain forest climate, the annual average temperature is 23 ∼ 26 degrees, the annual rainfall is about 1500 mm. From the chart of dengue fever in San Juan, Puerto Rico, we can see that dengue cases reached peak levels at 234th week in San Juan, Puerto Rico, as shown in Fig. 1. The number of dengue cases reached 461 people in this period, it is also the largest outbreak of dengue fever in this area. According to data trends, the incidence of this region is increasing.

2.2 Data

Dengue fever and meteorological data used in the method were collected from the website (https://predict.phiresearchlab.org/legacy/dengue/data.html). The daily meteorological factors includes: maximum temperature, minimum temperature, average temperature, dew point temperature, relative humidity, humidity, and precipitation.

Because the dengue fever cases of 1990–2013 was recorded in weekly, the daily weather data was converted into weekly. Finally, total 1, 196 samples were obtained. Figure 1 showed a sequence diagram of dengue and local climate data in San Juan, Puerto Rico.

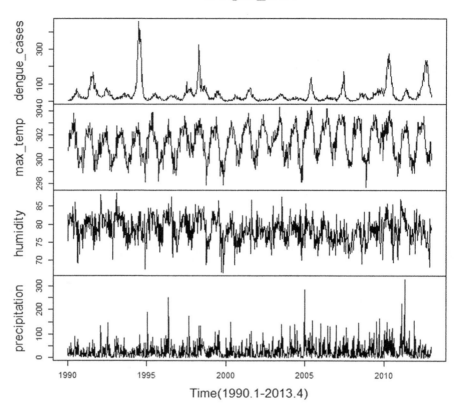

Fig. 1. Sequence of weekly dengue fever and meteorological data in San Juan, Puerto Rico region in 1990–2013.

3   Method

3.1   Normalization

Linear transformation of sample data was achieved by deviation Standardization, mapped the result value to [0,1]. Conversion function as:

$$x^* = \frac{x - x_{\min}}{x_{\max} - x_{\min}} \tag{1}$$

Among them: $x_{\max}$ is the maximum value of sample data; $x_{\min}$ is the minimum value of sample data.

3.2   Poisson Regression

In the Poisson regression model, yi represents the number of dengue fever cases in Ith month, assuming that the random variable Y is equal to the probability of y_i and consistent with the Poisson distribution of mean value λ_i. So the density function of the Poisson distribution is as follows (2):

$$P(Y_i = y_i | \lambda_i) = \frac{\exp(-\lambda_i)\lambda_i^{y_i}}{y_i!}, i = 0, 1, 2 \ldots \tag{2}$$

In the formula, λ_i can be estimated from the observed data, and the following structural equation can be deduced (3):

$$E(y_i) = Var(y_i) = \lambda_i = e^{\beta_i y_i} \tag{3}$$

Where $\beta' y_i = \theta_i$, β_i is a parameter variable, θ_i is got exponential to ensure that the parameter λ_i stays non negative. Meanwhile, λ_i is the mean value and variance of dengue fever.

To bring (3) into (2), the probability function is as follows (4):

$$P(y_i | \theta_i) = \frac{\exp(y_i \theta_i - e^{\theta_i})}{y_i!}, \quad i = 1, 2 \ldots, n \tag{4}$$

The logarithm to (4) can be obtained in another form as follows (5):

$$\theta_i = \log(\lambda_i) = X_i^T \beta, i = 1, 2 \ldots, n \tag{5}$$

(y_i, x_i^T) is the data point of group i, y_i represents the dengue case, $x_i = (x_{i1}, x_{i2}, \ldots x_{ip})'$ is a p*1 vector, it represents P factors affecting the onset of dengue fever (temperature, humidity, SOI, etc.) in month of I, $x_i^T = (x_{i1}, x_{i2}, \ldots x_{ip})$ is transpose of xi. β_i is a parameter variable for function settings, we can see β_i that the explanatory variable xi corresponds to the regression coefficient by (5).

3.3   KNN Regression Model

The nearest neighbor method (KNN) is a nonparametric regression method [26] which is different from the traditional regression method. The KNN regression does not have a fixed function form and estimates the data of a test sample by looking for similar testing state samples among historical data. KNN regression can achieve very good performance [25] when the feature dimension is low. The feature dimension of this study is less than nine dimensions. Thus, the KNN regression could be achieved a good performance. In this paper, the KNN algorithm is implemented in R language caret package. (https://cran.r-project.org/web/packages/caret/index.html).

4   Research and Discussion

4.1   Correlation Analysis Results

Dengue fever was positively related to temperature and humidity. The correlation coefficients of Dengue fever and Climatic factors listed in Table 1 were obtained according to the incidence data, Except daily temperature, dengue fever and other factors exhibited positive correlations. The largest correlation between dengue fever and weekly precipitation is 56.3%, but p value is 0.0514 ($p > 0.05$). Thus, the correlation analysis results are not significant. By analyzing the correlation of other factors, the correlations of dengue fever and the average maximum temperature, average minimum temperature, average temperature, average dew point temperature, absolute humidity are large than 17% ($p < 0.05$).

Table 1. Dengue fever and correlation analysis of influencing factors.

Dengue fever	Correlation coefficient	P value
Average maximum temperature	0.177	7.749e−10
Average minimum temperature	0.181	2.66e−10
Average temperature	0.174	1.478e−09
Daily temperature difference	−0.057	0.04719
Average dew point temperature	0.181	2.911e−10
Relative humidity	0.098	0.0006866
Absolute temperature	0.187	7.57e−11
Precipitation	0.563	0.0514

4.2   Poisson Regression

Table 2 showed that the estimation coefficients of regression results and prediction errors based on Poisson regression methods. By comparing with Table 1, the error of Poisson regression were obviously less than correlation analysis, and the P value is smaller. The error of Poisson regression is better than results of linear regression, and it is proved that precipitation in San, Juan, Puerto and Rico regions has quite significant influence on dengue fever [24]. According to the correlation analysis table, we can compare the effects of various factors on dengue fever, then the linear regression and Poisson regression were used to carry out regression analysis of the influencing factors.

4.3   KNN Regression Prediction Results

To enhance the regression performance, the KNN was applied to fit the dengue cases. The average minimum temperature, average maximum temperature, average temperature, average dew point temperature, temperature difference, relative humidity, absolute humidity, and Precipitation were choose as input data. The time series from 2000 to 2013 to fit the model. 20% duration as predicted data. Figure 2 demonstrated the real dengue instance and the predicted values and the actual incidence (testY) and

Table 2. Linear regression analysis about influencing factors of dengue fever.

Property	Coefficient estimate	Error	T value	Pr(> t)
Average maximum temperature	1.105e+01	4.408e+00	2.507	0.01231
Average minimum temperature	1.858e+00	5.070e+00	0.366	0.71413
Average temperature	−2.816e+01	2.027e+01	−1.389	0.1647
Daily temperature difference	−1.444e+01	5.089e+00	−2.838	0.00462
Average dew point temperature	−2.968e+01	2.705e+0	−1.097	0.27274
Relative humidity	−5.309e+00	4.013e+00	−1.323	0.18606
Absolute temperature	5.399e+01	1.830e+01	2.951	0.00323
Precipitation	−9.668e–03	5.282e–02	−0.183	0.85480

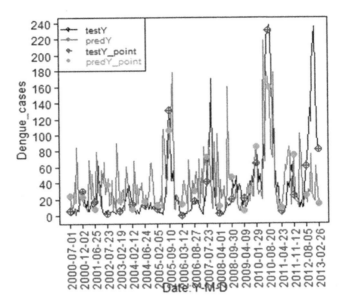

Fig. 2. Prediction curve of KNN regression

the incidence of dengue (predY) for KNN prediction model. The regression prediction error(RMSE) is 6.88 person/ week.

5   Conclusion

This paper predicated the dengue outbreak by compared linear correlation methods, Poisson regression model and KNN regression algorithm in San Juan, Puerto Rico. The results showed that there was a significant correlation between the average weekly peak temperature, the mean minimum temperature, the mean dew point temperature and dengue fever incidence. The predictive error of the established KNN regression prediction model was 6.88 per week. Because the San Juan, Puerto Rico region has an average more than 200 dengue fever per week, this paper result could be acceptable. Not only these results validate the author's previous conclusion [27] that dengue prediction could be archived 78.5% by using local weather data, but also provided potential an efficient method to predict dengue outbreak in other regions.

Acknowledgement. This study was supported by Guangxi cloud computing and big data Collaborative Innovation Center (No: YD16E18).

References

1. Bhatt, S., Gething, P.W., Brady, O.J., et al.: The global distribution and burden of dengue. Nature **496**(7446), 504–507 (2013)
2. Gubler, D.J.: Resurgent vector-borne diseases as a global health problem. Emerg. Infect. Dis. **4**(3), 442–450 (1998)
3. Beatty, M.E., Stone, A., Fitzsimons, D.W., et al.: Best practices in dengue surveillance: a report from the Asia-Pacific and Americas Dengue Prevention Boards. PLoS Negl. Trop. Dis. **4**(11), e890 (1935)
4. Guy, B., Saville, M., Lang, J.: Development of sanofi pasteur tetravalent dengue vaccine. Hum. Vaccines **46**(9), 696–705 (2009)
5. Hadinegoro, S.R., Arredondo-García, J.L., Capeding, M.R., et al.: Efficacy and long-term safety of a dengue vaccine in regions of endemic disease. N. Engl. J. Med. **373**(13), 1195–1206 (2015)
6. Villar, L., Dayan, G.H., Arredondogarcía, J.L., et al.: Efficacy of a tetravalent dengue vaccine in children in latin America. N. Engl. J. Med. **372**(2), 113–123 (2015)
7. Banu, S., Hu, W., Hurst, C., et al.: Dengue transmission in the Asia-Pacific region: impact of climate change and socio-environmental factors. Trop. Med. Int. Health **16**(5), 598–607 (2011). Tm & Ih
8. Descloux, E., Mangeas, M., Menkes, C.E., et al.: Climate-based models for understanding and forecasting dengue epidemics. PLoS Negl. Trop. Dis. **6**(2), e1470 (2012)
9. Ferreira, M.C.: Geographical distribution of the association between El Niño South Oscillation and dengue fever in the Americas: a continental analysis using geographical information system-based techniques. Geospatial Health **9**(1), 141–151 (2014)
10. Liao, C.M., Huang, T.L., Lin, Y.J., et al.: Regional response of dengue fever epidemics to interannual variation and related climate variability. Stoch. Env. Res. Risk Assess. **29**(3), 947–958 (2015)
11. Gagnon, A.S., Bush, A., Smoyer-Tomic, K.E.: Dengue epidemics and the El Niño southern oscillation. Clim. Res. **19**(1), 35–43 (2001)
12. Banu, S., Hu, W., Guo, Y., et al.: Projecting the impact of climate change on dengue transmission in Dhaka, Bangladesh. Environ. Int. **63**(3), 137–142 (2014)
13. Focks, D., Alexander, N., Villegas, E.: Multicountry study of Aedes aegypti pupal productivity survey methodology: findings and recommendations. Dengue Bull WHO **31**, 192–200 (2007)
14. Arcari, P., Tapper, N., Pfueller, S.: Regional variability in relationships between climate and dengue/DHF in Indonesia. Singap. J. Trop. Geogr. **28**, 251–272 (2007)
15. Bangs, M., Larasati, R., Corwin, A., et al.: Climatic factors associated with epidemic dengue in Palembang, Indonesia: implications of short-term meteorological events on virus transmission. Southeast Asian J. Trop. Med. Public Health **37**, 1103–1116 (2006)

16. Burattini, M., Chen, M., Chow, A., et al.: Modelling the control strategies against dengue in Singapore. Epidemiol. Infect. **136**, 309–319 (2007)
17. Chowell, G., Sanchez, F.: Climate-based descriptive models of dengue fever: the 2002 epidemic in Colima. Mexico. J Environ. Health **68**, 40–44 (2006)
18. Keating, J.: An investigation into the cyclical incidence of dengue fever. Soc. Sci. Med. **53**, 1587–1597 (2001)
19. Descloux, E., Mangeas, M., Menkes, C.E., et al.: Climate-based models for understanding and forecasting dengue epidemics. PLoS Negl. Trop. Dis. **6**, e1470 (2012)
20. Gharbi, M., Quenel, P., Gustave, J., et al.: Time series analysis of dengue incidence inGuadeloupe, French West Indies: forecasting models using climate variables as predictors. BMC Infect. Dis. **11**, 166 (2011)
21. Goto, K., Kumarendran, B., Mettananda, S., et al.: Analysis of effects of meteorological factors on dengue incidence in Sri Lanka using time series data. PLoS ONE **8**, e63717 (2013)
22. Chen, S.C., Liao, C.M., Chio, C.P., et al.: Lagged temperature effect with mosquito transmission potential explains dengue variability in southern Taiwan: insights from a statistical analysis. Sci. Total Environ. **408**, 4069–4075 (2010)
23. Antoine, A., Pascal, R., Morgan, M., et al.: Predicting dengue fever outbreaks in French guiana using climate indicators. PLoS Negl. Trop. Dis. **10**(4), e0004681 (2016)
24. Barrera, R., Amador, M., Mackay, A.J.: Population dynamics of aedes aegypti and dengue as influenced by weather and human behavior in San Juan, Puerto Rico. PLoS Negl. Trop. Dis. **5**(12), e1378 (2011)
25. Tipayamongkholgul, M., Fang, C.T., Klinchan, S., et al.: Effects of the El Niño-Southern Oscillation on dengue epidemics in Thailand, 1996–2005. BMC Public Health **9**(1), 422 (2009)
26. Earnest, A., Tan, S.B., Wildersmith, A.: Meteorological factors and El Niño southern oscillation are independently associated with dengue infections. Epidemiol. Infect. **140**(7), 1244–1251 (2011)
27. Jiang, Y., Zhu, G.: Prediction of dengue outbreak based on poisson regression and support vector machine. In: The 7th International Symposium on Computational Intelligence and Industrial Applications (ISCIIA2016), FM-GS1-01. Fuji Technology Press, Fuji (2016)

Identification of Nonlinear System Based on Complex-Valued Flexible Neural Network

Lina Jia, Wei Zhang, and Bin Yang[(✉)]

School of Information Science and Engineering, Zaozhuang University,
Zaozhuang, China
batsi@126.com

Abstract. Identification of nonlinear system could help to understand and model the internal mechanism of real complex systems. In this paper, complex-valued version of flexible neural tree (CVFNT) model is proposed to identify nonlinear systems. In order to search the optimal structure and parameters of CVFNT model, a new hybrid evolutionary method based on structure-based evolutionary algorithm and firefly algorithm is employed. Two nonlinear system identification experiments are used to test CVFNT model. The results reveal that CVFNT model performs better than the proposed real-valued neural networks.

Keywords: Nonlinear system · Complex-valued · Flexible neural network · Firefly algorithm

1 Introduction

Nonlinear system identification could help to understand and model the internal mechanism of real nonlinear systems. Artificial neural networks (ANNs) are powerful mathematical methods that can be used to learn complex linear and non-linear continuous functions, and have been successfully applied to identify nonlinear system in the past decades [1]. Nidhil et al. proposed a method to identify the system model, which involved use of back propagation (BP) neural network to predict the output of the system for a given input from the knowledge of past inputs & outputs [2]. Ayala et al. inserted the choice of the radial basis functions (RBF) neural networks model complexity and its inputs in the optimization procedure together with the model parameters, aiming at accuracy, model validity and regularization in a multiobjective approach [3]. Amin et al. proposed the use of an alternative Fuzzy Wavelet Neural Network (FWNN) to model the input-output maps of nonlinear dynamic systems [4]. Jiang et al. presented a dynamic fuzzy stochastic neural network model for nonparametric system identification using ambient vibration data [5].

Recently, complex-valued neural network (CVNN) has been proposed to predict and classify the time series data. Kitajima et al. proposed a wind speed prediction system using complex-valued neural network and frequency component of observed wind speed data as an input information [6]. Peker presented complex-valued neural network (CVANN) algorithm for automatic sleep scoring using EEG signals in psychiatry and neurology [7]. Hafiz et al. applied complex-valued neural networks as a

© Springer International Publishing AG 2017
H. Yin et al. (Eds.): IDEAL 2017, LNCS 10585, pp. 154–162, 2017.
https://doi.org/10.1007/978-3-319-68935-7_18

classification algorithm for the skeletal wireframe data that were generated from hand gestures [8]. Shamima et al. used a Fully Complex-valued Relaxation Network (FCRN) classifier to predict the secondary structure of proteins [9]. The results reveal that CVNN is more flexible and functional, and performs better than real-valued neural network.

Flexible neural tree (FNT) model can be seen as a flexible multi-layer feed-forward neural network. The FNT model was first proposed by Chen in year 2005, which has been applied widely in many areas. Chen et al. tried to build an effective early stage traffic identification model by applying flexible neural trees (FNT) [10]. Ojha proposed a computational intelligence (CI) technique named flexible neural tree (FNT) to predict die filling performance of pharmaceutical granules and to identify significant die filling process variables [11]. We proposed flexible neural tree model (FNT) to detect somatic mutations in tumor-normal paired sequencing data [12]. FNT model has over-layer connections and free parameters in activation functions, and could select the proper input variables or time-lags for constructing a model automatically, so it is more powerful and flexible than traditional neural networks, such as back propagation (BP) neural network, RBF neural network and wavelet neural network.

In this paper, a novel complex-valued neural network based on flexible neural tree (FNT), namely complex-valued flexible neural tree model (CVFNT) is proposed for identification of nonlinear system. The hierarchical structure of the CVFNT model is evolved using genetic programming (GP) like tree structure-based evolutionary algorithm with specific instructions. Firefly algorithm is used to optimize the parameters of CVFNT due to that it has few parameters and is easy to apply and implement. Two nonlinear systems are used to evaluate the performance of CVFNT.

2   Method

2.1   Structure of CVFNT

Complex-valued flexible neural tree (CVFNT) model is the extensions of real-valued FNT model. In a CVFNT, input layer, threshold values and weights are complex numbers. A tree-structural based encoding method with specific instruction set is selected for representing a CVFNT model. The used function set F and terminal instruction set T for generating a CVFNT model are described as follows:

$$S = F \cup T = \{+_2, +_3, \ldots, +_N\} \cup \{z_1, z_2, \ldots, z_n\}, \tag{1}$$

where $+_i (i = 2, 3, \ldots N)$ denotes non-leaf nodes' instructions and taking I arguments. $z_1, z_2, \ldots z_n$ ($z_i \in C^n$, $z_i = x_i + jy_i$ and j stands for the value of $\sqrt{-1}$.) are leaf nodes' instructions and taking no other arguments. The output of a non-leaf node is calculated as a flexible neuron model (see Fig. 1(left)). From this point of view, the instruction $+_i$ is also called a flexible neuron operator with i inputs.

In the creation process of neural tree, the operator is selected randomly from function set F and terminal instruction set T. If a non-terminal instruction, i.e., $+_i (i = 2, 3, \ldots, N)$ is selected, the i complex-valued weights $(w_1, w_2, \ldots w_i)$ are randomly generated and used for representing the connection strength between the node $+_i$ and its children.

The output of a flexible neuron $+_n$ can be calculated as follows. The total excitation of $+_n$ is

$$net_n = w_0 + \sum_{j=1}^{n} w_j z_j \tag{2}$$

Where w_0 is threshold value and $z_j (j = 1, 2, \ldots, n)$ are the inputs to node $+_n$. The output of the node $+_n$ is then calculated by

$$out_n = f(c, r, net_n) = \frac{net_n}{c + \frac{1}{r}|net_n|}. \tag{3}$$

Where $f(\cdot)$ is activation function, c and r are real variables, and $|net_n|$ is the modulus of complex net_n. The output of flexible activation function is complex-valued.

A typical complex-valued flexible neural tree model is shown as Fig. 1(right). This framework allows input variables selection and over-layer connections. So the structure of FNT model is sparse and flexible. The overall output of complex-valued flexible neural tree can be computed from left to right by depth-first method, recursively.

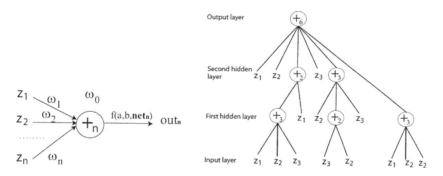

Fig. 1. A flexible neuron operator (left) and typical representation of neural tree (right) with function set $F = \{+_2, +_3, +_4, +_5, +_6\}$, and terminal set $T = \{z_1, z_2, z_3\}$.

2.2   Structure Optimization of CVFNT

Finding an optimal or near-optimal CVFNT is formulated as an evolutionary search process. In this paper, we use the following neural tree variation operators: mutation, crossover and selection [10].

2.3   Parameters Optimization of CVFNT

According to the optimal structure of CVFNT, tally the number of parameters. In a CVFNT model, weights w_i and threshold value w_0 are complex valued. In the optimization process, both real and imaginary parts need to be optimized. Firefly algorithm (FA) is an efficient optimization algorithm which was proposed by Xin-She Yang in 2009 [13]. It is very simple, has few parameters and easy to apply and implement, so this paper uses firefly algorithm to optimize the parameters of CVFNT.

Firefly algorithm is the random optimization method of simulating luminescence behavior of firefly in the nature. The firefly could search the partners and move to the position of better firefly according to brightness property. A firefly represents a potential solution, which represents $[\text{Re}(w_i), \text{Im}(w_i), \text{Re}(w_0), \text{Im}(w_0), c, r, \ldots]$. In order to solve optimization problem, initialize a firefly vector $[x_1, x_2, \ldots, x_n]$ (n is the number of fireflies). As attractiveness is directly proportional to the brightness property of the fireflies, so always the less bright firefly will be attracted by the brightest firefly. The brightness of firefly i is computed as

$$B_i = B_{i0} * e^{-\gamma r_{ij}} \tag{4}$$

Where B_{i0} represents maximum brightness of firefly i by the fitness function as $B_{i0} = f(x_i)$. γ is coefficient of light absorption, and r_{ij} is the distance factor between the two corresponding fireflies i and j.

The movement of the less bright firefly toward the brighter firefly is computed by

$$x_i(t+1) = x_i(t) + \beta_i(x_j(t) - x_i(t)) + \alpha \varepsilon_i \tag{5}$$

Where α is step size randomly created in the range [0, 1], and ε_i is gaussian distribution random number.

2.4   Fitness Function

Root mean squared error (RMSE) is used to search the optimal CVFNT according to actual data and predicted data.

$$RMSE = \sqrt{\frac{1}{N} \sum_{i=1}^{N} (y_{actual}^i - y_{predicted}^i)^2} \tag{6}$$

Where N is the number of data points, y_{actual}^i is the actual output of i-th data point and $y_{predicted}^i$ is the predicting output of i-th data point.

2.5   Flowchart of Our Method

1. The sample data of nonlinear system are real number, so input and output data need be transformed to complex values before being inputted into CVFNT model. The following transformation steps are used [14]:

- Suppose the input real numbers $[x_1, x_2, \ldots, x_m]$, and tally the maximum and minimum values of real numbers (*max* and *min*).
- i-th real number x_i is transformed into complex number as followed.

$$\varphi_i = \frac{x_i - min}{max - min}(2\pi - \delta),$$
$$z_i = e^{i\varphi_i} \tag{7}$$

Where δ is the shift angle and i stands for the value of $\sqrt{-1}$.

2. Train the CVFNT model according to the above complex numbers (input data) and firefly algorithm, which is introduced detailed in Sects. 2.2 and 2.3. The output of CVFNT is complex number, so the output value needs to be transformed into real number in order to evaluate model effectively. The inverse transformation shall be used:

$$\arg z = \varphi,$$
$$y = \frac{\varphi(max - min)}{2\pi - \delta} + min. \tag{8}$$

Where $\arg z$ is the argument of complex value z.

3 Experiments

Two nonlinear systems are used to evaluate the performance of CVFNT. We use five input variables to construct the complex-valued flexible neural tree model. The used instruction set $F = \{+_2, +_3, +_4\}$ and $T = \{z_1, z_2, z_3, z_4, z_5\}$. Shift angle δ is set as π.

3.1 The First Nonlinear System

The nonlinear system to be identified is described by

$$y(k) = 0.72y(k-1) + 0.025y(k-2)u(k-2) + 0.01u^2(k-3) + 0.2u(k-4). \tag{9}$$

Where $y(k)$ is the output of the system at the k-th time point and $u(k)$ is the plant input. We set input vector $\{y(k-1), y(k-2), u(k-2), u(k-3), u(k-4)\}$ as $\{z_1, z_2, z_3, z_4, z_5\}$.

In order to make the comparison fairly, the input signals used for training ALNN model are same as in literatures [15–17], which is an iid uniform sequence over $[-2,2]$ for about half of the 900 time points and the remaining data is given by $1.05\sin(\pi\frac{k}{45})$. To test the performance of identification model, the following input is used for test.

$$u(k) = \begin{cases} \sin(\pi \frac{k}{25}) & k < 250 \\ 1.0 & 250 \le k < 500 \\ -1.0 & 500 \le k < 750 \\ 0.3 \sin(\pi \frac{k}{25}) + 0.1 \sin(\pi \frac{k}{32}) + 0.6 \sin(\pi \frac{k}{10}) & 750 \le k < 1000 \end{cases} \quad (10)$$

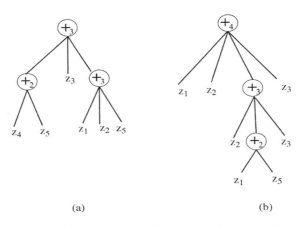

(a)   (b)

Fig. 2. The structure of optimized FNT model for Example 1 (a) and for Example 2 (b).

The training data contain 896 sample points and testing data have 1000 sample points. Figure 2(a) gives the structure of optimized CVFNT model. Figure 3 shows the identification performance between actual data and CVFNT model. Figure 3 also gives the identification error by CVFNT model. From Fig. 3, it can be seen that identification system is nearly as same as actual one, and the error by CVFNT is very tiny. The RMSE values of identification system using some different methods for testing data are listed in Table 1. From Table 1, it is clear that CVFNT model performs better than other real-valued neural network models.

3.2 The Second Nonlinear System

In this section, the plant to be identified is described by the following equation:

$$y(k+1) = f(y(k), y(k-1), y(k-2), u(k), u(k-1)), \quad (11)$$

where

$$f(x_1, x_2, x_3, x_4, x_5) = \frac{x_1 x_2 x_3 x_5 (x_3 - 1) + x_4}{1 + x_3^2 + x_2^2}. \quad (12)$$

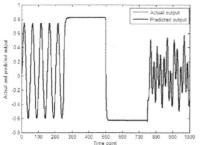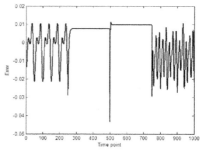

Fig. 3. Predicted data (left) and errors (right) by CVFNT for Example 1.

Table 1. Comparison of different models for Example 1.

Model	RMSE of testing
ERNN [17]	0.078
RSONFIN [18]	0.06
TRFN-S [15]	0.0313
FWNN [16]	0.0201
FWNN-inline-PSO [17]	0.0163
CVFNT	0.01007

In order to make the comparison fairly, the training and testing signals are same as literatures [15, 16]. The training and testing data both contain 1000 sample points. We set input vector $\{y(k), y(k-1), y(k-2), u(k), u(k-1)\}$ as $\{z_1, z_2, z_3, z_4, z_5\}$. Figure 2 (b) gives the structure of optimized FNT model. Figure 4 gives the predicted performance between actual data and CVFNT model. Figure 4 also gives the predicted error by CVFNT model. From Fig. 4, it can be seen that identification system of CVFNT is nearly as same as actual one. The RMSE values of identification system using six different methods for testing data are listed in Table 2. From Table 2, it is clear that CVFNT model performs better than other proposed real-valued neural network models.

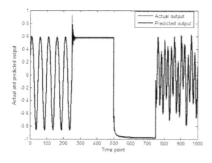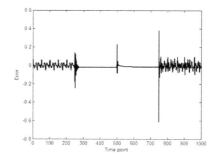

Fig. 4. Predicted data (left) and error (right) by CVFNT for Example 2

Table 2. Comparison of different models for Example 2.

Model	RMSE of testing
ERNN [17]	0.0575
RSONFIN [18]	0.0780
TRFN-S [15]	0.0346
FWNN [16]	0.0301
CVFNT	0.0289

4 Summary

In this paper, a novel complex-valued neural network based on flexible neural tree (FNT), namely complex-valued flexible neural tree model (CVFNT) is proposed for identification of nonlinear system. In a CVFNT model, inputs/outputs, weights, threshold values and activation functions are complex-valued. The hierarchical structure of the CVFNT model is evolved using genetic programming (GP) like tree structure-based evolutionary algorithm with specific instructions. Firefly algorithm is used to optimize the parameters of CVFNT. Two nonlinear system identification experiments are used to test CVFNT model. The results reveal that CVFNT model performs better than the proposed real-valued neural networks.

Acknowledgement. This work was supported by the PhD research startup foundation of Zaozhuang University (No. 2014BS13), Zaozhuang University Foundation (No. 2015YY02), and Shandong Provincial Natural Science Foundation, China (No. ZR2015PF007).

References

1. Angeline, P.J., Saunders, G.M., Pollack, J.B.: An evolutionary algorithm that constructs recurrent neural networks. IEEE Trans. Neural Netw. **5**, 54–65 (1994)
2. Nidhil, K.J., Sreeraj, S., Vijay, B.: System identification using artificial neural network. In: 2015 International Conference on Circuit, Power and Computing Technologies (ICCPCT), Cebu, Philippines, pp. 1–4 (2015)
3. Ayala, H., Coelho, L.: Multiobjective Cuckoo search applied to radial basis function neural networks training for system identification. IFAC Proc. **47**(3), 2539–2544 (2014)
4. Faijul Amin, M., Murase, K.: Single-layered complex-valued neural network for real-valued classification problems. Neurocomputing **72**(4–6), 945–955 (2009)
5. Jiang, X., Mahadevan, S., Yuan, Y.: Fuzzy stochastic neural network model for structural system identification. Mech. Syst. Signal Process. **82**, 394–411 (2016)
6. Kitajima, T., Yasuno, T., Ikeda, N.: Wind speed prediction system using complex-valued neural network and frequency component of wind speed. IEICE Tech. Rep. Neurocomput. **113**, 35–40 (2013)
7. Peker, M.: A new approach for automatic sleep scoring: combining Taguchi based complex-valued neural network and complex wavelet transform. Comput. Methods Programs Biomed. **129**, 203–216 (2016)

8. Hafiz, A.R., Al-Nuaimi, A.Y., Amin, M.F., Murase, K.: Classification of skeletal wireframe representation of hand gesture using complex-valued neural network. Neural Process. Lett. **42**(3), 1–16 (2015)
9. Shamima, B., Savitha, R., Suresh, S., Saraswathi, S.: Protein secondary structure prediction using a fully complex-valued relaxation network. In: International Joint Conference on Neural Networks, pp. 1–8 (2013)
10. Chen, Z., Peng, L., Gao, C., Yang, B., Chen, Y.: Flexible neural trees based early stage identification for IP traffic. Soft Comput., **21**(8), 1–12 (2017)
11. Ojha, V.K., Schiano, S., Wu, C.Y., Snášel, V., Abraham, A.: Predictive modeling of die filling of the pharmaceutical granules using the flexible neural tree. Neural Comput. Appl., 1–15 (2016)
12. Yang, B., Chen, Y.: Somatic mutation detection using ensemble of flexible neural tree model. Neurocomputing **179**, 161–168 (2016)
13. Yang, X.-S.: Firefly algorithms for multimodal optimization. In: Watanabe, O., Zeugmann, T. (eds.) SAGA 2009. LNCS, vol. 5792, pp. 169–178. Springer, Heidelberg (2009). doi:10. 1007/978-3-642-04944-6_14
14. Safi, S.K.: A comparison of artificial neural network and time series models for forecasting GDP in Palestine. Am. J. Theor. Appl. Stat. **5**(2), 58–63 (2016)
15. Juang, C.F.: A TSK-type recurrent fuzzy network for dynamic systems processing by neural network and genetic algorithms. IEEE Trans. Fuzzy Syst. **10**(2), 155–170 (2002)
16. Abiyev, R.H., Kaynak, O.: Fuzzy wavelet neural networks for identification and control of dynamic plants a novel structure and a comparative study. IEEE Trans. Ind. Electron. **55**(8), 3133–3140 (2008)
17. Cheng, R., Bai, Y.P.: A novel approach to fuzzy wavelet neural network modeling and optimization. Int. J. Electr. Power Energ. Syst. **64**, 671–678 (2015)
18. Juang, C.F., Lin, C.T.: A recurrent self-organizing neural fuzzy inference network. IEEE Trans. Neural Netw. **10**(4), 828–845 (1999)

Research on the Method of Splitting Large Class Diagram Based on Multilevel Partitioning

JinShuai Li$^{(\boxtimes)}$, XiaoFei Zhao, and BaoShan Sun

School of Computer Science and Software Engineering,
Tianjin Polytechnic University, Tianjin, China
785436853@qq.com

Abstract. The UML class diagrams generated in reverse engineering are often large and the readability is poor. In this paper, an automatic split algorithm based on multilevel partitioning technology is proposed. According to the coupling degree between classes, the classes with high coupling are divided into the same part and the class with low coupling are separated. Experimental results show that, compared with the traditional manual division, the class diagram obtained by the automatic splitting method is more readable and consumes less time. Using the automatic splitting method to split the large class diagram can save a lot of time, improving work efficiency greatly.

Keywords: Class diagram · UML · Multilevel partitioning · Automatic split · Coupling

1 Introduction

Software development model [1] refers to the whole process of software development, activities and tasks of the structure of the framework, which plays a vital role in the development and maintenance of the software. Compared with the source code, the software development model has the characteristics of high readability and high abstraction level. In software maintenance, software maintenance personnel may not participate in the software development process.

Despite the clear advantage of having software development model for the maintenance and evolution of a system, these artifacts are rarely maintained or may no longer exit [2]. As the development process is not standardized, many software systems are currently lack of software development model. The lack of software development model makes the software maintenance process very difficult. In order to solve this problem, the software development model can be generated from the source code by reverse engineering [3]. At present, the mainstream modeling tools have a certain degree of reverse engineering capabilities, ASTAH [4], JBOO [5] and so on can be generated from the source code unified modeling language (Unified modeling language, referred to as UML) [6] Software development model, and class diagram is one of the most basic and most important model. As the class diagram of the class, attributes and operations are

© Springer International Publishing AG 2017
H. Yin et al. (Eds.): IDEAL 2017, LNCS 10585, pp. 163–172, 2017.
https://doi.org/10.1007/978-3-319-68935-7_19

corresponding with the corresponding concept in the source code, so the reverse generation of class diagram is more direct.

Although the class diagram can be well supported in the reverse engineering, the class diagram generated by the mainstream reverse tools is often larger and less readable. The generation of large class diagram is related to the work of existing reverse tools. Existing reverse tools are usually organized according to the namespace of the class, and all classes in the same namespace are put into the same class diagram. In addition, some tools are organized according to the package or directory, and all the classes in a directory and subdirectories are organized into a class diagram. However, in practical engineering, a namespace or a directory may contain a large number of classes and interfaces. Some of the reverse engineering tools even include all the classes included in the project into a single class diagram, which is generally large and poorly readable. In accordance with the principle of 7 ± 2 [7], people in a certain period of time to be able to understand, the memory of the transaction is limited, generally not more than 7 ± 2. So, the large class diagram will be difficult to be read by software maintenance personnel. Therefore, it is urgent to split the class diagram to serve the software maintenance.

In view of the above situation, the basic idea of this paper is to classify the classes with higher coupling degree in the same class diagram according to the coupling between the classes. In particular, according to the different effects of inheritance, aggregation, association and dependency on the coupling degree between classes, put forward a method of calculating the coupling degree between classes, and then transforms the class diagram into the weighted graph of the reaction coupling strength, then compresses the weighted graphs several times, reduces the scale of the weighted graphs, makes it fit to divide, then divides the weighted graph into k parts. According to the order of compression to dismantle, and use the KL algorithm [8] to adjust the optimization, so that the classes with a large coupling degree can classified in a class diagram as far as possible.

2   Construction of Large Class Diagram and the Calculation of Coupling Degree

2.1   The Class Diagram Generated by Reverse Engineering

Class diagram, which is a type of UML diagram, is employed to model concepts in static views [9]. The main elements include classes, interfaces, and relationships between them (association, dependency, inheritance, composition, aggregation). In this paper, the class diagram split algorithm is used to split the class diagram, not to split internal structure of the class. Aggregate relationship represents the overall and partial relationship between classes. So we define the class diagram as:

$$Diagram = \{C, Ass, Dep, Inh, Agg\} \tag{1}$$

Where C denotes the class, Ass denotes the association between the classes, and if the classes $c_1, c_2 \in C$, then $< c_1, c_2 > \in Ass$. If it is a bidirectional association, then $< c_1, c_2 > \in Ass\ < c_1, c_2 > \in Ass$. Similarly, Dep represents the dependencies between classes, Inh represents the inheritance relationship between classes, Agg represents the aggregation relationship between classes.

2.2 Calculating the Coupling Degree

As one of the indicators that reflects the degree of interdependence between modules, the coupling degree has become an important basis for the partitioning module. In order to measure the coupling relationship of object-oriented systems, a large number of metrics and measurement frameworks [10] have been proposed, such as CBO, MPC, and DAC.

These metrics mainly measure the degree of interdependence between the classes and the surrounding environment. For the large class diagram splitting algorithm, we mainly care about the coupling strength between the classes. As a special view of the static structure of the system, class diagram mainly show the relationships among classes such as Inheritance, Association, Dependence and Aggregation. Depending on the degree of concern in class diagram, Inheritance, Aggregation, Association, Dependence, their influence on the partitioning of class diagrams is reduced in turn. In order to facilitate the split, the relationship between the classes is the following:

$$
\begin{aligned}
Coupleling(c_i, c_j) = Ancestor(c_i, c_j) \times 8 + Aggregated(c_i, c_j) \times 4 + \\
Associated(c_i, c_j) \times 2 + Depend(c_i, c_j) \times 1 \quad (2)
\end{aligned}
$$

If $< c_i, c_j > \in Inh$, then $Ancestor(c_i, c_j)$ is equal to one, otherwise it is equal to zero. The same principle, if $< c_i, c_j > \in Agg$, then $Aggregated(c_i, c_j)$ is equal to one, otherwise it is equal to zero. If $< c_i, c_j > \in Ass$, then $Associated(c_i, c_j)$ is equal to one, otherwise it is equal to zero. If $< c_i, c_j > \in Dep$, then $Depend(c_i, c_j)$ is equal to one, otherwise it is equal to zero. Here we set their influence to 8, 4, 2, 1 according to the degree of influence of Inheritance, Association, Aggregation, Dependence on the coupling between classes. So that the degree of coupling between classes with Inheritance is higher than the degree of coupling between classes without Inheritance. Similarly, the subsequent three quantification factors can also ensure the corresponding priority.

3 Class Diagram Splitting Algorithm

The partition of a graph is often called the cut of a graph. Given a division of P, if the two vertices of the edge are partitioned into different subsets, we call it edge cut. In this paper, we define the partition cost as the sum of the weights of all edges across different subsets. The higher the partition cost is, the higher the degree of correlation between different subsets is. In this paper, we must minimize the partition cost in the split of the large class diagram, which makes the readability and understandability of the sub graph as large as possible (Fig. 1).

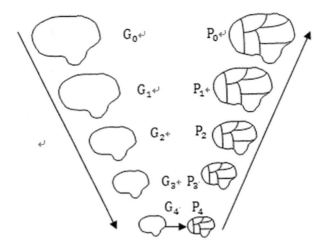

Fig. 1. Class diagram splitting process

The splitting algorithm is based on the coupling relationship in Sect. 2.2, putting the strongly coupled classes into the same class, and separating the classes with lower coupling degrees. This splitting method conforms to the high cohesion and low coupling design in object-oriented design in principle.

The algorithm is as follows:

1. Create vertices: Traverse all classes in class diagram of C, for each class $c_i \in C$, establish the vertex, name it with c_i.
2. Create a weighted graph: For any $c_i, c_j \in C$, calculate $Coupleling(c_i, c_j)$, if $Coupleling(c_i, c_j) > 0$, establish the edge of $e(i, j)$ between c_i and c_j, and let the weight of e(i, j) be equal to $Coupleling(c_i, c_j)$.
3. Coarsening stage: The weighted graph G_0 is compressed step by step, generate $G_1, G_2, G_3......G_n$ step by step, until the number of vertices in the G_n is less than 100.
4. Initial partitioning stage: The weighted graph G_n is divided $\log k$ times by 2-way partitioning algorithm, which is divided into partition P_n with k similar scale graphs.
5. Refining stage: The partition P_n of G_n is projected to G_0 through partitions of the $P_{n-1}, P_{n-2}...P_0$. After each restore, the KL optimization algorithm is used to process the partition edges so that the division of each region is minimized.

Due to the large scale of the class diagram, the scale of the weighted graph is relatively large. If we partition the weighted graph directly, the time complexity will be oversize, so the weighted graph will be processed before the division. Therefore, we compress the weighted graph in the coarsening stage to reduce the number of nodes to less than 100. After the coarsening stage, the number of nodes is greatly reduced. Now it is necessary to divide the graph with a much smaller scale. Compared with the direct division of the weighted graph, the time complexity of the partition is greatly reduced. At the end of the partition, each sub-graph is refined step by step until the original number of nodes is restored to obtain the original plot.

We expand the third to fifth step of the algorithm, and introduce them in the next three subsections.

3.1   Coarsening Stage

The weighted graph G_0 is compressed step by step, generate $G_1, G_2, G_3......G_n$ step by step, until the number of nodes in the G_n is less than 100. In the coarsening stage, we combine a set of vertices in the graph G_i into a large vertex, and use this newly formed vertex as a vertex in the coarser graph G_{i+1}. We make c_i^c as a set of all the vertices in the graph G_i that are merged into the new vertex, and let c be the vertex of its union. In order to get a good partitioning result in the initial partitioning stage, we set the weight of vertex c to the sum of all vertices's weights in c_i^c. In order to preserve the connection information between the vertices of the original graph in the coarser graph, c inherits the edge of c_i^c. If there are multiple vertices in c_i^c and are connected to the vertex h at the same time, then $w(c, h)$ equal to the sum of the weights of edge which consist of vertex h and all node in c_i^c.

Given the graph of the G_i, the corresponding coarser graph is obtained by combining the adjacent nodes in the graph. Several nodes are merged into node of G_{i+1}, and the edges between nodes are folded together. The edge folding method is mainly based on the matching, the matching of the graph is to find the set of edges in the graph. The vertices at both ends of the edges are not repeated. The coarser graph G_{i+1} is obtained by matching G_i, And the unmatched edges and vertices are directly copied into G_{i+1}. Since the purpose of combining vertices with matching is to reduce the size of graph G_i, the edges should be included as much as possible. So the match should be in the vertex to match the maximum. The maximum match is defined as: If any unmatched edge contains at least one matching vertex, then we call it the maximum match (Figs. 2 and 3).

Different matching methods will lead to different matching results, where we introduce two matching methods, Random matching and Heavy Edge matching. Random matching algorithm work process: access to the vertex in random order,

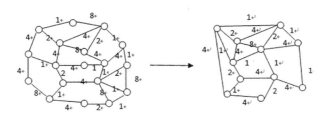

Fig. 2. Random matching coarsening

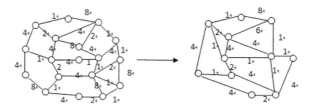

Fig. 3. Heavy edge matching coarsening

if the vertex u does not match, it randomly matches its adjacent unmatched vertex, if there is a matching adjacent vertex v, we put $e(u, v)$ included in the matched list, and then vertex u and vertex v are marked as matched; if the vertex u does not have an unmatched adjacent vertex v, then the vertex u still does not match.

Heavy Edge matching algorithm: We define M_i as the set of matching edges of the graph $G_i(C_i, E_i)$, then $W(M_i)$ is the sum of the weights of the matching edges. Obviously we can get the following formula:

$$W(E_{i+1}) = W(E_i) - W(M_i) \tag{3}$$

In this way, compared to $G_i(C_i, E_i)$, the decrement of the total edge-weight is equal to the sum of the weights of the matching edges. By selecting a matching M with a larger weight, the coarser graph has a smaller edge weight. So in initial partition stage, we can get a lower partition cost.

The same to Random matching algorithm, Heavy Edge matching algorithm use the random algorithm to traverse the vertex of graph. But here is not directly random matching the vertex u any adjacent vertex, but in all adjacent unmatched vertices to find the vertex makes the maximum edge-weight.

3.2   Initial Partition Stage

For the weighted graph $G_n(C_n, E_n)$, the recursion dichotomy is used to divide the vertices to obtain the k-way partition P_n, and the number of vertices per part is approximately equal. The Initial partitioning stage is to divide the coarser graph $G_n(C_n, E_n)$ into k parts, so that each part roughly contains the $1/k$ vertices of the original graph. We split the coarser graph into k parts by using the $logk$ recursive dichotomy. The process of dichotomy: randomly select a vertex on the coarser graph G_n, traverse it using the breadth-first algorithm until the trailing vertex weight is equal to half of the $G_n's$ total vertices weight and the remaining part as the other half. In this way, the coarser graph is divided into two parts. In the newly generated graph, a recursive dichotomy is used until the number of subgraphs reaches k.

3.3   Refining Stage

In the refining stage, the k-way partition P_n of the coarser graph G_n goes back to the original graph $G_{n-1}, G_{n-2}, ..., G_0$. Since the vertices in G_{i+1} are fused by one or several vertices in G_i, in order to obtain P_i, All the vertices in the partition P_i that are merged by the vertices of the upper-level coarser graph are dismantled, and the number of vertices is restored to the number of the upper-level coarser graph.

Although P_{i+1} is the local minimum division of the graph G_{i+1}, after the refinement, P_i is not necessarily the local minimum division of G_i. Compared with G_{i+1}, G_i is more specific, so there will be more space to optimize to reduce the partitioning cost. So every time we refine it, we optimize it. The basic idea of the optimization algorithm is to randomly select two adjacent partitions to select two sets of vertices, respectively. If the two vertex sets are exchanged, the new division can reduce the partition cost, then the new partition is adopted In our algorithm, we use KL optimization algorithm to optimize it.

After each refinement, we use the KL optimization algorithm, we define g_v for the vertex v from the partition to another partition after the division of the cost reduction, referred to as the cost gain. Let P be the partition of graph $G(C, E)$. g_v is obtained by the following formula:

$$ g_v = \sum_{(u,v) \in E \,\wedge\, P[v] \neq P[u]} w(u, v) - \sum_{(u,v) \in E \,\wedge\, P[v] = P[u]} w(u, v) \tag{4} $$

The formula w (u, v) is the weight of the edge (u, v). If g_v is positive, it means that moving the vertex v to another partition can reduce the cost of the division, and if g_v is negative, the division cost increases by the same amount. If the vertex v moves to another region, the cost gain of the vertex adjacent to the vertex v may change. So we need to update the cost gain of its neighboring nodes every time we move a vertex to another partition. And then until all the adjacent partition elements have used the optimization algorithm, then the end of this stage of optimization.

4   Experiment

4.1   Experimental Environment and Experimental Process

The experimental environment, Lenovo, y470 Intel (R) Core i5-2450M 2.50 GHZ 4G memory win7 64-bit operating system.

In order to split the large class diagram, first generate a class diagram from the source code through the reverse tool. At present, there are three ways to generate the class diagram from the source code. The first is to put all the categories and interfaces of the project into the same class diagram. The second way is to name the namespace and class names Space corresponds to a class diagram, the third according to the directory or package organization, each package or directory corresponds to a class diagram. The first type of generated class diagram scale is often too large, the second and third way to generate the class diagram is not only large scale, but also easy to lose critical information, it is because different namespace or different packages under the relevant class Tight coupling information can not be reflected in the class diagram.

The proposed splitting method in this paper is intended to split the class diagram, rather than merging the related classes scattered in different classes into a class diagram, so we choose to show all the categories and interfaces in the project A class diagram to fully demonstrate the coupling between categories (interfaces).

In order to make a more in-depth evaluation of the split effect, we manually split the large class graphs generated by the reverse engineering to create several smaller classes. By comparing the pure manual split and the proposed method, the effect of the automatic splitting algorithm on optimizing the software model of reverse engineering can be effectively measured.

The purpose of the workload comparison, we quantify the manual split and the algorithm to split the workload to do, in hours as a unit. For the division of the cost, after dividing the statistics of each method corresponding to the cost of the division can be. The quantification of readability and comprehensibility of class diagrams is difficult, and there is no widely accepted quantification standard. This experiment will evaluate the readability and comprehensibility of class diagrams by means of expert evaluation. By the five experts to independently assess the three kinds of split the way to get the class diagram, each expert on the three kinds of division of the way to score, very system. For each division method, the average score is counted.

4.2   Experimental Result

In order to facilitate description, we named the large class diagram split algorithm which use the random matching method as R&M, and named the large class diagram split algorithm which use the edge weight matching method as EW&M, named the traditional manual split algorithm as MANUAL. According to the principle of "7 ± 2", respectively, class 1, class 2, class 3, class 4, class 5, class 6 to split, k will be set to $29, 32, 44, 54, 57, 67$, which will be divided into $29, 32, 44, 57$ and 67 respectively (Table 1).

Table 1. Comparison of work time

Class diagram	R&M	HE&M	MANUAL
Class.Diagram1 with232 classes	1 h	1.2 h	25 h
Class.Diagram2 with256 classes	1.1 h	1.3 h	28 h
Class.Diagram3 with350 classes	1.3 h	1.7 h	35 h
Class.Diagram4 with427 classes	1.6 h	2.1 h	40 h
Class.Diagram5 with456 classes	1.7 h	2.2 h	42 h
Class.Diagram6 with534 classes	2 h	2.5 h	50 h

Table 2. Comparison of partition costs

Class diagram	R&M	HE&M	MANUAL
Class.Diagram1 with232 classes	323	278	288
Class.Diagram2 with256 classes	385	335	358
Class.Diagram3 with350 classes	496	420	465
Class.Diagram4 with427 classes	580	512	588
Class.Diagram5 with456 classes	732	594	690
Class.Diagram6 with534 classes	845	683	812

Table 3. Comparison of subjective evaluation results

Class diagram	R&M	HE&M	MANUAL
Class.Diagram1 with232 classes	6.2	9.2	9.0
Class.Diagram2 with256 classes	5.1	9.5	8.9
Class.Diagram3 with350 classes	7.4	9.4	8.4
Class.Diagram4 with427 classes	6.3	9.3	8.5
Class.Diagram5 with456 classes	5.5	9.4	8.4
Class.Diagram6 with534 classes	6.5	9.2	8.2

According to the experimental results we can find, Split the same size of the class diagram, R&M and HE&M with less time than the MANUAL, which R&M with the least time. As shown in the Table 2, for the partition cost, split the same size of the class diagram, HE&M have the least cost, MANUAL have the highest cost. With the scale of class diagram increase, the growth rate of the partition cost of MANUAL is higher than R&M and HE&M. From Table 3 we can see that the class diagram generated by HE&M has the highest readability, and the class diagram generated by R&M has the lowest readability. With the increase of the scale of the class diagram, the readability and comprehensibility of the class diagram generated by the R&M and the HE&M are relatively stable, conversely, the readability of the class diagram generated by MANUAL is lower. From what has discussed above, we can see that the proposed algorithm based on multi-level partitioning is superior to the traditional manual method, whether it is from the working time or the readability of the generated class diagram. Splitting the large class diagram using the large class diagram splitting algorithm can greatly improve the work efficiency.

5 Conclusions

The software development model plays a very important role in the maintenance of the software, but there are still many software systems lacking the software development model, which increases the difficulty of software maintenance.

In order to solve this problem, people use the reverse engineering tool to generate the corresponding UML class diagram from the source code, but the scale of the class diagram is often too large, which affects the readability and comprehensibility of the model.

Aiming at this problem, this paper proposes a split method based on coupling degree. The reverse engineering of the large class diagram is divided into a number of appropriate size of the smaller class diagram. The division of the class diagram improves the readability of the model and improves the efficiency of software maintenance. Through the application in several systems, the validity of the large class diagram splitting method proposed in this paper is verified. Compared with the manual split method, the automatic split method can greatly reduce the working time, and the readability is higher than the manual division of the class diagram.

Further research will continue to improve the effectiveness of class diagram division and reduce working hours. Using machine learning and other methods to understand and remember the experience of manual split may help to improve the split effect, but also help to adapt to different people in reading habits and reading ability differences.

References

1. Gousios, G., Pinzger, M., Van Deursen, A., et al.: An exploratory study of the pull-based software development model. In: 36th International Conference on Software Engineering, Hyderabad, pp. 345–355. ACM (2014)
2. Decker, M.J., Swartz, K., Collard, M.L., et al.: A tool for efficiently reverse engineering accurate UML class diagrams. In: IEEE International Conference on Software Maintenance and Evolution, pp. 607–609 (2016)
3. Jing, J., Wu, L.A., Sarandy, M.S., Muga, J.G.: Inverse engineering control in open quantum systems. Phys. Rev. A. **88**, 053422 (2013)
4. http://astah.net/editions/community
5. Zhiyi, M.A., Junfeng, Z., Xiangwen, M., Wenjuan, Z.: Research and implementation of Jade bird object-oriented software modeling tool. J. Softw. **14**, 97–102 (2003)
6. Booch, G., Rumbaugh, J., Vieweg, I., Werner, C., Wagner, K.P., et al.: Unified Modeling Language (UML). Einfhrung Wirtschaftsinformatik, pp. 367–377. Gabler Verlag (2012)
7. Miller, G.A.: The magical number seven, plus or minus two: some limits on our capacity for processing information. Psychol. Rev. **101**, 343–352 (1956)
8. Kernighan, B.W., Lin, S.: An efficient heuristic procedure for partitioning graphs. Bell Syst. Tech. J. **49**, 291–307 (1970)
9. Kaneiwa, K., Satoh, K.: On the complexities of consistency checking for restricted UML class diagrams. Theor. Comput. Sci. **411**, 301–323 (2010)
10. Briand, L.C., Daly, J.W., Wust, J.K., et al.: A unified framework for coupling measurement in object-oriented systems. Empirical Softw. Eng. **3**, 65–117 (1998)

Ford Motorcar Identification from Single-Camera Side-View Image Based on Convolutional Neural Network

Shui-Hua Wang[1,2], Wen-Juan Jia[1], and Yu-Dong Zhang[1,3(✉)]

[1] School of Computer Science and Technology, Nanjing Normal University,
Nanjing 210023, Jiangsu, China
yudongzhang@ieee.org
[2] Department of Mechanical and Control Engineering,
Kyushu Institute of Technology,
Kitakyushu 804-8550, Fukuoka Prefecture, Japan
[3] School of Computer Science and Technology, Henan Polytechnic University,
Jiaozuo 454000, Henan, People's Republic of China

Abstract. Aim: This study proposed an application of convolutional neural network (CNN) on vehicle identification of Ford motorcar. We used single camera to obtain vehicle images from side view. Method: We collected a 100-image dataset, among which 50 were Ford motorcars and 50 were non-Ford motorcars. We used data augmentation to enlarge its size to 3900-image. Then, we developed an eight-layer CNN, which was trained by stochastic gradient descent with momentum method. Results: Our CNN method achieves a sensitivity of 93.64%, a specificity of 93.13, and an accuracy of 93.38%. Conclusion: This proposed CNN method performs better than three state-of-the-art approaches.

Keywords: Vehicle identification · Ford motorcar · Side-view · Single camera · Convolutional Neural Network

1 Background

Vehicle identification is important in traffic control, factory production line [1], and accident site monitoring. There are many ways of identify vehicles, such as multiple spectrum model [2], multi-camera model [3], OCR and RFID model [4], sensor model, etc.

In this study, we used a single-camera model, which is easy to establish. There are already literatures discussing the classification algorithms from single-camera in frontal view. Ward et al. [5] developed a nonlinear autoregressive neural network (NARNet). Medeme et al. [6] used both Bayesian neural network and probabilistic neural network. Jia [7] proposed a wavelet entropy (WE), back propagation neural network (BPNN) and Levenberg-Marquardt (LM) algorithm.

Nevertheless, it is difficult to get the frontal view in realistic situations [8]. Hence, our team is the first time to capture the car images from the side-view. Besides, All above methods need to hand-design the features. Recently, the convolutional neural

© Springer International Publishing AG 2017
H. Yin et al. (Eds.): IDEAL 2017, LNCS 10585, pp. 173–180, 2017.
https://doi.org/10.1007/978-3-319-68935-7_20

network (CNN) achieved remarkable success since it does not need to extract feature manually [9, 10]. Therefore, we applied CNN into vehicle detection in this study.

The structure of the paper is organized as follows: Sect. 2 provides the materials. Section 3 presents the convolutional neural network methods. Section 4 offers the experiments and results. Finally, Sect. 5 gives the concluding remarks.

2 Materials

100 vehicle images are obtained, 50 are Ford vehicles, and the rest include Buick, Hyundai, Shanghai Volks, and Toyota. Two senior cameramen with more than five-year experiences were requested to shot the vehicles from side view. We removed the background and left the vehicle in the center of the picture by watershed algorithm [11] and manual revision. Each image was resized to 255 × 255. For simplicity, we converted the color images to gray-level images. Several samples of preprocessed dataset are listed in Fig. 1.

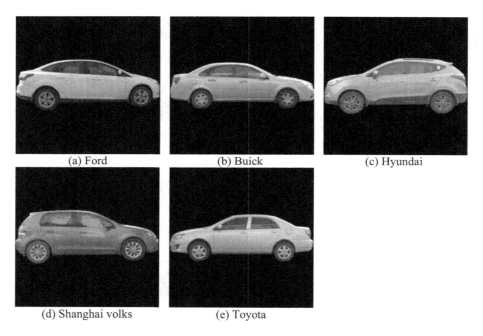

(a) Ford   (b) Buick   (c) Hyundai

(d) Shanghai volks   (e) Toyota

Fig. 1. Samples of dataset

Data augmentation [12] is used to enhance the size of our dataset. The rotation was used with rotation angle of −15 to 15 with increase of 5. The Gamma correction was employed with Gamma value of 0.7 to 1.3 with increase of 0.1. Gaussian noise [13] was added with variance of 0.01, and we generate 10 noise-corrupted image. Translation along horizontal was implemented with −15 to 15 with step of 5, and translation along vertical was carried out with −50 to 50 with step of 10. In total, each image can generate $6 + 6 + 10 + 6 + 10 = 38$ new images.

3   Convolutional Neural Network

Here the convolution and rectified linear unit (ReLU) [14] Layers are combined as one layer. We used max pooling [15] in this study. The parameters of the former six layers of proposed CNN is depicted below in Table 1. In the deep layer of the CNN, we create two layers with hidden numbers of 10 and 2, respectively. The sizes of their weights and biases are shown in Table 2. Finally, our CNN used softmax as the output function. In total, we created an 8-layer CNN.

Table 1. The former six layers of our CNN Structure

Name	Kernel size	No. of channels	No. filters	Stride	Padding
Conv + Relu_1	11 × 11	1	20	2 2	0
Pool_1	3 × 3			2 2	0
Conv + Relu_2	5 × 5	20	60	2 2	0
Pool_2	3 × 3			2 2	1 1
Conv + ReLu_3	3 × 3	60	60	2 2	0 0
Pool_3	3 × 3			2 2	0 0

Table 2. The latter two fully connect layers

Name	Weights	Biases
FC_1	10 × 540	10 × 1
FC_2	2 × 10	2 × 1

4   Experiments and Results

We used stochastic gradient descent with momentum (SGDM) method [16] as the training method. The maximum epoch was set as 50. Initial learning rate was 0.0001. The mini-batch size was set to 128. The l_2 regularization factor was 0.005.

4.1   Data Augmentation

Several samples of processed results after data augmentation are shown in Figs. 2, 3, 4 and 5. The 100-image dataset finally turn to a 3900-image dataset.

The 3900 dataset was split into two sets at random. One is 1950-image dataset used for training, and the other is 1950-image dataset used for test.

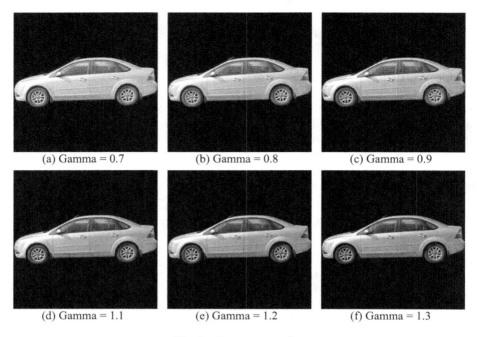

(a) Gamma = 0.7   (b) Gamma = 0.8   (c) Gamma = 0.9

(d) Gamma = 1.1   (e) Gamma = 1.2   (f) Gamma = 1.3

Fig. 2. Gamma correction

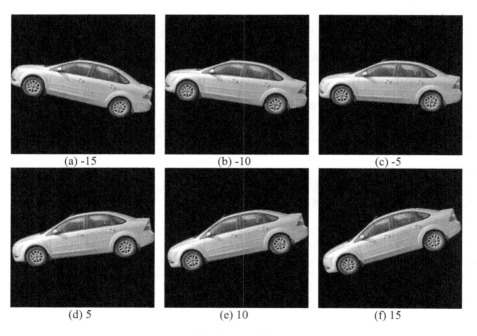

(a) -15   (b) -10   (c) -5

(d) 5   (e) 10   (f) 15

Fig. 3. Rotation

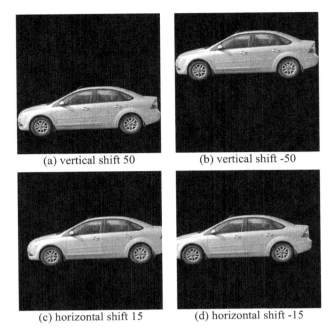

(a) vertical shift 50   (b) vertical shift -50

(c) horizontal shift 15   (d) horizontal shift -15

Fig. 4. Translation along vertical and horizontal directions

Fig. 5. Three noise corrupted image

4.2 CNN Convergence Analysis

The 50 epochs correspond to roughly 761 iterations. Here we plot the curves of training loss and training accuracy in Figs. 6 and 7, respectively. Note that the loss and accuracy was calculated over the randomly selected 128 samples every iteration.

4.3 Confusion Matrix

The confusion matrix over the test set was shown in Table 3. We can find that 913 Ford motors are correctly identified, and 62 Fords vehicles are misclassified as other brands.

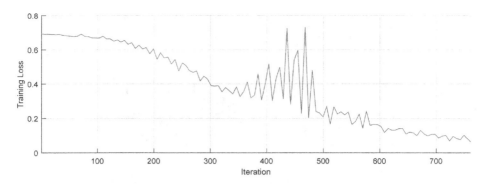

Fig. 6. Curve of training loss

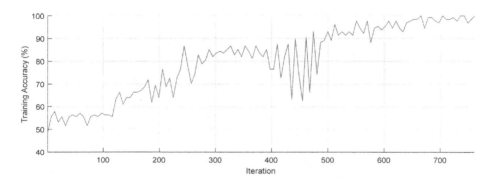

Fig. 7. Curve of training accuracy

In contrast, 908 Non-Ford vehicles are correctly identified, and 67 Non-Ford motorcars are misclassified as Ford motorcars. In all, the sensitivity of our method achieves 93.64%, the specificity is 93.13, and the accuracy is 93.38%.

4.4   Comparison to State-of-the-Art Approaches

We compared our method with three state-of-the-art methods: The comparison results were listed below in Table 4. We can observe that NARNet [5] achieves an accuracy of 76.97%, PNN [6] achieves an accuracy of 78.62%, WE + BPNN + LM [7] achieves an accuracy of 80.00%, and this proposed 8-layer CNN achieves an accuracy of 93.38%, which is significantly higher than state-of-the-art approaches.

We shall do more investigations in the future. For example, we shall compare different pooling techniques. We shall try to introduce the dropout layer and check its performance. Besides, we shall try to develop a method to optimize the structure of CNN automatically.

Nevertheless, this paper is a preliminary study work on Ford motorcar identification. Our results are based on laboratory condition. The performance cannot reflect in

Table 3. Confusion matrix over test set

	Ford	Non-Ford
Ford	913	62
Non-Ford	67	908

Table 4. Comparison to state-of-the-art approaches

Approach	Accuracy
NARNet [5]	76.97%
PNN [6]	78.62%
WE + BPNN + LM [7]	80.00%
8-layer CNN (Our)	93.38%

real situation where a mounted camera is used on a usual street scene. The 100-image dataset is limited. In the future, we shall collect more motorcar side-view images either from company or by ourselves. The experiment can be enhanced on multiple body colors, different motorcar models, and different years. Besides, we shall also preserve real background to try to enhance the identification ability of our method. Various CNNs with different structures will be explored to find the optimal network structure.

5 Conclusion

In this study, we developed an eight-layer convolutional neural network for Ford motorcar identification. The results show our method is superior to three state-of-the-art approaches. In the future, we shall make tentative tests of developing other-brand motorcar identification methods.

Acknowledgements. This paper is supported by Natural Science Foundation of China (61602250), Natural Science Foundation of Jiangsu Province (BK20150983), Open fund of Key Laboratory of Guangxi High Schools Complex System and Computational Intelligence (2016CSCI01), Open Fund of Fujian Provincial Key Laboratory of Data Intensive Computing (BD201607).

References

1. Xiang, L., et al.: Automatic vehicle identification in coating production line based on computer vision. In: International Conference on Computer Science and Engineering Technology, pp. 260–267. World Scientific Publication Co. Pvt. Ltd. (2016)
2. May, C.M., et al.: Multi-spectral synthetic image generation for ground vehicle identification training. In: Infrared Imaging Systems: Design, Analysis, Modeling, and Testing, vol. 27, pp. 496–503. SPIE-International Society Optical Engineering (2016)
3. Chen, H.T., et al.: Multi-camera vehicle identification in tunnel surveillance system. In: IEEE International Conference on Multimedia & Expo Workshops, pp. 1–6. IEEE (2015)

4. Jondhale, A., et al.: OCR and RFID enabled vehicle identification and parking allocation system. In: International Conference on Pervasive Computing (ICPC), pp. 4–11. IEEE (2015)
5. Ward, M.R., et al.: Vibrometry-based vehicle identification framework using nonlinear autoregressive neural networks and decision fusion. In: IEEE National Aerospace and Electronics Conference, pp. 180–185. IEEE (2014)
6. Medeme, N.R., et al.: Probabilistic vehicle identification techniques for semiautomated transportation security, Data Initiatives, pp. 190–198 (2005)
7. Jia, W.: Ford motor side-view recognition system based on wavelet entropy and back propagation neural network and Levenberg-Marquardt algorithm. In: Eighth International Symposium on Parallel Architectures, Algorithms and Programming (PAAP), pp. 11–17. IEEE (2017)
8. Vinoharan, V., et al.: A wheel-based side-view car detection using snake algorithm. In: 6th International Conference on Information and Automation for Sustainability (ICIAFS), pp. 185–189. IEEE (2012)
9. Lee, S.J., Kim, S.W.: Localization of the slab information in factory scenes using deep convolutional neural networks. Expert Syst. Appl. **77**, 34–43 (2017)
10. Tanaka, A., Tomiya, A.: Detection of phase transition via convolutional neural networks. J. Phys. Soc. Jpn. **86**, Article ID 063001 (2017)
11. Lu, S.Y.: A note on the marker-based watershed method for X-ray image segmentation. Comput. Meth. Prog. Biomed. **141**, 1–2 (2017)
12. Rao, Y., McCabe, B.: Is MORE LESS? The role of data augmentation in testing for structural breaks. Econ. Lett. **155**, 131–134 (2017)
13. Cameron, D., et al.: The effect of noise and lipid signals on determination of Gaussian and non-Gaussian diffusion parameters in skeletal muscle. NMR Biomed. **30**, Article ID: e3718 (2017)
14. Chen, Y.: Voxelwise detection of cerebral microbleed in CADASIL patients by leaky rectified linear unit and early stopping: a class-imbalanced susceptibility-weighted imaging data study. Multimed Tools Appl. Springer (2016)
15. Lai, S.X., et al.: Toward high-performance online HCCR: A CNN approach with DropDistortion, path signature and spatial stochastic max-pooling. Pattern Recogn. Lett. **89**, 60–66 (2017)
16. Mitliagkas, I., et al.: Asynchrony begets Momentum, with an application to deep learning. In: 54th Annual Allerton Conference on Communication, Control, and Computing, pp. 997–1004. IEEE (2016)

Predicting Personality Traits of Users in Social Networks

Zhili Ye[1,2]([✉]), Yang Du[1,2], and Li Zhao[1,2]

[1] Institute of Software Chinese Academy of Sciences, 4# South Fourth Street, ZhongGuanCun, Beijing 100190, People's Republic of China
{zhili2015,duyang2015}@iscas.ac.cn, zhaoli@nfs.iscas.ac.cn
[2] University of Chinese Academy of Sciences, 80# ZhongGuanCun East Road, Beijing 100190, People's Republic of China

Abstract. With the popularity of social media, an emerging interest is on predicting personality by mining users' digital footprints. Some researches have proven that social behaviors of users in social network are strongly influenced by their personality. However, current methods are mainly focused on selecting better features from user behaviors and then utilizing a classical classification model to separately predict different personality traits. Based on our observation and statistic analysis from the data, this paper proposes a unified semi-supervised method, Personality-dependent Variable Factor Graph model (PVFG), by considering not only the interrelations between a person's personality traits, but also the interrelations with their friends' personality traits. The experiment is carried on two real-world datasets and the results indicate that the proposed method outperforms several alternative methods.

Keywords: Personality prediction · Social network · Factor graph · Multi-label

1 Introduction

People leave considerable *digital footprints* on social network platforms, in the form of text information (e.g. status information on Facebook), pictorial information (e.g. pictures posted on Flicker), location information (e.g. visited places), etc. Recent researches suggest that people's digital footprints can be used to infer their *personalities*: for example, a person's music preferences [14], profile images [10] and even Facebook likes [17] have been proven to be good indicators of his/her personalities.

In psychology, personalities are often analysed from five dimensions: openness to experience, conscientiousness, extroversion, agreeableness and neuroticism (or emotion stability) [4]; this theory is known as the *Big Five* theory, and each dimension is termed as *personality trait*. Many existing works formulate the prediction of each personality trait as an independent regression (see, e.g. [8,17]) or classification problem (see, e.g. [5,11]). Some recent works formulate the personality prediction task as a *multi-label* classification problem [6], so as to explicitly

© Springer International Publishing AG 2017
H. Yin et al. (Eds.): IDEAL 2017, LNCS 10585, pp. 181–191, 2017.
https://doi.org/10.1007/978-3-319-68935-7_21

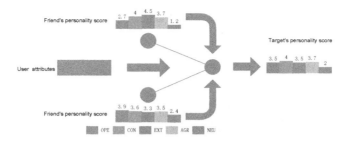

Fig. 1. A simplified illustration of inferring personality traits of a user. Friends' personality traits are used to help the inference.

consider the dependencies and correlations between personality traits and predict five personality traits using a unified model. However, all these methods fail to consider the *social influence* on a person's personality, which, intuitively, can be a considerable factor that affects a person's personality. Thus, the motivation is: Whether we can leverage the social relation information to more precisely predict the personalities?

To answer the above question, we first quantitatively study to what extent a person's personality is correlated with his/her close friends' personalities, and then we propose a unified semi-supervised *probabilistic graphical model* that explicitly considers (i) the target person's digital footprints, (ii) the inter-relations of the target person's personality traits, and (iii) the inter-relations between a friend's personality traits and the target person's personality traits, so as to more precisely predict the target person's personality. Figure 1 illustrates the intuition of the inference process of the model. We can get a result from the target person's attributes with softmax alike weights, and then refine the result by information propagation in the social network. Note that part of the nodes (person in the social network) are labelled. We utilize the unlabelled nodes by the network structure and use Loopy Belief Propagation (LBP) [12] to propagate until converge before the determination of each node's labels. To the best of our knowledge, this is the first work to model social connections to predict personalities, not just using social connections as features. We employ a semi-supervised learning algorithm to obtain the solution of our model by the way of personality contagion.

2 Problem Definition

Let $G = (V^L, V^U, E, \boldsymbol{T}^L)$ denotes a partial labelled network, where V^L is a set of labelled users with Big Five personality traits $\boldsymbol{T}^L$, and V^U is a set of unlabelled users with $V^L \cup V^U = V$; $\boldsymbol{T}^L = <T_1^L, T_2^L, \cdots, T_5^L>$ represents the five personality traits; E is a set of user relations $\{(u, v)|u \in V, v \in V\}$, indicating that u follows v; $\mathbf{X}$ is the attribute matrix, where each row x_i represents an $|x_i|$ dimension feature vector for user v_i. Given a partial labelled graph G and user attribute matrix $\mathbf{X}$, the intention is to learn a predictive function to infer the five personality traits simultaneously by leveraging the supervised information (labelled personality traits),

$$f : (G = (V^L, V^U, E, \boldsymbol{T}^L), \mathbf{X}) \rightarrow \boldsymbol{T}^U \qquad (1)$$

where $\boldsymbol{T}^U$ is the Big Five personality traits of unlabelled users V^U. Based on previous work [2, 6], we inferred each personality trait as a binary classification problem, i.e., high or low. The median was used to split self-reported scores.

3   Data and Observations

3.1   Data Collection

In this work, we use two datasets: Facebook dataset from myPersonality Project [7] and Friends and Family dataset [1]. MyPersonality[1] is a popular Facebook application that allows users to take real psychometric tests, and allows researchers to record users' psychological and Facebook profiles. We use the Big Five personality scores, demographic details, status updates, and Facebook social network. We first select the users who appear in all the above four categories of data and whose age ranging from 18 to 65 years old. Then, we remove the users whose total number of words in all status updates are less than 100 words. Finally, we obtain a dataset with 17,327 users and 12,104 social relations.

In Friends and Family dataset[2], the subjects are members of a young-family residential living community adjacent to a major research university in North America. There are 51 person with complete Big Five personality traits. The survey of friendship given tells that the subjects are all friends of each other, which is nearly fully connected. However, the call log and Bluetooth proximity are given. Considering the size of the dataset and the number of relationships, we use the Bluetooth proximity to generate social relationships. This dataset is got by scanning of nearby Bluetooth devices every 5 min, if two person appeared at the same time and the same device, there is a log. In such circumstances, these two people may be in the same office or the same home. This relation is more reliable, and finally we get 45 users and 709 relations. The data sizes of different ways of generating social relations are shown in Table 1.

Table 1. Different ways of generating network

	#nodes	#edges
Bluetooth	45	709
Call	35	92
Friendship survey	37	663

[1] http://mypersonality.org/.
[2] http://realitycommons.media.mit.edu/.

3.2   Observations

We perform statistical tests to examine the patterns between personality traits. Figure 2 illustrates the personality distribution of neighbours. Each row or column of the big matrix is one dimension of the Big Five personalty traits. Each dimension of the Big Five personalty traits is divided into two classes: low (label 0) and high (label 1). The column is the label of user v's personality traits and the row is the label of v's friends' personality traits. For each pair of personality traits, there is a 2×2 distribution matrix. For example, the pair (openness, openness), it's distribution matrix is in the top left corner. The value of each block is the frequency of co-occurrence of user v and v's friends on each pair of labels. When user v's label of openness is 0 and the probability that v's friend labels 0 in openness is 0.505, which is represented in the top left corner of the 2×2 matrix in Fig. 2(a). Overall, the correlation between a user's personality and his/her friends' personality in the dataset of Friends and Family is stronger than in myPersonality. There are some patterns in both dataset and we can utilize these differences for better inference. For example, in myPersonality, we find that a user and his/her friends tend to have the same label in the same personality trait, in other words, the color on the diagonal of the matrix is brighter than the other two cells in the same 2×2 matrix. Individuals who scored high in conscientiousness, extraversion and agreeableness are more unlikely to make friends who scored high in neuroticism. Part of the goal of our model is to learn the transition probability from data and then to help the inference of friends' labels.

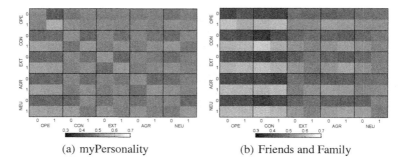

(a) myPersonality   (b) Friends and Family

Fig. 2. Plotmatrix for neighbors' personality distribution. Each column is the label of one trait of user v and each row is the label of one trait of v's friends. Each pair of traits has a 2×2 co-occurrence matrix.

4   Proposed Model

We propose a Personality-dependent Variable Factor Graph model (PVFG) to learn and predict a user's Big Five personality traits. The model aims to capture not only a user's attributes but also the correlations among users' five personality traits.

4.1 Personality-Dependent Variable Factor Graph

Using the definitions in Sect. 2, given a network $G = (V^L, V^U, E, \boldsymbol{T}^L)$ and user attributes matrix $\mathbf{X}$, each user (node) is associated with an attribute vector $\mathbf{x}_i$ and a label vector $\boldsymbol{t}_i = <t_i^1, t_i^2, \cdots, t_i^5>$ indicating the Big Five personality traits' labels. Let $\mathbf{X} = \{\mathbf{x}_i\}$ and $\boldsymbol{T} = \{\boldsymbol{t}_i\}$, the objective function is defined as follows:

$$P(\boldsymbol{T}|G, \mathbf{X}) \propto P(\mathbf{X}|\boldsymbol{T}) \cdot P(\boldsymbol{T}|G) \propto \prod_{v_i \in V} P(\mathbf{x}_i|\boldsymbol{t}_i) \prod_{e_{mn} \in E} P(\boldsymbol{t}_m, \boldsymbol{t}_n) \qquad (2)$$

where $P(\boldsymbol{T}|G)$ denotes the probability of labels given the structure of the network and $P(\mathbf{X}|\boldsymbol{T})$ is the probability of generating attributes $\mathbf{X}$ associated with all edges given their labels $\boldsymbol{T}$. According to the theory of factor graph [9], the "global" probability can be factorized into a product of "local" factor functions. Where $P(\boldsymbol{t}_m, \boldsymbol{t}_n)$ denotes the probability of labels associated with edge e_{mn}. e_{mn} represents a relationship between user m and user n. $\boldsymbol{t}_m$ denotes user m's five personality traits' labels.

We define two factors in this model. The first is an attribute factor $f(\mathbf{x}_i, \boldsymbol{t}_i)$ which models the correlations between users' personality traits and extracted features. The second is a pairwise-correlation factor $h(\boldsymbol{t}_m, \boldsymbol{t}_n)$ which captures the correlations between two friend's personality traits. By integrating these two factor, Eq. 2 can be further factorized as:

$$P(\boldsymbol{T}|G, \mathbf{X}) = \prod_{v_i \in V} [f(\mathbf{x}_i, \boldsymbol{t}_i)] \times \prod_{e_{mn} \in E} [h(\boldsymbol{t}_m, \boldsymbol{t}_n)] \qquad (3)$$

Figure 3 shows an illustration of our proposed model. In factor graph, there are two kind of nodes: variable node (represented as white circle) and factor node (represented as black square). The variable nodes connecting to the same factor node are interrelated with each other. The node t_i^j is the latent variable node of user v_i in the j-th dimension of the Big Five personality traits. We model the correlation between attributes $\mathbf{x}_i$ and the Big Five personality traits $\boldsymbol{t}_i$ of user v_i by the factor node $f(\mathbf{x}_i, \boldsymbol{t}_i)$. The interrelations between user v_m's personality traits and the interrelations with his/her friend v_n's personality traits are modeled in the factor node $h(\boldsymbol{t}_m, \boldsymbol{t}_n)$.

Specifically, we define $f(\mathbf{x}_i, \boldsymbol{t}_i)$ as

$$f(\mathbf{x}_i, \boldsymbol{t}_i) = \frac{1}{W_v} exp\{\boldsymbol{\alpha_{t_i}} \cdot \mathbf{x_i}\} \qquad (4)$$

where $\boldsymbol{\alpha}$ is a real valued vector of parameters and W_v is a normalization term. As $\boldsymbol{t}_i$ is a 5-length vector, the output of $f(\mathbf{x}_i, \boldsymbol{t}_i)$ is also a 5-length vector, and the value of dimension j is the probability of labelling 1 in personality trait j ($t_i^j = 1$) considering the feature $\mathbf{x}_i$.

The next factor is the pairwise-correlation factor. The intention is to consider the correlation between two users' personality traits. Given an edge $e_{mn} \in E$,

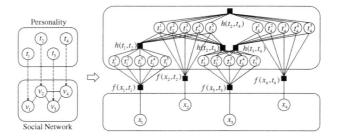

Fig. 3. An illustration of the proposed model. $f(\mathbf{x}_i, \boldsymbol{t}_i)$ and $h(\boldsymbol{t}_i, \boldsymbol{t}_j)$ respectively represent attribute factor and pairwise-correlation factor in the proposed model.

which indicates the friendship between user v_m and v_n, we can define the correlation between their personality traits $\boldsymbol{t}_m$ and $\boldsymbol{t}_n$ as:

$$
h(\boldsymbol{t}_m, \boldsymbol{t}_n) = \begin{cases}
\frac{1}{W_{e_1}} exp\{\beta_1 \cdot I(t_m^1, t_n^1)\} \\
\cdots \\
\frac{1}{W_{e_{25}}} exp\{\beta_{25} \cdot I(t_m^5, t_n^5)\} \\
\cdots \\
\frac{1}{W_{e_{35}}} exp\{\beta_{35} \cdot I(t_m^4, t_m^5)\} \\
\frac{1}{W_{e_{36}}} exp\{\beta_{36} \cdot I(t_n^1, t_n^2)\} \\
\cdots \\
\frac{1}{W_{e_{45}}} exp\{\beta_{45} \cdot I(t_n^4, t_n^5)\}
\end{cases}
\tag{5}
$$

where β_p indicates the transition probability in personality traits pair p. $I(\cdot)$ is an indicator function, and W_{e_p} is a normalization term. There are $5 \times 5 = 25$ different "cross" correlation pairs from $(t_m^1, \cdots, t_m^5, t_n^1, \cdots, t_n^5)$ for the interrelations between user v_m's personality and his/her friend v_n's personality, and $\binom{5}{2} = 10$ different combinations of five personality traits of user v_m and v_n respectively for the interrelations between a person's own personality traits.

By combining these two factor functions together, we obtain the following log-likelihood of the proposed model as:

$$
\mathcal{O}(\alpha, \beta) = \sum_{v_i \in V} \boldsymbol{\alpha}_{\boldsymbol{t}_i} \cdot \mathbf{x_i} + \sum_{e_{mn} \in E} \sum_{p=1}^{45} \beta_p \cdot h_p^{'}(\cdot) - logW
\tag{6}
$$

where $W = W_v W_e$ is the global normalization term, $W_e = \prod_{e_p=1}^{45} W_{e_p=1}$.

Overall, different from the most state-of-the-art works, this work defines the problem of personality inference as a semi-supervised multi-label classification problem and built a unified factor graph model to handle this problem. Similar to most multi-label algorithms, we consider the correlations between target labels (i.e., personality traits in this case). But the difference is that we model the correlations between two users' personality traits' labels by utilizing the social network relations.

4.2   Feature Definition

Based on the previous literature in this field [5, 15], we define three categories of features.

Profile features. User profile is easy to obtain on the social networks. We use two features, age and gender, which are the only information we can get from user's demographic dataset.

Graph features. Graph features consist of some statistical characteristics of the network structure. Based on [5], we extract 9 graph features: degree, betweenness, closeness, authority centrality, hub centrality degree, clustering Coefficient, eigenvector Centrality, embeddedness and neighborhood Connectivity.

Status features. Similar to [5, 15], we define two types of status features, including 11 statistical features of status updates and Latent Dirichlet Allocation (LDA) [3] features of each user. The 11 statistical features consist of the number of status updates, frequency of status updated per day, 4 statistical features (sum, avg, max, min) about the length of status updates, and 5 time distribution counts of status updates. As for LDA features, we set up 100 topics to get users' status updates' topic distribution.

4.3   Model Learning

To learn the proposed model, we maximize the log-likelihood object function $\mathcal{O}(\theta)$. The gradient for each parameter is calculated as:

$$\frac{\partial \mathcal{O}(\theta)}{\partial \alpha} = E[\sum_{v_i \in V} f(\mathbf{x}_i, t_i)] - E_{P_\alpha(\mathbf{T}|\mathbf{X})}[\sum_{v_i \in V} f(\mathbf{x}_i, t_i)]$$

$$\frac{\partial \mathcal{O}(\theta)}{\partial \beta} = E[\sum_{e_{mn} \in E} h(t_m, t_n)] - E_{P_\beta(\mathbf{T}|\mathbf{X}, G)}[\sum_{e_{mn} \in E} h(t_m, t_n)] \tag{7}$$

where $E[\sum_{v_i \in V} f(\mathbf{x}_i, t_i)]$ is the expection of the summation of the attribute factor functions given the data distribution over $\mathbf{T}$ and $\mathbf{X}$, and $E_{P_\alpha(\mathbf{T}|\mathbf{X})}[\sum_{v_i \in V} f(\mathbf{x}_i, t_i)]$ is the expection of the summation of the attribute factor function given by the estimated model. The expection term in second Equation of Eq. 7 has similar meaning.

To learn the factor graph model, we adopt Loopy Belief Propagation (LBP) [12] to calculate the marginal probability of $P(\mathbf{T})$ and compute the exception terms. After we proceed LBP twice to get the two components of each Equation in Eq. 7, we update θ by Eq. 8. Finally, with the estimated parameter θ, we now assign the value of unknown labels $\mathbf{T}$ by looking for a label configuration that will maximize the objective function, as shown in Eq. 9.

$$\theta_{new} = \theta_{old} + \eta \cdot \frac{\partial \mathcal{O}(\theta)}{\partial \theta} \tag{8}$$

$$\mathbf{T}^* = \arg \max \mathcal{O}(\mathbf{T}|G, \mathbf{X}, \theta) \tag{9}$$

5   Experiments

Data and Evaluation. We conduct the experiments on myPersonality and Friends and Family datasets. Detailed data information is introduced in Sect. 3. In the second dataset, we only use the graph feature, as the rest two kinds of feature information are not provided. Next, we define the evaluation method as most multi-label works [13] do: 0/1 Loss, Hamming Loss, Accuracy mean and F1-score mean. The first two are define as:

$$0/1 \quad Loss = 1 - \frac{1}{N} \sum_{i=1}^{N} 1_{\boldsymbol{y}^i = \widehat{\boldsymbol{y}}^i} \tag{10}$$

$$Hamming \quad Loss = 1 - \frac{1}{NL} \sum_{i=1}^{N} \sum_{j=1}^{L} 1_{y_j^i = y_j^i} \tag{11}$$

where N is the number of instances, L is the number of targets (five personality traits in this work), $\boldsymbol{y}^i$ is vector of real label of instance i and $\widehat{\boldsymbol{y}}^i$ is the label inferred. Accuracy mean and F1-score mean are the average of the corresponding measures of five personality traits.

Comparison Methods. We compare the proposed PVFG with different classification algorithms, including Logistic Regression (LR), Support Vector Machine (SVM), Random Forest (RF), Ensembles of Classifier Chains (ECC), Conditional Random Field (CRF) and Factor Graph Model (FGM). For LR and RF, we employ Weka[3]. For SVM, we use LIBSVM[4]. For ECC, we utilize Classifier Chains (CC) in MEKA[5] as base learner and ensembles 20 different classifier chains. We use the model in [16] for the FGM.

Table 2. Performance comparison of different methods for personality classification.

Method	myPersonality				Friends and Family			
	0/1 Loss	Hamm. Loss	Accu. Mean	F1-score Mean	0/1 Loss	Hamm. Loss	Accu. Mean	F1-score Mean
LR	0.9223	0.4060	0.5940	0.5937	1.0000	0.5231	0.4769	0.4473
SVM	0.9159	0.4057	0.5943	0.5940	1.0000	0.5077	0.4923	0.3495
RF	0.9312	0.4157	0.5843	0.5827	1.0000	0.4923	0.5077	0.4783
ECC	0.9098	0.4176	0.5824	0.5819	**0.9231**	0.5185	0.4815	0.4689
CRF	**0.8883**	0.4252	0.5748	0.5745	**0.9231**	0.4923	0.5077	0.4817
FGM	0.9264	0.4049	0.5951	0.5952	1.0000	0.4769	0.5231	0.4791
PVFG	0.9142	**0.3994**	**0.6006**	**0.6003**	1.0000	**0.4308**	**0.5692**	**0.5101**

---

[3] http://www.cs.waikato.ac.nz/ml/weka/.
[4] https://www.csie.ntu.edu.tw/~cjlin/libsvm/.
[5] http://meka.sourceforge.net/.

Note that LR, SVM and RF are traditional methods which do not consider the correlations among personality traits. CRF and ECC consider the correlations among one user's own personality traits but do not utilize social relations. FGM uses social relations but does not model the five personality traits in a unified model, i.e. not a multi-label model. In contrast, our proposed PVFG model considers the correlation among personality traits of two friends and models the five personality traits in a unified model.

Table 2 lists the performances of different methods on myPersonality and Friends and Family dataset. We randomly select 70% users to create training set (labelled set) and denote the rest as testing set (unlabelled set) in both datasets. As the proposed PVFG model is a semi-supervised model, we use the training set and test set together in the process of model learning.

When infering one trait of a user, LR, SVM and RF do not consider the influence of his/her other traits and the influence of his/her friends' traits. They are the baselines of this task. ECC and CRF only consider the influence of his/her other traits, and they get better 0/1 loss but worse result in other metrics. FGM only consider the influence of his/her friends' traits, which gets slightly bad 0/1 loss and better result in other metrics. The PVFG taking into account these two factors. In both datasets, the PVFG is superior to other models in terms of Hamming Loss, Accuracy mean and F1-score mean. In myPersonality dataset, the PVFG also achieves improvement in 0/1 loss comparing to baselines. The results of PVFG in Friends and Family dataset achieve significance improvement comparing to other methods. As PVFG utilizes the information of friends, the more distinctive pattern, the better the result of inference. The distribution of friends' personality traits is more distinguishable in the dataset of Friends and Family, as shown in Fig. 2. The network in myPersonality may be incomplete due to the process of collecting questionnaire. Many users have only one or two friends in that dataset. The information of friends may be not enough for good inference. Table 3 show the effect of degree on the performance in test set of myPersonality. The degree of all users in Friends and Family dataset dataset is greater than 20, which is large enough, thus we do not conduct the same analysis in that dataset. Considering both the sample ratio and performance, PVFG get best result when the user's degree is 4 or 5. This may fit the intuitive understanding that only a few good friends have a greater impact on us.

Table 3. PVFG performance of different degree in the test data of myPersonality dataset

Degree	1	2	3	4	5	6	7	>7
Ratio	47.98%	28.80%	13.83%	5.10%	1.62%	0.73%	0.65%	1.29%
0/1 Loss	0.9056	0.9157	0.9298	0.8889	0.9000	0.7778	1.0000	0.9375
Hamming loss	0.3987	0.4140	0.4246	0.3619	0.3600	0.4444	0.4250	0.4125

6   Conclusion

In this paper, we study the correlation between two friends' personality and find some distinctive pattern in the distribution of friends' personality traits. Based on the observations, we define the inference of Big Five personality traits as a multi-label classification problem and propose a unified semi-supervised factor graph models to utilize the social network structure and unlabelled data. In our proposed model, a users personality traits are determined by not only footprint features, but also the personality traits of his/her friends. There are two technical points in the proposed model. First, it is a graph-based semi-supervised node classification method. Unlike the previous methods which used social network structure as features, we directly build model on the network structure. Second, it is a multi-label method. We infer the labels of five traits in a unified model rather than in five independent models.

In online social network, friends' informations can help the inference of user's label. In this work, we directly utilize friends' labels. For future work, some other graph-based methods can be explored and validated for this problem, and we can also utilize friends' features besides friends' labels.

References

1. Aharony, N., Pan, W., Ip, C., Khayal, I., Pentland, A.: Social fMRI: investigating and shaping social mechanisms in the real world. Pervasive Mob. Comput. **7**(6), 643–659 (2011)
2. Bai, S., Zhu, T., Cheng, L.: Big-five personality prediction based on user behaviors at social network sites. arXiv preprint arXiv:1204.4809 (2012)
3. Blei, D.M., Ng, A.Y., Jordan, M.I.: Latent Dirichlet allocation. J. Mach. Learn. Res. **3**, 993–1022 (2003)
4. Costa, P.T., McCrae, R.R.: The revised NEO personality inventory (NEO-PI-R). In: The SAGE Handbook of Personality Theory and Assessment, vol. 2, pp. 179–198 (2008)
5. Farnadi, G., Zoghbi, S., Moens, M.F., De Cock, M.: Recognising personality traits using facebook status updates. In: Proceedings of the Workshop on Computational Personality Recognition (WCPR 2013) at the 7th International AAAI Conference on Weblogs and Social Media (ICWSM 2013). AAAI (2013)
6. Iacobelli, F., Culotta, A.: Too neurotic, not too friendly: structured personality classification on textual data. In: Proceedings of Workshop on Computational Personality Recognition. AAAI Press, Melon Park, pp. 19–22 (2013)
7. Kosinski, M., Matz, S.C., Gosling, S.D., Popov, V., Stillwell, D.: Facebook as a research tool for the social sciences: opportunities, challenges, ethical considerations, and practical guidelines. Am. Psychol. **70**(6), 543 (2015)
8. Kosinski, M., Stillwell, D., Graepel, T.: Private traits and attributes are predictable from digital records of human behavior. Proc. Natl. Acad. Sci. **110**(15), 5802–5805 (2013)
9. Kschischang, F.R., Frey, B.J., Loeliger, H.A.: Factor graphs and the sum-product algorithm. IEEE Trans. Inf. Theory **47**(2), 498–519 (2001)

10. Liu, L., Preotiuc-Pietro, D., Samani, Z.R., Moghaddam, M.E., Ungar, L.: Analyzing personality through social media profile picture choice. In: Tenth International AAAI Conference on Web and Social Media (2016)
11. Markovikj, D., Gievska, S., Kosinski, M., Stillwell, D.: Mining facebook data for predictive personality modeling. In: Proceedings of the 7th International AAAI Conference on Weblogs and Social Media (ICWSM 2013), Boston, MA, USA (2013)
12. Murphy, K.P., Weiss, Y., Jordan, M.I.: Loopy belief propagation for approximate inference: an empirical study. In: Proceedings of the Fifteenth Conference on Uncertainty in Artificial Intelligence, pp. 467–475. Morgan Kaufmann Publishers Inc. (1999)
13. Read, J., Pfahringer, B., Holmes, G., Frank, E.: Classifier chains for multi-label classification. Mach. Learn. **85**(3), 333–359 (2011)
14. Rentfrow, P.J., Gosling, S.D.: The Do Re Mi's of everyday life: the structure and personality correlates of music preferences. J. Pers. Soc. Psychol. **84**(6), 1236 (2003)
15. Schwartz, H.A., Eichstaedt, J.C., Kern, M.L., Dziurzynski, L., Ramones, S.M., Agrawal, M., Shah, A., Kosinski, M., Stillwell, D., Seligman, M.E., et al.: Personality, gender, and age in the language of social media: the open-vocabulary approach. PLoS ONE **8**(9), e73791 (2013)
16. Tang, J., Sun, J., Wang, C., Yang, Z.: Social influence analysis in large-scale networks. In: Proceedings of the 15th ACM SIGKDD International Conference on Knowledge Discovery and Data Mining, pp. 807–816. ACM (2009)
17. Youyou, W., Kosinski, M., Stillwell, D.: Computer-based personality judgments are more accurate than those made by humans. Proc. Natl. Acad. Sci. **112**(4), 1036–1040 (2015)

Face Anti-spoofing Algorithm Based on Gray Level Co-occurrence Matrix and Dual Tree Complex Wavelet Transform

Xiaofeng Qu, Hengjian Li, and Jiwen Dong[✉]

Shandong Provincial Key Laboratory of Network Based Intelligent Computing,
University of Jinan, Ji'nan, China
ise_quxf@163.com, {ise_lihj,ise_dongjw}@ujn.edu.cn

Abstract. By analyzing the difference of facial texture features between living face and photo, we propose a novel face anti-spoofing algorithm based on gray level co-occurrence matrix (GLCM) and dual-tree complex wavelet tree (DT-CWT). Firstly, inspired by the co-occurrence matrix, we extract five texture features including angle second moment, entropy, contrast, correlation and local uniformity to represent the gray direction, interval and amplitude information for the face texture information. Secondly, DT-CWT has the advantages of approximate translation invariance and good direction selectivity. Therefore, the coefficients of DT-CWT can enhance the texture information and edge information in the frequency domain. At last, the SVM classification is used to distinguish between true and fake face. Our algorithm is demonstrated on the published NUAA database. Compared with the existing methods, the feature dimension is reduced. The experimental results show that the proposed algorithm improves the detection accuracy.

Keywords: Anti-spoofing · Liveness detection · Gray level co-occurrence matrix · Dual tree complex wavelet transform

1 Introduction

In recent years, biometric identification technology has developed rapidly, and fingerprint identification, face recognition, iris recognition and other identification technologies have been widely used. Among them, face feature is widely welcomed because of its advantages such as not easy to be lost, stolen and user-friendly. As an effective method of identity authentication, face recognition has brought great convenience to people's daily life. However, the acquisition and forgery of biometric information has become easier and faster than before because of the development of biotechnology. N.M Duc [1] and others succeeded in passing the certification system using only a printed face picture, a variety of spoofed counterfeit technologies continue to emerge which pose a great challenge to the application of liveness detection at the

---

This work was supported by the Shandong Provincial Key R&D Program (2016ZDJS01A12).

H. Yin et al. (Eds.): IDEAL 2017, LNCS 10585, pp. 192–200, 2017.
https://doi.org/10.1007/978-3-319-68935-7_22

same time. In order to make authentication more secure, it is important to detect whether the identified object is true or fake face. Given the a face recognition(FR) with spoofing detection module (see Fig. 1). In the real application scenario, the identity authentication system is mainly faced with three common ways to deceive: Photo spoofing, Replay spoofing and 3D mask spoofing.

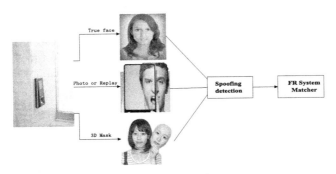

Fig. 1. A face recognition (FR) with spoofing detection module.

In recent years, many domestic and foreign research institutions of scholars are committed to the study of in liveness detection, there are many liveness detection algorithm involved in many international conferences and periodicals. The methods that have been applied to face detection today include:

(1) Based on motion information analysis: From the perspective of static observation, the difference between true face and fake photo is that the former is a three-dimensional object, the latter is a two-dimensional planar structure, and the motion of the two is quite different. T. Choudhary [2] and others used the eye, nose and mouth as the tracking point of the estimated feature points, the three-dimensional depth coordinates of the feature points are estimated from the head motion. K. Kollreider [3] proposed the optical flow method, it can detect dynamic changes in the face to carry out in liveness detection. This method has some disadvantages such as it is difficult to obtain the characteristics required for liveness detection and the method requires the users to active cooperate. Moreover, when counterfeiters deliberately bend the photo. At the same time, this method is susceptible to noise and lighting conditions with the reliability is not strong, the high cost of computer defects.

(2) Based on the analysis of true face image features information: For example, some people analyze facial features. Such as thermal infrared images, blinking, etc. Diego A [4] and others shot the visible face and the hot infrared face to detect, according to the relevance of the two in living judgment. Hyung-Keun Jee [5] and others detected the eye in the input continuous picture sequence and calculate the motion of each eye area to determine whether the input face is true or fake. These methods require additional equipment, there are unstable factors. Therefore, these methods can't be widely promoted in the ordinary equipment.

(3) Based on texture information analysis: The researchers analyze the difference between 3D human face and human face image, and they extracted the relevant texture features for liveness detection. A.K. Jain [6] and others proposed the method that analyzed Fourier spectrum, definiting the frequency dynamic descriptors to distinguish true face and photo face. Andre Anjos [7], Jukka Maatta [8] et al. analyzed the validity of the LBP feature; Jukka Maatta [8] used $LBP^{16,2}$, $LBP^{8,1}$, $LBP^{8,2}$ to extract face features, classificating training SVM to recognize living face. Jukka Maatta [9] and others improved the algorithm in 2012, they combined the fusion method of LBP features, Gabor wavelet and HOG three characteristics. And D. Wen [10] proposed the state-of-the-art methods in spoof detection in 2015. These methods have large computational complexity due to the large dimension of feature extraction.

In this paper, we propose a face anti-spoofing algorithm based on GLCM and DT-CWT, extracting texture features of true and fake face images and training SVM classifier to make true judgment verification on the face database. The method does not need the user's active cooperation and extracts low feature dimension, reducing the time cost and the complexity of the algorithm. This paper demonstrates DT-CWT in Section 2. Then we will describe the method of living face detection in Sect. 3. In Sect. 4, some experimental results will be compared. Finally, we summarize this work in Sect. 5.

2   Dual Tree Complex Wavelet Transform

Wavelet transform [11] had good localization performance of time domain and frequency domain. DT-CWT decomposition adopts two parallel discrete wavelet trees, which named tree-a and tree-b respectively. Each tree is composed of a series of low-pass filter and high pass filter stack, and the filter length of tree-a and tree-b are odd number and even number. When the two-dimensional image signal, the tree-a acts on the rows of the image and generates the real part and the tree-b acts on the columns of the image and generate the imaginary part.

Not only overcomes the transform shift sensitivity and direction selectivity of traditional wavelet transform, but DT-CWT provides more choices for the direction decomposition of each scale space and produces high frequency sub-band of six direction and two low-frequency sub-bands. In this paper, our method is aimed at two-dimensional DT-CWT. As mentioned earlier, DT-CWT has good directional selectivity, and the image is transformed into DT-CWT transform, and each scale is decomposed into a high frequency subband of $\pm15°$, $\pm45°$, $\pm75°$, which 6 complex coefficients in total. we use the mean and variance of the low-frequency sub-band coefficients matrix are extracted as the feature vectors for the SVM classification.

3 Face Anti-spoofing Algorithm Based on GLCM and DT-CWT

Based on the analysis of the difference of texture between true face and photo, therefore, this paper presents a face detection method that it based on GLCM and DT-CWT. The angle second moment, entropy, contrast, correlation and local uniformity of five texture features were extracted on the basis of GLCM. We use 2D DT-CWT of face image is decomposed into four layers of complex wavelet transform at the same time, extracting the low frequency sub-band coefficients as the feature vectors for training SVM classification. The algorithm framework is shown in Fig. 2.

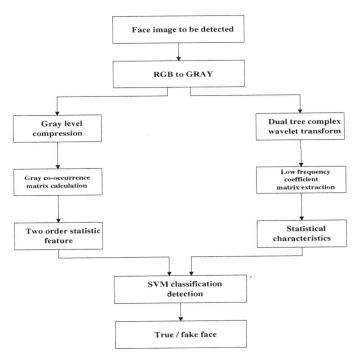

Fig. 2. Framework of face anti-spoofing algorithm based on GLCM and Dual tree complex wavelet transform.

(1) Firstly, inspired by the co-occurrence matrix, we extract five texture features including angle second moment, entropy, contrast, correlation and local uniformity to represent the gray direction, interval and amplitude information for the face texture information. And the mean value and standard deviation of the five texture features are taken as the feature vectors.

The spatial co-occurrence estimation of gray scale is the two order statistical feature of image, and the corresponding statistical features can be used to describe various texture features of images. Set $f(x,y)$ as a two-dimensional digital image. Its size is $M * N$ and gray level is Ng, and the Co-occurrence Matrix which must satisfies certain spatial relations is

$$p(i,j) = \#\{(x_1,y_1),(x_2,y_2) \in M * N | f(x_1,y_1) = i, (x_2,y_2) = j\} \tag{6}$$

Where, $\#(x)$ is the number of elements in set x and p is a matrix of $Ng * Ng$.

The statistical features in this paper include Angle second moment, Entropy, Contrast, Correlation and Local uniformity based on the co-occurrence matrix characteristics:

(a) Angle second moment (energy):

$$ASM = \sum_{i=0}^{L-1}\sum_{j=0}^{L-1} P^2(i,j) \tag{7}$$

It is the square sum of the elements of the gray level co-occurrence matrix, so it is also called energy. It reflects the uniformity of the image gray distribution and the texture roughness.

(b) Entropy:

$$H = -\sum_{i=0}^{L-1}\sum_{j=0}^{L-1} P(i,j) \log P(i,j) \tag{8}$$

It is a measure of the image which has the amount of information. The texture information, which is a measure of randomness, belongs to the information of the image.

(c) Moment of inertia (contrast):

$$CON = \sum_{i=0}^{L-1}\sum_{j=0}^{L-1} (i-j)^2 P(i,j) \tag{9}$$

It reflects the image definition and the depth of the texture groove. The deeper the texture trench, the greater the contrast and the visual effect is more clear.

(d) Relevance:

$$COR = \frac{\sum_i \sum_j ijP(i,j) - \mu_X \mu_y}{\sigma_X^2 \sigma_y^2} \tag{10}$$

It measures the similarity of the spatial gray-level co-occurrence matrix elements in the row or column direction, so it reflects the correlation of local gray in the image. When the matrix element values are even equal, the larger the value.

(e) Local uniformity (inverse gap):

$$L = \sum_{i=0}^{L-1} \sum_{j=0}^{L-1} \frac{P(i,j)}{1 + (i-j)^2} \tag{11}$$

It reflects the homogeneity of the image texture, and it measures the texture locality changes of the image.

(2) Secondly, DT-CWT has the advantages of approximate translation invariance and good direction selectivity. Therefore, the coefficients can enhance the texture information and edge information in the frequency domain. In this paper, we extract the mean and variance of the low-frequency coefficients of DT-CWT.

(3) At last, the SVM classification is used to distinguish between true and fake face. We use SVM to train the feature of true and fake face respectively in time domain and frequency domain to distinguish true or fake face.

4   Experimental Results

In order to analyze and test the proposed algorithm's ability to detect Client and Imposter, we test cases in the open face database NUAA. It is difficult for us to accurately distinguish the images from true faces and fake faces, as shown in Fig. 3.

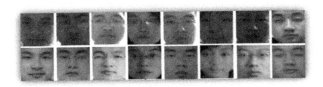

Fig. 3. Examples of images captured from true faces (upper row) and fake faces (lower row) in NUAA database.

The distance d is 1 and the angles are $0°$, $45°$, $90°$ and $135°$ respectively. As can be seen from Fig. 4, the two order statistical features of 100 client faces and 100 corresponding imposter faces are obviously different at four angles in GLCM.

In the database, the positive example is the sequence of 15 human face images collected using webcam and the counterexample sample is the sequence of images captured using a photograph taken by the 15. The samples taken from the counterexample sample are standardized grayscale images of the 64×64. After face detection and eye location, We can obtain standardized positive samples for 15 sets of

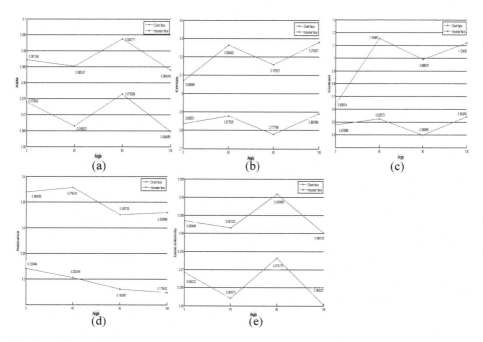

Fig. 4. Differences in the two order statistical characteristics of GLCM with client faces and imposter faces.

face picture sequences, totaling 5105 real face images. And we can obtain standardization of counterexample samples for 15 groups photos of face image sequences, totaling 7509 pictures of face images. 5 groups were randomly selected from 15 groups of live pictures, a total of 1200 photos are taken as examples of training samples. And we selected group of photos of face images randomly, a total of 1200 photos as an example of training samples to train the SVM. Besides the rest, which are never participated in the training of all the samples as a test set, and the number of pictures of the training and test sets are shown in Table 1.

Table 1. The face images in training set and test set of NUAA database The face images in training set and test set of NUAA database

Client face photos	Training set			
	300	800	1000	1200
Imposter face photos	Training set			
	300	800	1000	1200
Client face photos	Test set			
	119	490	1190	1200
Imposter face photos	Test set			
	300	800	1000	1200

In order to test the effectiveness of the algorithm, the LBP feature extraction and face classification for training SVM, the other were tested based on GLCM in wavelet feature detection of true and false after adding face accuracy as shown in Table 2.

Table 2. Comparison of basic LBP features and all LBPV [14] features and Uniform LBPV [14] features and CDD (Component Dependent Descriptor) [13] and GLCM and wavelet analysis features with GLCM wavelet analysis features with GLCM features

Sample	Detection accuracy	Characteristic dimension
Basic LBP features	93.87%	59
All LBPV [14]	88.03%	256
Uniform LBPV [14]	86.95%	59
CDD [13]	97.7%	512
GLCM	94.27%	8
GLCM + Haar [12]	96.97%	12
GLCM + DWT	97.31%	12
GLCM + DT-CWT (**proposed**)	**99.52%**	**12**

Table 2 shows that using LBP and GLCM as the basic characteristics, Wavelet features are added and compared with them, and their influence on the detection rate of the algorithm is tested. Compared with All LBPV, Uniform LBPV and the basic feature of LBP, GLCM and wavelet features have obvious advantages, which the 12 dimensional feature with respect to the 256 and 59 dimensions, not only the algorithm complexity is greatly reduced, and the detection accuracy is greatly improved. In especial, the detection of GLCM and DTCWT wavelet transform is proposed in this paper, which the correct rate of 99.52% compared with the basic LBP features significantly improve the accuracy rate of 93.87%. Although the accuracy of GLCM and wavelet feature detection is decreased, the feature dimension is greatly reduced, and the complexity of the algorithm is decreased obviously compared with the CDD [13] features. The algorithm proposed in this paper not only reduces the feature dimensionality obviously, but also improves the detection accuracy. In the next step, we will try to reduce the feature dimension to reduce the computational complexity and gain higher detection rate.

5   Summary

In this paper, an algorithm is proposed based on GLCM and DT-CWT. We mainly analyze the difference of facial texture features between true face and photo. And we analyze the texture in two aspects: spatial domain and frequency domain. Among them, GLCM is adopted in the spatial domain, and DT-CWT is adopted in the frequency domain. And we use support vector machine (SVM) classification finally. DT-CWT has the advantages of approximate shift invariance and multi directional selectivity and it can enhance the texture and edge information of the image especially. The experimental results show that this method can improve the detection accuracy on the basis of reducing the feature dimension.

References

1. Duc, N.M., Minh, B.Q.: Your face is not your password. In: Black Hat Conference, pp. 1–16 (2009)
2. Choudhury, T., Clarkson, B., Jebara, T., Pentland, A.: Multimodal person recognition using unconstrained audio and video. In: AVBPA 99, Washington DC, pp. 176–181 (1999)
3. Kollreider, K., Fronthaler, H., Bigun, J.: Evaluating liveness by face images and the structure tensor. In: Fourth IEEE Workshop on Automatic Identification Advanced Technologies, pp. 75–80, October 2005
4. Socolinsky, D.A., Selinger, A., Neuheisel, J.D.: Face recognition with visible and thermal infrared imager. CVIU **91**(1–2), 72–114 (2003)
5. Jee, H.-K., Jung, S.-U., Yoo, J.-H.: Liveness detection for embedded face recognition system. Int. J. Med. Sci. **1**, 235–238 (2006)
6. Li, J., Wang, Y., Tan, T., Jain, A.: Live face detection based on the analysis of fourier spectra. In: Proceedings of SPIE Biometric Technology for Human Identification, vol. 5404, pp. 296–303 (2004)
7. Chingovska, I., Anjos, A., Marcel, S.: On the effectiveness of local binary patterns in face anti-spoofing. In: Bromme, A., Busch, C., (eds.) BIOSIG, pp. 1–7. IEEE (2012)
8. Maatta, J., Hadid, A., Pietikainen, M.: Face spoofing detection from single images using micro-texture analysis. In: International Joint Conference on Biometrics (IJCB2011), Washington DC, USA, pp. 10–17 (2011)
9. Maatta, J., Hadid, A., Pietikainen, M.: Face spoofing detection from single images using texture and local shape analysis. IET Biometrics **1**(1), 3–10 (2012). The Institution of Engineering and Technology 2012
10. Wen, D., Han, H., Jain, A.K.: Face spoof detection with image distortion analysis. IEEE Trans. Inf. Forensics Secur. **10**, 1–16 (2015)
11. Dong, Z., Tang, X., Zhang, S., et al.: Image restoration combined with wavelet transform and texture synthesis. Chin. J. Image Graph. **20**(7), 882–894 (2015)
12. Cao, Yu., Ling, T., Lifang, W.: Live face gray co-occurrence matrix and wavelet analysis in authentication detection algorithm. J. Sig. Process. **30**(7), 830–835 (2014)
13. Tan, X., Li, Y., Liu, J., Jiang, L.: Face liveness detection from a single image with sparse low rank bilinear discriminative model. In: Daniilidis, K., Maragos, P., Paragios, N. (eds.) ECCV 2010. LNCS, vol. 6316, pp. 504–517. Springer, Heidelberg (2010). doi:10.1007/978-3-642-15567-3_37
14. Kose, N., Dugelay, J.-L.: Classification of captured and recaptured images to detect photograph spoofing. In: ICIEV 2012, pp. 1027–1032 (2012)

High-Accuracy Deep Convolution Neural Network for Image Super-Resolution

Wen'an Tan[1,2] and Xiao Guo[1(✉)]

[1] Nanjing University of Aeronautics and Astronautics,
Nanjing, People's Republic of China
{wtan,njguoxiao}@foxmail.com
[2] Shanghai Polytechnic University, Shanghai, People's Republic of China

Abstract. Higher performance is the eternal purpose for super-resolution (SR) methods to pursue. Since the deep convolution neural network is introduced into this issue successfully, many SR methods have achieved impressive results. To further improve the accuracy that current SR methods have achieved, we propose a high-accuracy deep convolution network (HDCN). In this article, deeper network structure is deployed for reconstructing images with a fixed upscaling factor and the magnification becomes alternative by cascading HDCN. L_2 loss function is substituted by a more robust one for reducing the blurry prediction. In addition, gradual learning is adopted for accelerating the rate of convergence and compacting the training process. Extensive experiment results prove the effectiveness of these ingenious strategies and demonstrate the higher-accuracy of proposed model among state-of-the-art SR methods.

Keywords: Super-resolution · Gradual learning · Cascade · Charbonnier loss function

1 Introduction

Limited by imaging equipment, digital images tend to be low-resolution (LR) for missing high-frequency constituent. Single image super-resolution (SISR) is introduced to recover the high-resolution (HR) from its LR input. This technique has been widely adopted in security monitoring and medical imaging where requires extra image details.

In recent years, Convolutional neural network (CNN) provides impetus for SISR methods. Dong et al. [2] certify the validity of learning a nonlinear LR-to-HR mapping in an end-to-end manner. They further deploy a shrink layer and a expand layer [3] in their model, termed FSRCNN, to downsize the scale of parameters and maintain the accuracy. Kim et al. [6] substitute a plain structure composed of 20 convolutional layers for anterior shallow network. This very deep CNN (VDSR) achieves impressive results for its substantial increase in accuracy. Subsequent articles demonstrate various ingenious methods, only a few of them, however, could surpass VDSR in accuracy.

© Springer International Publishing AG 2017
H. Yin et al. (Eds.): IDEAL 2017, LNCS 10585, pp. 201–210, 2017.
https://doi.org/10.1007/978-3-319-68935-7_23

In order to further improve the accuracy of current SR methods, we propose a high-accuracy deep convolution neural network to reconstruct image in a fix upscaling factor. By cascading HDCN, we can achieve scalable super-resolution and larger magnification. The merits of proposed HDCN are stated as follows:

(1) For the purpose of reducing the blurry or over-smooth prediction, a robust penalty function is adopted to substitute for L_2 loss function.
(2) During training process, gradual training method is introduced to accelerate the rate of convergence. Training samples are divided into individual parts via calculating the average local gray value difference (ALGD) of their origin and residual patches.
(3) Experimental results on benchmark datasets demonstrate the superior accuracy of proposed HDCN to some state-of-art SR methods.

2   Related Work

Cascade network for SISR: Original neural network based SISR methods can only reconstruct HR images in a fixed upscaling factor [2,3]. Altering the scale factor would be accompanied by fine-tuning the whole model. It's quite an ordeal for those non-professionals, their computers would cost extensive time. In order to augment practicability, researchers dedicate to implement models with feasible upscaling factors. Some scalable SISR methods, such as VDSR, utilize the similarity that images across different scales share some common structures and textures. Thus, images with different scales are combined into a large dataset to train a multi-scale model. Other methods cascade a fixed scale network until the desired size is reached. Wang et al. observe that a cascade of sparse coding based network (CSCN) trained for small scaling factors performs better than the single sparse coding based network trained for a large scaling factor. This strategy is simple but suffers the risk of error accumulation during repetitive upscaling. Fortunately, this risk could be reduced by enhancing the performance of each fixed-scale network. For instance, Zhen et al. employ non-local self-similarity (NLSS) search and shrink the upscaling factor to a small degree in their experiment. In contrast to the multi-scale model trained directly, cascade network could simplify training period and achieve better accuracy. In addition, larger upscaling factor can be achieved by cascading more sub-models.

Residual-like learning: Before introducing residual-like learning, NN based SISR methods tend to implement narrow network and prudent step during iteration. Gradient exploding/vanishing problem hampers deeper network structure and rate of convergence. Since He et al. accomplish a very deep residual network for image recognition and achieve impressive results, subsequent SISR methods tend to adopt residual-like structure to improve the performance of their models. On behalf of them, VDSR attempts to learn the residual image, defined as the difference between input and output image, rather than origin ground-truth image and substantial increases the accurate of reproduce results. From the perspective of network structure, VDSR, inspired by the merits of VGG net, deploys a deep and plain network. In order to approximate the initial structure of residual network, Yang et al. carry out identity mapping shortcuts as a projection to change feature dimensions. Considering the deep structure deployed in proposed model, residual learning is also utilized to improve the efficiency of training process.

3 Experiment

3.1 Model Structure

Inspired by 'the deeper, the better' [6], proposed network, as shown in Fig. 1, deploys a plain structure with d layers ($d = 25$). The first layer casts as the feature extraction and representation operating on the input patches. 64 filters with size of 3×3 are utilized in this layer. Subsequent layers, except the last, dedicate to learn the end-to-end mapping from LR patch to its relative HR patch. 64 filters with size $3 \times 3 \times 64$ are deployed in these layers. The last layer is aimed to reconstruct the HR image. One $3 \times 3 \times 64$ filter is implemented to reproduce the image.

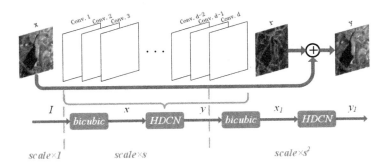

Fig. 1. HDCN and cascaded HDCN structure.

A single HDCN can only upscale the image with a fix factor s. To achieve other upscaling factors, HDCN can be cascaded as shown in Fig. 1. Considering the practicability and flexibility, s is set to 2. Thus, 4×magnification can be implemented by cascading two HDCN models and images with upscaling factor 3 can be reconstructed from 4×magnification via using bicubic interpolation. In addition, in Fig. 1, I represents the input LR image and x is the bicubic result of I. We denote the residual image as $r = y - x$ where y is the output of single HDCN. Obviously, x_1 and y_1 are the input and output image of the next HDCN.

3.2 Enhancement Strategies

Loss function: When an original NN based SR model is utilized to reconstruct images, it can be observed that the reproduce results tend to contain blurry prediction or over-smooth edge. The primary cause is one LR image patch may correspond different similar HR patches while L_2 loss function fails to recognize the complex potential relation [8]. Hence, a robust penalty function, proposed by Charbonnier et al., is introduced in proposed model to substitute for L_2 loss function.

The loss function is defined as (1):

$$L(x, R; f) = \frac{1}{N} \sum_{i=1}^{N} \sqrt{(R - f(x))^2 - \xi^2} \tag{1}$$

In (1), we denote R as the residual image, computed as the difference between the x and ground truth image. In addition, N is the amount of patches in each mini-batch and ξ is set to $1e - 3$ empirically.

Gradual training: The chosen of training images is a crucial factor of model's performance. With the structure of network going deeper, more training images are required to suffice the training process, specifically, improve the model performance and overcome overfitting. However, extensive training images arise attendant problem, retarding the rate of convergence. Meanwhile, convergence curve of training process tends to be wavery. Those small undulations indicate instability of the model in some degree. The primary cause is the inhomogeneous distribution of edge-like patterns. Patches with sharp edges are easier to be learned than other kinds of patches. Considering the extensive patches (over 700000) prepared for the experiment, gradual learning is implemented to improve the efficiency of training process. In contrast to the gradual upsampling network (GUN) proposed by Zhao et al. [14], residual learning and gradient clipping strategies are retained to expedite rate of convergence. In addition, edge-like patches are classified by means of the ALGD, computed as (2):

$$V_{ALGD} = \sum_{p=1}^{N_p} |G_p - \bar{G}| \tag{2}$$

where N_p denotes the amount of pixels in an image patch and G_p ($p = 1, 2, \cdots, N_p$) is the gray value of according pixel while $\bar{G}$ represents the average gray value of that patch. The ALGD of the whole training samples is also calculated and denoted as $\overline{V_{ALGD}}$. Hence, the evaluation parameter δ of each patch can be calculated as (3):

$$\delta = V_{ALGD}/\overline{V_{ALGD}} \tag{3}$$

The essential of the gradual learning is the learning process from easy to difficult. In other words, the model would achieve better performance gradually. The rate of convergence will be accelerated due to the reduction of instable undulations. While selecting edge-like patches, the ALGD of residual images is also taken into consideration to ensure the simpleness. We denote δ_r as the ALGD of residual patches and δ_o presents the ALGD of original patches. The training

Table 1. Discriminant condition on
division of training set

Subset	Discriminant condition
Part 1	$\delta_o \geqslant 1.2 > \delta_r \geqslant 0.8$
Part 2	$\delta_o, \delta_r \geqslant 1.2$ or $1.2 \geqslant \delta_o \geqslant 0.8 \geqslant \delta_r \geqslant 0.5$
Part 3	Reminder

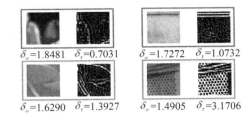

$\delta_o=1.8481$ $\delta_r=0.7031$ $\delta_o=1.7272$ $\delta_r=1.0732$

$\delta_o=1.6290$ $\delta_r=1.3927$ $\delta_o=1.4905$ $\delta_r=3.1706$

Fig. 2. δ_o and δ_r of training samples

set is divided into three parts in accordance with the discriminant condition as
shown in Table 1. These parameters are set by referring to [14].

In Fig. 2, we demonstrate several image samples. Patches in blue border are
easier to be learned by model than patches in green one, whereas their δ_o are
both greater than 1.2. It indicates the significance of deploying the ALGD of
residual patches.

Those images with small parameters are combined in one subset for their
tiny contribution on improving performance. During gradual learning, sharp-
edge patches of first part are prior utilized to form the initial training set. After
temporary convergence, we then feed second subset to the network and it can be
observed in Fig. 3 that the reproduce image becomes brilliant. Ultimately, those
remainder with small parameters are added to fine-tune the model. However,
the improvement is not perceptible for human visual. Details are demonstrated
in the next chapter.

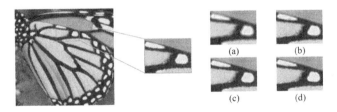

Fig. 3. 2×magnification results in each state. (a) Bicubic (b) training with Part 1(c)
training with Part 2 (d) training with remainder.

4   Implement Details

Training dataset: The origin training set is comprised of two datasets, 91
images from Yang et al. [12] and 200 images from Berkeley Segmentation Dataset
[9]. VDSR and RFL [10] employ the identical dataset. Data augmentation is
a crucial measure of overcoming the overfitting problem. The training data is
augmented through three approaches. (1) Rotation: images can be rotated by
90, 180 or 270. Other degrees are not adopted for unnecessary imresize process.

(2) Flipping: flip images horizontally and vertically. (3) Scaling: randomly down-scale between [0.5, 1.0]. Limited by hardware, we operate the augmentation on Yang's dataset and part of Berkeley's. Ultimately, training set consists of approximately 2000 images.

Test dataset: SISR experiments are carried on four datasets: Set5 [1], Set14 [13], urban100 [4], B100. Set5 and Set14 are two benchmarks used for years. Urban100 and B100 are adopted extensively among recent state-of-the-art papers.

Parameters settings: As other mature models, momentum parameter is set to 0.9 and the weight decay is $1e-4$. The initial learning rate is set to 0.1 and then decreases by a factor of 10 every 10 epochs (not less than 0.001).

The model converges very quickly with the first part of training set, approximately 3–5 epochs. We then feed the second part, consists of over 200000 image patches, to the network and decrease the learning rate to 0.01 during the training process and it costs 12–15 epochs to converge. Ultimately the remainder is utilized to fine-tune the model and the learning rate drops to 0.001 after 20 epochs. The total training process requires 30–35 epochs and roughly costs 30 h on a personal computer using a GTX 970. In addition, image patches with $\delta_r<0.5$ are deserted due to their residual images are almost dark completely. It can be observed in Fig. 4 that the model with gradual learning converges faster and performs better than origin model.

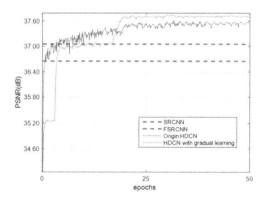

Fig. 4. Convergence analysis on the gradual learning.

Proposed model is implemented underwith Caffe[5]. Each mini-batch consists of 64 sub-images with size of 51*51. An epoch has 7812 iterations.

Experiment results: Proposed model is compared to 6 state-of-the-art SR methods: A+[11], RFL, SRCNN, FSRCNN, DRCN [7], VDSR.

In Figs. 5 and 6, it can be observed that proposed model reduces blurry prediction, which proves the effectiveness of deploying robust loss function. In addition, Table 2 shows that proposed model can achieve higher accuracy in most

Table 2. Average PSNR/SSIM for 2×, 3× and 4×magnification. Red indicates the best and blue indicates the second best performance.

Set	Scale	A+ PSNR/SSIM	RFL PSNR/SSIM	SRCNN PSNR/SSIM	FSRCNN PSNR/SSIM	DRCN PSNR/SSIM	VDSR PSNR/SSIM	Ours PSNR/SSIM
Set 5	2×	36.54/0.9544	36.54/0.9537	36.66/0.9542	36.99/0.9553	37.63/0.9588	37.53/0/9576	37.67/0.9613
	3×	32.58/0.9088	32.43/0.9057	32.75/0.9090	33.16/0.9169	33.82/0.9226	33.66/0.9213	33.84/0.9232
	4×	30.28/0.8603	30.14/0.8548	30.48s/0.8628	30.71/0.8654	31.53/0.8854	31.35/0.8821	31.54/0.8858
Set 14	2×	32.28/0.9056	32.26/0.9040	32.42/0.9063	32.73/0.9092	33.04 /0.9118	33.03/0.9124	33.14/0.9156
	3×	29.13/0.8188	29.05/0.8164	29.28/0.8209	29.43/0.8272	29.79/0.8311	29.77/0.8314	29.87/0.8345
	4×	27.32/0.7491	27.24/0.7451	27.49/0.7503	27.70/0.7557	28.02/0.7670	28.01/0.7674	28.10/0.7690
B100	2×	31.22/0.8872	31.16/0.8853	31.36/0.8875	31.51/0.8914	31.85/0.8942	31.90/0.8960	31.92/0.8965
	3×	28.29/0.7835	28.22/0.7806	28.41/0.7863	28.58/0.7905	28.80/0.7963	28.82/0.7976	28.81/0.7973
	4×	26.82/0.7103	26.75/0.7071	26.91/0.7124	26.97/0.7136	27.23/0.7233	27.29/0.7254	27.25/0.7246
urban 100	2×	29.23/0.8937	29.13/0.8905	29.52/0.8949	29.87/0.9007	30.75/0.9133	30.76/0.9137	30.83/0.9139
	3×	26.03/0.7973	25.86/0.7900	26.24/0.8088	26.40/0.8117	27.15/0.8276	27.14/0.8279	27.18/0.8286
	4×	24.34/0.7201	24.20/0.7113	24.53/0.7238	24.61/0.7273	25.14/0.7510	25.18/0.7534	25.20/0.7535

of test set. Meanwhile, experiment results in B100 also indicate the existence of error accumulation during repetitive upscaling. However, cascaded model is more flexible to reconstruct images with larger magnification. Considering the Laplacian Pyramid Networks model [8], termed LapSRN, can also reproduce image with larger scale, proposed model is compared to LapSRN and VDSR. In Fig. 7, proposed model and LapSRN can both reconstruct image with less blurry prediction in virtue of the robust loss function, while proposed model performs better due to the model structure and gradual learning strategy.

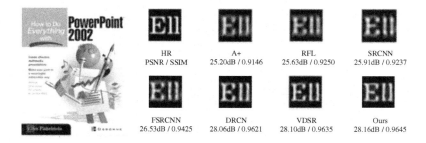

Fig. 5. Super-resolution results of 'ppt3' (Set14) with scale factor×3.

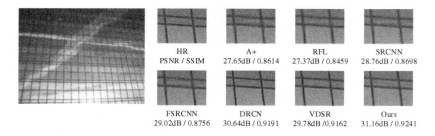

Fig. 6. Super-resolution results of 'img_055' (urban100) with scale factor×4.

Fig. 7. Super-resolution results of '3096' (B100) with scale factor×8.

5   Conclusions and Future Perspectives

In this article, we present a high accuracy SR method using deep convolution neural network. Proposed model can reduce the blurry prediction and reconstruct images with larger scale. After a more stable training process, proposed model achieves higher accuracy than some state-of-the-art works. Our further work is to combine other image process techniques, such as object recognition and optical character recognition (OCR), into SISR model. We believe that the more useful information is involved, the more brilliant reconstruct images we can obtain.

Acknowledgment. The paper is supported in part by the Natural Science Foundation of China (No. 61672022 and No. 61272036), the Key Discipline Foundation of Shanghai Second Polytechnic University (No. XXKZD1604), and the Graduate Innovation Program (No. A01GY17F022).

References

1. Bevilacqua, M., Roumy, A., Guillemot, C., Alberi Morel, M.L.: Low-complexity single-image super-resolution based on nonnegative neighbor embedding. In: British Machine Vision Conference (BMVC), pp. 1–12 (2012)
2. Dong, C., Loy, C.C., He, K., Tang, X.: Learning a deep convolutional network for image super-resolution. In: Fleet, D., Pajdla, T., Schiele, B., Tuytelaars, T. (eds.) ECCV 2014. LNCS, vol. 8692, pp. 184–199. Springer, Cham (2014). doi:10.1007/978-3-319-10593-2_13
3. Dong, C., Loy, C.C., Tang, X.: Accelerating the super-resolution convolutional neural network. In: Leibe, B., Matas, J., Sebe, N., Welling, M. (eds.) ECCV 2016. LNCS, vol. 9906, pp. 391–407. Springer, Cham (2016). doi:10.1007/978-3-319-46475-6_25
4. Huang, J.B., Singh, A., Ahuja, N.: Single image super-resolution from transformed self-exemplars. In: 2015 IEEE Conference on Computer Vision and Pattern Recognition (CVPR), pp. 5197–5206 (2015)
5. Jia, Y., Shelhamer, E., Donahue, J., Karayev, S., Long, J., Girshick, R.B., Guadarrama, S., Darrell, T.: Caffe: convolutional architecture for fast feature embedding. CoRR abs/1408.5093 (2014)
6. Kim, J., Lee, J.K., Lee, K.M.: Accurate image super-resolution using very deep convolutional networks. CoRR abs/1511.04587 (2015)
7. Kim, J., Lee, J.K., Lee, K.M., undefined, undefined, undefined, undefined: deeply-recursive convolutional network for image super-resolution. In: 2016 IEEE Conference on Computer Vision and Pattern Recognition (CVPR), pp. 1637–1645 (2016)
8. Lai, W., Huang, J., Ahuja, N., Yang, M.: Deep laplacian pyramid networks for fast and accurate super-resolution. CoRR abs/1704.03915 (2017)
9. Martin, D., Fowlkes, C., Malik, J., Tal, D.: A database of human segmented natural images and its application to evaluating segmentation algorithms and measuring ecological statistics. In: IEEE International Conference on Computer Vision, vol. 02, p. 416 (2001)
10. Schulter, S., Leistner, C., Bischof, H.: Fast and accurate image upscaling with super-resolution forests. In: 2015 IEEE Conference on Computer Vision and Pattern Recognition (CVPR), pp. 3791–3799 (2015)

11. Timofte, R., De Smet, V., Van Gool, L.: A+: adjusted anchored neighborhood regression for fast super-resolution. In: Cremers, D., Reid, I., Saito, H., Yang, M.-H. (eds.) ACCV 2014. LNCS, vol. 9006, pp. 111–126. Springer, Cham (2015). doi:10.1007/978-3-319-16817-3_8
12. Yang, J., Wright, J., Huang, T.S., Ma, Y.: Image super-resolution via sparse representation. Trans. Img. Proc. **19**(11), 2861–2873 (2010)
13. Zeyde, R., Elad, M., Protter, M.: On single image scale-up using sparse-representations. In: Boissonnat, J.-D., Chenin, P., Cohen, A., Gout, C., Lyche, T., Mazure, M.-L., Schumaker, L. (eds.) Curves and Surfaces 2010. LNCS, vol. 6920, pp. 711–730. Springer, Heidelberg (2012). doi:10.1007/978-3-642-27413-8_47
14. Zhao, Y., Wang, R., Dong, W., Jia, W., Yang, J., Liu, X., Gao, W.: GUN: gradual upsampling network for single image super-resolution. CoRR abs/1703.04244 (2017)

An Improved Density Peak Clustering Algorithm

Jian Hou[1]([✉]) and Xu E[2]

[1] College of Engineering, Bohai University, Jinzhou 121013, China
dr.houjian@gmail.com
[2] College of Information Science, Bohai University, Jinzhou 121013, China

Abstract. Density based clustering is an important clustering approach due to its ability to generate clusters of arbitrary shapes. Among density based clustering algorithms, the density peak (DP) based algorithm is shown to a potential one with some attractive properties. The DP algorithm calculates the local density of each data, and then the distance of each data to its nearest neighbor with higher density. Based on these two measurements, the cluster centers can be isolated from the non-center data. As a result, the cluster centers can be identified relatively easily and the non-center data can be grouped into clusters efficiently. In this paper we study the influence of density kernels on the clustering results and present a new kernel. We also present a new cluster center selection criterion based on distance normalization. Our new algorithm is shown to be effective in experiments on ten datasets.

Keywords: Density peak · Cluster center · Distance normalization

1 Introduction

Clustering refers to the task of grouping data into clusters based on their similarity, so that the similar data are in the same cluster and dissimilar ones are in different clusters. While there are already a large amount of clustering algorithms of different types, in this paper our work is based on the density peak (DP) based clustering algorithm presented in [26].

The DP algorithm is a density based clustering approach proposed recently and is shown to be attractive in clustering tasks. However, the original DP algorithm cannot be regarded as a reliable clustering approach. First, the density kernel and involved parameters have a significant influence on the density calculation, and then on the final clustering results. In [26] the cutoff kernel and Gaussian kernel are tested, and it is found that both the kernel types and cutoff distance impact on the clustering results. This means that we may need a careful tuning process to obtain satisfactory clustering results. Second, while cluster centers can be isolated from non-center data based on ρ and δ in theory, the identification of cluster centers in practical tasks is much more difficult than it seems. In order to avoid determining thresholds for both ρ and δ, [26] proposes

© Springer International Publishing AG 2017
H. Yin et al. (Eds.): IDEAL 2017, LNCS 10585, pp. 211–221, 2017.
https://doi.org/10.1007/978-3-319-68935-7_24

to use $\gamma = \rho\delta$ as the sole criterion of cluster center selection, and data with large γ have larger probability to be identified as cluster centers. This method is found to be afflicted by three major problems. The first is that by multiplying ρ by δ, some non-center data with very large ρ or very large δ may have a large γ and therefore be recognized as cluster centers by mistake. Even if the non-center data have smaller γ than cluster centers, their difference may not be evident enough and the cluster centers cannot be identified automatically. In this case, we need the number of clusters to select the cluster centers. A third problem is that this criterion fails to take the density difference among clusters into account, and the centers of small-density clusters have a small chance to be identified correctly.

In this paper we improve the original DP algorithm to obtain better clustering results. In order to solve the problems caused by density difference among clusters, [15] proposes to normalize the density ρ based on the density of neighboring data. In this paper we resort to a different approach to solve this problem. By studying the relationship between δ and cluster centers, we propose to use the normalized δ as the criterion of cluster center selection, in contrast to [15] where $\gamma = \rho\delta$ is still the final criterion. In addition, we present a new density kernel in density calculation. We experiment with ten datasets and validate the effective of the proposed approach.

2   Related Works

By grouping a set of data into a number of clusters, the implicit data distribution pattern can be identified and utilized in further processing. Data clustering has wide application in various fields including data mining, pattern recognition, machine learning and image processing. Typical examples include the application in social networks [8,17], music [7], virtual worlds [9] and bioinformatics [23].

A vast amount of clustering algorithms have been proposed and some of them are reviewed briefly here. The k-means algorithm is one of the most simple and most commonly used clustering approach, and a lot of variants [18,19,21,22] have been proposed to improve on the original one. The DBSCAN [5] algorithm is one of the most popular density based clustering algorithms. The normalized cuts (NCuts) algorithm [27] is a popular spectral clustering approach and has been used widely as a benchmark of image segmentation methods. Another important work in the field builds robust graphs to make use of the data similarity information [32]. Similar to NCuts, many graph-based algorithms have been proposed to utilize the information encoded in the pairwise data similarity matrix [1,17,20]. The affinity propagation (AP) [2] algorithm passes among the data the affinity message encoded in the pairwise similarity and identifies cluster centers and cluster members. In contrast to the above-mentioned algorithms, the dominant sets (DSets) algorithm [25] defines a dominant set as a graph-theoretic concept of a cluster, and extracts the clusters sequentially. Based on its nice properties, this algorithm has obtained many successful applications and further works include [11–14,24,28,29].

The DP algorithm requires as input the pairwise data distance matrix, which contains the major information needed by this algorithm. At the beginning, we calculate the local density ρ of each data based on the distance matrix. This step can be accomplished with the cutoff kernel, Gaussian kernel or other density kernels. Then For each data, we compute its distance δ to the nearest neighbor with higher density. By intuition, the cluster centers are usually the data with large density, and they are relatively far away from each other. This implies that cluster centers are often with both large ρ and large δ. In contrast, the non-center data are usually with either small ρ or small δ. This observation enables us to isolate cluster centers from non-center data relatively easily. Another important contribution of the DP algorithm is an efficient method of grouping non-center data. It assumes that each data should be in the same cluster as its nearest neighbor with higher density. While this method has no sound theoretical foundation, it is shown to be effective in experiments.

3 Density Peak Clustering

The DP algorithm is based on the observation that cluster centers are usually high-density data surrounding by low-density ones. This observation has the following implications. First, a cluster center is the density peak in its cluster since its density is larger than those of the non-center data in its cluster. Second, a cluster center usually has a large δ as its nearest neighbor with higher density is in a different cluster. Here we see that cluster centers have both large ρ and large δ. On the contrary, the non-center data are often close to their nearest neighbor with larger density in the same cluster and their δ's are small. While the data far away from others may have large δ's, their density ρ's are usually small due to their isolation from the majority of data. In summary, the non-center data cannot have large ρ and large δ in the same time in general. This difference between cluster centers and non-center data can be utilized to identify cluster centers. In the next step, we group non-center data into clusters based on the assumption that a data has the same label as its nearest neighbor with higher density. This assumption can be supported by the observation that in a cluster non-center data have smaller density than the cluster center.

In the first step we need to calculate the local density ρ of each data. In [26] two density kernels, i.e., the cutoff kernel and Gaussian kernel, are used for this task. The cutoff kernel calculates the density of one data i as

$$\rho_i = \sum_{j \in S, j \neq i} \chi(d_c - d_{ij}), \tag{1}$$

where S denotes the set of data to be clustered, $d_c \in R$ is the cutoff distance which is determined beforehand, $d_{ij} \in R$ stands for the distance between i and j, and

$$\chi(x) = \begin{cases} 1, & x > 0, \\ 0, & x < 0. \end{cases} \tag{2}$$

Evidently, with the cutoff kernel the density is measured by the number of data in the neighborhood of radius d_c. While the cutoff kernel considers only the data in a neighborhood, the Gaussian kernel takes all the data into account and calculates the density by

$$\rho_i = \sum_{j \in S, j \neq i} exp(-\frac{d_{ij}^2}{d_c^2}). \tag{3}$$

Noticing that with both kernels the parameter d_c is involved, [26] proposes to determine d_c to include 1% to 2% of all data into the neighborhood on average. With the local density of each data, the distance δ_i to the nearest neighbor with higher density is calculated by definition as

$$\delta_i = \min_{j \in S, \rho_j > \rho_i} d_{ij}. \tag{4}$$

In order to identify cluster centers based on the ρ's and δ's of the data, [26] represents the data in a so-called $\rho - \delta$ decision graph. We use the R15 [30] dataset to illustrate this graph in Fig. 1, where Fig. 1(a) shows the $\rho - \delta$ decision graphs with the cutoff kernel. In this paper we determine d_c by including 1.6% of the data in the neighborhood on average. It can be observed in the $\rho - \delta$ decision graphs that there are 15 data points far away from the others. Evidently the 15 data should be regarded as cluster centers and all the others are non-center data. Since the $\rho - \delta$ decision graph involves two thresholds in identifying cluster centers, [26] further proposes to use $\gamma_i = \rho_i \delta_i$ as the single criterion and represents the data in the γ decision graph, as illustrated in Fig. 1(b). In the γ decision graph the data are sorted in the decreasing order according to γ, and those with the largest γ's will be identified as cluster centers. Finally, the clustering result is shown in Fig. 1(c), which is quite close to the ground truth.

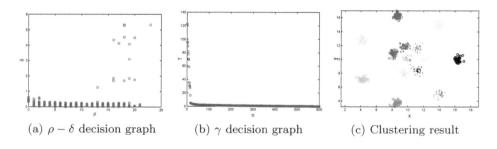

(a) $\rho - \delta$ decision graph   (b) γ decision graph   (c) Clustering result

Fig. 1. The $\rho - \delta$ decision graph, γ decision graph and clustering results with the cutoff kernel on the R15 dataset.

In theory we can identify cluster centers from either the $\rho - \delta$ decision graph or γ decision graph as those data with both large ρ's and large δ's. However, in practical tasks there is no clear distinction between the "large" and "small". If we do not know the number of clusters, we are actually not sure if all the 15 data or a subset of them in Fig. 1 should be regarded as cluster centers.

In other words, while in theory the cluster centers can be identified automatically, this problem is actually not solved in [26].

Considering the difficulty in identifying cluster centers automatically, in this paper we assume that the number of clusters is determined beforehand. In this condition, we can use γ as the single criterion to select cluster centers. This method may work well in some cases, as illustrated on the R15 dataset. However, in more complex conditions, including data with very large ρ or very large δ, and different clusters have quite different density, this simple criterion may not work well. In order to improve this algorithm, in this paper we propose to normalize the δ of data, and use the normalized δ as the single criterion to identify cluster centers. In addition, we also study a new density kernel. The details of these works are presented in the next section.

4   Our Algorithm

In the last section we see that on the R15 dataset, the cutoff kernel is able to generate very good clustering results, on condition that the number of clusters is given. However, this does not mean that it performs well on other datasets too. For example, we apply the cutoff kernel to the Spiral dataset [3], and show the selected cluster centers and corresponding clustering results in Fig. 2(a) and (d). On this dataset the selected cluster centers are in different clusters. However, the clustering result is not satisfactory. Our explanation of this observation is that this kernel generates unsatisfactory ρ's, δ's and γ's, and then lead to the unsatisfactory clustering result.

4.1   New Density Kernel

The cutoff kernel uses the number of clusters in a neighborhood to measure the density of one data. This kernel ignores the distance to the neighboring data and may cause information loss. In contrast, the Gaussian kernel takes all the data into account and assigns a weight to each data. Between these two extremes, we propose to measure the density by the average distance to a number of neighboring data. In implementation, we firstly calculate the average distance to five nearest neighbors, and uses the reverse of this distance as the density ρ. With this new density kernel, the cluster centers and clustering results on Spiral and Flame datasets are reported in Fig. 2(b) and (e). It is evident to see that with the new density kernel, the clustering results are improved significantly.

4.2   Distance Normalization

In the original DP algorithm we use $\gamma = \rho\delta$ as the cluster center selection criterion, and regard cluster centers as with both large ρ and large δ. However, in a cluster, it is very likely that the cluster center and its nearest neighbors are all with large density. As a result, the density ρ seems not very useful in differentiate cluster centers from non-center data.

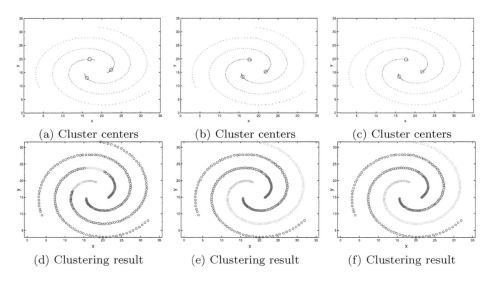

(a) Cluster centers   (b) Cluster centers   (c) Cluster centers

(d) Clustering result   (e) Clustering result   (f) Clustering result

Fig. 2. The selected cluster centers and clustering result with different kernels and criterions on the Spiral dataset. The left column belongs to the cutoff kernel, and the middle column belongs to the new density kernel. The right column is generated with the new density kernel and normalized δ.

In this paper, we present a new criterion based on the observation that δ is more informative than ρ. We have a look at a cluster center and its neighboring data. Normally, a cluster center and its neighboring data are all with large density. Therefore rho is not very informative in differentiating between cluster centers and non-center data. In fact, we observe in Fig. 1 that the ρ's of all data are distributed rather evenly in the whole local density range. In contrast, only a small number of data are with large δ's, and the δ's of most data are very small. This means that δ is much more effective than ρ in differentiating between cluster centers and non-center data.

In order to improve the discriminative ability of δ further, we study the distribution of δ's of the data around the cluster centers. The cluster centers are with large δ's, and its neighbors are non-center data with smaller δ's. In contrast, the non-center data are with small δ's, and the δ's of their neighbors are usually slightly different. If we normalize the δ of one data based on the δ's of its neighboring data, the difference between cluster centers and non-center data can be enlarged. Specifically, we normalize the δ_i of each data i by

$$\delta_i' = \frac{\delta_i}{\frac{1}{|D_{inn}|} \sum_{j \in D_{inn}} \delta_j},$$ (5)

and then use δ' as the single criterion of cluster center selection. Evidently with this criterion the cluster centers are with large $delta'$'s, whereas those of non-center are usually small. With this new criterion, the cluster center selection and clustering results on the Spiral and Flame datasets are shown in Fig. 2(c) and (f). By comparing with the case with only the new density kenrel, we find that the normalized δ criterion generates basically the same results as γ. This means that the normalized δ can be used as an alternative to γ in selecting cluster centers.

5   Experiments

In this section we do experiments to validate the effectiveness of the proposed algorithms. The experiments are conducted on ten datasets, including Aggregation [6], Compound [31], Pathbased [3], Spiral, D31 [30], R15 [30], Jain [16], Flame and two UCI datasets Iris and Glass. We use Normalized Mutual Information (NMI) to evaluate the clustering results.

First, we compare the new density with the cutoff and Gaussian kernels. In this part, we still use γ to select the cluster centers. The clustering result comparison among these three kernels are reported in Fig. 3. It can be observed that on the majority of datasets, our kernel performs better than or comparably to the better-performing one in the other two kernels. At the same time, we notice that our kernel is outperformed by both the cutoff and Gaussian kernels on D5 and D7. We attribute this observation to the fact the our kernel uses a fixed number of nearest neighbors to calculate local density on different datasets. This may cause unsatisfactory results in some cases, for example, the D5 and D7 datasets in our experiments. Consequently, we plan to study the possibility of an adaptive neighborhood in the future work.

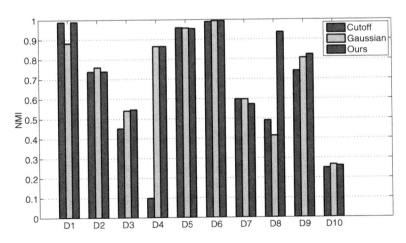

Fig. 3. The comparison of three density kernels in clustering on ten datasets. Here D1, D2, $\cdots$, D10 represent the ten datasets in the order of Aggregation, Compound, Pathbased, Spiral, D31, R15, Jain, Flame, Iris and Glass.

We then compare the whole algorithm with several other algorithms, including k-means, NCuts, DBSCAN, AP, DSets, DSets-DBSCAN [10], DP-c (DP with cutoff kernel) and DP-G (DP with Gaussian kernel). For k-means and NCuts, we set the number of clusters as ground truth. In DBSCAN, we use $MinPts = 3$ and determine Eps with the method proposed in [4]. With AP algorithm, the preference value p is determined in the following way. We firstly calculate the range of p $[p_{min}, p_{max}]$ with the code published by the authors of [2]. Then p is selected to be $p_{min} + 9.2step$, where $step = (p_{max} - p_{min})/10$. With the DSets algorithm the data similarity is calculated by $s(i, j) = exp(-d(i, j)/\sigma)$ and σ is set as $20\bar{d}$, where $d(i, j)$ refers to the Euclidean distance and $\bar{d}$ stands for the mean of all pairwise distances. The DSets-DBSCAN is a parameter free algorithm and no parameter tuning is needed. With both DP-c and DP-G, the parameter d_c is determined by including 1.6% of all data in the neighborhood on average. The comparison of these algorithms is reported in Table 1.

Table 1. Comparison of different algorithms on ten datasets with NMI.

	DSets	k-means	NCuts	DBSCAN	AP	DSets-DBSCAN	DP-c	DP-G	Ours
Aggregation	0.86	0.85	0.76	0.92	0.82	0.89	0.99	0.88	0.99
Compound	0.75	0.72	0.63	0.89	0.81	0.92	0.74	0.76	0.74
Pathbased	0.76	0.55	0.50	0.64	0.54	0.82	0.45	0.54	0.55
Spiral	0.14	0.00	0.00	0.71	0.00	0.66	0.10	0.86	0.86
D31	0.85	0.94	0.96	0.84	0.59	0.67	0.96	0.96	0.95
R15	0.83	0.92	0.99	0.87	0.74	0.91	0.99	0.99	0.99
Jain	0.43	0.37	0.33	0.73	0.46	0.87	0.60	0.60	0.57
Flame	0.60	0.40	0.42	0.83	0.57	0.90	0.49	0.41	0.94
Iris	0.65	0.76	0.74	0.75	0.79	0.60	0.74	0.81	0.82
Glass	0.31	0.37	0.37	0.40	0.33	0.37	0.25	0.27	0.26
Average	0.62	0.59	0.57	0.76	0.57	0.76	0.63	0.71	0.77

We arrive at some conclusions from Table 1. For each algorithm, the clustering results vary significantly on different datasets, and there is no one algorithm which outperform another consistently. This shows the difficulty in obtaining an universal clustering algorithm. In this case, our algorithm performs fairly evenly on different datasets, and the average clustering quality is better than those of other algorithms. This validates the effectiveness of the proposed algorithm.

As one kind of graph-based clustering algorithms, our approach requires the pairwise similarity (distance) matrix as the input, indicting a large computation load. On the basis of the original DP algorithm, our approach introduces a distance normalization step, which adds to the computation burden. Future work will be devoted to developing more efficient alternatives to the current normalization method.

In this paper we validate the proposed algorithm with a number of relatively small datasets. Noticing that the original DP algorithm has been employed in some real world problems, we believe that our improved algorithm can be applied to real world problems as well. This will be one of the directions in our future work.

6   Conclusions

On the basis of the promising density peak based clustering algorithm, in this paper we present an improved algorithm for better performance. First, we study the influence of density kernels on the clustering results, and present a new density which is based on the distance to neighboring data. Then, we investigate the criterions used in cluster center selection, and proposed to use normalized δ distance to replace the original criterion. Finally, we conduct experiments on ten datasets and compare our algorithm with eight algorithms. Experimental results validate the effectiveness of our algorithm.

Acknowledgement. This work is supported in part by the National Natural Science Foundation of China under Grant No. 61473045, and by the Natural Science Foundation of Liaoning Province under Grant Nos. 20170540013 and 20170540005.

References

1. Bello-Orgza, G., Camacho, D.: Evolutionary clustering algorithm for community detection using graph-based information. In: IEEE Congress on Evolutionary Computation, pp. 930–937 (2014)
2. Brendan, J.F., Delbert, D.: Clustering by passing messages between data points. Science **315**, 972–976 (2007)
3. Chang, H., Yeung, D.Y.: Robust path-based spectral clustering. Pattern Recogn. **41**(1), 191–203 (2008)
4. Daszykowski, M., Walczak, B., Massart, D.L.: Looking for natural patterns in data: Part 1. density-based approach. Chemometr. Intell. Lab. Syst. **56**(2), 83–92 (2001)
5. Ester, M., Kriegel, H.P., Sander, J., Xu, X.W.: A density-based algorithm for discovering clusters in large spatial databases with noise. In: International Conference on Knowledge Discovery and Data Mining, pp. 226–231 (1996)
6. Gionis, A., Mannila, H., Tsaparas, P.: Clustering aggregation. ACM Trans. Knowl. Discov. Data **1**(1), 1–30 (2007)
7. González-Pardo, A., Granados, A., Camacho, D., de Borja Rodríguez, F.: Influence of music representation on compression-based clustering. In: IEEE Congress on Evolutionary Computation, pp. 1–8 (2010)
8. González-Pardo, A., Jung, J.J., Camacho, D.: Aco-based clustering for ego network analysis. Future Generat. Comput. Syst. **66**, 160–170 (2017)
9. González-Pardo, A., Ortíz, F.B.R., Pulido, E., Fernández, D.C.: Influence of music representation on compression-based clustering. In: ACM Workshop on Surreal Media and Virtual Cloning, pp. 9–14 (2010)

10. Hou, J., Gao, H., Li, X.: DSets-DBSCAN: a parameter-free clustering algorithm. IEEE Trans. Image Process. **25**(7), 3182–3193 (2016)
11. Hou, J., Gao, H., Xia, Q., Qi, N.: Feature combination and the kNN framework in object classification. IEEE Trans. Neural Netw. Learn. Syst. **27**(6), 1368–1378 (2016)
12. Hou, J., Liu, W., Xu, E., Cui, H.: Towards parameter-independent data clustering and image segmentation. Pattern Recogn. **60**, 25–36 (2016)
13. Hou, J., Pelillo, M.: A simple feature combination method based on dominant sets. Pattern Recogn. **46**(11), 3129–3139 (2013)
14. Hou, J., Qi, X., Qi, N.M.: Experimental study on dominant sets clustering. IET Comput. Vision **9**(2), 208–215 (2015)
15. Hou, J., Pelillo, M.: A new density kernel in density peak based clustering. In: International Conference on Pattern Recognition, pp. 463–468 (2016)
16. Jain, A.K., Law, M.H.C.: Data clustering: a user's dilemma. In: Pal, S.K., Bandyopadhyay, S., Biswas, S. (eds.) PReMI 2005. LNCS, vol. 3776, pp. 1–10. Springer, Heidelberg (2005). doi:10.1007/11590316_1
17. Menéndez, H.D., Barrero, D.F., Camacho, D.: Adaptive k-means algorithm for overlapped graph clustering. Int. J. Neural Syst. **22**(5), 1250018 (2012)
18. Menéndez, H.D., Barrero, D.F., Camacho, D.: A multi-objective genetic graph-based clustering algorithm with memory optimization. In: IEEE Congress on Evolutionary Computation, pp. 3174–3181 (2013)
19. Menéndez, H.D., Barrero, D.F., Camacho, D.: A co-evolutionary multi-objective approach for a k-adaptive graph-based clustering algorithm. In: IEEE Congress on Evolutionary Computation, pp. 2724–2731 (2014)
20. Menéndez, H.D., Barrero, D.F., Camacho, D.: A genetic graph-based approach for partitional clustering. Int. J. Neural Syst. **24**(3), 1430008 (2014)
21. Menéndez, H., Camacho, D.: A genetic graph-based clustering algorithm. In: Yin, H., Costa, J.A.F., Barreto, G. (eds.) IDEAL 2012. LNCS, vol. 7435, pp. 216–225. Springer, Heidelberg (2012). doi:10.1007/978-3-642-32639-4_27
22. Menéndez, H.D., Otero, F.E.B., Camacho, D.: MACOC: a medoid-based ACO clustering algorithm. In: Dorigo, M., Birattari, M., Garnier, S., Hamann, H., Montes de Oca, M., Solnon, C., Stützle, T. (eds.) ANTS 2014. LNCS, vol. 8667, pp. 122–133. Springer, Cham (2014). doi:10.1007/978-3-319-09952-1_11
23. Menéndez, H.D., Plaza, L., Camacho, D.: Combining graph connectivity and genetic clustering to improve biomedical summarization. In: IEEE Congress on Evolutionary Computation, pp. 2740–2747 (2014)
24. Zemene, E., Pelillo, M.: Interactive image segmentation using constrained dominant sets. In: Leibe, B., Matas, J., Sebe, N., Welling, M. (eds.) ECCV 2016. LNCS, vol. 9912, pp. 278–294. Springer, Cham (2016). doi:10.1007/978-3-319-46484-8_17
25. Pavan, M., Pelillo, M.: Dominant sets and pairwise clustering. IEEE Trans. Pattern Anal. Mach. Intell. **29**(1), 167–172 (2007)
26. Rodriguez, A., Laio, A.: Clustering by fast search and find of density peaks. Science **344**, 1492–1496 (2014)
27. Shi, J., Malik, J.: Normalized cuts and image segmentation. IEEE Trans. Pattern Anal. Mach. Intell. **22**(8), 167–172 (2000)
28. Tripodi, R., Pelillo, M.: A game-theoretic approach to word sense disambiguation. Comput. Linguist. **43**(1), 31–70 (2017)
29. Vascon, S., Mequanint, E.Z., Cristani, M., Hung, H., Pelillo, M., Murino, V.: Detecting conversational groups in images and sequences: a robust game-theoretic approach. Comput. Vis. Image Underst. **143**, 11–24 (2016)

30. Veenman, C.J., Reinders, M., Backer, E.: A maximum variance cluster algorithm. IEEE Trans. Pattern Anal. Mach. Intell. **24**(9), 1273–1280 (2002)
31. Zahn, C.T.: Graph-theoretical methods for detecting and describing gestalt clusters. IEEE Trans. Comput. **20**(1), 68–86 (1971)
32. Zhu, X., Loy, C.C., Gong, S.: Constructing robust affinity graphs for spectral clustering. In: IEEE International Conference on Computer Vision and Pattern Recognition, pp. 1450–1457 (2014)

Consensus-based Parallel Algorithm for Robust Convex Optimization with Scenario Approach in Colored Network

Fan Feng and Feilong Cao$^{(\boxtimes)}$

College of Sciences, China Jiliang University,
Hangzhou 310018, Zhejiang Province, China
feilongcao@gmail.com

Abstract. This paper mainly proposes an parallel distributed learning algorithm for the robust convex optimization (RCO). Firstly, the scenario approach is used to transform RCO into its probabilistic approximation Scenario Problem (SP), which is distributively solved by multiprocessors to lighten the computational burden. Secondly, each processor (node) of the colored network processes the local optimization via a primal-dual subgradient algorithm (PDSA) to obtain an optimal solution called a local variable. Finally, a consensus method named the Colored Distributed Average Consensus (CDAC), which is based on Distributed Average Consensus (DAC), is proposed to act on the whole local variables to obtain the global optimal solution. Experimental results show that CDAC has an advantage in terms of computational time over DAC, while they have the same results.

Keywords: Distributed learning · Robust convex optimization · Scenario approach · Primal-dual subgradient algorithm · Distributed average consensus · Colored network

1 Introduction

1.1 Background

A problem named Robust Optimization (RO) is that its constraints are parameterized with uncertainties. The aim of RO is to find a solution which is not affected by the uncertainty. There are some research studies on RO. See, e.g., [1,2]. RO has been widely applied in many areas such as economic management [3], communication systems [4], and other areas [2]. In fact, RO is a semi-infinite optimization problem and cannot be efficiently solved numerically. Some traditional ways are no longer applicable, which may cause that the optimal solution becomes feasible solution even infeasible solution under the affection of the uncertainties.

This study was supported by the National Natural Science Foundation of China under Grant no. 61672477.

Robust convex optimization (RCO) is a special case of RO that uses a convex objective function. The primary method for solving RCO is transforming it into an equivalent deterministic tractable convex problem under some mild assumptions [2,5] at present. The transformed optimization can be solved by using existing convex optimization methods. However, a drawback of this primary approach is that the transformed convex optimization is often not as scalable as the nominal optimization.

This paper will apply a method called the scenario approach to solve RCO. This was studied in [6–8], where it was shown that most of the constraints of an original optimization are satisfied if the number of sampled constraints is sufficiently large. This approach randomly samples, with independent and identically distributed (i.i.d.) conditions, finite constraints from the constraints of RCO according to an arbitrary, absolutely continuous distribution to obtain a standard convex optimization called the Scenario Problem (SP). SP is a probabilistic approximation of RCO. The distinguishing feature of SP is that the number of sampled constraints can be computed in advance, which can guarantee that the solution to the SP is optimal with some given confidence [9].

1.2   Motivation and Contribution

Although RCO can be probabilistically approximated by SP, it is a computational challenge when there are a large number of constraints in SP. In particular, the more the number of sampling constraints is, the higher the computational complexity with only one processor is. Thus, in this paper, an efficient distributed learning framework with many processors (nodes) interconnected in a colored network is proposed to alleviate the computational burden of only one processor.

Our main idea can be described in three main steps:

Sampling and distributing. First, N finite constraints from the infinite constraints of RCO are randomly sampled. As how many constraints should be sampled, the theory guarantees that it is a probabilistic approximation, as given in [7,11]. Then, the N constraints are distributed to m processors (nodes) according to the computational capability. This means that processors with more computational power have more constraints distributed to them.

Solving optimization. After sampling and distribution, each processor (node) starts to solve the standard convex optimization with local constraints (called local optimization). The primal-dual subgradient algorithm [12] (PDSA) is adopted to handle the local optimizations. Actually, the algorithm is developed to invest the drawbacks of typical subgradient algorithms, and generates a pair of primal-dual iterates that are demonstrated to converge to an optimal pair of primal-dual solutions.

Although the distributed alternating direction method of multipliers (ADMM) [13,18], which has been a common method of distributed machine learning to solve general optimization and improved in terms of convergence rate by many literatures like [19,20], can be applied to solve the SP, this paper does not adopt the method. This decision was made because the variables are updated according to an optimization of the augmented Lagrangian function, which will cause that an optimization must be solved per iteration unless the updates is an explicitly expression. Obviously, it must be time-consuming and complex with regard to computation. Thus, the PDSA is much simpler and suitable.

Consensus. After each processor (node) solves its own local optimization and obtains the optimal solution, namely, local variable. The last step is to apply the Colored Distributed Average Consensus (CDAC) to local variables to obtain a global optimal solution. CDAC is based on Distributed Average Consensus (DAC). DAC is an algorithm that obtains the global average with the values of all nodes of a network [8,10]. This is actually a linear iterative algorithm. The iterative form is $x_i^{k+1} = \sum_{j \in N_i, j=i} C_{ij} x_j^k$, where C is the a coefficient matrix, and satisfies some conditions. Compared with DAC, CDAC aims to the colored network, which will result that variables of nodes with the same color can update in parallel rather must wait all nodes to finish their updates. This can save waiting time and ensure the convergence at the same time. The idea is originated from [17]. In a word, CDAC is efficient in saving computational time.

In conclusion, the proposed CDAC can not only deal with RCO with large-scale constraints in a distributed manner, but can also shorten the computational time in an efficient way.

1.3   Organization

The rest of this paper is organized as follows. Section 2 outlines RCO and the scenario approach. Section 3 deals with SP in an efficient distributed framework. Section 3.1 introduces the PDSA to solve subproblem optimization. Section 3.2 is devoted to designing the efficient learning framework in a colored network with CDAC. Section 3.3 summarizes the proposed algorithm and analyzes the convergence. Section 4 presents the results of a numerical simulation, and Sect. 5 concludes the text.

2   Preliminaries and Problem Statement

This section introduces the RCO and its probabilistic approximation (SP).

2.1   Robust Convex Optimization (RCO)

Consider a robust convex optimization (RCO) of the following form: [1]

$$\begin{aligned} \underset{x \in \Theta}{\text{minimize}} \quad & f_0(x) \\ \text{subject to} \quad & h(x,q) \le 0, \forall q \in Q, \end{aligned} \tag{1}$$

where $\Theta \subseteq \mathbb{R}^n$ is a convex, closed, and non-empty set, $f_0(x) : \mathbb{R}^n \to \mathbb{R}$ is convex, and $h(x,q) : \mathbb{R}^n \times Q \to \mathbb{R}$ is convex with respect to x for any $q \in Q \subseteq \mathbb{R}^l$. Assuming that the uncertainty q in $h(x,q)$ has no structure, except for the Borel measurability of $h(x,q)$, with any fixed x.

2.2 RCO with Scenario Approach (SP)

RCO is a semi-infinite optimization problem, and cannot be efficiently solved numerically. Instead of satisfying all constraints in (1), the core of the scenario approach is to replace the infinite constraints with finite ones. This is a probabilistic approximation and is called Scenario Problem (SP).

$$\begin{aligned}
&\underset{x \in \Theta}{\text{minimize}} && f_0(x) \\
&\text{subject to} && h(x, q^i) \le 0, \forall i = 1, 2, \ldots, N.
\end{aligned} \tag{2}$$

where $\{q^i\} \subseteq Q$ are randomly sampled from Q according to an arbitrary absolutely continuous distribution, and meet the i.i.d. conditions.

Furthermore, in order to guarantee the probabilistic approximation, the following explanations are necessary:

Assumption 1 *(see [12]). The scenario problem* (2) *is feasible, and there is no duality gap between* (2) *and its dual problem.*

Note that the Assumption 1 guarantees the existence of some dual variable of a dual problem of (2) such that the difference between the primal objective value and the dual objective value is 0. Determining the number of sampled constraints is a problem. However,[7,11] give the answer.

Lemma 1. *Assume that there exists a unique solution to* (2). *If* $\eta, \delta \in (0,1)$ *and* N *satisfy the following inequality*

$$N \ge \frac{e}{\eta(1-e)}\left(\ln \frac{1}{\delta} + n - 1\right), \tag{3}$$

where e is Euler's number and the violation probability $V(x)$ *is defined as* $V(x) = \mathbb{P}\{q \in Q | h(x, q) > 0\}$, *then the solution* x *to* (2) *satisfies* $V(x) \le \eta$ *with a probability of at least* $1 - \delta$, *i.e.,* $\mathbb{P}\{V(x) \le \eta\} \ge 1 - \delta$.

Lemma 1 indicates that the minimum constraint number N can be computed once the parameters δ, η are fixed and (2) can probabilistically approximate (1) with a absolutely small violation probability.

3 Parallel Algorithm for Scenario Problem with Colored Network

In this section, the parallel distributed learning framework for solving (2) is designed. As introduced above, solving (2) will approximately solve (1) under the proposed conditions, but the constraint number N will be very large with quite small parameters δ, η according to (3). Consequently, it will be time-consuming and result in a high computational cost. It is also a burden to use only one processor to solve (2). For these reasons, an parallel distributed framework to solve (2) is put forward.

Instead of solving (2) with one processor, utilizing m processors cooperatively to solve (2) in a distributed way is adopted. Let $G(V, E)$ be an undirected graph, $V = \{1, 2, \ldots, m\}$ be the set of nodes, and E be the set of edges of the graph G. We denote the neighbors of node i by the set $N_i = \{j | (i, j) \in E\}$. If node i and node j are directly connected, which means that they can communicate their information with each other, then $(i, j) \in E, \forall i, j \in V$. It's necessary to make an assumption to ensure the communication between nodes.

Assumption 2 *The network $G(V, E)$ is strongly connected, that is, $\forall i, j \in V$, there exists at least one path from the node i to the node j, whose length is at least 1.*

3.1 Subproblem Optimization

The main purpose of this section is to solve the subproblem with primal-dual subgradient algorithm. Let n_i denote the constraint number of nodes (processors) i. Obviously, the sum of all nodes' constraint numbers must be no less than the total constraint number N, that is $\sum_{i=1}^{m} n_i \leq N$.

Each processor needs to solve the subproblem optimization (4) to obtain its optimal solution x_i^* (called a local variable)

$$
\begin{aligned}
\underset{x_i \in \Theta}{\text{minimize}} \quad & f_0(x_i) \\
\text{subject to} \quad & h(x_i, q^{s_i}) \leq 0, \forall s_i = 1, 2, \ldots, n_i,
\end{aligned}
\tag{4}
$$

The PDSA is adopted to reach this goal. The augmented Lagrangian is

$$
L(x_i, \lambda_i) = f_0(x_i) + \lambda_i^\top H(x_i, q_i) + \frac{\rho}{2} \|H(x_i, q_i)\|_2^2,
$$

where $H(x_i, q_i) = (h(x_i, q^1)_+, \ldots, h(x_i, q^{n_i})_+), \rho > 0$, and $h(\cdot)_+ = \max(0, h(\cdot))$, $\lambda_i \in \mathbb{R}^{n_i}$ is the Lagrangian multiplier corresponding to the constraints. (x_i, λ_i) is primal-dual optimal if and only if

$$
0 \in T_i(x_i, \lambda_i) = \partial_{x_i} L(x_i, \lambda_i), \quad 0 \in P_i(x_i, \lambda_i) = -\nabla_{\lambda_i} L(x_i, \lambda_i)
$$

Then, the update of primal-dual subgradient method is

$$
x_i^{k+1} = x_i^k - \alpha_i^k T_i(x_i^k, \lambda_i^k) = x_i^k - \alpha_i^k (\partial f_0(x_i^k) + \sum_{s_i=1}^{n_i} (\lambda_{i,s_i}^k + \rho h(x_i^k, q^{s_i})_+) \partial h(x_i^k, q^{s_i})_+)
$$

$$
\lambda_i^{k+1} = \lambda_i^k - \alpha_i^k P_i(x_i^k, \lambda_i^k) = \lambda_i^k + \alpha_i^k H(x_i^k, q_i)
$$

with the step size $\alpha_i^k = \frac{\gamma^k}{\|[T_i^k; P_i^k]\|_2^2}$, where λ_{i,s_i} is the s_ith element of λ_i, γ^k satisfies the following conditions:

$$
\gamma^k > 0, \sum_k \gamma^k = \infty, \sum_k (\gamma^k)^2 < \infty.
$$

In this paper, let $\gamma^k = \frac{1}{k}$, which satisfies the conditions above.

3.2 Consensus with Colored Distributed Average Consensus

CDAC bases on DAC. DAC computes the global average with the values of all nodes. Let every node have a vector $x_i, i = 1, 2, \ldots, m$. The goal is to converge to the global average with x_is of all nodes:

$$\hat{x} = \frac{1}{m} \sum_{i=1}^{m} x_i. \tag{5}$$

DAC is an linear iterative procedure to compute (5). Let x_i^k denote the estimate of the average of node i at iteration k, and $x_i^0 = x_i$. Then, every node updates its variable according to the communication with its neighbors, i.e.,

$$x_i^{k+1} = C \cdot x_i^k = \sum_{j \in N_i, j=i} C_{ij} x_j^k, \tag{6}$$

which is the weighted sum of the previous value of its neighbors and itself. Where $C_{i,j} \neq 0$ if node i and node j are directly connected, and 0 otherwise

Connectivity matrix C is a symmetric weight matrix that is determined beforehand. It must satisfy some properties [14] to achieve asymptotic average consensus:

$$C \cdot 1 = 1, 1^T \cdot C = 1^T, \rho(C - \frac{1 \cdot 1^T}{m}) < 1,$$

where $1 \in \mathbb{R}^m$ is a column vector all ones, $\rho(\cdot)$ denotes the spectral radius of a matrix given by $\rho(M) = \max_r \{|\lambda_r(M)|\}$, $\lambda_r(M)$ is the rth eigenvector of matrix M. Only then can the DAC converge to the global average in (5), i.e., $x_i^k \to \hat{x}(i = 1, 2, \ldots, m)$ as $k \to \infty$.

To ensure convergence, a simple choice of the real-valued connectivity matrix C referred to in [15] is given by

$$C_{ij} = \begin{cases} \frac{1}{d+1} & \text{if } j \in N_i, \\ 1 - \frac{d_i}{d+1} & \text{if } j = i, \\ 0 & \text{otherwise,} \end{cases}$$

where d_i is the degree of node i, and d is the maximum degree of the graph.

The CDAC has similar procedures to those of DAC. The difference is that CDAC is focused on the colored network. Since each node communicates only with its neighbors, nodes with different colors have no communication in a colored network. Thus, if nodes with the same color update their variables in parallel, the computation time will be shortened and the efficiency of the algorithm will be improved.

Obviously, once the topology of a network is determined, the coloring scheme is fixed. Set R indicate the color number of the network, and R_r be the set of nodes with the same color $r, r = 1, 2, \ldots, R$. For simplicity but without generality, nodes are numbered such that the first $|R_1|$ nodes are in R_1, the next $|R_2|$ nodes are in R_2, and so on, where $|R_r|$ denotes the number of nodes in R_r. A simple example illustrates the above assumption and is shown in Fig. 1.

The network in Fig. 1 with $m = 11$ nodes and 17 edges is colored with $R = 4$ colors, and $R_1 = \{1, 2, 3\}, |R_1| = 3, R_2 = \{4, 5, 6, 7\}, |R_2| = 4, R_3 = \{8, 9\}, |R_3| = 2, R_4 = \{10, 11\}, |R_4| = 2$.

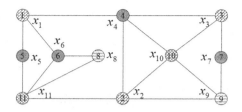

Fig. 1. Example of connected color network.

Similarly, in CDAC, every node has a measurement vector $(x_i, i = 1, 2, \ldots, m)$. From the color 1 to the color R, nodes in R_r with color $r (r = 1, 2, \ldots, R)$ update their variables in parallel by

$$x_i(t + 1) = \sum_{j \in N_i, i > j} C_{ij} x_j(t + 1) + \sum_{j \in N_i, i \leq j} C_{ij} x_j(t).$$

The coefficient C is identical to that of DAC.

3.3 Algorithm and Convergence Analysis

The processes and details of the proposed method are summarized as the following algorithm.

The proposed Algorithm 1 is ensured to be convergent for two reasons. First, both the PDSA and DAC are convergent. This was proven in [14] and [12], respectively. Second, the convergence of CDAC can also be guaranteed by [14]. In particular, DAC and CDAC are equivalent to some extent, thus the convergence is also the same.

4 Simulation and Experiment

We adopt a linear system with a single input and output. This is the same as the example in [9]. Let $u \in \mathbb{R}^n$ be input, $y \in \mathbb{R}^n$ be output, and $y = Ux$, where $U \in \mathbb{R}^{n \times n}$ is the Toeplitz matrix of input u. Suppose that the actual input is $u + \Delta u$ and the output is $y + \Delta y$. Considering that the perturbations Δu and Δy are unknown, the usual optimization models are no longer appropriate for such a problem. To simplify, let $q = (\Delta u^T, \Delta y^T) \in \mathbb{R}^{2n}$. This can be reformulated into a RCO problem as (1) to solve, and is listed as follows:

$$\min_x \|(y + \Delta y) - (U + \Delta U)x\|_2^2, \tag{7}$$

Algorithm 1 (Efficient distributed learning algorithm for RCO with scenario approach.)

1: Initialization: for all node i, set $x_i^0 = 0, \lambda_i^0 = 0, k = t = 0$.
2: **for** $i = 1, 2, \ldots, m$ [in parallel] **do**
3:    **repeat**
4:       $k = k + 1, \gamma_k = \frac{1}{k}$.
5:       compute the subgradient
6:       $T_i^k = \partial f_0(x_i^k) + \sum_{s_i=1}^{n_i} (\lambda_{i,s_i}^k + \rho h(x_i^k, q^{s_i})_+) \partial h(x_i^k, q^{s_i})_+, \; P_i^k = -H(x_i^k, q_i)$.
7:       compute the step size $\alpha_i^k = \frac{\gamma^k}{\|[T_i^k; P_i^k]\|_2^2}$.
8:       update primal and dual variable $x_i^{k+1} = x_i^k - \alpha_i^k T_i^k, \lambda_i^{k+1} = \lambda_i^k - \alpha_i^k P_i^k$.
9:    **until** a predefined stopping criterion is met. Then, obtain local variable x_i^*s.
10: **end for**
11: **repeat**
12:    **for** $c = 1, 2, \ldots, C$ **do**
13:       **for** $i \in C_c$ [in parallel] **do**
14:          every node i with color c receives the variables from node $j \in N_i$
15:          and parallelly updates its variable
16:          $x_i^*(t+1) = \sum_{j \in N_i, i > j} C_{ij} x_j^*(t+1) + \sum_{j \in N_i, i \leq j} C_{ij} x_j^*(t)$,
17:          and then broadcasts its updated variable to its neighbors.
18:       **end for**
19:    **end for**
20: **until** some stopping criterion is met. global optimal solution $\hat{x} = x_{\text{cons}}^*$.

for all $q = (\Delta u^T, \Delta y^T) \in Q$, where $Q = \{q | q \in \mathbb{R}^{2n}, \|q\|_2 \leq \rho\}$ and ρ denotes the uncertainty of data.

We reformulate (8) in the form of (2)

$$\begin{aligned} \underset{x,t}{\text{minimize}} \quad & t \\ \text{subject to} \quad & \|(y + \Delta y) - (U + \Delta U)x\|_2^2 \leq t, \forall q \in Q. \end{aligned} \tag{8}$$

Then, (8) can be solved by our method.

We set $\eta = 0.001$ and $\delta = 10^{-4}$. Thus, the minimum sample number N is 17735 according to (3). Hence, we adopt $N = 20000$ and a network with 200 nodes. Thus, every node handles 100 samples. The network is randomly generated with the nodes connected with a probability of $p = 0.5$. The uncertainty varies from 0 to 2 with a step size of 0.1. Then, similar to [16], we set the input as $u = (1, 2, 3)^T$ and the output as $y = (4, 5, 6)^T$. Given a fixed uncertainty, we define that the maximum of the residual with DAC or CDAC is

$$\text{residual} = \max_{i=1,2,\ldots,N} \|(y + \Delta y^i) - (U + \Delta U^i)x\|_2,$$

and the standard maximum residual is the maximum absolute difference of the standard solution $x = (U + \Delta U)^{-1}(y + \Delta y)$ and the optimal solution $x = U^{-1}y$. The simulation results are shown in Fig. 2.

With our method, the total computing time is around 13 h for 21 scenario. Thus, the average time in each node for a scenario is around $\frac{3600 \times 13}{200 \times 21} \approx 11$ s.

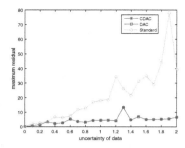

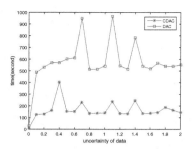

Fig. 2. Maximum residual with respect to uncertainty of data.

Fig. 3. Computing time of CDAC and DAC over uncertainty.

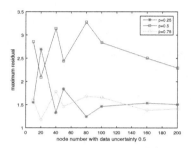

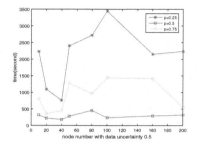

Fig. 4. Maximum residual of CDAC over node number with different network connect probability.

Fig. 5. Computing time of CDAC over node number with different network connect probability.

Figure 2 shows that DAC and CDAC have similar maximum residuals with varying uncertainty. As noted previously, CDAC has an advantage over DAC in terms of time. We show the average time over uncertainty in Fig. 3.

Moreover, we also consider that how the network connect probability p and node number m influence the maximum residual and the computing time of CDAC. As shown in Figs. 4 and 5, it must be a trade-off between lower maximum residual and less computing time, the probability and the node number is not as high as possible or as low as possible.

5   Conclusion

RCO has received considerable attention in recent decades and is widely applied in many real-world areas. In this paper, an efficient distributed learning framework to solve RCO with a scenario approach was proposed. Firstly, RCO was probabilistically approximated by SP. Secondly, each processor (node) processes the local optimization via a PDSA to obtains its own optimal solution (called a local variable). Thirdly, the improved CDAC was acted on all the local variables to obtain the global optimal solution. Compared with DAC, CDAC has an advantage in terms of computational time while has the same results.

References

1. Ben-Tal, A., Nemirovski, A.: Robust convex optimization. Math. Oper. Res. **23**, 769–805 (1998)
2. Bertsimas, D., Brown, D., Caramanis, C.: Theory and applications of robust optimization. SIAM Rev **53**, 464–501 (2011)
3. Bertsimas, D., Thiele, A.: A robust optimization approach to inventory theory. Oper. Res. **54**, 150–168 (2006)
4. Bertsimas, D., Sim, M.: Robust discrete optimization and network flows. Math. Program. **98**, 49–71 (2003)
5. Ben-Tal, A., Hertog, D., Vial, J.: Deriving robust counterparts of nonlinear uncertain inequalities. Math. Program. **149**, 265–299 (2015)
6. Calafiore, G., Campi, M.: Uncertain convex programs: randomized solutions and confidence levels. Math. Program. **102**, 25–46 (2005)
7. Campi, M., Garatti, S.: The exact feasibility of randomized solutions of uncertain convex programs. SIAM J. Optim. **19**, 1211–1230 (2008)
8. Farias, D., Roy, B.: On constraint sampling in the linear programming approach to approximate dynamic programming. Math. Oper. Res. **29**, 462–478 (2004)
9. You, K., Tempo, R.: Networked parallel algorithms for robust convex optimization via the scenario spproach. arXiv preprint. arXiv: 1607.05507 (2016)
10. Scardapane, S., Wang, D., Panella, M., Uncini, A.: Distributed learning for random vector functional-link networks. Inf. Sci. **301**, 271–284 (2015)
11. Alamo, T., Tempo, R., Luque, A., Ramirez, D.: Randomized methods for design of uncertain systems: sample complexity and sequential algorithms. Automatica **52**, 160–172 (2013)
12. Sen, S., Sherali, H.: A class of convergent of primal-dual subgradient algorithms for decomposable convex program. Math. Program. **35**, 279–297 (1986)
13. Boyd, S., Parikh, N., Chu, E., Peleato, B., Eckstein, J.: Distributed optimization and statistical learning via the alternating direction method of multipliers. Found. Trends® Mach. Learn. **3**, 1–122 (2011)
14. Xiao, L., Boyd, S.: Fast linear iterations for distributed averaging. Syst. Control Lett. **53**, 65–78 (2004)
15. Olfati-Saber, R., Fax, J., Murray, R.: Consensus and cooperation in networked multi-agent systems. Proc. IEEE **95**, 215–233 (2007)
16. Ghaoui, L., Lebret, H.: Robust solutions to least-squares problems with uncertain data. SIAM J. Matrix Anal. A. **18**, 1035–1064 (1997)
17. Mota, J., Xavier, J., Aguiar, P., Püschel, M.: D-ADMM: A communication-efficient distributed algorithm for separable optimization. IEEE Trans. Sig. Process. **61**, 2718–2723 (2013)
18. Wang, H., Gao, Y., Shi, Y., Wang, R.: Group-based alternating direction method of multipliers for distributed linear classification. IEEE Trans. Cybern. **99**, 1–15 (2016)
19. Wang, H., Gao, Y., Shi, Y., Wang, H.: A fast distributed classification algorithm for large-scale imbalanced data. In: 16th IEEE International Conference on Data Mining, pp. 1251–1256. IEEE, New Orleans (2016)
20. Goldstein, T., O'Donoghue, B., Setzer, S., Baraniuk, R.: Fast alternating direction optimization methods. SIAM J. Imaging. Sci. **7**, 1588–1623 (2014)

Heterogeneous Context-aware Recommendation Algorithm with Semi-supervised Tensor Factorization

Guoyong Cai[1,2(✉)] and Weidong Gu[1(✉)]

[1] Guilin University of Electronic Technology, Guilin, China
ccgycai@gmail.com, guweidong007@163.com
[2] Guangxi Key Lab of Trusted Software, Guangxi, China

Abstract. Data sparsity is one of the most challenging problems in recommender systems. In this paper, we tackle the data sparsity problem by proposing a heterogeneous context-aware semi-supervised tensor factorization method named HASS. Firstly, heterogeneous context are classified and processed by different modeling approaches. We use a tensor factorization model to capture user-item interaction contexts and use a matrix factorization model to capture both user attributed contexts and item attributed contexts. Secondly, different context models are optimized with semi-supervised co-training approach. Finally, the two sub-models are combined effectively by an weight fusing method. As a result, the HASS method has several distinguished advantages for mitigating the data sparsity problem. One is that the method can well perceive diverse influences of heterogeneous contexts and another is that a large number of unlabeled samples can be utilized by the co-training stage to further alleviate the data sparsity problem. The proposed algorithm is evaluated on real-world datasets and the experimental results show that HASS model can significantly improve recommendation accuracy by comparing with the existing state-of-art recommendation algorithms.

Keywords: Recommender systems · Heterogeneous Context-aware · Data sparsity · Semi-supervised learning

1 Introduction

In recommendation systems, traditional collaborative filtering algorithms usually are impacted easily by data sparsity problems. In fact, contextual information (time, location and so on) also influence recommendation [1]. For example, a user may tend to read a book in the morning. Thus, incorporating contextual information into recommendation, can mitigate data sparsity problems and increase recommender accuracies.

Contextual information is usually divided into two types. One is the attribute contexts of user and item [2] (age, sex, occupation, genre and so on). The other is the user-item interaction context [3] (time, location and so on). As contextual information becomes increasing important in recommender systems, many different kinds of context-aware recommender methods have been developed.

© Springer International Publishing AG 2017
H. Yin et al. (Eds.): IDEAL 2017, LNCS 10585, pp. 232–241, 2017.
https://doi.org/10.1007/978-3-319-68935-7_26

The works in [4, 5] represent contextual information as one or several dimensions which is similar to representation of users and items, which captures the common effect of various contexts to users and items but without taking into account the specific impact of contextual information. To deal with this problem, Liu [3] proposed a contextual operating tensor recommender model (COT) which contained contextual operating tensor to model the specific operation of contexts. Nevertheless this method neglect the influence of attribute contexts of users and iterms. Based on COT model in [3], Wu [6] proposed three contextual operating tensors for user attribute context, item attribute context and interaction context, which captured specific contextual influences on the latent vector of users or items. But this method has problem in dealing with large amount of contextual information because multiple tensors bring too many parameters. Liu et al. [7] proposed context-aware sequential recommender model in the circumstance containing rich users history behavior.

Facing the data sparsity problem in recommender systems, only using sparse labeled samples to train models is not enough. Therefore, the unlabeled samples are also important for training. Zhang et al. [2] proposed to construct two matrix factorization models for the user attribute context (age, sex and occupation) and the item attribute context (genre). Then, the two models are optimized by applying the semi-supervised co-training. Qu et al. [8] proposed a semi-supervised co-training recommender algorithm to incorporate views of movie texts, images and audios. However, these two semi-supervised recommender methods ignore the effect of both interaction contexts and attribute contexts.

To construct a model to overcome the shortages mentioned above and mitigate the data sparsity problem, we propose a heterogeneous context-aware semi-supervised model named HASS. First, we use different methods to model different type of contexts. We construct not only the contextual operating tensor model based on method in [3] for the interaction context but also the matrix factorization model based on method in [2] for the attribute context. Then, the two models are optimized by semi-supervised co-training and fused into a whole HASS model.

2   Heterogeneous Context-aware Semi-supervised Method

The proposed HASS model is graphically depicted in Fig. 1 and contains four steps given as follows:

Step 1: Constructing Interaction Context-aware Model. In the first step, a variant of contextual operating tensor model based on the work in [3] is applied to denote the complicated user-item interaction contexts.

Step 2: Constructing Attribute Context-aware Model. In the second step, a variant of matrix factorization model based on the work in [2] is applied to denote the attribute contexts of user and item.

Step 3: Semi-supervised Co-training. In the third step, two heterogeneous context-aware models are optimized by the way of semi-supervised co-training.

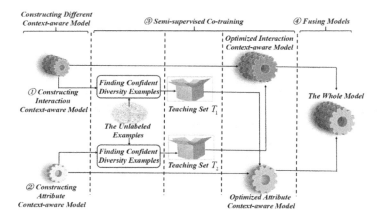

Fig. 1. Overview of our proposed HASS method.

Step 4: Fusing Models. In the final step, two heterogeneous context-aware models are fused into a whole model to predict ratings.

2.1 Constructing Interaction Context-aware Model

As shown in Fig. 2, the overview of the interaction context-aware model can be treated as a multi-layer matrix factorization model which is fit for complicated features of user-item interaction contexts.

The interaction context-aware model consists of three steps given as follows:

Step 1: Perceiving Interaction Contexts. In the first step, influences of the interaction contexts are perceived by fusing contextual information into the contextual operating tensor to form the corresponding contextual operating matrix.

Step 2: Contextual Operation to Latent Vectors. In the second step, the latent vector of users or items can be processed by these contextual operating matrix.

Step 3: Predicting Rating. In the final step, this model can predict the rating by the corresponding product of operated latent vectors of users and items.

As shown in the Fig. 2, users and items are denoted by $U = \{u_1, u_2, \ldots\}$ and $V = \{v_1, v_2, \ldots\}$.

The d dimensional latent vectors of user i and item j are denoted by $u_i \in R^d$ and $v_j \in R^d$. The rating that user i grade item j can be denoted by $r_{i,j}$. Various user-item interaction contexts can be denoted by $C_1, C_2, \ldots, C_n.c$ is a variable of the context $C.(c_{1,k}, c_{2,k}, \ldots, c_{n,k})$ is a feature vector indicating the current context circumstance k. Then each context variable c can be denoted by d_c dimensional latent vector h. Thus, the interaction context circumstance k also can be represented as a corresponding latent matrix $H_k = [h_{1,k}, h_{2,k}, \ldots, h_{n,k}] \in R^{d_c \times n}$. Besides, in the first step, Perceiving Interaction Contexts, contains the follow several components: The contextual operating tensors for users $T_U^{[1:d]}$ and items $T_V^{[1:d]}$ respectively indicate the common context effects on users and items. The latent matrix of the context circumstance H_k indicates the specific

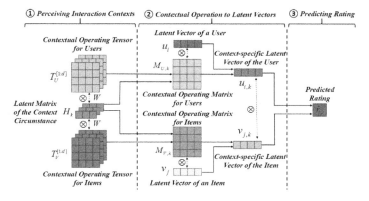

Fig. 2. Constructing Interaction Context-aware Model.

properties of contexts. W denotes the weights of each context. $\otimes$ denotes the matrix multiplication. In the second step, there are several components. The contextual operating matrix for users $M_{U,k}$ and items $M_{V,k}$ as two kinds of context operator can correspondingly process latent vectors of user i and item j to form the context-specific latent vectors of the user $u_{i,k}$ and item $v_{j,k}$. In the final step, the predicting rating $\hat{r}_{i,j}$ is influenced by user $u_{i,k}$ and item $v_{j,k}$.

Perceiving Interaction Contexts. To consider the specific contextual influence to users and items, the way of perceiving interaction contexts is constructing two contextual operating matrices respectively to users and items.

Different interaction contexts have similar or common contextual effects (for example, both weekend and being at the cinema can make you would like to see movies).

Therefore, contextual operating matrix is constructed by the product of contextual latent vector which captures contextual specific properties and contextual operating tensor which indicates common contextual influences:

$$M_{U,k} = a_k^t T_U^{[1:d]}, \ M_{V,k} = a_k^t T_V^{[1:d]} \tag{1}$$

Where $M_{U,k}$ and $M_{V,k}$ respectively denote $d \times d$ dimensional contextual operating matrices of users and items under the context circumstance k. $T_U^{[1:d]}$ and $T_V^{[1:d]}$ denote both $d_c \times d \times d$ dimensional contextual operating tensors. $[1:d]$ denotes that the tensor contains d slices. t denotes the transpose of a matrix. a_k is a d_c dimensional contextual latent vector which is a weighted combination of each context latent vector:

$$a_k = H_k W \tag{2}$$

Where H_k is a $d_c \times n$ dimensional latent matrix of context values under the context circumstance k. Where W is a n dimensional vector, indicating weights of each context.

Contextual Operation to Latent Vectors. After perceiving interaction contexts, the contextual operating matrix describes how a context circumstance affects the properties of entities of users and items. This model can process the latent vector of users or items by the way of contextual operations. This process of the contextual operation is described in Eq. (3) as the corresponding product of the contextual operating matrix and the original latent vectors of users and items:

$$u_{i,k} = M_{U,k}u_i, \ v_{j,k} = M_{V,k}v_j \tag{3}$$

Where u_i and v_j are original latent vectors of users and items. Hence, substituting the Eqs. (1) and (2) into Eq. (3), we have:

$$u_{i,k} = (H_kW)^t T_U^{[1:d]}u_i, \ v_{j,k} = (H_kW)^t T_V^{[1:d]}v_j \tag{4}$$

Where $u_{i,k}$ and $v_{j,k}$ are both d dimensional latent vectors of users and items operated by the contextual circumstance k.

Predicting Rating. After latent vectors of users and items are influenced by the contextual operating matrix. The predicting function of this model can be written as:

$$\hat{r}_{i,j} = w_0 + w_i + w_j + \sum_{m=1}^{n} w_{m,k} + u_{i,k}^t v_{j,k} \tag{5}$$

Where the rating is factorized into six components: Global average bias w_0, user i bias w_i, item j bias w_j, the bias $w_{m,k}$ of the context value, $u_{i,k}$ and $v_{j,k}$ denoting the latent vector of user and item influenced by context circumstance k.

Then, substituting Eq. (4) into Eq. (5), we have the objective function of this model as the model $h_1(i,j)$:

$$h_1(i,j) = \hat{r}_{i,j} = w_0 + w_i + w_j + \sum_{m=1}^{n} w_{m,k} + \left[(H_kW)^t T_U^{[1:d]}u_i\right]^t (H_kW)^t T_V^{[1:d]}v_j \tag{6}$$

To optimize the parameters, we define the following objective function J_1:

$$\min_{U,V,H,T,W} J_1 = \sum_{r_{i,j}\in\Omega} (r_{i,j} - \hat{r}_{i,j})^2 + \frac{\lambda}{2}(\|U\|^2 + \|V\|^2 + \|H\|^2 + \|T\|^2 + \|W\|^2) \tag{7}$$

Where Ω denotes the train set, λ denotes regularization factor. Parameters can be optimized by the following correspondingly stochastic gradient descent (SGD) equation: $\theta = \theta - \frac{\partial J_1(\theta)}{\partial \theta}$ Where θ denotes each parameter $u_i, v_j, H_k, T_U^{[1:d]}, T_V^{[1:d]}, W$.

2.2 Constructing Attribute Context-aware Model

In the recommender system, except the interaction contexts, attribute contexts of user and item also contain preference information. Therefore, in this section, we use a variant of matrix factorization model proposed in paper [2] which captures the influence of attribute contexts of user and item to improve performance. We have:

$$\hat{r}_{i,j} = w_0 + w_i + w_j + \sum_{m \in user\_attributes} w_m + \sum_{n \in item\_attributes} w_n + v_j^t u_i \qquad (8)$$

Here, the predicted rating is factorized into seven components: Global average bias w_0, user i bias w_i, item j bias w_j, the latent vector u_i of user, the latent vector v_j of item, user attributes bias w_m, item attributes bias w_n.

To obtain a further optimization, we can not only connect the user with the biases of item contexts but also relate the item to the biases of user contexts as below to construct the model $h_2(i,j)$:

$$h_2(i,j) = \hat{r}_{i,j} = w_0 + w_i + w_j + \sum_{m \in user\_attributes} w_{jm} + \sum_{n \in item\_attributes} w_{in} + v_j^t u_i$$
$$(9)$$

Where w_{jm} denotes item j's biases to samples which are related with the user attribute contexts of a specific kind m (such as age, sex and so on), w_{in} is user i's preferences to samples which are related with the item attribute context of a the specific categories n (such as genre and so on).

Based on this further optimization, using stochastic gradient descent (SGD) algorithm to learn the parameters. w_i is optimized merely when the current rating is related with user $i.w_j$ is similar to $w_i.w_{in}$ is optimized only when the current rating is related with user i and the items belonging the specific categories n of item attribute contexts. w_{jm} is similar to w_{in}.

And the objective function J_2 of this model can define the following equation:

$$\min_{u_*, v_*, w_*} J_2 = \sum_{r_{i,j} \in \Omega} (r_{i,j} - \hat{r}_{i,j})^2 + \lambda(\|u_i\|^2 + \|v_j\|^2 + \sum w_*^2) \qquad (10)$$

Where Ω denotes the train set, λ denotes regularization factor and w_* denotes all bias parameters w_0, w_i, w_j, w_{jm} and w_{in}. The parameters of the attribute context-aware model can be optimized by the following stochastic gradient descent (SGD) equation:

$$\Delta\theta = -\gamma \frac{\partial e_{i,j}^2}{\partial \theta} - \lambda\theta = 2\gamma e_{i,j} \frac{\partial \hat{r}_{i,j}}{\partial \theta} - \lambda\theta, \ \theta = \theta + \Delta\theta$$

Where $e_{i,j}^2 = (r_{i,j} - \hat{r}_{i,j})^2$, γ is the learning rate, θ denotes each parameter u_i, v_j and w_*.

2.3 Semi-supervised Co-training

To ensure the efficiency of the semi-supervised learning, there are two key steps:

Constructing Teaching Set. Different models have different advantages in predicting ratings. Therefore, constructing the teaching set T_m is a significant step. One challenge is how to formulate the criteria for selecting confident unlabeled samples from the subset U' of unlabeled samples so as to build the teaching set. The criteria we used in this work is the model's confidence as follow.

We know that if some kinds of samples account for the large proportion in training set, the model will have more confidence in predicting these same kinds of unlabeled samples. We construct two confidence criteria for these heterogeneous context-aware models. The confidence criteria of the first model $h_1(i,j)$ is measured by following $C_1(x_{i,j})$ confidence coefficient equation:

$$C_1(x_{i,j}) = \frac{d_i^{(1)} \times d_j^{(1)} \times \prod_{c \in interaction} d_c^{(1)}}{N} \qquad (11)$$

Where $x_{i,j}$ denotes a unlabled sample that user i grade item j. $d_i^{(1)}$ and $d_j^{(1)}$ correspondingly denote the proportion that the first model training sample related with user i and item j accounts for the total training sample. Similarly, $d_c^{(1)}$ denotes the ratio that the training sample related with corresponding interaction context kind c accounts for the total training sample and N is a normalization factor.

The confidence criteria of the second model $h_2(i,j)$ is measured by following $C_2(x_{i,j})$:

$$C_2(x_{i,j}) = \frac{d_i^{(2)} \times d_j^{(2)} \times \prod_{c \in attribute} d_c^{(2)}}{N} \qquad (12)$$

Similarly, where $d_i^{(2)}$ and $d_j^{(2)}$ correspondingly denote the proportion that the second model training sample related with user i and item j accounts for the total training sample. In our work, we train two models by the same data set, which results in that $d_i^{(1)}$ equals to $d_i^{(2)}$ and $d_j^{(1)}$ equals to $d_j^{(2)}$. $d_c^{(2)}$ denotes the proportion that the training sample related with corresponding attribute context kind c accounts for total training samples.

To detect the more confident unlabled samples for each model, the confidence of each unlabeled sample should be normalized by the following confidence probability:

$$\Pr(x_{i,j}, m) = \frac{C_m(x_{i,j})}{\sum x_{k \in U'} C_m(x_k)} \qquad (13)$$

Where $m = 1$ denotes the first model $h_1(i,j)$ and $m = 2$ denotes the second model $h_2(i,j)$. $\Pr(x_{i,j}, m)$ denotes the confidence probability that the confidence coefficient $C_m(x_{i,j})$ of $x_{i,j}$ under the model $h_m(i,j)$ accounts for all confidence coefficients of unlabeled samples from the unlabeled samples subset U'.

The higher confidence probability of unlabled sample $x_{i,j}$ denotes the more accurate in recommender predicting. Nonetheless, only utilizing confidence probability can not guarantee constructing the appropriate teaching sets because a big diversity between the outputs of the two models is important for the optimization. We set a threshold to meet the diversity of the teaching set and model. Therefore, after selecting unlabeled samples with high confidence probability $\Pr(x_{i,j}, m)$ from U' and model $h_m(i,j)$, predicts these unlabeled samples ratings ($\hat{r}_1$ and $\hat{r}_2$) by two models. If exist an unlabeled sample $x_{i,j}$ meeting $|\hat{r}_1 - \hat{r}_2| > \tau$, the model then fuse this sample into teaching set T_m of model $h_m(i,j)$. Finally, U' should remove this T_m by $U' = U' - T_m$.

Mutually Training. The mutually training is divided into two steps: Teaching and Optimizing. In Teaching step:$L_1 = L_1 \cup T_2$; $L_2 = L_2 \cup T_1$; Where L_1 and L_2 denotes the training set of model $h_1(i,j)$ and $h_2(i,j)$. Teaching step is that teaching set T_1 and T_2 respectively are fused into the training set L_2 and L_1 of the peer model.

Optimizing step is that model $h_m(i,j)$ can be optimized by learning the own new fused training set L_m. Optimize the Two Models:$h_1(i,j) \leftarrow L_1$ and $h_2(i,j) \leftarrow L_2$.

2.4 Fusing the Models

We adopt the weight fusing method to fuse $h_1(i,j)$ and $h_2(i,j)$ models into our final whole model $h(i,j)$ by a weight coefficient α between 0 and 1.

$$h(i,j) = \alpha h_1(i,j) + (1 - \alpha)h_2(i,j) \tag{14}$$

3 Experiment

Datasets. Our experiments are conducted on real datasets, MovieLens-100 K, consisting of 100,000 ratings (1-5) from 943 users on 1,682 movies.

Except ratings, This MovieLens dataset also contains rich context information, such as attribute context (the age, sex and occupation of users and the genre of films) and interaction context (the day and hour of timestamps).

We randomly sample about 20% of the ratings from the original dataset to create the test set, and the remaining 80% ratings are treated as the train set.

Compared Methods. Our HASS model captures not only this dataset's attribute context to construct attribute context-aware sub-model but also this dataset's interaction context to construct interaction context-aware sub-model.

We compare HASS model with three traditional collaborative filtering recommendation algorithms and three state-of-the-art context-aware recommender models.

Traditional Collaborative Filtering Algorithms:
UB k-NN: The user-based k-nearest neighbor collaborative filtering algorithms.
IB k-NN: The item-based k-nearest neighbor collaborative filtering algorithms.
FactCF: The factorization-based collaborative filtering algorithm.

Context-aware Recommender Models:
CONTEXT: The attribute context-aware model [2], given by Eq. 9.
CSEL: The attribute context-aware model [2], which not only divides CONTEXT model into two sub-model (attribute context models of user and item) but also fuses these sub-model into a whole model with co-training algorithm.
COT: The contextual operating tensor model [3] for interaction context-aware recommender systems, given by Eq. 6.

Evaluation Metrics. To measure the performance of rating prediction, we use the most popular metric, Root Mean Square Error (RMSE):

$$RMSE = \sqrt{\frac{\sum\limits_{(i,j)\in\Omega_{test}} (r_{i,j} - \hat{r}_{i,j})^2}{n_{test}}} \qquad (15)$$

Experiment Results and Analysis. In Fig. 3, experiment performances are represented by different colors indicating various kinds of compared methods.

- Blue: Traditional Collaborative Filtering Algorithms.
- Orange: State-of-the-art Context-aware Algorithms.
- Green: Our HASS Algorithm.

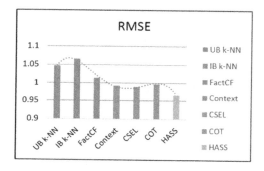

Fig. 3. The overall RMSE Performance.

The overall performance on the various compared method are given in Fig. 3. The traditional collaborative filtering algorithms neglects the influence of the contextual information on recommender system. Therefore, the performance of this kind (Blue) is usually inferior to the performance of context-aware models (Orange).

We find that the performance of COT is slightly inferior to CSEL's performance, which is maybe resulted from that the attribute context information is more plentiful than the interaction context information in the Movielens-100 K dataset.

Comparing with UB k-NN, IB k-NN, FactCF, CONTEXT, CSEL and COT, our proposed HASS improves RMSE values by 7%, 9%, 4%, 2%, 2% and 3% respectively.

Our proposed HASS algorithm can be used for those cases that containing both attribute contexts and interaction contexts (such as movie, book, food recommendation and so on). Comparing others algorithms, HASS model not only perceives the influence of the heterogeneous context but also alleviates the data sparsity problem by the way of semi-supervised co-training algorithm.

4   Conclusion

This paper not only perceives the different influence of the attribute and interaction context but also proposes a heterogeneous context-aware semi-supervised method HASS to solve the data sparsity problem. First, we construct two sub-models of heterogeneous context-aware by the contextual operating tensor and the matrix factorization. Second, we fuse two sub-models into a whole model by semi-supervised co-training. The experiment results show that our strategy outperforms the various state-of-the-art context-aware models and makes a significant progress in solving the data sparsity problem. In the future, we would like to construct three or more appropriate context-aware models to improve recommender performance by semi-supervised co-training.

Acknowledgments. This work is supported by Chinese National Science Foundation (#61763007), Guangxi Key Lab of Trusted Software under project Kx201503 and Innovation Project of GUET Graduate Education (#2017YJCX44).

References

1. Adomavicius, G., Tuzhilin, A.: Context-aware recommender systems. In: Recommender Systems Handbook, pp. 191–226. Springer, Boston (2015)
2. Zhang, M., Tang, J., Zhang, X., et al.: Addressing cold start in recommender systems: A semi-supervised co-training algorithm. In: Proceedings of the 37th International ACM SIGIR Conference on Research & Development in Information Retrieval, pp. 73–82. (2014)
3. Liu, Q., Wu, S., Wang, L.: COT: contextual operating tensor for context-aware recommender systems. In: AAAI Conference on Artificial Intelligence (AAAI), pp. 203–209 (2015)
4. Karatzoglou, A., Amatriain, X., Baltrunas, L., Oliver, N.: Multiverse recommendation: n-dimensional tensor factorization for context-aware collaborative filtering. In: Proceedings of the Fourth ACM Conference on Recommender Systems - RecSys 2010 (2010)
5. Rendle, S., Gantner, Z., Freudenthaler, C., Schmidt-Thieme, L.: Fast context-aware recommendations with factorization machines. In: Proceedings of the 34th International ACM SIGIR Conference on Research and Development in Information - SIGIR 2011, pp. 635–644 (2011)
6. Wu, S., Liu, Q., Wang, L., Tan, T.: Contextual operation for recommender systems. IEEE Trans. Knowl. Data Eng. **28**(8), 2000–2012 (2016)
7. Liu, Q., et al.: Context-aware sequential recommendation. In: IEEE 16th International Conference on Data Mining (ICDM) (2016)
8. Qu, W., Song, K.-S., Zhang, Y.-F., Feng, S., Wang, D.-L., Yu, G.: A novel approach based on multi-view content analysis and semi-supervised enrichment for movie recommendation. J. Comput. Sci. Technol. **28**(5), 776–787 (2013)

Object Detection with Proposals in High-Resolution Optical Remote Sensing Images

Huoping Ding[1,2], Qinhan Luo[3], Zhengxia Zou[4(✉)], Cuicui Guo[1,2], and Zhenwei Shi[4(✉)]

[1] Space Star Technology CO., Ltd., Beijing, China
[2] State Key Laboratory of Space-Ground Integrated Information Technology, Beijing, China
[3] The 28th Research Institute of China Electronics Technology Group Corporation, Nanjing, China
[4] Image Processing Center, School of Astronautics, Beihang University, Beijing 100191, China
{zhengxiazou,shizhenwei}@buaa.edu.cn

Abstract. Detecting object in remote sensing images remains a challenge due to multi-scale objects, complex ground environment and large image size despite of the fast development of machine learning and computer vision technology in recent years. The primary difficulty lies in the fast and accurate location of candidate bounding boxes from a large-size remote sensing image. In this letter, we propose a novel remote sensing object detection method inspired by the recent-popular technique, Object Proposals, to quickly generate high-quality object bounding box locations in remote sensing images. A simple but effective objectness measurement, based on the image gradients and its variants, is proposed. Moreover, to evaluate the effectiveness of our method, we complete the subsequent detection flow based on the convolution neural networks as a standard detection baseline. Experiments show that our method is able to produce high-quality proposals with a desirable computational speed.

1 Introduction

Automatic detection of objects in remote sensing images (RSI) consists in determining whether objects exist in the image, and if so where in the image they occur. Although a variety of approaches exist, it remains a challenge due to the huge searching space of locating the objects and the variation of intra class variations, multi-scale objects and the complex background distributions.

The most popular approach for detection task in RSIs over the past decade [2–4,6,10,11,13,15,17] can be grouped in two stages: candidate region extraction and object identification. This approach involves applying some simple tests to each hypothesized object location and extracting the candidate regions that are likely to contain the objects quickly. Then the sophisticated classifier is trained on

H. Yin et al. (Eds.): IDEAL 2017, LNCS 10585, pp. 242–250, 2017.
https://doi.org/10.1007/978-3-319-68935-7_27

some well-designed features and applied on these candidate regions to complete detection task.

Among previous methods for candidate region extraction, the image segmentation techniques [3,4,10] have been commonly used. The threshold values are usually carefully manual chosen in the grayscale [3] or the gradient magnitude space [4] of input image. Generally, the experimental threshold may be successful to a certain extent, but such skill is not robust and may lack the physical meaning. Different from the above works, in [10], the threshold value is adaptive acquired by analyzing the information of the related region. However, this method is only effective in very limited background conditions. Apart from these segmentation based methods, the shape information of objects has also been utilized frequently. Some image processing techniques like Hough transform [11], Ellipse and Line Segment Detector (ELSD) [16], Circle Frequency Filter [2,12] have been employed for some specific requirements such as airports, oil tanks, airplanes and ships. More recently, some researchers [6,15] consider the objects to be salient and the visual saliency mechanism is applied for candidate region extraction. Unfortunately, these saliency based methods still lack convincing performance in complex remote sensing scenes.

There is no doubt that the quality of extracted candidate region have important effects on the final detection results [9]. Although these previous works have achieved some good results, they share three common drawbacks. First, they are task-specific that usually only support single object category and is not easy to generalize to other types of objects. Second, and more critical, most of these methods concentrate more on candidate regions rather than candidate bounding-boxes. These methods utilize the threshold segmentation technique and subsequent connectivity domain analysis, which lack the scale representation of an object and may subject to the complex environmental impact. Third, these methods are usually unsupervised. As the prior information of object plays an important role for candidate region extraction stage, the absence of prior information will further influence the quality of the extracted candidate regions and detection performance.

Object Proposals, a new computer vision technique, has received increasing interest from the research community in recent years. It aims to generate a relatively small set of bounding box proposals that are likely to contain objects of interest. A comprehensive survey on this problem can be found in [9]. Our method is inspired by the works [1,5]. Motivated by its success in [8,9], this technique emerges as an alternative to the traditional object detection paradigm in remote sensing images.

In this paper, we propose a novel Object Proposals method for RSIs, aiming to produce a set of bounding boxes proposals with high recall rate and accurate locations in complex ground environments. Comparing to previous works, our method has the following advantages.

- **Being adaptive to the object's scales changes.** A bounding box proposal not only identifies the location of an object, but also specifies its corresponding scale, where both of these factors are crucial for high-quality detection performance.

- **With high recall rate.** Candidate regions should cover all potential targets since any undetected targets at this stage cannot be recovered later.
- **Computational efficiency.** This is especially important for large size RSIs.

To evaluate the effectiveness of our proposed method, we use a recent-popular CNN based processing flow [8] for object identification stage as a verification baseline. The remainder of this letter is organized as follows. In Sect. 2 we present detailed description of our proposed Object Proposals method. Section 3 reports the experimental results. Conclusions are drawn in Sect. 4.

2   Object Proposals for Remote Sensing Images

In this section, we begin with an overview of our method. Then we discuss the design of our features. Finally we explain how we learn the objectness measure and the implementation details for full image detection.

The goal of our approach is to compute an objectness map quickly that highlights the objects and suppresses the background. The flowchart of our method is shown in Fig. 1. To get the objectness map, we scan the whole image on its multiple feature channels over a predefined window size. Each window $\mathbf{F}_l \in \mathcal{R}^{D \times 1}$ of the multiple feature channels is then scored with a linear model $\mathbf{w} \in \mathcal{R}^{D \times 1}$:

$$s_l = \langle \mathbf{w}, \mathbf{F}_l \rangle, \tag{1}$$
$$l = (i, x, y), \tag{2}$$

where, $s_l, \mathbf{F}_l, i, (x, y)$ are filter score, the feature representation, scale and position of a window, and D is the dimension of the filter. Using non-maximal suppress, we are able to select the object candidate windows from each scale i of the objectness map. Here we further squash the object score into the interval (0,1) as a probability of containing an object by using the sigmoid function

$$o_l = \frac{1}{1 + \exp(-s_l)}. \tag{3}$$

To learn the objectness measure of image windows, we follow the linear SVM [7] for its good capability of generalization and less computational cost at inference stage. The features of the ground truth object windows and random sampled background windows are used as positive and negative training samples.

For feature design, we focus on both computational efficiency and representation ability. Our study is based on the assumption that the gradient gives an informative representation of an image. Concretely, we compute the normalized gradients firstly, then several channels include the magnitude and the local histograms are accumulated. Our features are then single lookups in these channels without further computation can be computed very efficiently. The detailed features are listed as follows.

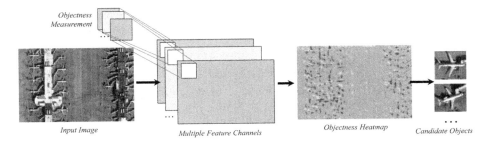

Fig. 1. Flowchart of our proposal generation method.

Gray pixel value: The gray scale version of the image acts as the simplest feature channel.

Gradient magnitude: Gradients are computed by the discrete derivative mask $[-1, 0, 1]$. For color images a common trick is to compute the gradient on the 3 color channels separately and use the maximum response.

$$\mathbf{G}(x, y) = ((\mathbf{I}_{x+1,y} - \mathbf{I}_{x-1,y})^2 + (\mathbf{I}_{x,y+1} - \mathbf{I}_{x,y-1})^2)^{1/2}, \qquad (4)$$

where the $I, (x, y)$ are the gray image and the location.

Gradient orientation histogram: The gradient orientation histogram is a weighted histogram where bin index is determined by gradient angle and weighed by gradient magnitude. The channels are given as

$$\Theta(x, y) = \arctan \frac{\mathbf{G}_y}{\mathbf{G}_x}, \qquad (5)$$

$$\mathbf{Q}_\theta(x, y) = \mathbf{G}(x, y) \cdot \mathbf{1}[\Theta(x, y) = \theta], \qquad (6)$$

where θ and $\mathbf{1}$ represents the range of the gradient angles and the indicator function, $\mathbf{G}(x, y)$ and $\Theta(x, y)$ are the gradient magnitude and gradient angle at $\mathbf{I}(x, y)$.

2.1   Other Implementation Details

Some other implementation details of our method are listed as follows.

Scales: We deal with multiple scales detection problem by building an image feature pyramids before detection.

Orientations: We train several objectness model $\mathbf{w}$ with individual samples of corresponding orientations. The response of each orientation will be considered independently to extract the candidate bounding boxes.

Speed up: For a 5000×5000 pxl remote sensing image, the feature channels can be computed efficiently within seconds on a standard PC. To further reduce the sliding window inference time, we split and convert $\mathbf{w}$ into several parts with corresponding feature channels and replacing the inner product operation with convolution

$$\langle \mathbf{w} \cdot \mathbf{F}_i \rangle = \langle \mathbf{w}_1 \cdot \mathbf{F}_{i_1} \rangle + \ldots + \langle \mathbf{w}_m \cdot \mathbf{F}_{i_m} \rangle, \tag{7}$$
$$= \mathbf{w}'_1 * \mathbf{F}_{i_1} + \ldots + \mathbf{w}'_m * \mathbf{F}_{i_m}, \tag{8}$$

where j represents the feature channel id and $\mathbf{F}_{i_j}(j = 1, 2, \ldots m)$ represent the feature of the window.

3   Experimental Results

In this section, the datasets with airplanes and oil tanks are utilized to evaluate the effectiveness of our method. Moreover, following [8], we will show its good performance with the standard CNN features to complete the detection task. Another remote sensing object candidate location method in [6] is compared against our method.

3.1   Datasets and Experimental Setup

We collected several optical remote sensing images from google earth with the resolution of 0.8 m for evaluation. It consists of two parts:

- **Oil tank dataset** which consists of 22 images for training and 5 images for test. The image size varies from 1500×1500 to 3000×3000.
- **Airplane dataset** which consists of 72 images for training and 5 images for test. The image size varies from 5000×5000 to 12000×12000.

All the objects in these images have been manually annotated with bounding boxes. The size of the objectness filter $\mathbf{w}_j(j = 1, 2, \ldots m)$ is set to 16×16.

3.2   Approach Variants

Following [5], we evaluate the DR-#WIN on our test set, which means counting the detection rate (DR) given #WIN proposals. The test data set consists of 10 large size RSIs with bounding box annotations of objects. The variety of the objects in scale, orientation, position and shape make it very suitable for our evaluation. The evaluation result is shown in Fig. 2. An object is considered as being detected if the intersection over union (IoU) score with our proposed bounding box location is no less than the predefined threshold. We compare with our feature to its baseline: a single gradient magnitude feature as is used in [5], as shown in Fig. 2. Our method gives a better performance than that of [5]. In fact, in remote sensing images, sometimes the gradient lines of an object are not always closed, even not salient. Detecting object candidates only based on the gradient map may result in some miss alarms. Instead, our method is designed based on richer feature maps with better discriminative ability.

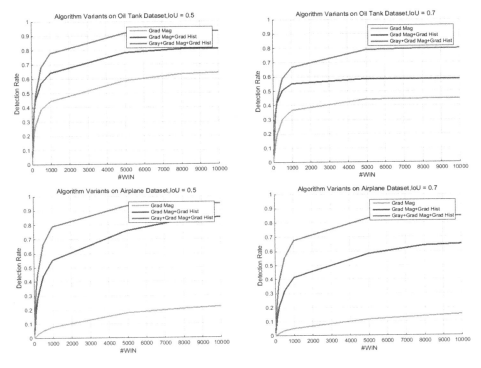

Fig. 2. Comparison of our method with different feature channels.

3.3   Comparison with Other Method

We compare our method against Diao's method [6]. The result is shown in Table 1. The predicted bounding box is considered correct only if it overlaps more than 50 percent with the ground-truth bounding box. Unfortunately, as mentioned in its own work, their proposed saliency-based method is not applicable for multi-scale object [6] thus some of their output candidate region may contain part of the objects yet have a low IoU score with the ground truth.

Table 1. Comparisons of recall and time performance

Method	Recall (oil tank)	Time (oil tank)	Recall (plane)	Time (plane)
Ours	93.07%	3.84 s	94.84%	44.05 s
Method in [6]	34.35%	2.65 s	32.05%	16.50 s

Table 2 list the processing time to detect airplanes with 5000 × 5000 images and oil tanks with 2000 × 2000 images. We test our algorithm on an Intel i7 PC with 16G RAM and Matlab 2015b.

Fig. 3. Some examples of our detection results. **Blue boxes:** detected objects. **Red boxes:** missed objects. **The first row:** candidate bounding box locations generated by Object Proposals. **The second row:** final detection results. (Color figure online)

Table 2. Comparisons of final detection result

Method	Recall	Precision
Ours	88.25%	87.33%
Method in [6]	17.40%	36.20%

We use a recent-popular CNN based processing flow [8] for object identification stage as a verification baseline. Each candidate location is resized to a fixed size image and is fed into the vgg-f net [14] that extracts a high quality fixed-length feature vector. A linear SVM is then trained to complete the detection task. The final detection result is shown in Table 2.

At test time, the algorithm will output a set of bounding boxes with corresponding scores. The predicted bounding box is considered correct when the IoU with the ground truth is more than 0.5. Multiple detection are penalized. The results show that our approach is effective for further detection with a cascaded sophisticated classifier (Fig. 3).

4   Conclusion

In this letter, we have proposed new a object proposal based candidate bounding location method for object detection in high resolution RSIs. The proposed method is able to generate a small sets of bounding boxes efficiently that have high IoU with the ground truth regardless of the change of scales, orientations and appearances. A CNN based more sophisticated classifier is employed on these candidate bounding boxes and show good performance for detection task.

Acknowledgement. The work was supported by the funding project of the State Key Laboratory of Space-Ground Integrated Information Technology (SKL-SGIIT), under the Grant 2016-SGIIT-KFJJ-YG-03. The authors also would like to thank Piotr Dollar for his image processing toolbox.

References

1. Alexe, B., Deselaers, T., Ferrari, V.: What is an object? In: 2010 IEEE Conference on Computer Vision and Pattern Recognition (CVPR), pp. 73–80 (2010)
2. An, Z., Shi, Z., Teng, X., Yu, X., Tang, W.: An automated airplane detection system for large panchromatic image with high spatial resolution. Optik - Int. J. Light Electron Opt. **125**(12), 2768–2775 (2014)
3. Bo, S., Jing, Y.: Region-based airplane detection in remotely sensed imagery. In: International Congress on Image and Signal Processing, pp. 1923–1926 (2010)
4. Chen, X., Xiang, S., Liu, C.L., Pan, C.H.: Aircraft detection by deep belief nets. In: 2013 2nd IAPR Asian Conference on Pattern Recognition, pp. 54–58, November 2013
5. Cheng, M.M., Zhang, Z., Lin, W.Y., Torr, P.: BING: binarized normed gradients for objectness estimation at 300fps. In: IEEE Conference on Computer Vision and Pattern Recognition, pp. 3286–3293 (2014)
6. Diao, W., Sun, X., Zheng, X., Dou, F.: Efficient saliency-based object detection in remote sensing images using deep belief networks. IEEE Geosci. Remote Sens. Lett. **13**(2), 137–141 (2016)
7. Fan, R.E., Chang, K.W., Hsieh, C.J., Wang, X.R., Lin, C.J.: LIBLINEAR: a library for large linear classification. J. Mach. Learn. Res. **9**(9), 1871–1874 (2008)
8. Girshick, R., Donahue, J., Darrell, T., Malik, J.: Region-based convolutional networks for accurate object detection and segmentation. IEEE Trans. Patt. Anal. Mach. Intell. **38**(1), 1–1 (2015)
9. Hosang, J., Benenson, R., Dollar, P., Schiele, B.: What makes for effective detection proposals? IEEE Trans. Patt. Anal. Mach. Intell. **38**(4), 814–830 (2016)
10. Liang, Y.L., Colgan, W., Lv, Q., Steffen, K., Abdalati, W., Stroeve, J., Gallaher, D., Bayou, N.: A decadal investigation of supraglacial lakes in west greenland using a fully automatic detection and tracking algorithm. Remote Sens. Environ. **123**(6), 127–138 (2012)
11. Liu, D., He, L., Carin, L.: Airport detection in large aerial optical imagery. In: IEEE International Conference on Acoustics, Speech, and Signal Processing, 2004, Proceedings, (ICASSP 2004), vol. 5, pp. V-761-4, May 2004
12. Shi, Z., Yu, X., Jiang, Z., Li, B.: Ship detection in high-resolution optical imagery based on anomaly detector and local shape feature. IEEE Trans. Geosci. Remote Sens. **52**(8), 4511–4523 (2014)
13. Shi, Z., Zou, Z.: Can a machine generate human-like language descriptions for a remote sensing image? IEEE Trans. Geosci. Remote Sens. **55**(6), 3623–3634 (2017)
14. Simonyan, K., Zisserman, A.: Very deep convolutional networks for large-scale image recognition. CoRR abs/1409.1556 (2014)
15. Wang, G., Chen, Y., Yang, S., Gao, M., Shan, G.: Salient target detection in remote sensing image via cellular automata. In: 2015 Sixth International Conference on Intelligent Control and Information Processing (ICICIP), pp. 417–420, November 2015

16. Zhang, L., Shi, Z., Wu, J.: A hierarchical oil tank detector with deep surrounding features for high-resolution optical satellite imagery. IEEE J. Sel. Top. Appl. Earth Observations Remote Sens. **8**(10), 4895–4909 (2015)
17. Zou, Z., Shi, Z.: Ship detection in spaceborne optical image with SVD networks. IEEE Trans. Geosci. Remote Sens. **54**(10), 5832–5845 (2014)

Towards Spectral-Texture Approach to Hyperspectral Image Analysis for Plant Classification

Ali AlSuwaidi$^{(\boxtimes)}$, Bruce Grieve, and Hujun Yin

School of Electrical and Electronic Engineering, The University of Manchester, Manchester M13 9PL, UK
ali.bghalsuwaidi@postgrad.manchester.ac.uk,
{bruce.grieve,hujun.yin}@manchester.ac.uk

Abstract. The use of hyperspectral imaging systems in studying plant properties, types, and conditions has significantly increased due to numerous economical and financial benefits. It can also enable automatic identification of plant phenotypes. Such systems can underpin a new generation of precision agriculture techniques, for instance, the selective application of plant nutrients to crops, preventing costly losses to soils, and the associated environmental impact to their ingress into watercourses. This paper is concerned with the analysis of hyperspectral images and data for monitoring and classifying plant conditions. A spectral-texture approach based on feature selection and the Markov random field model is proposed to enhance classification and prediction performance, as compared to conventional approaches. Two independent hyperspectral datasets, captured by two proximal hyperspectral instrumentations with different acquisition dates and exposure times, were used in the evaluation. Experimental results show promising improvements in the discrimination performance of the proposed approach. The study shows that such an approach can shed a light on the attributes that can better differentiate plants, their properties, and conditions.

Keywords: Feature selection · Hyperspectral imaging · Markov random field · Spectral analysis · Texture analysis

1 Introduction

Hyper-spectral imaging (HSI), a branch of multivariate imaging [1], gathers optical properties of a target with several spectral representations using a mixture of spectroscopy and imaging technologies [2]. HSI has been utilised in an increasing number of applications, for instance, remote sensing [3], proximal sensing [4], industrial processes [5], medical imaging [6] and chemical processes [7]. Moreover, several configurations have been used to capture hyperspectral images: point, line, area, and single shot scanning [8].

Texture characteristics and spectral information are the fundamental properties of hyperspectral images. Textural information is described as attributes

H. Yin et al. (Eds.): IDEAL 2017, LNCS 10585, pp. 251–260, 2017.
https://doi.org/10.1007/978-3-319-68935-7_28

representing texture arrangement of grey levels [9]. This information is associated with many image properties such as coarseness, smoothness, orientation, depth, etc. Whilst, spectral information defines the measured spectrum of the corresponding texture images, where each image represents a unique spectral signature [2]. It is worth noting that the spectrum information covers single or several parts of the electromagnetic spectrum.

Several spectral analysis techniques have been introduced to analyse hyperspectral images. These techniques have played an important role in several domains such as agriculture, medicine, and industry for many tasks - especially image classification [10]. Since hyperspectral imaging senses a wider range of the electromagnetic spectrum, effective and efficient approaches are needed for analysing the images [10]. These approaches include feature extraction and feature selection, used to reduce the dimensionality of hyperspectral images as well as the requested processing time, thus analysing only the information relevant to the investigated problem.

Texture analysis provides insight about texture properties, which are important basis to recognition and description. Generally, texture analysis techniques can be broadly categorised into statistical, structural, transform-based, and model-based [9]. The first two use statistics of the grey levels and arrangement rules of the grey levels to describe the texture. The characteristics of transformed-based are the use of transforms to describe texture properties in the transform-based techniques. While the model-based approach uses models and (estimated) model parameters to define textures. Several studies published in the past have shown that the Markov random field (MRF) is one of the most powerful models for describing various textures [11]. MRF, in which inter-pixel dependency is modelled probabilistically, has been utilised in many applications such as image de-noising, image compression, image segmentation and super-resolution.

This work focuses on classifying different plant conditions (e.g. stressed vs. normal; diseased vs. healthy) using a spectral-texture approach and compares the results with those from using conventional methods, individual spectral or texture approaches. The relevant features (i.e. spectral wavelengths) along with the texture representation (i.e. estimated MRF parameters) are used in the classification stage. Furthermore, support vector machines (SVM) are used as classifier for two reasons [12]: (1) it is considered as a state of the art classification algorithm, (2) it reduces the risk of overfitting (in order to deal efficiently with the dimensionality problem). The approach has been evaluated on real-world datasets with promising and improved discrimination achieved.

The remainder of the paper is organised as follows: the background on feature selection and MRF is reviewed in Sect. 2. Section 3 presents HSI systems, HSI datasets, and the proposed spectral-texture approach. Results and discussions are given in Sect. 4, followed by the concluding remarks in Sect. 5.

2   Background

An overview of feature selection is presented in the first subsection, while the Markov random field (MRF) model is highlighted in the second.

2.1   Feature Selection

HSI systems gather large amounts of information; however, not all the data collected is necessarily relevant to the problem investigated. The problem of high dimensionality can be alleviated by using a feature selection process. Feature selection is the process of choosing a relevant subset of features (in this study, wavelengths) and discarding the remaining ones (e.g. irrelevant and redundant) [13,14]. The process of feature selection can be described in four steps: search organisation, subset evaluation, stopping criteria, and result validation. The first step is responsible for generating several subsets of features and that includes determining search direction and procedure. The second step involves evaluation of the relevance of the generated subsets, based on certain criteria, in order to select the optimal one (i.e. the one that maximises the evaluation criteria). The last two steps determine when the process should be halted and the significance of the selection parameters to the investigated problem.

Feature selection models can be separated - based on certain evaluation criteria - into the following categories: filter, wrapper, and embedded [13,14]. The discrimination capability depends solely on data characteristics in the first model, while it depends on the mining algorithms used to assess the relevancy of the features in latter models. It should be noted that the embedded model was introduced to utilise both filter and wrapper models, i.e. to rank features based on their data characteristics and evaluate their goodness through classification algorithms. In addition, the filter model can produce acceptable to good performances in short time, while the wrapper and embedded models are easy to implement.

Various feature selection algorithms have been introduced in the past. The correlation-based feature selection (CFS) [15] algorithm has been shown to be particularly powerful due to its ability to discard irrelevant and redundant features, as well as producing good discrimination compared to other selection algorithms [12]. It uses Shannon's entropy $H(\boldsymbol{x}) = -\sum_{i=1}^{n} P(x_i) \log_2 P(x_i)$ and information gain $I(\boldsymbol{x}, \boldsymbol{y}) = H(\boldsymbol{x}) - H(\boldsymbol{x}|\boldsymbol{y})$ to minimise feature bias and then measures the correlation between the features and the classes. The measured correlation is then used to evaluate the feature heuristically:

$$Merit_S = \frac{n(\overline{r_{cf}})}{n + (n + n(n-1)\overline{r_{ff}})} \tag{1}$$

where $\overline{r_{cf}}$ denotes the average feature-class correlation, $\overline{r_{ff}}$ represents the average feature-feature correlation, and $Merit_S$ is the heuristic merit of a subset containing features.

2.2   MRF Model

MRF is an extension of the Markov chain model [16]. It has a set of nodes, each of which corresponds to a variable or set of variables (an example of MRF neighbour structure and corresponding parameters is shown in Fig. 1). It is also termed as undirected graph model since it is more natural for modelling certain problems, such as spatial statistics and image analysis [17]. Moreover, the orientation of the texture features is not required, unlike the directed graph model. The main advantages of MRF models compared to directed graph models are: (1) more natural for certain domains (i.e. symmetric) and (2) the discrimination of former models work better than the latter one due to the normalisation process (i.e. globally vs. locally). In contrast, the major disadvantages are: (1) less interoperable and (2) parameter estimation can be computationally more expensive (e.g. maximum likelihood estimate).

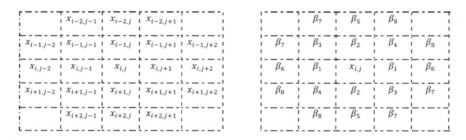

Fig. 1. Markov random field: neighbour structure (left) and corresponding parameters of $x_{i,j}$ (right)

The MRF can be mathematically described using the equivalent Gibbs distribution with regard to the same graph [18]. Let $P(x)$ denotes a Gibbs distribution for realisation x, $\mathcal{N}$ represents a neighbouring system, Ω denotes a finite lattice, $\mathcal{C}$ represents all possible cliques, i.e. the subset of a lattice consists of single and/or set of pixels which are neighbours to each other, then the distribution can be represented as:

$$P(x) = \frac{1}{Z} e^{-U(x)/T} \tag{2}$$

where T is a constant and stands for temperature; $U(x)$ represents the energy function that depends only on clique potential V_C on the lattice and can be written as:

$$U(x) = \sum_{c \in C} V_c(x) \tag{3}$$

Z denotes a normalising constant, also termed as partition function, and is defined as:

$$Z = \sum_x e^{-U(x)/T} \tag{4}$$

In terms of estimation of texture parameters, the least square (LS) and maximum likelihood (ML) are two estimates that are widely used with the texture [11]. The former is simple and it has low computational requirements compared to the latter, which is why it is more preferable in practice. For the

LS estimate, the parameters over a finite lattice Ω can be estimated using the following equation:

$$\hat{\beta} = [\sum_{m\in\Omega} \beta_m\beta_m^T]^{-1}[\sum_{m\in\Omega}\sum_{i,j\in\Omega} \beta_m x_{i,j}], m = 1, 2, \ldots, M \tag{5}$$

where $x_{i,j}$ represents the middle pixel and β_m denotes the neighbouring pixels that can be represented as:

$$\beta_m = col[x_{i+u,j+v}], (u,v) \neq 0, (u,v) \in \mathcal{N} \tag{6}$$

where u, v represent the location of the neighbouring pixels horizontally and vertically respectively and *col* stands for column. It is worth noting that LS is not consistent for non-causal neighbour sets [11,17]. However, it is more preferable compared to the ML estimate since ML is computationally expensive. In addition, the ML result is not always guaranteed (if not impossible) and requires an alternative function, i.e. iterative and computationally expensive, such as pseudo likelihood (MPL).

3 Materials and Methods

This section describes the materials and the spectral-texture approach. It first emphasises the specifications of the HSI systems used to capture the hyperspectral datasets and then describes the datasets used in the experiment, followed by the description of the spectral-texture approach.

3.1 HSI Systems and Datasets

Two HSI systems were used to collect the hyperspectral images: The University of Manchester (UoM) HSI system and The University of Bonn (Bonn) HSI system. The key specifications of both systems are given in Table 1. Both systems operate in controlled environments (dark room vs. dark chamber) in order to minimise the effect of unwanted noise. Furthermore, the dynamic range of both systems is managed to prevent saturation. In addition, three images (scene, dark noise, and flat field) are captured by both systems and then used to spectrally normalise the scene images to enhance the quality of the image. More information about both systems can be found in [19,20].

Two HSI datasets (called UoM and Bonn for simplicity) captured with different acquisition dates and exposure times were used for analysis purposes. The scene images of the UoM dataset consisted of six Arabidopsis leaf samples, while the Bonn dataset consisted of four sugar leaf samples [20] placed flattened on the sample plate in both cases (shown in Fig. 2). Moreover, the former dataset consists of two normal and four stressed (cold and heat) leaves (i.e. top and bottom left: normal, middle top and bottom: cold stress, and top and bottom right: heat stress), while the latter consists of four leaves under one condition; either healthy (controlled) or unhealthy (Cercospora). 648 samples were extracted from

Table 1. Key specifications of UoM and Bonn HSI systems

Specification	HSI systems	
	UoM	Bonn
Sensor type	Area	Line
Effective pixels	1024×1344	1600
Spectral range	$400-700\,nm$	$400-1000\,nm$
Spectral resolution	$10\,nm$	$2.8\,nm$
Spatial resolution	High	$0.19\,mm$
Radiometric resolution	12-bits	12-bits
Dispersion device	Liquid crystal tunable filter	Image spectrograph

the UoM dataset and divided into two groups: normal and stressed. The normal group was represented by 216 samples and the remaining represented stressed samples. The Bonn dataset yielded 196 samples: 98 samples of controlled and Cercospora conditions. It should be stated that only green areas of both datasets were considered for sample extraction. In addition, 50% of the samples was used for training purposes with 10-fold cross validation with the remaining ones used for testing.

Fig. 2. Samples of UoM and Bonn scene images: UoM Arabidopsis samples captured on February 2017 (left), and Bonn Sugar samples captured on March 2013 (right)

3.2 Spectral-Texture Approach

The proposed spectral-texture approach (is illustrated in Fig. 3) and can be described in four steps: spectral signature extraction, significant wavelengths selection, texture parameters estimation, and classification. The spectral signature is extracted from the pixel value of the small leaf region and then averaged over the entire wavelengths spectrum. This averaged signature is used to reduce the variation in pixel intensities across the selected leaf region. CFS is used in the wavelengths selection step in order to simplify the dataset and select the most significant wavelengths. The LS estimate is used to estimate the second order parameters of the MRF model and then average them either over the entire wavelength spectrum or over the list of selected wavelengths. In the final step, the selected wavelength and the estimated texture parameters are combined and passed to a conventional SVM classifier with a radial basis function (RBF) kernel for classification. The SVM uses a quadratic programming routine to solve the following quadratic problem with the regards to training set in order to find the best hyperplane [21]:

$$\min_{\omega \in \mathbb{R}^d, \xi_i \in \mathbb{R}^+} \frac{\|\omega\|^2}{2} + C \sum_i^N \xi_i \tag{7}$$

$$\text{subject to} : y_i(\omega.\boldsymbol{x_i} + b) \geq 1 - \xi_i$$

where ω, b, $\boldsymbol{x_i}$, y_i, ξ_i represent the weight vector, bias, training set, desired class label, and a non-zero slack variable respectively. Moreover, C is the regularisation parameter and it is used to mark the misclassified samples, thus determining the flexibility of the decision boundary. In this case, the decision function y can be solved using the weight vector as well as the bias:

$$y = \text{sign}(\omega.\boldsymbol{x_i} + b) \tag{8}$$

The value of decision function $y \in \{\pm1\}$, where 1 denotes one class and -1 the other class. It should be mentioned that false positive and negative errors have to be reduced in order to obtain a good classification result. In addition, the RBF kernel was used to employ the nonlinear hyperplane and can be defined as the following exponential function:

$$K(\boldsymbol{x}, \boldsymbol{y}) = e^{-\gamma\|\boldsymbol{x}-\boldsymbol{y}\|^2}, \gamma = \frac{1}{2\sigma^2} \tag{9}$$

4   Results and Discussions

The experimental results assessed the usefulness of the spectral-texture approach in analysing and classifying plant hyperspectral images under different conditions. Both UoM and Bonn datasets were used in the analysis. The final results were then compared with the classification results of existing spectral analysis approach, texture analysis approach, and the combination of all the wavelengths and the estimated texture parameters. 50% of the samples was used for training with 10-fold cross validation and the remaining 50% was used for testing. Table 2 displays the average classification rates of 100 runs with the standard deviations.

What stands out in this table is that the average classification rate of the spectral-texture approach outperforms other approaches (i.e. the all wavelengths, the selected wavelengths, the estimated texture parameters, and the combination of all the wavelengths and the estimated texture parameters). Moreover, a positive correlation is found between the selected wavelengths (e.g. 550 nm and 710 nm in UoM dataset and 513 nm and 698 nm in Bonn dataset) and the wavelengths used in the previous empirical studies [22,23]. These results suggest the proposed method is a valid approach for studying and analysing different plant

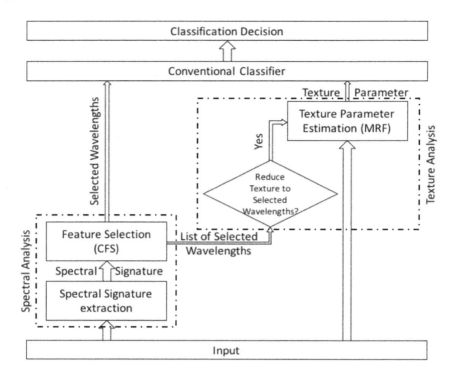

Fig. 3. Schematic of the proposed spectral-texture analysis approach

Table 2. Average classification rates

Technique	Average classification rates (%) (standard deviation)	
	UoM	Bonn
All wavelengths	87.35 (0.019)	98.54 (0.019)
CFS	88.40 (0.014)	98.85 (0.010)
Estimated texture parameters	74.64 (0.019)	65.42 (0.037)
All wavelengths + Estimated texture parameters	83.42 (0.020)	88.51 (0.042)
CFS + Estimated texture parameters	**92.87 (0.009)**	**99.36 (0.007)**

types and plant conditions. Further statistical tests revealed that the improvements are significant compared to the other approaches at a significance level of 1%, especially the wavelengths-texture and texture approaches (p-value$<10^{-5}$). On the other hand, the proposed approach tends to be slower compared to the spectral approach. It should be noted that using single feature selection algorithm might affect the prediction performance as well as the robustness since there is no single feature selection algorithm can deal with all situations. A more robust and effective selection framework was proposed in our previous work [10].

5 Conclusions

This paper presented a spectral-texture approach for analysis and classification of hyperspectral plant types and conditions. The experimental results from this approach have shown marked improvements in discrimination performance compared to other approaches. The improvements are statistically significant. The findings suggest that such approach seems valid and applicable for the study of different plant properties, types, and conditions. Future study can explore the effect of the estimated parameters with different orders (e.g. first and third orders MRF) to identify the optimal neighbouring system, as well as different classification routines, such as novelty detection to best identify plant properties, types, and conditions. In addition, different computational platforms can be explored to improve the speed of the proposed approach, thus reducing the time complexity.

References

1. Geladi, P.L.M., Grahn, H.F., Burger, J.E.: Multivariate images, hyperspectral imaging: background and equipment. In: Techniques and Applications of Hyperspectral Image Analysis, pp. 1–15. John Wiley and Sons, Ltd. (2007)
2. ElMasry, G., Sun, D.W.: Principles of Hyperspectral Imaging Technology. In: Sun, D.W. (ed.) Hyperspectral Imaging for Food Quality Analysis and Control, pp. 3–43. Academic Press, San Diego (2010)
3. Campbell, J., Wynne, R.: Introduction to Remote Sensing, 5th edn. Guilford Publications, New York (2011)
4. Liu, H., Lee, S.H., Chahl, J.S.: Development of a proximal machine vision system for off-season weed mapping in broadacre no-tillage fallows. J. Comput. Sci. **9**(12), 1803–1821 (2013)
5. Duchesne, C., Liu, J., MacGregor, J.: Multivariate image analysis in the process industries: a review. Chemometr. Intell. Lab. Syst. **117**, 116–128 (2012)
6. Lu, G., Fei, B.: Medical hyperspectral imaging: a review. J. Biomed. Opt. **19**(1), 010901 (2014)
7. Geladi, P., Bengtsson, E., Esbensen, K., Grahn, H.: Image analysis in chemistry i. Properties of images, greylevel operations, the multivariate image. TrAC Trends Anal. Chem. **11**(1), 41–53 (1992)
8. Qin, J.: Hyperspectral Imaging Instruments. In: Sun, D.W. (ed.) Hyperspectral Imaging for Food Quality Analysis and Control, pp. 129–172. Academic Press, San Diego (2010)
9. Bharati, M.H., Liu, J., MacGregor, J.F.: Image texture analysis: methods and comparisons. Chemometr. Intell. Lab. Syst. **72**(1), 57–71 (2004)
10. AlSuwaidi, A., Veys, C., Hussey, M., Grieve, B., Yin, H.: Hyperspectral feature selection ensemble for plant classification. In: Hyperspectral Imaging and Applications (HSI 2016), October 2016
11. Yin, H., Allinson, N.M.: Self-organised parameter estimation and segmentation of MRF model-based texture images. In: Proceedings of the IEEE International Conference on Image Processing, ICIP 1994, vol. 2, pp. 645–649. IEEE (1994)
12. AlSuwaidi, A., Veys, C., Hussey, M., Grieve, B., Yin, H.: Hyperspectral selection based algorithm for plant classification. In: 2016 IEEE International Conference on Imaging Systems and Techniques (IST), pp. 395–400, October 2016

13. Liu, H., Motoda, H.: Feature Selection for Knowledge Discovery and Data Mining. Kluwer Academic Publishers, Norwell (1998)
14. Liu, H., Yu, L.: Toward integrating feature selection algorithms for classification and clustering. IEEE Trans. Knowl. Data Eng. **17**(4), 491–502 (2005)
15. Hall, M.A., Smith, L.A.: Feature selection for machine learning: Comparing a correlation-based filter approach to the wrapper. In: Proceedings of the Twelfth International Florida Artificial Intelligence Research Society Conference, pp. 235–239 (1999)
16. Blake, A., Kohli, P., Rother, C.: Markov Random Fields for Vision and Image Processing. The MIT Press, Cambridge (2011)
17. Murphy, K.P.: Machine Learning: A Probabilistic Perspective. The MIT Press, Cambridge (2012)
18. Geman, S., Geman, D.: Stochastic relaxation, gibbs distributions, and the Bayesian restoration of images. IEEE Trans. Pattern Anal. Mach. Intell. **PAMI–6**(6), 721–741 (1984)
19. Foster, D.H., Amano, K., Nascimento, S.M.C.: Color constancy in natural scenes explained by global image statistics. Vis. Neurosci. **23**(3–4), 341–349 (2006)
20. Mahlein, A.K., Hammersley, S., Oerke, E.C., Dehne, H.W., Goldbach, H., Grieve, B.: Supplemental blue led lighting array to improve the signal quality in hyperspectral imaging of plants. Sensors **15**(6), 12834–12840 (2015)
21. Kulkarni, S., Harman, G.: An Elementary Introduction to Statistical Learning Theory, 1st edn. Wiley Publishing, New Jersey (2011)
22. Gitelson, A., Merzlyak, M.N.: Spectral reflectance changes associated with autumn senescence of aesculus hippocastanum l. and acer platanoides l. leaves. spectral features and relation to chlorophyll estimation. J. Plant Physiol. **143**(3), 286–292 (1994)
23. Mahlein, A.K., Rumpf, T., Welke, P., Dehne, H.W., Plmer, L., Steiner, U., Oerke, E.C.: Development of spectral indices for detecting and identifying plant diseases. Remote Sens. Environ. **128**, 21–30 (2013)

Face Attributes Retrieval by Multi-Label Contractive Hashing

Xuan Zhao$^{(\boxtimes)}$, Xin Jin, and Xiao Guo

Nanjing University of Aeronautics and Astronautics,
Nanjing, People's Republic of China
{zhaoxuan16,x.jin}@nuaa.edu.cn, njguoxiao@foxmail.com

Abstract. Substantial increase of Internet data requires efficient storage and rapid retrieval strategy. Hence, supervised hashing method is introduced in this issue. By mapping high dimensional data to compact binary codes, supervised hashing methods could downsize data while preserving semantic similarity based on labels. However, most of these hashing methods are designed for simple binary similarity, therefore they fail to manage the complex multi-level semantic structure of multi-label images. In this work, we propose a novel Multi-Label Contractive Hashing (MLCH) to preserve multi-level semantic similarity of face attributes images. To improve the efficiency of training process, an optimized triplet selection algorithm is implemented. Gradual learning is adopted to accelerate the rate of convergence and enhance the performance of proposed model. Meanwhile, contractive constraint is introduced to obtain more saturated binary codes. The proposed MLCH is evaluated on datasets CelebA and PubFig. Experimental results prove the validity of these ingenious strategies and demonstrate superiority of MLCH to the state-of-the-art hashing methods in large-scale image retrieval.

Keywords: Hashing method · Multi-label · Contraction · Face attribute

1 Introduction

The ever-growing amount of big data arises attendant problems. For instance, larger storage and faster query speed are required on demand. Hashing techniques could reduce storage cost and accelerate retrieval speed in virtue of mapping high-dimensional data to compact binary codes, and simultaneously preserve the similarity of the original data. Hence, hashing has been widely used in computer vision, machine learning, image retrieval and related areas.

Hashing methods can be divided into two main categories: unsupervised and supervised methods. As a representative among early unsupervised hashing methods, Locality-Sensitive Hashing (LSH) [10] learns hashing functions by random projections without exploring the data distribution. LSH and its variants tend to require long binary codes for high precision and recall. Some methods attempt to analysis metric similarity from distribution information, such

© Springer International Publishing AG 2017
H. Yin et al. (Eds.): IDEAL 2017, LNCS 10585, pp. 261–269, 2017.
https://doi.org/10.1007/978-3-319-68935-7_29

as Spectral Hashing (SH) [11], Iterative Quantization (ITQ) [2]. In contrast to unsupervised methods, supervised hashing applies to measure semantic similarity [8,11]. To improve the saturation of binary codes, Auto-encoder Jacobian Binary Hashing (Auto-JacoBin) [1] introduces a contractive constraint, proposed by Contractive Auto-Encoders (CAE) [8], to enhance the efficiency of encoding.

Most of hashing methods tend to utilize single class labels or pairwise relationships as supervisory information. In practice, however, extensive images are associated with multiple labels. These complex potential relations could be handled by triplet loss effectively. Whereas, hashing methods for multi-label tasks have not been sufficient researched.

Attribute labels, which contain human-comprehensible semantic information, are mid-level representation of images. Thus they can effectively bridge the Semantic Gap between low-level features and high-level semantics. Recently, attribute learning has drawn substantial attention in computer vision tasks. For instance, police need to retrieve suspect with similar attribute labels described by witnesses from massive surveillance videos. However, current researches are mainly devoted to face attribute prediction rather than face attribute retrieval.

Considering multi-label retrieval containing more information than single-label, we propose a novel Multi-Label Contractive Hashing (MLCH) to achieve efficient storage and rapid retrieval of multi-label face images. Our main contributions can be summarized as follows:

(1) To improve the efficiency of training process, an optimized algorithm for selecting triplet is introduced and gradual learning strategy is adopted to accelerate the rate of convergence.
(2) A contractive constraint term is deployed to improve the saturation of binary codes, thereby lessens the idle coding bits.
(3) Experimental results demonstrate that proposed method outperforms many state-of-the-art hashing methods on face attribute datasets CelebA and Pub-Fig.

The rest of this paper is organized as follows: Related work is briefly discussed in Sect. 2. In Sect. 3, we will introduce the details of our MLCH model. Experimental evaluations are presented in Sect. 4. Finally, we conclude the paper in Sect. 5.

2   Related Work

Multi-label hashing method: Iterative Quantization with Canonical Correlation Analysis (CCA-ITQ) [3] utilizes multi-labels to decrease the dimensionality of input data by CCA, and encodes the outcome into binary codes through minimizing the quantization error. However, it learns hash function with point-wise labels, which cannot deal with complex semantic similarity relationships very well. Deep Semantic Ranking Based Hashing (DSRH) [12] deploys deep convolutional neural network to jointly learn feature representations and mappings.

Meanwhile, a ranking list, obtained with multi-level semantic similarity information, is introduced to guide the foundation of hash functions. Considering DSRH suffers from huge computational cost brought by the triplet ranking loss for hashing, gradual learning is adopted to address this issue.

Attribute learning: Most of face attribute relevant works focus on attribute prediction. The most representative work of face attribute retrieval is [4], where 65 facial attributes are defined, and a text-based method is implemented to retrieve face images with similar attributes. However, it constructs binary classifiers for every attribute instead of training a classifier for multiple attributes. In contrast, proposed method concentrates on establishing a multi-label classifier, which is more suitable for practical situation.

3   The Proposed Method

3.1   Problem Definition

Suppose that we have N training images organized as $X = [x_1, x_2, \cdots, x_N] \in R^{d \times N}$. Each image possesses a d-dimensional descriptor, denoted as x, and corresponding attribute labels $a \in \{-1, +1\}^t$, where $+1$ and -1 represent presence or absence of that attribute respectively. Proposed model aims for establishing a hash function h that maps descriptors to k-dimensional ($k \ll d$) binary codes in Hamming space, while retaining semantic similarity of origin images.

3.2   Enhancement Strategies

Optimization triplet selection algorithm: A triplet consists of anchor point, similar point and dissimilar point, denoted as x^a, x^s, x^d. The criterion of choice is $\delta_a(x^a, x^s) > \delta_a(x^a, x^d)$, where δ represents the description of semantic similarity between two samples, computed by the number of common labels they share. Our purpose is learning a hashing function h to ensure that each triplet satisfies the following formula:

$$d_H(h(x^a), h(x^s)) + \alpha \leqslant d_H(h(x^a), h(x^d)) \tag{1}$$

In Eq. (1), α represents the minimum margin and $d_H(h(x_1), h(x_2))$ is Hamming distance between $h(x_1)$ and $h(x_2)$, computed as $-h(x_1)^T h(x_2)$.

Traditional triplet selections tend to choose triplets randomly among the whole training set. However, the entirely stochastic process neglects to handle some optimizable situation. As shown in Fig. 1, (a) can be utilized for learning, whereas (b) will be ignored during training process. Hence, we propose an optimization triplet selection to improve the efficiency of training process.

Gradual learning: Considering some images could not be cast as representative samples in classes they belong to, the training set is divided into two parts, the optimized triplet selection would be implemented on subset consists of representative samples and the remainder is utilized to fine-tune model for retaining robustness.

Fig. 1. Two different cases while selecting triplets. (a) $d_H(h(x^a), h(x^s)) > d_H(h(x^a), h(x^d))$ (b) $d_H(h(x^a), h(x^s)) < d_H(h(x^a), h(x^d))$

Algorithm 1. Optimized Triplets selection

Input:
   Binary codes of training data: $h(x_1), h(x_2), \cdots, h(x_n)$;
   Attribute labels: $a_1, a_2, ..., a_n$;
Output:
   Triplet Set: $\{x^a, x^s, x^d\}^m$
   while $i < m$ **do**
      1.Choose 3 samples from training data randomly;
      2.Calculate the similarity δ_a between any two of the 3 samples, select the maximum and minimum; if 3 similarities are the same, **go to 1**;
      3.Pick out x^d according to similarities (remove the two samples with the highest similarity and the remaining sample is x^d);
      4.Calculate the distances of x^d with the other two samples, denote the smaller one as x^a and the other as x^s;
      5.If $d_H(h(x^a), h(x^s)) + \alpha > d_H(h(x^a), h(x^d))$,then select $\{x^a, x^s, x^d\}$ into triplets set and $i \Leftarrow i + 1$; Else, **go to 1**;
   end while

In problem definition, t has been denoted as the number of attributes. In order to emphasize the advantages of multi-attributes, we have tried different values of t. Limited by hardware device, t is set to 10. Multiclass linear discriminant analysis is utilized to decrease the inner-class while enlarge the inter-class distance. Subsequently, we attempt to seek out the cluster center of each class by the following steps:

(1) Set initial cluster center as the average of total samples in the class.
(2) Calculate the distance between each sample and current cluster center, remove 2% samples with longest distance.
(3) Update cluster center by computing the average of remaining samples.
(4) If the distance between new cluster center and the anterior one is large than threshold, which is set to 80, go to (2); else end.

Hence, first training subset consists of remaining images and second part contains those samples which removed during step (2). First part is prior utilized to train model for their preferable representation of classes they belong to, then second part will fine-tune the model in a low learning rate to maintain the robustness of model. For instance, the first part of CelebA consists of over 32,000 images and the initial learning rate is set to 0.01. The model converges very fast, around 6-8 epochs and we then feed remainder images to fine-tune the model with decreased learning rate, set to 0.002 in particular. The whole training process costs almost 15 epochs (750 iterations).

3.3 The Contractive Constraint

In order to obtain more saturated binary codes, a contractive constraint is introduced, computed as the Jacobian matrix of input data in Frobenius norm. This constraint term dedicates to contract the manifold of input data, thereby improving the balance of encoding result. The contractive constraint can be calculated as:

$$\|J_f(x)\|_F^2 = \sum_{ij} (\frac{\partial h(x_j)}{\partial x_i})^2 \tag{2}$$

In addition, the implementation of contractive constraint can retain more information of original images while the bit of hash code is fixed and achieve better performance.

3.4 The Objective Function

The initial loss function is defined as:

$$L(x^a, x^s, x^d) = max\{d_H(h(x^a), h(x^s)) + \alpha - d_H(h(x^a), h(x^d)), 0\} \tag{3}$$

To impose different penalties for the multi-label problem, a penalty coefficient, according to the different similarity, is introduced into triplet loss:

$$\gamma(x^a, x^s, x^d) = 2^{\delta_a(x^a, x^s)} - 2^{\delta_a(x^a, x^n)} \tag{4}$$

Then Eq. (3) can be rewritten as:

$$L(x^a, x^s, x^d) = \gamma(x^a, x^s, x^d) max\{d_H(h(x^a), h(x^s)) + \alpha - d_H(h(x^a), h(x^d)), 0\} \tag{5}$$

The hashing function is defined as follows:

$$h(x; W) = 2\sigma(W^T x) - 1 \tag{6}$$

In Eq. (6), $W \in R^{d \times k}$, $\sigma(x) = \frac{1}{(1+exp(-x))}$ (images are encoded to $\{-1, +1\}^k$ during training process and mapped to $\{0, 1\}^k$ ultimately).

We denote D as the set of triplets, then the objective function is

$$F(W) = \sum_{(x^a, x^s, x^d) \in D} L(x^a, x^s, x^d) + \mu \|J_f(X)\|_F^2 + \frac{\alpha}{2} \|mean_x(h(x; W))\|_2^2 + \frac{\beta}{2} \|W\|_2^2 \tag{7}$$

In Eq. (7), $L(x^a, x^s, x^d)$ is the loss function defined in Eq. (5), μ, α and β are set to 0.01,1 and $5e^{-4}$ respectively. The third term of Eq. (7) encourages each binary bit averaged over the training data to be mean-zero and the last term of constraint is the L_2 weight decay to penalize large weights.

4   Experiments

4.1   Datasets and Evaluation Standards

Proposed model MLCH has been evaluated on two face attribute datasets, Celeb-Faces Attributes Dataset (CelebA) [7] and Public Figures Face Database (Pub-Fig) [5]. Images of both datasets are taken in completely unconstrained scenes and cover large variations in lighting and imaging conditions, etc. *CelebA* has more than 200,000 celebrity images and each sample possesses 40 attribute annotations. 50,000 images are randomly selected from training set for training and retrieval, 1,000 images are sampled from testing set as testing queries. *PubFig* consists of 58,797 images of 200 people collected from the internet. This dataset can be divided into development set and evaluation set, containing 16,336 images and 42,461 images respectively. During experiment, 1,000 images are randomly sampled from evaluation set as queries and the entire development set is utilized as training set. Those attribute labels in float have been normalized to ±1. After extracting image features by VGG-FACE [9], 4096-dimensional vectors are utilized as input data of subsequent hashing process.

For the better evaluation of multi-label problem, Normalized Discounted Cumulative Gains (NDCG) and Average Cumulative Gain (ACG) are chosen as benchmark evaluation of experimental results.

4.2   Evaluation of Enhancement Strategies

To verify the effectiveness of proposed strategies implemented on training set, we illustrate the convergence rate of MLCH and MLCH-NS (MLCH without enhancement strategies). MLCH converges very fast due to the optimized triplet selection implemented on divided subset, while MLCH-NS need more than 30 epochs to improve its performance. In addition, smaller undulations in MLCH indicate the more stable training process than MLCH-NS and the ultimate result of MLCH surpasses MLCH-NS achieved (Fig. 2).

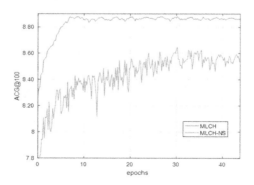

Fig. 2. Convergence analysis of MLCH and MLCH-NS.

4.3   Evaluation of Contractive Constraint

MLCH without contractive constraint, termed MLCH-NC, is also trained to verify the significance of deploying such constraint. The saturation analysis of binary codes is illustrated in Fig. 3. The vertical axis indicates the percentage of code bits that less than 0.01 or greater than 0.09, and the horizontal axis is epochs in training process. It can be observed that binary codes of MLCH is closer to 0 or 1 than MLCH-NC, in other word, more saturated (Table 1).

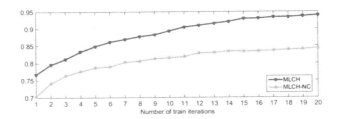

Fig. 3. Saturation analysis of MLCH and MLCH-NC.

Table 1. NDCG and ACG of MLCH and MLCH-NC.

Evaluation criteria	Methods	CelebA					Pubfig				
		16 bits	32 bits	48 bits	64 bits	128 bits	16 bits	32 bits	48 bits	64 bits	128 bits
NDCG@100	MLCH-NC	0.8784	0.8798	0.8790	0.8804	0.8926	0.8164	0.8287	0.8352	0.8411	0.8517
	MLCH	0.8799	0.8842	0.8879	0.8911	0.8946	0.8214	0.8310	0.8360	0.8429	0.8586
ACG@100	MLCH-NC	8.8050	8.8471	8.8618	8.8713	8.8840	7.8884	8.0875	8.1841	8.1998	8.2237
	MLCH	8.8320	8.8672	8.8851	8.8913	8.9106	7.9013	8.1003	8.1970	8.2161	8.2480

As shown in Fig. 3, introducing contractive constraint achieves higher NDCG and ACG as theoretical expected. The primary reason is that the increase of saturation enables binary codes to carry more semantic information.

4.4   Method Comparison

Proposed method is compared with ITQ, SH, LSH and Supervised Hashing with Kernels (KSH) [6], CCA-ITQ[1]. To ensure the fair comparison with competing approaches, feature vectors obtained from VGG-FACE are utilized as input data and parameters in those approaches are set to optimal settings declared in their literatures. Five groups of experiments are conducted while each group contains 10 attributes selected randomly as labels. Ultimate result is computed as the average of each group.

As illustrated in Table 2, MLCH shows superior performance of NDCG and ACG on both datasets. CCA-ITQ also utilizes multi-labels information, but with

---

[1] Proposed model does not compared to DSRH because the training code is not available.

Table 2. Average NDCG and ACG for various bits. Red indicates the best and blue indicates the second best performance.

Evaluation criteria	Methods	CelebA					Pubfig				
		16 bits	32 bits	48 bits	64 bits	128 bits	16 bits	32 bits	48 bits	64 bits	128 bits
NDCG@100	LSH	0.7867	0.7921	0.7926	0.7999	0.8103	0.7607	0.7679	0.7740	0.7754	0.7840
	SH	0.7919	0.8013	0.8049	0.8140	0.8146	0.7649	0.7707	0.7752	0.7824	0.7842
	CCA-ITQ	0.8395	0.8316	0.8284	0.8248	0.8175	0.8052	0.7939	0.7907	0.7864	0.7886
	ITQ	0.8080	0.8157	0.8203	0.8250	0.8267	0.7914	0.7990	0.8004	0.8019	0.8041
	KSH	0.8285	0.8359	0.8367	0.8387	0.8380	0.7763	0.7822	0.7871	0.7929	0.7957
	MLCH	0.8803	0.8841	0.8876	0.8908	0.8944	0.8196	0.8283	0.8375	0.8444	0.8454
ACG@100	LSH	8.0689	8.1228	8.1488	8.1943	8.2953	7.2798	7.3764	7.4424	7.4544	7.5049
	SH	8.1416	8.1994	8.2333	8.3212	8.3253	7.3786	7.4240	7.4667	7.5631	7.5642
	CCA-ITQ	8.5737	8.5045	8.4618	8.4273	8.3489	7.6804	7.6383	7.6029	7.5845	7.5749
	ITQ	8.2894	8.3586	8.4202	8.4448	8.4767	7.6240	7.6766	7.7220	7.7647	7.8042
	KSH	8.4851	8.5347	8.5524	8.5669	8.5873	7.5773	7.5981	7.6381	7.6520	7.6766
	MLCH	8.8300	8.8652	8.8852	8.8947	8.9046	7.8500	8.0860	8.1776	8.2093	8.2409

the increase of code bits, its performance is decreased, even worse than KSH and unsupervised ITQ. We can see that MLCH can achieve a better retrieval result in virtue of higher learning efficiency and more saturated codes obtained by enhancement strategies and the contractive constraint.

5   Conclusion

In this paper, we propose a novel MLCH to retrieve the face images with multi attributes. An optimized triplets selection and gradual learning strategy are proposed to accelerate the rate of convergence and suffice the training process. A contractive constraint term is introduced to obtain more saturated binary codes. Thus we can preserve the multi-level semantic similarity of face images while improve retrieval speed and reduce storage space. Experimental results demonstrate the superiority of the proposed method over many state-of-the-art hashing methods.

References

1. Fu, X., McCane, B., Mills, S., Albert, M., Szymanski, L.: Auto-Jacobin: auto-encoder Jacobian binary hashing. CoRR abs/1602.08127 (2016)
2. Gong, Y., Lazebnik, S.: Iterative quantization: a procrustean approach to learning binary codes. In: IEEE Conference on Computer Vision and Pattern Recognition, pp. 817–824 (2011)
3. Hardoon, D.R., Szedmak, S.R., Shawe-taylor, J.R.: Canonical correlation analysis: an overview with application to learning methods. Neural Comput. **16**(12), 2639–2664 (2004)
4. Kumar, N., Belhumeur, P., Nayar, S.: FaceTracer: a search engine for large collections of images with faces. In: Forsyth, D., Torr, P., Zisserman, A. (eds.) ECCV 2008. LNCS, vol. 5305, pp. 340–353. Springer, Heidelberg (2008). doi:10.1007/978-3-540-88693-8_25

5. Kumar, N., Berg, A.C., Belhumeur, P.N., Nayar, S.K.: Attribute and simile classifiers for face verification (2010)
6. Liu, W., Wang, J., Ji, R., Jiang, Y.G.: Supervised hashing with kernels. In: IEEE Conference on Computer Vision and Pattern Recognition, pp. 2074–2081 (2012)
7. Liu, Z., Luo, P., Wang, X., Tang, X.: Deep learning face attributes in the wild. In: IEEE International Conference on Computer Vision, pp. 3730–3738 (2015)
8. Rifai, S., Vincent, P., Muller, X., Glorot, X., Bengio, Y.: Contractive auto-encoders: explicit invariance during feature extraction. In: Getoor, L., Scheffer, T. (eds.) Proceedings of the 28th International Conference on Machine Learning (ICML 2011), pp. 833–840. ACM, New York (2011)
9. Simonyan, K., Zisserman, A.: Very deep convolutional networks for large-scale image recognition. CoRR abs/1409.1556 (2014). http://arxiv.org/abs/1409.1556
10. Slaney, M., Casey, M.: Locality-sensitive hashing for finding nearest neighbors [lecture notes]. IEEE Signal Process. Mag. **25**, 128–131 (2008)
11. Weiss, Y., Torralba, A., Fergus, R.: Spectral hashing. In: Conference on Neural Information Processing Systems, Vancouver, British Columbia, Canada, pp. 1753–1760, December 2008
12. Zhong, Y., Sullivan, J., Li, H.: Face attribute prediction with classification CNN. CoRR abs/1602.01827 (2016). http://arxiv.org/abs/1602.01827

Trajectory Similarity-Based Prediction with Information Fusion for Remaining Useful Life

Zhongyu Wang, Wang Tang, and Dechang Pi[(✉)]

College of Computer Science and Technology,
Nanjing University of Aeronautics and Astronautics, 29 Jiangjun Road,
Nanjing 211106, Jiangsu, People's Republic of China
1109763309@qq.com, 46015316@qq.com, nuaacs@126.com

Abstract. Prediction of remaining useful life (RUL) has widely application in industrial domain, especially for aircraft where safety and reliability are of high importance. RUL Prediction can provide the time of failure for a degrading system, so that there are high requirements of its accuracy. In this paper, we propose a new trajectory similarity-based RUL prediction approach with an information fusion strategy (named IF-TSBP) in the similarity measure step. The novel information fusion strategy allows us to get more precise trajectory similarity degree than traditional similarity measure strategy which contributes to the prediction result. The experimental results show that the prediction accuracy of our proposed algorithm IF-TSBP outperforms the traditional trajectory similarity-based prediction approach and some common machine learning algorithms.

Keywords: Remaining useful life · Similarity measure · Prognostics

1 Introduction

In modern aircraft industry, the safety of aircraft equipment plays an important role for it is tightly related with the safety of passengers. Among all equipment, aircraft engine is the most important. In this situation, prognostics and health management (PHM) and Condition-Based Maintenance (CBM) are popular in research domain.

Prediction of remaining useful life (RUL) is the main task of PHM whose purpose is to provide, as early as possible, accurate prediction of the time when the system comes to failure. In general, prediction of RUL can be divided into three kinds of strategies: model-based, data-driven and hybrid approaches. For the lack of priori physical information which is fundamental to model-based and hybrid approaches, this paper focuses on the data-driven approaches.

There are many algorithms can be used in data-driven approach to estimate the system RUL such as neural network, support vector machine, decision tree and so on.

However, these machine learning methods all need to find a degradation model before estimation that may have a negative effect on the result. In this paper, we choose

© Springer International Publishing AG 2017
H. Yin et al. (Eds.): IDEAL 2017, LNCS 10585, pp. 270–278, 2017.
https://doi.org/10.1007/978-3-319-68935-7_30

the trajectory similarity-based prediction (TSBP) approach that the data set must meet the assumptions:

i. The health state of system can be collect from sensors;
ii. There are enough historic run-to-failure data;
iii. If two systems of same kind have similar degradation situation, we can consider that they have similar RUL. Also, the similarity and RUL are positively related.

Wang et al. [1] first proposed the TSBP approach and got top three in PHM2008 Challenge. After that, Racha K et al. [2] proposed an enhanced TSBP approach, whose prediction results are as good as common machine learning methods and have satisfying efficiency. In this paper, we propose an approach named IF-TSBP. This approach uses a new similarity measure method with information fusion, which increases the final result of RUL estimation.

The rest paper is organized as follows. In Sect. 2, algorithm IF-TSBP is proposed. The experimental dataset was shown in Sect. 3. The analysis of the experimental results is provided in Sect. 4. Finally, the Sect. 5 offers conclusion of the paper.

2 IF-TSBP Approach

The overall workflow of IF-TSBP approach is shown in Fig. 1. There are two fundamental processes: health modeling of system state and RUL estimation based on IF-TSBP. First, we build health indicator (HI) model of history run-to-failure data [3]. Then, we apply the same model to the to-be-predict data collected from sensory. Finally we get the estimated RUL by matching the to-be-predict HI data with historic HI data.

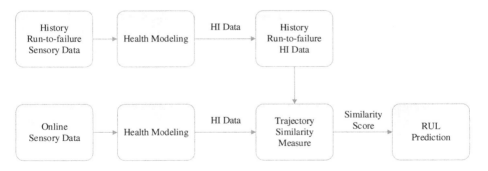

Fig. 1. Overall approach workflow

2.1 Health Modeling of System State

The data of system state is a high-dimension data collected by several sensors. It is not wise to use the high-dimension data directly in similarity measure process. We need to build a model of system state data at first. The common approach is to establish health indicator (HI) model, which allows us change the high-dimension health state data into one-dimension HI data. HI model is widely accepted in RUL estimation. Wang et al. [1] used regression method to build HI model after sensor selection. Benkedjouh et al. [4] apply ISOMAP to calculate HI of data.

According to [1], regression model gets better results than other methods such as Kalman filter and PCA, and linear regression model is the most suitable model among regression model. In this case, we choose linear regression to build the HI model. The algorithm flow is shown in Algorithm 1.

Algorithm 1: Health modeling of system state

Function: Build health indicator model by linear regression

Input: Run-to-failure data set S_{train} (including n units), still operating data set S_{test}

(including m units), max health threshold $Tmax$, min health threshold $Tmin$

Output: HI of each unit

01. *let set F={}*   //F is the set of feature points involved in regression model training

02. *for i from 1 to n*

03.   *Unit=S_{train}(i)* //Get this unit of data set

04.   *if lifecycle of Unit>Tmax* //if there are some point whose life larger than Tmax

05.   *F=F ∪{(p,1)|Unit.life(p)> Tmax }*//set the points HI=1

06.   *end if*

07.   *F=F ∪{(p,0)|Unit.life(p)<Tmin}*//set the points whose life smaller than *Tmin* HI=0

08. *end for*

09. *Reg=Linear Regression result of point set F* //Reg is a function changing point into HI

10. *use Reg to calculate the HI of all data points in S_{train} and S_{test}*

The results we get in Algorithm 1 is discrete data of HI. In order to build efficient trajectory model, quadratic exponential regression is selected to operate curve fitting method, as is shown in formula (1).

$$y = ae^{bx} + ce^{dx} \qquad (1)$$

The fit curve after quadratic exponential regression is shown in Fig. 2.

As shown in Fig. 2, discrete data of HI have been fit by a quadratic exponential curve which is the result of health indicator curve fitting process.

Through this process, an efficient system health trajectory model has been established which is fundamental to the following trajectory similarity match process.

2.2   RUL Estimation with IF-TSBP

According to assumption iii, we can change the RUL estimation problem to the problem of finding the best match between current to-be-estimated HI trajectory and history HI trajectory. Then we can estimate RUL by the chosen history HI trajectory. However, the results will be seriously affected by noise in data if just one history HI

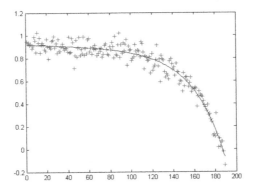

Fig. 2. Example of health indicator curve fitting

trajectory is chosen. In this case, we proposed a novel information fusion similarity measure method.

2.2.1 Information Fusion Similarity Measure

In order to get better similarity measure between degradation trajectories, in this paper, we proposed a novel information fusion similarity measure which allows us get more valuable information than normal similarity measure method.

Assume that two degradation trajectories with the same length, as formula (2).

$$
\begin{aligned}
T_a &= \{a_1, a_2, \ldots, a_n\} \\
T_b &= \{b_1, b_2, \ldots, b_n\}
\end{aligned}
\tag{2}
$$

Their local degradation trend T'_a and T'_b is shown in formula (3).

$$
\begin{aligned}
T'_a &= \{a_2 - a_1, a_3 - a_2, \ldots, a_n - a_{n-1}\} \\
T'_b &= \{b_2 - b_1, b_3 - b_2, \ldots, b_n - b_{n-1}\}
\end{aligned}
\tag{3}
$$

The similarity score $Score_{IF}(T_a, T_b)$ is defined in formula (4).

$$
Score_{IF} = e^{-\mu Score_d(T_a, T_b) - \lambda Score_t(T'_a + T'_b)}
\tag{4}
$$

$Score_d$ is the global numerical similarity score, $Score_t$ is the local degradation trend similarity score. They are defined in formula (5) and (6):

$$
Score_d = \sum_{i=1}^{m} W_i sim(seg_i(T_a), seg_i(T_b))
\tag{5}
$$

$$
Score_t = \sum_{i=1}^{m} W_i sim(seg_i(T'_a), seg_i(T'_b))
\tag{6}
$$

In formula (5) and (6), we divide the trajectory A and B into m segments, then calculate the similarity between each two segments which have their own weight. Euclidean distance is chosen to calculate the similarity. The weight is proportional to the series time, because the late segments are more important in similarity measure.

If the similarity score is closed to 1, it means the two trajectories are similar. If the similarity score is closed to 0, it means the two trajectories are different.

2.2.2   Trajectory Match

Trajectory match is the main task of TSBP. The purpose of trajectory match is to find one or more than one history trajectories which are most similar with to-be-estimated trajectory. Under the assumption iii, we can use these history trajectories to estimate the to-be-estimated RUL.

It is not easy to measure the similarity between the trajectory to be measured and the history trajectory because their lengths are generally different. In this case, we use slide window method that compares the to-be-estimated trajectory to the same length window on history trajectories. Then the top K best window will be chosen to calculate the best average similarity score and the best match offset point. The algorithm is shown in Algorithm 2.

Algorithm 2: Trajectory Match with Information Fusion

Function: Get similar score of two trajectory

Input: Two trajectories T_{test} and T_{train}

Output: Best match point P_{best} and best average similarity score SC_{best}

01. *for offset from 1 to length(T_{train})-length(T_{test})*

02. $T_a = T_{test}(i)$

03. $T_b = \{T_{train}(i) | offset <= i <= offset + length(T_{train})\}$ //resize history trajectory

04. $SC(offset) = Score_{IF}(T_a, T_b)$ //calculate IF-Score of this offset

05. *end for*

06. *find top-K best SC and offset*

07. SC_{best}=*average of top-K best SC*, P_{best}= *average of top-K best offset*

SC_{best} and P_{best} are obtained by Algorithm 2. The prediction of RUL in this history trajectory was calculated by P_{best}, as is shown in Fig. 3. The RUL calculation method is formula (7).

$$RUL_{predict} = Lifecycle_{train} - Length_{test} - P_{best} \qquad (7)$$

As shown in Fig. 3, the RUL of test curve is calculated by run-to-failure length of history curve $Lifecycle_{train}$, best match point P_{best} and the length of the test curve $Length_{test}$.

2.2.3   RUL Estimation

The prediction RUL of to-be-estimated trajectory on one history trajectory and the score of prediction are obtained through approach above. In the same way, the

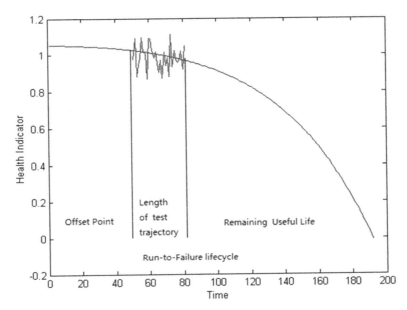

Fig. 3. Use Trajectory match to RUL estimation

prediction RUL and prediction score on each history trajectory will be obtained and constitute a set of prediction results $\{R_i | 1 \leq i \leq N\}$, N is the number of history trajectories.

When the RUL is comprehensively estimated, not all the forecast results can be taken into account, because some of the predictions score is too low to be considered.

To avoid the negative effect caused by these predictions with low score, an unsupervised classifier K-means is used to pick them out.

Finally, the prediction RUL is obtained by formula (8).

$$RUL_{final} = \sum_{i=1}^{m} \frac{Score_{IF}(i)}{Sum} RUL_{predict}(i) , \ Sum = \sum_{i=1}^{m} Score_{IF}(i) \tag{8}$$

3   Experimental Data and Performance Evaluation

3.1   Turbofan Data Set

The experimental data used in this paper is the data of aircraft turbofan, which is the Dataset#FD001 of NASA Prognostic Data Repository. The data set consists of one training set and one test set, both of which have 100 engine degradation time series. Each time series contains 26 dimensions, of which 21 dimensions are collected by the sensors. The data in the training set are run-to-failure time series, and the end time of the sequence is the exact time when the engine went wrong. The data in the test set is the time series of the engine which is running without problems.

3.2 Performance Evaluation

To evaluate the forecast results, we use the competition criteria of PHM Challenge 2008 [5], which is one of the universal standards for aircraft turbine engine evaluation as well. The criteria is shown in formula (9).

$$
S = \begin{cases} \sum_{i=1}^{n} e^{-\frac{d}{a_1}} - 1, & \text{for } d < 0 \\ \sum_{i=1}^{n} e^{\frac{d}{a_2}} - 1, & \text{for } d \geq 0 \end{cases} \tag{9}
$$

S is the calculated score, the smaller the value of S, the better the result of the evaluation. N represents on behalf of the amount of the test set. D is the difference between the estimated value and the actual value. According to literature [6], we set $a_1 = 10, a_2 = 13$.

As shown in (9), when d > 0, S grows faster. This is because if the prediction RUL is longer than the actual RUL, it may cause the system to fail before the expected dead time, which could lead to serious consequences. Therefore, the evaluation score will make a penalty for the late prediction.

4 Experimental Results and Analysis

This section uses the proposed IF-TSBP algorithm to predict turbofan's remaining useful life based on Dataset#FD001 of NASA. Software environment: Intel(R) Core (TM) i5-3470 3.2 GHZ CPU, 8G main memory, 1T hard disk and Microsoft Windows7 system. Experiment platform: Matlab. Language: Matlab.

4.1 HI Modeling Result

According to the prior knowledge in [1, 2], we select the 2nd, 3rd, 4th, 7th, 11th, 12th and 15th sensors, modeling the data of these seven sensors by the system health model method proposed in Sect. 2.1 to establish HI model for all these experimental data. We set the maximum health threshold $T_{max} = 300$ and the minimum health threshold $T_{min} = 5$. The modeling results are shown in Fig. 4.

In Fig. 4 we can find that, all the training data and most of the test data have shown a significant degradation. A small part of the test data, especially the data which have a too short sampling time, did not show significant degradation. Since this type of data has a small amount, there is no much effect on the testing results.

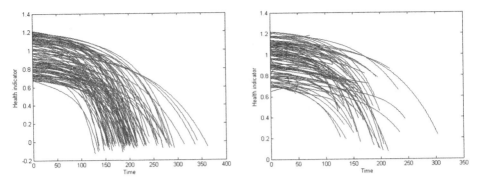

Fig. 4. HI modeling result of all training and testing data

4.2 RUL Predict Result

In this part, we use the HI modeling results in Sect. 4.1 for the RUL prediction by the proposed IF-TSBP algorithm. In calculating the similarity, we selected $\mu = 0.8, \lambda = 0.2$, which distance metric is the Euclidean distance.

The results of the prediction are calculated using the evaluation criteria proposed in Sect. 3.2, and compared with the scores which test on the same data set using approaches in [1, 2, 6], the results are shown in Table 1.

Table 1 shows that, IF-TSBP has a better prediction than the basic TSBP, and is also superior to some common machine learning algorithms.

Table 1. RUL predict score

Algorithm	RUL predict score
SVR [4]	1382
CNN [6]	1287
TSBP- Basic Euclidean Distance [1]	2534
TSBP-Weighed Euclidean Distance [2]	1426
IF-TSBP	**974**

5 Conclusion

In this paper, an enhanced approach named IF-TSBP is proposed, which optimizes the original TSBP from the perspective of similarity measure. The IF-TSBP approach takes both the global numerical similarity and local trend similarity among trajectories into account, making the trajectory similarity measure more accurate and improving the effect of the RUL prediction as well. Based on the experiment results, our approach gets higher accuracy than the traditional TSBP, and it is superior to some common machine learning algorithms in accuracy and practicability.

Acknowledgments. The research work is supported by National Natural Science Foundation of China (U1433116), and the Fundamental Research Funds for the Central Universities (NP2017208) and Foundation of Graduate Innovation Center in NUAA (kfjj20171603).

References

1. Wang, T., Yu, J., Siegel, D., et al.: A similarity-based prognostics approach for remaining useful life estimation of engineered systems. In: International Conference on Prognostics and Health Management, PHM 2008, pp. 1–6. IEEE (2008)
2. Khelif, R., Malinowski, S., Chebel-Morello, B., et al.: RUL prediction based on a new similarity-instance based approach. In: 2014 IEEE 23rd International Symposium on Industrial Electronics (ISIE), pp. 2463–2468. IEEE (2014)
3. Lam, J., Sankararaman, S., Stewart, B.: Enhanced trajectory based similarity prediction with uncertainty quantification. In: PHM 2014 (2014)
4. Benkedjouh, T., Medjaher, K., Zerhouni, N., et al.: Remaining useful life estimation based on nonlinear feature reduction and support vector regression. Eng. Appl. Artif. Intell. **26**(7), 1751–1760 (2013)
5. Saxena, A., Goebel, K., Simon, D., et al.: Damage propagation modeling for aircraft engine run-to-failure simulation. In: International Conference on Prognostics and Health Management, PHM 2008, pp. 1–9. IEEE (2008)
6. Sateesh Babu, G., Zhao, P., Li, X.-L.: Deep convolutional neural network based regression approach for estimation of remaining useful life. In: Navathe, S.B., Wu, W., Shekhar, S., Du, X., Wang, X.S., Xiong, H. (eds.) DASFAA 2016, Part I. LNCS, vol. 9642, pp. 214–228. Springer, Cham (2016). doi:10.1007/978-3-319-32025-0_14

Co-clustering with Manifold and Double Sparse Representation

Fang Li[1,2,3] and Sanyuan Zhang[1(✉)]

[1] College of Computer Science and Technology,
Zhejiang University, Hangzhou, China
`syzhang@zju.edu.cn`
[2] The Key Laboratory of Image and Graphic Intelligent Processing
of Higher Education in Guangxi, Guilin University
of Electronic Technology, Guilin, China
[3] Guangxi Key Laboratory of Trusted Software,
Guilin University of Electronic Technology, Guilin, China

Abstract. Clustering is a fundamental tool that has been applied in dealing with huge volumes of text documents and images. For extracting relevant information from the enormous volumes of available data, some co-clustering algorithms have been proposed and shown to be superior to traditional one-side clustering. In this paper, we proposed a novel co-clustering approach called double sparse manifold learning (DSML). We based our formulation on double sparse constraints and manifold learning which use a modified version of mutual k-nearest neighbor graph to capture the underlying structure, modeled sample-feature relationship from the data reconstruction perspective. We developed an iterative procedure to get the solution. Our method preserves local geometrical structure better. Experiments on three benchmark datasets show that our method can get more promising performance on all analyzed data-sets.

Keywords: Co-clustering · Sparse representation · Manifold learning · Double sparse · Mutual kNN

1 Introduction

In data mining and information retrieval field, complex data such as text documents and images have been receiving more and more attention. Traditional clustering is one side clustering that emphasizes clustering along the sample or feature dimensions. More and more applications, such as text analysis and image clustering, are often stacked in columns and rows of a dyadic data matrix, and require to co-cluster both the samples and features. To overcome the drawback that focus on clustering along unilateral dimension, some co-clustering methods have been proposed [1–4]. These works have shown that using the dual interdependence between samples and features to discover hidden clustering structures in data can enhance the clustering performance. From a geometrical point of view, a data set is a set of discrete samplings on continuous manifold, and

© Springer International Publishing AG 2017
H. Yin et al. (Eds.): IDEAL 2017, LNCS 10585, pp. 279–286, 2017.
https://doi.org/10.1007/978-3-319-68935-7_31

clustering aims at finding intrinsic structures of the manifold. Some manifold learning based co-clustering methods have been proposed [5–8].

In this paper, we propose an approach which based on manifold learning and sparse co-clustering, models the sample-feature relationship from the data reconstruction perspective. After mapping Laplacian weighted graph to the sparsity-inducing representations and imposing the l_1-norm sparsity on both dictionary and coefficients, a more compact and/or robust dictionary will be learned from the original data sets, and then a more expressive coefficient can be used to represent the text or image. The learned sparse dictionary and coefficient preserve the local manifold, and can be used for co-clustering the samples and features. Since the local geometrical structure of the sample space can be better preserved, more promising co-clustering results can reasonably be expected.

The rest of the paper is organized as follows. Section 2 briefly subsumes sparse representation clustering methods and introduce manifold embedding formulation. Proposed double sparse manifold learning algorithm is presented in Sect. 3. Section 4 conducts experiments to validate the proposed method. Section 5 gives our conclusions.

2   Related Work

2.1   Sparse Representation for Clustering

The sparse representation method is one of the most representative linear representation methods and has shown huge potential capabilities in handling problems in the field of data mining, knowledge management, and clustering. In this section, we discuss sparse representation as a tool for clustering.

Let there be a set of vectors $\{x_j \in \mathbf{R}\}_{j=1}^{J}$ in K clusters. The goal of clustering is to find $K \ll J$ cluster means $\{\mu_k \in \mathbf{R}_{k=1}^{K}\}$ and an assignment of each x_j to a best-matching cluster $k^*(j)$ such that $\sum_j |x_j - \mu_{k^*(j)}|^2$ (or other suitable mismatch cost) is minimized [9]. K-means algorithm is commonly used procedure for clustering. It can be written as $\min \|X - DS\|_F^2$ where X is dataset, D is cluster mean matrix, and S is coefficient matrix. K-SVD is a generalization of the K-Means clustering process for adapting dictionaries so as to represent signals sparsely [10]. The objective function of K-SVD can be written as: $\|X - DS\|_F^2$ subject to $\forall i, \|s_i\|_0 \leq T$ where T is a predetermined small positive constant. D is clustering mean matrix called dictionary and S is coefficient matrix. Some existing biclustering algorithms use the SVD directly or have a strong association with it. [11–13] based on sparse SVD (SSVD) method to find biclusters in data. Its objective function can be written as: $\|X - suv^T\|_F^2 + \lambda_u P_1(su) + \lambda_v P_2(sv)$ where $P_1(su)$ and $P_2(sv)$ are sparsity-inducing penalty terms.

Double sparsity enables larger dictionary to be trained and shows much more stable performance. So in our proposed method, we will impose the l_1-norm sparsity on both dictionary and coefficients.

2.2   Manifold Learning

Recent studies show that many real world data are actually sampled from a nonlinear low dimensional manifold which is embedded in the high dimensional ambient space. Yet existing clustering algorithms fail to consider the geometric structure in the data which is essential for clustering data on manifold. This limits the application of clustering for the data lying on manifold. Data points are discrete samplings from data manifold. We hope that the cluster labels of data points are smooth with respect to the intrinsic data manifold. A common way for the data sampled from a manifold is to construct a graph to approximate the manifold. The vertices of the graph correspond to the data samples, while the edge weight represents the affinity between the data points. One common assumption about the affinity between data points is cluster assumption, which says if two samples are close to each other in the input space, then their labels (or embeddings) are also close to each other.

Let a data graph whose vertices correspond to $[x_1, x_2, ..., x_n]$. If data points x_i and x_j are close to each other, then their sparse coefficient s_i and s_j should be close as well. This is formulated as $\frac{1}{2} \sum_{i,j} \|s_i - s_j\|^2 W_{ij}$ where W_{ij} is the affinity matrix with weights characterizing the similarity of s_i and s_j. This formula can be rewritten as follows:

$$\frac{1}{2} \sum_{i,j} \|s_i - s_j\|^2 W_{ij} = \sum_{i,j} s_i B_{ii} s_j^t - \sum_{i,j} s_i W_{ij} s_j^t = Tr(S(B-W)S^T) = Tr(SLS^T)$$

(1)

where $B_{ii} = \Sigma_j W_{ij}$ is the diagonal degree matrix, and $\mathrm{L} = \mathrm{B} - \mathrm{W}$ is the graph Laplacian of the data graph G. Equation (1) reflects the smoothness of the data points.

3   Double Sparse Manifold Learning Algorithm

3.1   Objective Function

Inspired by recent progress in manifold learning, co-clustering, and double sparsity, in this paper, we propose a novel clustering algorithm, called double sparse manifold learning (DSML). This method explicitly considers the local geometrical structure of the data, builds a Laplacian weighted graph to encode the geometrical information in the data. The Laplacian weighted graph is incorporated into the double sparse coding objective function as a regularizer.

Let $X = [x_1, x_2, ..., x_n] \in \mathbf{R}^{m \times n}$ the training samples of the ith object class, with each column of X is a sample vector. Sparse coding tries to find a dictionary and a sparse coefficient matrix whose product can best approximate the original data matrix. The column vectors of D can be regarded as the basis vectors and each column of S is the new representation of each data point in this new space. Where each column is a sparse representation for a data point.

By incorporating the manifold embedding regularizer into the original sparse coding and impose the l_1-norm sparsity on the dictionary and coefficients, our objective function is defined as follows:

$$\min_{D,S} \|X - DS\|_F^2 + \alpha\|D\|_1 + \beta\|S\|_1 + \gamma Tr(SLS^T) \quad s.t. \quad d_j^T d_j = 1, \forall j \quad (2)$$

where S is the representation matrix of X over the manifold embedding sparse dictionary D, and parameter α, β is a positive scalar number that balances the F-norm term and the l_1-norm term. Parameter γ is Laplacian graph coefficient and it is also a positive scalar number.

It is important to select a good Laplacian graph representation. Laplacian graph tries to capture the underlying structure of the data by defining a neighborhood for each data-point in terms of a neighborhood graph. Strategies for improving these affinity matrices mainly consist of two different steps: normalizing the affinity matrix W and capturing the underlying structure of the data points by building nearest neighbor graphs.

Nearest neighbor graphs usually model the local relation between a data point i (k nearest neighbor) or all points within a pre-defined distances (ε-neighborhood). Nevertheless, both graph structures have obvious drawbacks. One drawback of ε-neighborhood graphs is that the data points have to form non-elongated, tight clusters. The disadvantage of k nearest neighbor graphs is that only the absolute neighborhood ranking is considered for the connection of two points. Since our method have to deal with asymmetry neighbor affinity matrix, we propose to use mutual kNN graph (mkNN). The mkNN adds an additional constraint to the kNN graph, which requires that two connected data points belong to each other's k-neighborhood [14]. The mutual k nearest neighbor graph can be considered as hybrid form of ε-neighborhood graphs and k nearest neighbor graphs [15]. mkNN can provide information that is not available form ε-neighborhood or k nearest neighbor. We use a non-symmetric neighborhoods criterion as follows:

$$W_{ij} = \begin{cases} exp(-\|x_i - x_j\|^2/t) & x_j \in kNN(x_i) \quad and \quad x_i \in ckNN(x_j) \\ 0 & \text{otherwise} \end{cases} \quad (3)$$

The asymmetry coefficient c control the range of neighborhood incorporation during the mutual k nearest neighbor graph construction.

3.2 Double Sparse Manifold Learning

The optimization problem for the proposed formulation (2) is a joint optimization problem of the dictionary D and the representation coefficient matrix S. This problem is non-convex, like in some clustering algorithm, such as K-SVD, we employ two stages iteration: one for sparse coding and the other for updating the dictionary. The optimization procedures of the proposed method are described in the following Algorithm 1.

S is the coefficient matrix, where each column is a sparse representation for a data point. After learning a new representation for the data, we then apply k-means in the new representation space.

Algorithm 1. Double sparse manifold learning

Input:
    $X = [x_1, x_2, ..., x_n] \in \mathbf{R}^{m \times n}$ the training samples of the ith object class
Output:
    sparse dictionary D and coefficient S.
1: Initialize each column of D to have unit l_2-norm.
2: Construct W by using (3) and let $B_{ii} = \Sigma_j W_{ij}$, then let L=B-W
3: Fix D and solve S using convex optimization techniques to solve the following
    function
$$\min_{S} \|X - DS\|_F^2 + \alpha \sum_i \|s_i\|_1 + \gamma Tr(SLS^T)$$
4: Fix S and update D. We update all the d_j one by one by solving the function
$$\min_{D} \|X - DS\|_F^2 + \beta \sum_j \|d_j\|_1 \quad s.t. \quad d_j^T d_j = 1, \forall j$$
5: Go back to 3. The iterative minimization process is continued until the stopping
    criterion is met.
6: Output D,S.

---

4 Experiments

To facilitate objective comparison, we evaluate our approach on real world data sets to investigate the clustering performance of DSML. Three datasets are used: COIL20 image database, TDT2 corpus, MNIST dataset.

    Our method is compared with six other methods for data clustering: K-means clustering algorithm (K-means); graph regularized sparse coding (GSC) [16]; sparse coding (SC) [17]; KSVD [10]; Non-negative Matrix Factorization (NMF) [18]; Graph Regularized Non-negative Matrix Factorization (GNMF) [19].

4.1 Evaluation Metrics

The clustering result is evaluated by comparing the obtained label of each sample with the label provided by the data set. Two standard clustering metrics, the accuracy (AC) and the normalized mutual information metric (NMI) are used to measure the clustering performance [16]. Clustering Accuracy discovers the one-to-one relationship between clusters and classes, measures the extent to which each cluster contained data points from the corresponding class. NMI is used for evaluating the quality of clusters. The larger the NMI is, the better the clustering result will be.

4.2 Data Set

1. TDT2 corpus: The TDT2 corpus (Nist Topic Detection and Tracking corpus) consists of data collected during the first half of 1998 and taken from 2 newswires, 2 radio programs and 2 television programs sources. We use a subset of Top 10 Categories with 7456 samples.

2. MNIST dataset: The MNIST database of handwritten digits from Yann LeCun's page has a training set of 60,000 examples, and a test set of 10,000

Fig. 1. Examples of MNIST

Fig. 2. Examples of COIL20

examples. Figure 1 shows some examples of MNIST. We use a subset of 4,000 examples.

3. COIL20 image database: The COIL20 image library contains 1,440 32×32 gray scale images of 20 objects, 72 images per object. Figure 2 shows some examples of COIL20.

4.3 Results

Table 1 summarizes the characteristics of the data sets used in experiment. Tables 2 and 3 show the clustering performance of the seven algorithms. Table 2 shows the clustering accuracy on three benchmark data sets. Table 3 shows the clustering normalized mutual information on three benchmark data sets. The clustering performances of seven algorithms are listed in the table. As can be seen, our DSML algorithm outperforms many proposed image clustering algorithm. These results demonstrate that by encoding geometrical information into

Table 1. Description of the data sets

Data sets	Samples	Features	Classes
TDT2	7456	36771	10
MNIST	4000	784	10
COIL20	1440	1024	20

Table 2. Clustering accuracy on the data sets.

Data sets	K-means	KSVD	SC	GSC	NMF	GNMF	DSML
TDT2	0.7743	0.7543	0.7940	0.9008	0.7110	0.8910	0.9010
MNIST	0.5420	0.5095	0.5705	0.6657	0.4490	0.5478	0.6758
COIL20	0.6048	0.6437	0.6049	0.7229	0.6673	0.7222	0.7431

Table 3. Normalized mutual information on the data sets

Data sets	K-means	KSVD	SC	GSC	NMF	GNMF	DSML
TDT2	0.7872	0.7324	0.7345	0.8466	0.7528	0.7986	0.8475
MNIST	0.4804	0.4311	0.5560	0.6670	0.4135	0.5898	0.6690
COIL20	0.7386	0.7185	0.7073	0.8067	0.7436	0.8023	0.8440

the double fused sparse representations, the clustering performance can be significantly enhanced.

5 Conclusion

In this paper, we proposed a double sparse manifold leaning (DSML) method that considers the manifold structure and double fused sparse representation for co-clustering. By combining a graph Laplacian regularizer with double sparse representation and using a modified version of a mutual k-nearest neighbor graph to capture the underlying structure, our method can obtain sparse representations which capture the intrinsic geometrical information in the data and can be used in sparse representation based data clustering. Our comparative experiments on three data sets demonstrate that the proposed DSML algorithm has high accuracy and can get more normalized mutual information.

Acknowledgements. This research is supported by National Natural Science Foundation of China, under Grant No. 61272304. Supported by National Nature Science Foundation under Grant No. 61363029. Supported by The Key Laboratory of image and graphic intelligent processing of higher education in Guangxi (No. GIIP201606). Supported by Guangxi Key Laboratory of Trusted Software (No. kx201628)

References

1. Rohe, K., Qin, T., Bin, Y.: Co-clustering directed graphs to discover asymmetries and directional communities. Proc. Nat. Acad. Sci. **113**(45), 12679–12684 (2016)
2. Wang, S., Huang, A.: Penalized nonnegative matrix tri-factorization for co-clustering. Expert Syst. Appl. **78**, 64–73 (2017)
3. Del Buono, N., Pio, G.: Non-negative matrix tri-factorization for co-clustering: an analysis of the block matrix. Inf. Sci. **301**, 13–26 (2015)

4. Busygin, S., Prokopyev, O., Pardalos, P.M.: Biclustering in data mining. Comput. Oper. Res. **35**(9), 2964–2987 (2008)
5. Gu, Q., Zhou, J.: Co-clustering on manifolds. In: Proceedings of the 15th ACM SIGKDD International Conference on Knowledge Discovery and Data Mining, pp. 359–368. ACM (2009)
6. Li, P., Bu, J., Chen, C., He, Z.: Relational co-clustering via manifold ensemble learning. In: Proceedings of the 21st ACM International Conference on Information and Knowledge Management, pp. 1687–1691. ACM (2012)
7. Li, P., Jiajun, B., Chen, C., He, Z., Cai, D.: Relational multimanifold coclustering. IEEE Trans. Cybern. **43**(6), 1871–1881 (2013)
8. Allab, K., Labiod, L., Nadif, M.: Multi-manifold matrix decomposition for data co-clustering. Pattern Recogn. **64**, 386–398 (2017)
9. Papalexakis, E.E., Sidiropoulos, N.D., Bro, R.: From k-means to higher-way co-clustering: multilinear decomposition with sparse latent factors. IEEE Trans. Signal Process. **61**(2), 493–506 (2013)
10. Aharon, M., Elad, M., Bruckstein, A.: K-SVD: an algorithm for designing over-complete dictionaries for sparse representation. IEEE Trans. Sig. Process. **54**(11), 4311–4322 (2006)
11. Sill, M., Kaiser, S., Benner, A., Kopp-Schneider, A.: Robust biclustering by sparse singular value decomposition incorporating stability selection. Bioinformatics **27**(15), 2089–2097 (2011)
12. Lee, M., Shen, H., Huang, J.Z., Marron, J.S.: Biclustering via sparse singular value decomposition. Biometrics **66**(4), 1087–1095 (2010)
13. Ji, S., Zhang, W., Liu, J.: A sparsity-inducing formulation for evolutionary co-clustering. In: Proceedings of the 18th ACM SIGKDD International Conference on Knowledge Discovery and Data Mining, pp. 334–342. ACM (2012)
14. Kontschieder, P., Donoser, M., Bischof, H.: Improving affinity matrices by modified mutual KNN-graphs. In: 33rd Workshop of the Austrian Association for Pattern Recognition (AAPR/OAGM) (2009)
15. Donoser, M.: Replicator graph clustering. In: BMVC (2013)
16. Zheng, M., Jiajun, B., Chen, C., Wang, C., Zhang, L., Qiu, G., Cai, D.: Graph regularized sparse coding for image representation. IEEE Trans. Image Process. **20**(5), 1327–1336 (2011)
17. Lee, H., Battle, A., Raina, R., Ng, A.Y.: Efficient sparse coding algorithms. In: Advances in Neural Information Processing Systems, vol. 19, p. 801 (2007)
18. Cai, D., He, X., Wu, X., Han, J.: Non-negative matrix factorization on manifold. In: Eighth IEEE International Conference on Data Mining, ICDM 2008, pp. 63–72. IEEE (2008)
19. Cai, D., He, X., Han, J., Huang, T.S.: Graph regularized nonnegative matrix factorization for data representation. IEEE Trans. Pattern Anal. Mach. Intell. **33**(8), 1548–1560 (2011)

Artifact Removal Methods in Motor Imagery of EEG

Yanlong Zhu, Zhongyu Wang, Chenglong Dai, and Dechang Pi[✉]

College of Computer Science and Technology,
Nanjing University of Aeronautics and Astronautics, 29 Jiangjun Road,
Nanjing 211106, Jiangsu, China
1109763309@qq.com, 917730526@qq.com,
{zhuyanlong,dc.pi}@nuaa.edu.cn

Abstract. EEG reflects the strength of the neuronal activity in the brain. Since EEG signal is weak, noisy and mixed with a large number of artifacts, which causes interference to the processing and identification of the EEG signal. Using EEG related pretreatment can effectively remove artifact, noise, and improve EEG signal-noise ratio and efficient, which provides more accurate data for feature extraction and classification. In this paper, we introduce several methods including PCA, ICA and CSP. Based on these methods, the complete process of EEG signal de-noising, feature extraction and classification are established, which can complete the classification and recognition of the motor imagery signals. We use a combination of a lot of pretreatment methods to analysis and process motor imagery of EEG and propose an improved algorithm named CS-CSP. The experimental results show that the Chebyshev type II filter is superior to the conventional pre-treatment methods and the recognition accuracy of CS-CSP is higher than CSP.

Keywords: Brain computer interface · Motor imagery · Common spatial patterns · Artifact removal

1 Introduction

With the development of the science and technology, researchers often amplify the bio-electric signal of the brain and record them into a graph [1], which is called EEG. The flow chart of the recognition of EEG signal is shown in Fig. 1. First, pre-process the EEG signal and do feature extraction and feature recognition on the signal. Then convert the signal and use the Brain-Computer Interface (BCI) transmitting them to the external device. In this way, we achieve interact of the EEG signal and the external devices [2]. Due to the ease of the operation and non-invasive, this method has received much concern and has thought to be one of the hot issues of brain and neuroscience research by numerous international authorities.

The EEG is very weak and is susceptible to various factors, such as ECG, EMG and power frequency interference. Thus, it will influence the accuracy of the EEG signal

© Springer International Publishing AG 2017
H. Yin et al. (Eds.): IDEAL 2017, LNCS 10585, pp. 287–294, 2017.
https://doi.org/10.1007/978-3-319-68935-7_32

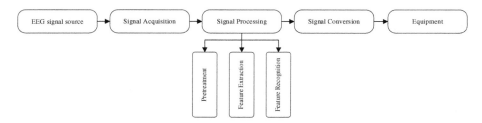

Fig. 1. EEG signal recognition process

classification [3]. Therefore, how to accurately and efficiently remove the artifacts in the original EEG signal is important to the implementation of the BCI system.

The EEG signal can be divided into different categories depending on the source [4], and the motor imagery (MI) signal is one of them. Motor image refers to the movement of a body in the mind but the body itself does not perform and it is a kind of EEG signal that the brain sends out as it imagines movements. Commonly used MI signals are left and right hands, feet, tongue etc.

The rest of this paper is organized as follows. We introduce classical algorithms in Sect. 2. An improved algorithm named CS-CSP is proposed in Sect. 3. Experiment results are shown in Sect. 4. Finally, the Sect. 5 offers concluding remarks.

2   The Classical Algorithms

This section introduction some methods about EEG signal processing, which could remove the artifacts that are unrelated to MI and improve the recognition accuracy.

Principal Components Analysis (PCA) is a classical technique for analyzing simplifying datasets. It is often used to reduce the dimensionality of datasets, and to preserve the features of the worst contributions of the dataset. At present, PCA has become one of the most widely used algorithms in data dimensionality reduction [5].

Independent component analysis (ICA) is a computational method for separating a multivariate signal into additive subcomponents. This is done by assuming that the subcomponents are non-Gaussian signals and that they are statistically independent from each other. EEG signals can be regarded as a mixture of independent, multi-channel brain waves that are mixed in different proportions, and artifact signals and brain waves are produced in different parts of the brain [6].

The frequency response of the Butterworth filter is maximally flat in the passband and rolls off towards zero in the stopband. The Butterworth filter is a type of signal processing filter designed to have as flat frequency response as possible in the passband.

Chebyshev Type II Filter also known as inverse Chebyshev filters. It has no ripple in the passband, but does have equiripple in the stopband. This feature preserves the signal feature in the passband, but because of equiripple in the stopband, it is possible to cause some interference to the boundary. We use Chebyshev Type II filter to process EEG signals, in order to prevent pretreatment method disturb the EEG signal.

3   The CS-CSP Algorithm

3.1   Common Spatial Patterns

The Common Spatial Patterns (CSP) algorithm is proposed by Fukunaga et al. [7]. And applied to the BCI, which used to extract EEG signal feature. It is the mainstream algorithm for feature extraction of MI signals. The CSP algorithm is to find the optimal spatial projection for the maximum power of the two classes of signals.

The classical CSP algorithm is less efficient than expected in the small dataset because of the influence of noise and overfitting. Regularized Common Spatial Patterns (RCSP) is an improvement over the small dataset of the classical CSP algorithm to improve performance on small datasets.

The feature vectors extracted by RCSP algorithm have the same dimensions as the original number of channels. When the number of channels is larger, the vector dimension is too high, and the training classifier may have an overfitting phenomenon.

3.2   CS-CSP Algorithm

In order to solve the over fitting phenomenon, to avoid the influence of artifact to classifier, a new improved algorithm is proposed in this paper. On the basis of CSP, the CS-CSP algorithm is adopted to reduce the dimension of the feature vectors.

Algorithm 1: Roulette wheel

Input: probability of being selected $P(i)$, the number of point in the dataset num

Output: selected number id $result$

```
01.  m:=0, r:=rand
02.  for(i:=1;i ≤ num;i++)// num: the number of point in the dataset
03.      m:=m+P(i)
04.      if m > r then
05.          result := i
06.          break
07.      end if
08.  end for
```

The process of CS-CSP is shown as Algorithm 2.

Algorithm 2: CS-CSP algorithm

Input: population size P, probability of performing crossover Pc, probability of mutation Pm, evolution number G

Output: Matrix after channel selection

01. *while (t<G)   //t: Loop variable*

02.   *for(i:=1;i $\leq$ P;i++)*

03.   *if rand < Pc*

04.   *find two object A and B   //by roulette wheel selection*

05.   *C := Crossover(A,B)*

06.   *if rand < Pm then*

07.   *C := Mutation(C)   // Mutation: Randomly change some values*

08.   *end if*

09.   *end if*

10.   *end for*

11. *t++*

12. *end while*

The crossover means get half of the data from A and B, while Mutation means randomly change some values in C.

4   Experiments and Analysis

This section introduces the data source, the experimental process, the experimental scheme and the evaluation index. 5-fold cross validation method is used to verify the accuracy of each experiment. Repeat 10 times to eliminate the error caused by random division of data, then the average recognition accuracy is obtained by averaging.

4.1   Dataset

The EEG MI dataset used during the experiment was taken from BCI Competition III [8]. The dataset I is provided by University of Tübingen and Universität Bonn. The dataset IVa is provided by the German Berlin-BCI research group. The detailed attributes of the datasets are listed in Table 1.

4.2   Compared Methods

In this experiment, the passband of the Butterworth filter is 0–35 Hz, the passband of the Chebyshev Type II Filter is 0–30 Hz, PCA and ICA select the first 30 components (Fig. 2).

Table 1. Experiment dataset information

DataSet		Type	Trials	Electrode number	Frequency
DataSet I		Tongue, Left pinkie	278	64	1000 Hz
DataSet IVa	aa	Right foot, Right hand	168	118	100 Hz
	al	Right foot, Right hand	224	118	100 Hz
	av	Right foot, Right hand	84	118	100 Hz
	aw	Right foot, Right hand	56	118	100 Hz
	ay	Right foot, Right hand	28	118	100 Hz

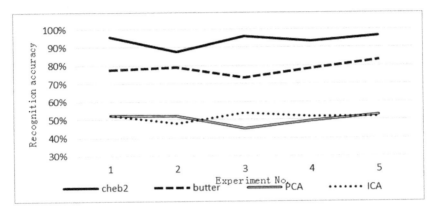

Fig. 2. Compared each method with test data I

It can be seen that the Chebyshev Type II Filter average recognition accuracy is basically over 90%. The results of PCA and ICA are not well, this is because these algorithms are required experience.

4.3 Compared Filtering Methods with Different Parameters

According to the results of 4.2, Chebyshev Type II Filter and Butterworth filter work better. We know that the main component of the brain wave is alpha wave (8–15 Hz), containing small theta wave (4–7 Hz) and beta wave (12.5–28 Hz) (Table 2).

Table 2. Recognition accuracy of different pass-band frequency in dataset I

Methods	Pass-band Frequency				
	20 Hz	25 Hz	30 Hz	35 Hz	40 Hz
Butterworth filter	94.62%	**95.68%**	65.47%	78.35%	71.45%
Chebyshev type II filter	85.64%	85.24%	93.87%	**95.70%**	74.81%

It can be seen that filter passband cut-off frequency is closely related to the recognition accuracy. The two filter methods achieve the best effect while the passband cut-off frequency is not same. This is because Butterworth filter has not ripple at the stopband cut-off frequency, and the frequency response curve is very smooth.

4.4   Experimental Results of Classical CSP and RCSP Algorithms

The result is shown in Table 3. It can be seen that the result of classic CSP is worse while data samples are becoming smaller.

Table 3. Experimental results of classical CSP algorithm

DataSet		TEST1	TEST2	TEST3	TEST4	TEST5	Average
DataSet I		89.603	95.330	93.888	92.791	97.129	93.948
DataSet IVa	aa	72.720	68.382	66.764	76.213	71.360	71.088
	al	61.600	52.747	65.098	65.237	57.312	60.399
	av	59.166	52.361	57.916	54.583	64.305	57.666
	aw	52.000	53.666	50.333	51.333	53.333	52.133
	ay	35.000	35.000	36.666	36.666	36.666	36.000

Then we tried the RCSP algorithm, there are two parameters, β and γ $(0 \leq \beta, \gamma \leq 1)$, need to be defined. We set $\beta \in \{0, 0.01, 0.1, 0.2, 0.4, 0.6\}$, when β is fixed, we set $\gamma \in \{0, 0.001, 0.01, 0.1, 0.2\}$.

From the Fig. 3, it can be seen that When both β and γ are 0.1, RCSP has the highest accuracy at 74.8%, while classic CSP only get 61.9%. When dataset becomes smaller, the accuracy of RCSP decreases slightly, but more stable than the classic CSP.

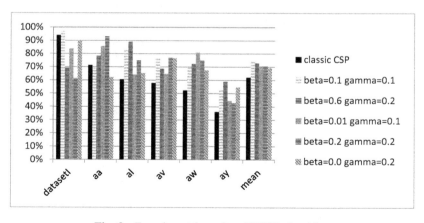

Fig. 3. Experimental results of RCSP algorithm

4.5 Experimental Results of CSP and CS-CSP Algorithms

From Table 3 and Fig. 3, we can see that classic CSP and RCSP are not good enough to handle small dataset, this paper provides an improved algorithm named CS-CSP. We choose the al data filtered by Chebyshev type II. We set the population size is 100, iterations is 20, both probability of performing crossover and probability of mutation are 0.5, minimum criteria is recognition accuracy higher than 97%, both β and γ are 0.1.

As we can see from Table 4, the CS-CSP algorithm has some advantages over the original algorithm. The accuracy of the classical CSP algorithm is improved by nearly 5%, while the accuracy of RCSP is improved by about 2%.

Table 4. Experimental results of classical CSP algorithm

	Classic CSP	RCSP
Original	60.399%	83.277%
CS-CSP	65.269%	85.249%

5 Conclusion

This paper focus on how to improve the accuracy of recognition of MI signal. We implemented a set of experiments with the public dataset. The effects of some methods are not ideal. The effects of PCA and ICA are not as good as Butterworth filter and Chebyshev Filter. The classical CSP algorithm has better effect on large datasets than under small dataset. RCSP can solve this problem, but it doesn't perform well with high feature dimensions. Our proposed algorithm CS-CSP is better than others.

Acknowledgments. The research work is supported by National Natural Science Foundation of China (U1433116) and the Fundamental Research Funds for the Central Universities (NP2017208).

References

1. Lan, Z., Liu, Y., Sourina, O., et al.: Real-time EEG-based user's valence monitoring. In: 2015 10th International Conference on Information, Communications and Signal Processing (ICICS), pp. 1–5. IEEE (2015)
2. Stewart, A.X., Nuthmann, A., Sanguinetti, G.: Single-trial classification of EEG in a visual object task using ICA and machine learning. J. Neurosci. Methods **228**, 1–14 (2014)
3. Lan, Z., Sourina, O., Wang, L., et al.: Real-time EEG-based emotion monitoring using stable features. Visual Comput. **32**(3), 347–358 (2016)
4. Stikic, M., Johnson, R.R., Tan, V., et al.: EEG-based classification of positive and negative affective states. Brain-Comput. Interfaces **1**(2), 99–112 (2014)
5. Subasi, A., Gursoy, M.I.: EEG signal classification using PCA, ICA, LDA and support vector machines. Expert Syst. Appl. **37**(12), 8659–8666 (2010)

6. Hou, X., Liu, Y., Sourina, O., et al.: EEG based stress monitoring. In: 2015 IEEE International Conference on Systems, Man, and Cybernetics (SMC), pp. 3110–3115. IEEE (2015)
7. Fukunaga, K.: Instruction to Statistical Pattern Recognition. Elsevier, Orlando (1972)
8. Blankertz, B., Dornhege, G., Müller, K.R., et al.: Results of the BCI Competition III. BCI Meeting (2005)

Clustering by Searching Density Peaks via Local Standard Deviation

Juanying Xie$^{(\boxtimes)}$, Weiliang Jiang, and Lijuan Ding

School of Computer Science, Shaanxi Normal University,
Xi'an 710062, People's Republic of China
`xiejuany@snnu.edu.cn`

Abstract. To solve the problem of DPC (Clustering by fast search and find of Density Peaks) that it cannot find the cluster centers coming from sparse clusters, a new clustering algorithms is proposed in this paper. The proposed clustering algorithm uses the local standard deviation of point i to define its local density ρ_i, such that all the cluster centers no matter whether they come from dense clusters or sparse clusters will be found as the density peaks. We named the new clustering algorithm as SD_DPC. The power of SD_DPC was tested on several synthetic data sets. Three data sets comprise both dense and sparse clusters with various number of points. The other data set is a typical synthetic one which is often used to test the performance of a clustering algorithm. The performance of SD_DPC is compared with that of DPC, and that of our previous work KNN-DPC (K-nearest neighbors DPC) and FKNN-DPC (Fuzzy weighted K-nearest neighbors DPC). The experimental results demonstrate that the proposed SD_DPC is superior to DPC, KNN-DPC and FKNN-DPC in finding cluster centers and the clustering of a data set.

Keywords: Clustering · Density peaks · Local standard deviation · DPC · KNN-DPC · FKNN-DPC · SD_DPC

1 Introduction

Clustering analysis is to discover the group structure of a data set, and disclose the knowledge, patterns and rules hidden in the data [5,7,14,16]. It is an unsupervised learning process and implemented by grouping similar objects into same clusters and dissimilar ones in other clusters [1,3–5,10,12,13,16]. Its applications range from astronomy to bioinformatics, bibliometrics, biomedical, pattern recognition [1,3–5,11,15]. With the emerging of big data from various areas in the real world, there have been more and more experts focusing on studying the clustering techniques to try to understand and summarize the complex data automatically as far as possible and find the potential knowledge and rules and patterns embedded in the big data without any previous domain knowledge [1–3,9,11,15].

There are many kinds of clustering algorithms such as partitioning, hierarchical, density-based, etc. [1,3–5,12,13,16]. The novel density-based clustering

H. Yin et al. (Eds.): IDEAL 2017, LNCS 10585, pp. 295–305, 2017.
https://doi.org/10.1007/978-3-319-68935-7_33

algorithm was proposed by Alex Rodríguez and Alessandro Laio in [1]. The clustering algorithm can find the clustering of a data set by finding the density peaks as cluster centers and assign each point except for the density peaks to its nearest neighbor with higher density. We call the clustering algorithm as DPC for short. DPC is powerful except for its weaknesses in calculating the densities of points and its one step assignment strategy [1,12,13]. There are several advanced density peaks based clustering algorithms [1,8,12,13]. The K nearest neighbors based density peaks finding clustering (KNN-DPC) [12] and the fuzzy weighted K nearest neighbor based density peaks finding clustering (FKNN-DPC) [13] were proposed to remedy the deficiencies of DPC. The extensive experiments demonstrated the excellent performance of KNN-DPC and FKNN-DPC. However, because DPC uses the arbitrary *cutoff* distance d_c to define the local density of a point, which makes the cluster centers from sparse clusters may not be found for they cannot become density peaks. KNN-DPC and FKNN-DPC use the K nearest neighbors of a point to define its local density, and can solve the problem of DPC to some extent, but they did not thoroughly solve it. It is the common phenomena that there are both dense and sparse clusters in a date set simultaneously, which make it difficult for the aforementioned clustering algorithms to find the proper cluster centers and even the clustering as well.

To try to let the cluster centers can become density peaks no matter they come from dense or sparse clusters by absorbing the distributive information of points in a data set as far as possible, we propose the heuristic clustering algorithm by adopting the local standard deviation of a point to define its local density because it is well known that the local standard deviation of a point embodies the information of how dense the local area is around the point. As a result we introduce the new local standard deviation based density peaks finding clustering algorithm named SD_DPC. We tested its power on some special synthetic data sets which are composed of dense and sparse clusters simultaneously. We also test SD_DPC on a typical synthetic dataset from [6]. The experimental results demonstrate that SD_DPC is powerful in finding the cluster center by finding the local standard deviation based density peaks and the clustering of a data set, and its performance is superior to DPC, KNN-DPC and FKNN-DPC.

This paper is organized as follows: Sect. 2 introduces the new SD_DPC in detail. Section 3 tests the power of it by special synthetic data sets, and compares its performance with that of DPC, KNN-DPC and FKNN-DPC in terms of several typical criteria for testing a clustering algorithm, namely clustering accuracy (Acc), adjusted mutual information (AMI), and adjusted rand index (ARI). Section 4 is some conclusions.

2   The Main Idea of the Proposed Clustering Algorithm

DPC [1] has become the hotspot in machine learning for its strong power in detecting cluster centers and exclude outliers and recognize the clusters with any arbitrary shapes and dimensions. The basic assumption in DPC is that the ideal cluster centers are always surrounded by neighbors with lower local density, and they are at a relatively large distance from any other points with higher local density. To find the ideal cluster centers, DPC introduces the *local density metric* of point i in (1) and the *distance* δ_i of point i in (2).

$$\rho_i = |\{d_{ij}|d_{ij} < d_c\}| \tag{1}$$

$$\delta_i = \begin{cases} max_j \{d_{ij}\}, & \rho_i = max_j\{\rho_j\} \\ min(\{d_{ij}|\rho_j > \rho_i\}), & otherwise \end{cases} \tag{2}$$

where d_{ij} is Euclidean distance between points i and j, and d_c the *cutoff* distance given manually. We can see from (1) that the local density ρ_i of point i is the number of points j that are closer to i than d_c. DPC is robust to d_c for large data sets [1]. The definition in (2) disclose that δ_i is the maximum distance from point i to any other point j when point i has got the highest density, otherwise δ_i is the minimum distance between point i and any other point j with higher density [1]. It can be seen from (2) that δ_i is much larger than the nearest neighbor distance for the points with local or global maxima density. The points with anomalously large value of δ_i are to be chosen as cluster centers by DPC.

The most important contribution of DPC is that it proposed the idea of decision graph which is the collection of points (ρ_i, δ_i) in a 2-dimension space with ρ and δ to be x-axis and y-axis respectively. Cluster centers are the points with high δ and relatively high ρ, that is the cluster centers are the ones at the top right corner of the decision graph. The second innovation of DPC is its one step assignment that it assign each remaining points except for those density peaks to the same cluster as its nearest neighbor with higher density. This one step assignment contribution leads the DPC's efficient execution.

However, everything has two sides. The local density definition of DPC in (1) may results in the lower density for those cluster centers from sparse clusters for the arbitrary *cutoff* distance d_c. The very efficient one step assignment of DPC for remaining points may lead to the similar "Domino Effect" [12,13], that is once a point is assigned erroneously, then there may be many more points subsequently assigned incorrectly [12,13], especially in the case where there are several overlapping clusters.

$$\rho_i = \sum_{j \in KNN_i} exp(-d_{ij}) \tag{3}$$

KNN-PDC [12] and FKNN-DPC [13] introduced new definition of density ρ_i for point i by (3). The new definition can be used to any size of data set to calculate the density ρ_i of point i. The KNN_i in (3) means the data set composed of the points of the K-nearest neighbors of point i. Furthermore, KNN-DPC and FKNN-DPC respectively proposed their own two-step assignment strategy, where FKNN-DPC introduced fuzzy weighted K nearest neighbors theory in its second step of assignment strategy other than that of KNN-DPC only the K nearest neighbors theory is used in its two-step assignment strategy. It was demonstrated that both KNN-DPC and FKNN-DPC are superior to DPC [12, 13]. Due to the contribution of fuzzy weighted K-nearest neighbors that FKNN-DPC is more robust than KNN-DPC [13].

Although KNN-DPC and FKNN-DPC have been demonstrated very powerful in finding the patterns of data sets, they did not solve the problem of DPC thoroughly, such as finding the cluster centers from the sparse clusters by finding density peaks.

To detect the cluster centers from sparse clusters by finding the density peaks, we try to absorb the distribution information of points in a dataset as far as possible, so we introduce the new local density ρ_i for point i in (4) by introducing the local standard deviation of point i with the knowledge that the local standard deviation of a point embodying the local information of how dense the local area is around point i. The KNN_i in (4) is the same as that in (3), that is the set of K nearest neighbors of point i, and d_{ij} the Euclidean distance between data points i and j. Then the new local standard deviation based density peaks finding clustering algorithm is proposed in this paper and named as SD_DPC. The assignment strategy in SD_DPC is the same as that of DPC only to demonstrate that the local density definition ρ_i of point i in (4) can solve the problem existing in DPC that the cluster center of sparse cluster may not be detected by finding density peaks.

$$\rho_i = \frac{1}{\sqrt{\frac{1}{K-1}\sum_{j \in KNN_i} d_{ij}^2}} \tag{4}$$

Here are the main steps of SD_DPC.

step1 calculate the density ρ_i of point i in (4), and its distance δ_i in (2);

step2 plot all of points in the 2-dimensional space with their densities ρ and δ as their x and y coordinates respectively;

Table 1. Description of synthetic data sets

Datasets	Number of records	Number of attributes	Number of clusters
a3	7,500	2	50
dataset1	678	2	3
dataset2	10,900	2	8
dataset3	20,000	2	4

Table 2. The parameters for generating dataset1

Parameter	cluster1	cluster2	cluster3
Mean	$[3, 2]$	$[8, 2]$	$[6, 5.5]$
Covariance	$\begin{bmatrix} 0.7, & 0 \\ 0, & 0.7 \end{bmatrix}$	$\begin{bmatrix} 0.7, & 0 \\ 0, & 0.7 \end{bmatrix}$	$\begin{bmatrix} 1, 0 \\ 0, 1 \end{bmatrix}$
Number of points	300	300	78

Table 3. The parameters for generating dataset2

Parameter	cluster1	cluster2	cluster3	cluster4	cluster5	cluster6	cluster7	cluster8
Mean	[1, 8]	[1, 3]	[3, 6]	[8, 10]	[9, 2]	[13, 6]	[16, 1]	[16, 12]
Covariance	$\begin{bmatrix} 0.2, & 0 \\ 0, & 0.2 \end{bmatrix}$	$\begin{bmatrix} 0.2, & 0 \\ 0, & 0.2 \end{bmatrix}$	$\begin{bmatrix} 0.2, & 0 \\ 0, & 0.2 \end{bmatrix}$	$\begin{bmatrix} 1, & 0 \\ 0, & 1 \end{bmatrix}$	$\begin{bmatrix} 1, & 0 \\ 0, & 1 \end{bmatrix}$	$\begin{bmatrix} 1, & 0 \\ 0, & 1 \end{bmatrix}$	$\begin{bmatrix} 0.2, & 0 \\ 0, & 0.2 \end{bmatrix}$	$\begin{bmatrix} 0.2, & 0 \\ 0, & 0.2 \end{bmatrix}$
Number of points	2,000	2,000	2,000	300	300	300	2,000	2,000

Table 4. The parameters for generating dataset3

Parameter	cluster1	cluster2	cluster3	cluster4
Mean	[2, 2]	[9, 2]	[6, 5.5]	$x \in [0.5, \ 12.5]$
Covariance	$\begin{bmatrix} 0.2, & 0 \\ 0, & 0.2 \end{bmatrix}$	$\begin{bmatrix} 1, & 0 \\ 0, & 1 \end{bmatrix}$	$\begin{bmatrix} 1.5, & 0 \\ 0, & 1.5 \end{bmatrix}$	$y \in [8, \ 9.5]$
Number of points	8,000	3,000	3,999	5,001

step3 select those points at the top right corner in the 2-dimensional space, that is density peaks with relatively higher densities and distances, as cluster centers of the data set;

step4 assign the remaining points except for the cluster centers to its nearest neighbor with higher density.

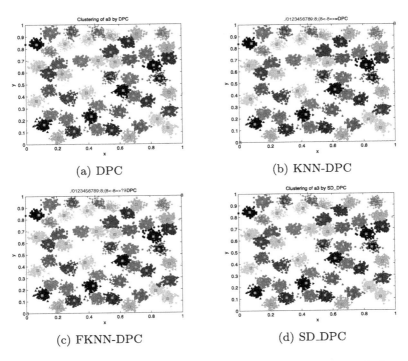

(a) DPC

(b) KNN-DPC

(c) FKNN-DPC

(d) SD_DPC

Fig. 1. The clusterings of a3 by 4 clustering algorithms, respectively.

3   Experiments and Analysis

This section will display the experimental results of SD_DPC and the analysis. In order to test the power of SD_DPC, especially its power to detect the cluster centers from sparse clusters and find the clustering of a data set which has got both dense and sparse clusters simultaneously, we synthetically generated data sets with both dense and sparse clusters. In addition, we test the power of SD_DPC by the typical data set a3 from other reference. It should be explained that for the page limitation we cannot display more experimental results on any other typical bench mark data sets. The Subsect. 3.1 will display the data sets used in this paper. The experimental results of SD_DPC and the analysis are shown in Subsect. 3.2. We compared the performance of SD_DPC with that of DPC, KNN-DPC and FKNN-DPC. For the page limitation, the decision graph and some other experimental results such as the comparison with other clustering algorithms cannot be included in this paper.

We normalize the data using a min-max normalization given by (5), where x_{ij} is the value of attribute j of point i, and $min(x_j)$ and $max(x_j)$ are the minimum and maximum values of attribute j, respectively. The min-max normalization in (5) can preserve the original relationship in data [4], and reduce the influence on experimental results from different metrics for attributes and

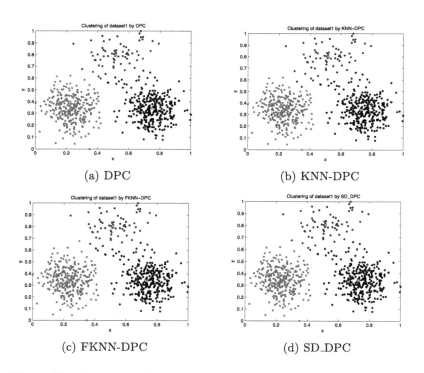

(a) DPC   (b) KNN-DPC

(c) FKNN-DPC   (d) SD_DPC

Fig. 2. The clusterings of dataset1 by 4 clustering algorithms, respectively

reduce the runtime of algorithms' as well.

$$x_{ij} \leftarrow \frac{x_{ij} - min(x_j)}{max(x_j) - min(x_j)} \tag{5}$$

3.1 Data Sets Description

Table 1 shows the informations of the synthetic data sets used in our experiments. The data set a3 comes from [6], and the other three data sets are synthetically generated. The parameters for generating the synthetical data sets are shown in Tables 2, 3 and 4, respectively. We designed these various size of data sets with dense and sparse clusters simultaneously only to test the ability and the scalability of SD_DPC in finding the cluster center from a sparse cluster and the clustering of a data set as well.

3.2 Results and Analysis

Figures 1, 2, 3 and 4 respectively display the clusterings of data sets from Table 1 by 4 clustering algorithms. The square point in each cluster is the cluster center detected by the related algorithm. Table 5 displays the quantity results of the

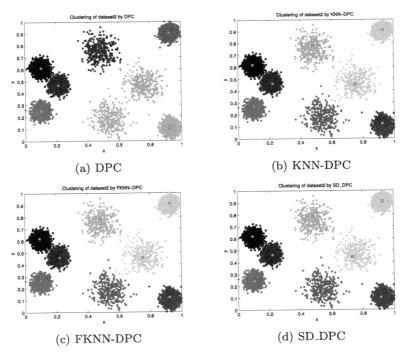

(a) DPC

(b) KNN-DPC

(c) FKNN-DPC

(d) SD_DPC

Fig. 3. The clusterings of dataset2 by 4 clustering algorithms, respectively

aforementioned clustering algorithms in terms of Acc, AMI and ARI, and the number of clusters discovered by each algorithm and the number of clusters covering the cluster centers by F/P, where F refers to the number of cluster centers found by algorithms and P the number of clusters in which the cluster centers lie. The parameters pre-specified for each algorithm are also shown in Table 5 by *Par*.

From the clusterings of a3 and dataset1 by 4 clustering algorithms respectively shown in Figs. 1 and 2, we can say that all of the 4 clustering algorithms can detect all of the cluster centers and the clustering of a3 and dataset1. The difference between the clusterings by 4 algorithms only comes from the border points between clusters. The quantity comparison between the clustering results of 4 clustering algorithms on a3 and dataset1 will be displayed in Table 5 in terms of Acc, AMI and ARI.

The clusterings shown in Fig. 3 disclose that the cluster centers of three sparse clusters in dataset2 cannot be detected by DPC, while the other three algorithms can detect all the cluster centers by finding density peaks. However, the cluster centers found by 4 clustering algorithms are not completely same, and border points between cluster5 and cluster6 are grouped into different clusters by KNN-DPC, FKNN-DPC and SD_DPC. The clustering by SD_DPC is the

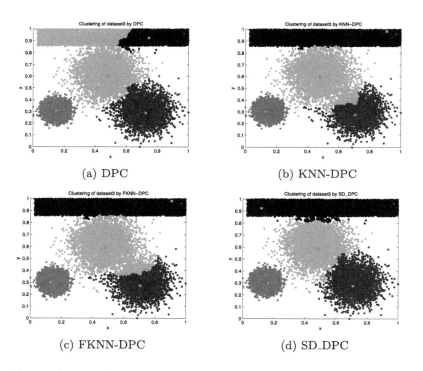

(a) DPC

(b) KNN-DPC

(c) FKNN-DPC

(d) SD_DPC

Fig. 4. The clusterings of dataset3 by 4 clustering algorithms, respectively

Table 5. Comparison of 4 clustering algorithms on 4 synthetic data sets.

Algorithm	a3					dataset1				
	Acc	AMI	ARI	F/P	Par	Acc	AMI	ARI	F/P	Par
DPC	**0.989**	**0.986**	0.977	50/50	1.25	0.978	0.896	0.946	3/3	5
KNN-DPC	**0.989**	**0.986**	0.978	50/50	10	**0.981**	**0.907**	**0.953**	3/3	12
FKNN-DPC	0.987	0.985	0.975	50/50	10	0.973	0.877	0.934	3/3	12
SD_DPC	**0.989**	**0.986**	**0.979**	50/50	6	0.978	0.896	0.946	3/3	7
Algorithm	dataset2					dataset3				
	Acc	AMI	ARI	F/P	Par	Acc	AMI	ARI	F/P	Par
DPC	0.917	0.864	0.895	5/5	2	0.848	0.810	0.779	4/4	2
KNN-DPC	**0.999**	**0.997**	**0.999**	8/8	10	0.959	0.908	0.931	4/4	12
FKNN-DPC	**0.999**	0.996	0.998	8/8	12	0.968	0.918	0.947	4/4	12
SD_DPC	**0.999**	**0.997**	**0.999**	8/8	13	**0.982**	**0.940**	**0.965**	4/4	8

best one compared to original pattern of dataset2, which is not displayed for the page limitations.

The clusterings of dataset3 by 4 clustering algorithms shown in Fig. 4 demonstrate the power of our SD_DPC. From the results in Fig. 4, it can be seen that DPC cannot detect the clustering of dataset3, while the other three algorithms can. KNN-DPC and FKNN-DPC are much better than DPC in finding the clustering of dataset3, but they are not as good as our SD_DPC for the mistakes they made in finding the cluster2 and cluster3 of dataset3. The detail quantity evaluation of the 4 clustering algorithms on dataset3 will be shown in Table 5.

The clustering results in Table 5 reveal that our SD_DPC has got the best performance among the 4 clustering algorithms while it was defeated by KNN-DPC on dataset1. Our previous study KNN-DPC is also a good clustering algorithm. DPC only has got comparable performance on a3.

The overall analysis demonstrate that our proposed SD_DPC is powerful in finding the cluster centers and the clustering of a data set no matter the clusters are dense or sparse and with any arbitrary shapes. So we can conclude that the proposed local density of a point based on its local standard deviation is valid.

4   Conclusions

This paper proposed to adopt the local standard deviation of point i to define its local density ρ_i, so that the distribution information of points in a data set can be absorbed as much as possible to overcome the problem of DPC which may not detect the cluster center of a sparse cluster by finding density peaks. As a consequence the heuristic clustering algorithms named SD_DPC has been introduced. The performance of SD_DPC was tested on several synthetic data sets and compared with that of DPC, KNN-DPC, FKNN-DPC. The experimental results demonstrate that the proposed SD_DPC is superior to DPC, KNN-DPC and FKNN-DPC.

Acknowledgments. We are much obliged to those who provide the public data sets for us to use. This work is supported in part by the National Natural Science Foundation of China under Grant No. 61673251, is also supported by the Key Science and Technology Program of Shaanxi Province of China under Grant No. 2013K12-03-24, and is at the same time supported by the Fundamental Research Funds for the Central Universities under Grant No. GK201701006, and by the Innovation Funds of Graduate Programs at Shaanxi Normal University under Grant No. 2015CXS028 and 2016CSY009.

References

1. Alex, R., Alessandro, L.: Clustering by fast search and find of density peaks. Science **344**(6191), 1492–1496 (2014)
2. Dan, F., Melanie, S., Christian, S.: Turning big data into tiny data: constant-size coresets for k-means, PCA and projective clustering. In: Proceedings of the Twenty-Fourth Annual ACM-SIAM Symposium on Discrete Algorithms, SODA 2013, pp. 1434–1453. SIAM (2013), http://dl.acm.org/citation.cfm?id=2627817.2627920
3. Frey, B.J., Dueck, D.: Clustering by passing messages between data points. Science **315**(5814), 972–976 (2007)
4. Han, J., Kamber, M., Pei, J.: Data Mining: Concepts and Techniques, 3rd edn. The Morgan Kaufmann Series in Data Management Systems, Morgan Kaufmann (2011)
5. Jain, A.K.: Data clustering: 50 years beyond k-means. Pattern Recogn. Lett. **31**(8), 651–666 (2010)
6. Karkkainen, I., Franti, P.: Dynamic local search for clustering with unknown number of clusters. In: Proceedings of the 16th International Conference on Pattern Recognition, vol. 2, pp. 240–243. IEEE (2002)
7. MacQueen, J., et al.: Some methods for classification and analysis of multivariate observations. In: Proceedings of the Fifth Berkeley Symposium on Mathematical Statistics and Probability, vol. 1, Statistics, Oakland, CA, USA, pp. 281–297 (1967)
8. Mehmood, R., EI-AShram, S., Bie, R., Dawood, H., Kos, A.: Clustering by fast search and merge local density peaks for gene expression microarray data. Sci. Rep. **7**, 45602 (2017)
9. Tong, H., Kang, U.: Big data clustering. In: Aggarwal, C.C., Reddy, C.K. (eds.) Data Clustering: Algorithms and Applications, chap. 11, pp. 259–276. CRC Press (2013)
10. Von Luxburg, U., Williamson, R.C., Guyon, I.: Clustering: science or art? J. Mach. Learn. Res. Proc. Track **27**, 65–80 (2012)
11. Xie, J., Gao, H.: Statistical correlation and k-means based distinguishable gene subset selection algorithms. J. Softw. **25**(9), 2050–2075 (2014)
12. Xie, J., Gao, H., Xie, W.: K-nearest neighbors optimized clustering algorithm by fast search and nding the density peaks of a dataset. SCIENTIA SINICA Informationis **46**(2), 258–280 (2016)
13. Xie, J., Gao, H., Xie, W., Liu, X., Grant, P.W.: Robust clustering by detecting density peaks and assigning points based on fuzzy weighted k-nearest neighbors. Inf. Sci. **354**, 19–40 (2016)

14. Xie, J., Jiang, S., Xie, W., Gao, X.: An efficient global K-means clustering algorithm. J. Comput. **6**(2), 271–279 (2011)
15. Xie, J., Li, Y., Zhou, Y., Wang, M.: Differential feature recognition of breast cancer patients based on minimum spanning tree clustering and F-statistics. In: Yin, X., Geller, J., Li, Y., Zhou, R., Wang, H., Zhang, Y. (eds.) HIS 2016. LNCS, vol. 10038, pp. 194–204. Springer, Cham (2016). doi:10.1007/978-3-319-48335-1_21
16. Xu, R., Wunsch, D.I.: Survey of clustering algorithms. IEEE Trans. Neural Netw. **16**(3), 645–678 (2005)

Sparse Representation Based on Discriminant Locality Preserving Dictionary Learning for Face Recognition

Guang Feng[1,2], Hengjian Li[1,2(✉)], Jiwen Dong[1,2], and Xi Chen[3]

[1] School of Information Science and Engineering, University of Jinan,
Jinan 250022, China
ise_lihj@ujn.edu.cn
[2] Shandong Provincial Key Laboratory
of Network Based Intelligent Computing, Jinan 250022, China
[3] School of Bigdata and Computer Science, Guizhou Normal University,
Guiyang, China

Abstract. A novel discriminant locality preserving dictionary learning (DLPDL) algorithm for face recognition is proposed in this paper. In order to achieve better performance and less computation, dimensionality reduction is applied on original image samples. Most of the proposed dictionary learning methods learn features and dictionary, however, the inner structure of feature is hardly considered. Therefore, by incorporating discriminant locality preserving criteria into dictionary learning, the margin of coefficients distance between between-class and within-class is encourage to be large in order to enhance the classification ability and gain discriminative information. What is more, the local structure of the feature is also preserved, which is very vital in face recognition performance. Our experiments on Extended Yale B, AR and CMU face database demonstrated the proposed algorithm has higher recognition performance than other dictionary learning based classification methods.

Keywords: Dictionary learning · Sparse representation · Discriminant locality preserving criterion · Face recognition

1 Introduction

As a basic research of signal representation, the sparse representation of signal is widely used in the field of signal processing because of its good robustness, strong generalization ability and strong anti-interference ability. Sparse coding theory and algorithm have been used in the field of image restoration and compressed sensing. In recent years, the sparse representation technology has made a great progress in image classification [1–7], such as face recognition and digit classification. The success of the image classification based on sparse representation is due to the fact that a high-dimensional image can be represented or coded by several representative samples from the same class in a low-dimensional manifold.

The pattern classification method based on dictionary learning has attracted wide attention because the method of sparse representation classification has the advantages

H. Yin et al. (Eds.): IDEAL 2017, LNCS 10585, pp. 306–314, 2017.
https://doi.org/10.1007/978-3-319-68935-7_34

of good robustness, strong generalization ability and strong anti-interference ability. It can adaptively select the sparsest one of all representations to represent the test sample and reject all other non-sparse representations. Therefore, the classification method can be designed by combining the classification information and the discriminant of sparse representation. For example, Wright et al. [1] presents a sparse representation classifier (SRC), which linearly represents a probe image by all the training images under the L1-norm based sparsity constraint. The experimental results show that SRC has a high recognition rate and a strong robustness to illumination transformation and occlusion when the number of training samples is sufficient. Inspired by SRC, many scholars continue to study the method of face recognition based on sparse representation. Yang et al. [2] regard as the sparse coding problem is a robust regression problem with sparse constraints. A fast-sparse representation model is constructed by using the maximum likelihood estimation of sparse coding. Yang et al. [3] makes use of the Gabor feature to compress the dictionary. In [4], Yang et al. proposed a sparse representation dictionary learning method based on Fisher criterion (FDDL). This method makes use of the discriminant information between the training samples of different classes so that the dictionary atoms of different categories have the smallest within-class dispersion and maximal between-class dispersion, so that the dictionary has stronger ability of discrimination. Zhu et al. [5] proposed multiple maximum scatter difference discrimination dictionary learning, this method enhances the ability to minimize within-class scatter and maximize between-class scatter, so that the learned dictionary has a stronger representation of the relevant class. The face recognition method based on dictionary learning has received extensive attention [6, 7].

The purpose of dictionary learning is to represent the original image with fewer atoms, and to make the dictionary more discriminative. While, the Discriminant Locality Preserving Projections (DLPP) [8–12] is a classical model of data dimensionality reduction. DLPP can not only make the training data have the discrimination, but also keep the local structure of the original data. This paper presents a novel sparse representation dictionary learning (DLPDL) based on discriminant locality preserving criterion, the discriminant locality preserving constraint term is added to the dictionary learning model, and the validity of the proposed DLPDL method is verified by experiments. The rest of this paper is organized as follows. The DLPDL model and optimization procedure are described in Sect. 2. Section 3 presents the DLPDL based classifier. The experimental results are described in Sect. 4. Lastly, Sect. 5 concludes this paper.

2   Discriminant Locality Preserving Dictionary Learning

In order to improve the performance of existing dictionary learning methods, this paper propose a novel discriminant locality preserving based dictionary learning (DLPDL) scheme. Suppose that the sparse representation dictionary $D = [D_1, D_2, \ldots D_c]$, where D_i is the corresponding i-th sub-dictionary, and assume that the training sample set is A, $A = [A_1, A_2, \ldots A_c]$, Y is the sparse coding coefficient matrix of the training sample on the dictionary D, where $Y = [Y_1, Y_2, \ldots Y_c]$, that is $A \approx DY$. This requiring the dictionary D to have a strong ability to reconstruct the sample A. Therefore, we proposed the following DLPDL model:

$$J_{(D,Y)} = \arg\min_{D,Y}\left\{r(A, D, Y) + \lambda_1\|Y\|_1 + \lambda_2 f(Y)\right\} \tag{1}$$

Where $r(A, D, Y)$ is the discrimination fidelity term; $\|Y\|_1$ is the sparsity constraint term; $f(Y)$ is a discriminant locality preserving constraint imposed on the coefficient matrix X; and parameters $\lambda_1 > 0$, $\lambda_2 > 0$. This paper mainly discusses the design problem of discrimination fidelity term $r(A, D, Y)$ and discriminant locality preserving constraint term $f(Y)$.

2.1 Discriminative Fidelity Term

From the previous definition, Y_i is the representation of A_i on D, denoted as $Y_i = \left[Y_i^1; \ldots; Y_i^j; \ldots; Y_i^c,\right]$, where Y_i^j is the coding coefficient of A_i over the sub-dictionary D_j. Define the representation of A_i on the sub-dictionary D_j as $R_j = D_j Y_i^j$. First, A_i can be represented by D, that is $A_i \approx DY_i = D_1 Y_i^1 + \cdots + D_j Y_i^j + \cdots + D_c Y_i^c = R_1 + \cdots + R_j + \cdots + R_c$, second, we hope that A_i can be better represented by D_i, rather than by $D_j (j \neq i)$, which means that the $Y_i^j (j \neq i)$ part of the coefficient is close to zero. Thus, all the effective coefficients are in Y_i^i, we will define the discriminative fidelity term as follows:

$$r(A_i, D, Y_i) = \|A_i - DY_i\|_F^2 + \|A_i - D_i Y_i^i\|_F^2 + \sum_{j=1, j\neq i}^c \left\|D_i Y_i^j\right\|_F^2 \tag{2}$$

Where $\|A_i - DY_i\|_F^2$ reflects the linear reconstruction ability of the training sample A_i on the dictionary D, $\|A_i - D_i Y_i^i\|_F^2$ reflects the discrimination ability of dictionary D_i.

2.2 Discriminative Locality Preserving Constraint Term $f(X)$

Discriminant locality preserving projections combines the Fisher criterion and the locality preserving criterion [8]. Define a training set $A = \left[A_{1,1}, \ldots, A_{i,j}, \ldots, A_{C,n}\right]$ of $N \times C$ face images belonging to C different persons. The objective function of DLPP is as follows:

$$a = \arg\max_a \frac{\sum_{i,j=1}^C B_{i,j}\left(a^T m_i - a^T m_j\right)^2}{\sum_{c=1}^C \sum_{i=1}^N W_{i,j}^c\left(a^T A_{c,i} - a^T A_{c,j}\right)^2} \tag{3}$$

Where the parameter m_i is derived from $m_i = \frac{1}{N}\sum_{k=1}^N A_{i,k}$, $A_{c,k}$ is the k-th sample of class c. The parameter $W_{i,j}^c$ is the weight between $A_{c,i}$ and $A_{c,j}$, $W_{i,j}^c = exp\left(-\|A_{c,i} - A_{c,j}\|^2/t\right)$, $B_{i,j}$ is the weight between m_i and m_j, $B_{i,j} = exp\left(-\|m_i - m_j\|^2/t\right)$. This function not only uses the discriminant information between classes, but also maintains the local information of the original training samples.

Based on discriminant locality preserving criterion, we can define discriminant locality preserving term $f(X)$ as:

$$f(X) = trace\left(\sum_{c=1}^{C}\sum_{y_i,y_j\in Y_c} W_{i,j}^c\left(y_c^i - y_c^j\right)\left(y_c^i - y_c^j\right)^T\right)$$
$$- trace\left(\sum_{i,j=1}^{C} B_{i,j}\left(m_i - m_j\right)\left(m_i - m_j\right)^T\right) + \eta\|Y\|_F^2 \qquad (4)$$

Where y is the sparse coefficient of original training samples, m_i and m_j are the mean vector of within-class sparse coefficient, $W_{i,j}^c$ and $B_{i,j}$ are the weight matrix of within-class sparse coefficient and between-class sparse coefficient mean respectively.

2.3 The DLPDL Model

By incorporating Eqs. (4) and (6) into Eq. (3), we can get the following function model:

$$J_{(D,Y)} = \underset{D,Y}{argmin}\left\{\sum_{i=1}^{c} r(A_i, D, Y_i) + \lambda_1\|Y\|_1 + \lambda_2\left(trace\left(\sum_{c=1}^{C}\sum_{y_i,y_j\in Y_c} W_{i,j}^c\left(y_c^i - y_c^j\right)\left(y_c^i - y_c^j\right)^T\right)\right.\right.$$
$$\left.\left. - trace\left(\sum_{i,j=1}^{C} B_{i,j}\left(m_i - m_j\right)\left(m_i - m_j\right)^T\right) + \eta\|Y\|_F^2\right)\right\} \qquad (5)$$

Although the objective function $J_{(D,Y)}$ in the Eq. (7) is not always convex, it is convex when D or Y is fixed to solve the other. Thus, the dictionary D and the sparse coefficient matrix Y can be alternately optimized, that is, the objective function (Eq. (7)) is divided into two sub-problems: fixed D to update Y and fixed Y to update D.

2.4 Optimization of DLPDL

In this paper, we use the following methods to solve the dictionary D and the sparse coefficient matrix Y.

First, fix the dictionary D and update the coefficient matrix Y. Equation (7) is reduced to a sparse coding problem to compute $Y = [Y_1, Y_2, \ldots, Y_c]$. We successively computed each class of sparse coding Y_i, when compute Y_i, all Y_j $(j \neq i)$ are fixed. Therefore, the objective function of Eq. (7) can be further simplified as:

$$J_{(Y_i)} = \arg\underset{Y_i}{\min}\left\{r(A_i, D, Y_i) + \lambda_1\|Y_i\|_1 + \lambda_2 f_i(Y_i)\right\} \qquad (6)$$

With

$$f_i(Y_i) = trace\left(\sum_{y_i,y_j\in Y_i} W_{i,j}\|y_i - y_j\|_2^2\right) - trace\left(\sum_{i,j=1}^{C} B_{i,j}\|m_i - m_j\|_2^2\right) + \eta\|Y_i\|_F^2$$

$f_i(Y_i)$ is a strictly convex function for Y_i, and all of its items (except $\eta\|Y_i\|_F^2$) can be differentiated. Therefore, we use the iterative preserving projection algorithm (IPM) [13] to solve Eq. 8.

Second, when the coefficient Y is fixed, we update dictionary D_i class by class. When update D_i, all D_j $(j \neq i)$ are fixed. Therefore, the objective function can be simplified as:

$$J_{(D_i)} = \arg\min_{D_i} \left\{ \left\| A - D_i Y^i - \sum_{j=1, j\neq i}^{c} D_i Y^j \right\|_F^2 + \left\| A - D_i Y_i^i \right\|_F^2 + \sum_{j=1, j\neq i}^{c} \left\| D_i Y_i^i \right\|_F^2 \right\} \tag{7}$$

In general, it is required that each column of the dictionary D is a unit vector. Equation (9) is a quadratic programming problem that can be solved by an algorithm in MLSRC [14].

The whole DLPDL iterative algorithm is optimized as follows:

3   The Classification Scheme

Yang et al. [4] proposed a global classifier (GC) scheme which uses the reconstruction error of sparse coding coefficients to classify. We code the testing sample y based on the global dictionary D. Define the objective function as:

$$\hat{\alpha} = \arg\min_{\alpha} \left\{ \|y - D\alpha\|_2^2 + \gamma\|\alpha\|_1 \right\} \tag{8}$$

Where γ is a constant. Denote by $\hat{\alpha} = [\hat{\alpha}_1; \hat{\alpha}_2; \ldots; \hat{\alpha}_c]$, where $\hat{\alpha}_i$ is the sparse coefficient vector corresponding to the sub-dictionary D_i. We define the classification criteria:

$$e_i = \|y - D_i\hat{\alpha}_i\|_2^2 + w\|\hat{\alpha}_i - m_i\|_2^2 \tag{9}$$

Where the first term is the reconstruction error of class i, the second term is the distance between the coefficient vector $\hat{\alpha}$ of the training sample y and the learned training samples mean coefficient vector m_i of class i, and w is a weight parameter. The final classification result is $l(y) = \arg\min_{i}\{e_i\}$.

4   Experimental Results and Analysis

In order to verify the performance of our proposed DLPDL algorithm, we apply the algorithm on the Extended Yale B [15], AR [16] and CMU PIE [17] face database, and we compare DLPDL with other methods (Nearest Neighbor Classification with Euclidean distance (NN), Sparse Representation Classification (SRC) [1], Support Vector Machines (SVM), Fisher Discrimination Dictionary Learning (FDDL) [4]). In this paper, we use the SPGL1 [18] method to solve the reconstruction problem in SRC, the SVM toolbox version is libsvm-3.11 [19]. In our experiment, Eigenfaces (PCA) [20] was first applied to reduce the dimensionality of face images, and the performance of multiple methods under different dimensions was tested as a comparison.

4.1   Extended Yale B Database

The Extended Yale B database contains 2414 images from 38 people (about 64 images per subject). In the experiment, we randomly selected 20 images of each class for training, and the remaining images of each class for testing. In experiment, the image is scaled to 32×32. The parameters of DLPDL are set as follows: $\lambda_1 = 0.005$, $\lambda_2 = 0.05$, $\gamma = 0.001$, $w = 0.05$. The recognition rate of various methods is listed in Table 2. It can be seen that the DLPDL method achieves the best classification performance in any dimension. When the dimension of training samples is relatively small, DLPDL achieved a higher performance improvement.

Table 1. Alg. 1. discriminant locality preserving dictionary learning

Input: training set A, parameter λ_1, λ_2.
Output: dictionary D, sparse coefficient Y.
(1) Initialization dictionary D: initialize each atom in the dictionary D_i with a random vector.
(2) Updating sparse coefficient matrix: fixed dictionary D, and use IPM [13] algorithm to solve the sparse coefficient Y.
(3) Updating dictionary D: fixed sparse coefficient Y and use MLSRC [14] algorithm to update the dictionary D_i ($i = 1, 2, \ldots c$) class by class.
(4) If the value of $J_{D,X}$ in adjacent iterations is sufficiently similar, or the maximum number of iterations has been reached, the algorithm stops, otherwise, return to the second step. Output D and Y.

4.2   AR Database

The AR database consists of over 4000 frontal images from 126 individuals. In the experiment, we select a subset of the AR database, which included 50 male subjects and 50 female subjects. Each of people has 14 images, we choose 7 images of them for training, and the remaining 7 images are used for testing. The image mainly includes changes in light and expression and the size of original image is 32×32. The parameters of DLPDL are set as follows: $\lambda_1 = 0.005$, $\lambda_2 = 0.005$, $\gamma = 0.001$, $w = 0.05$. In the following experiments, we tested the performance of the various methods. As can be seen from the table, FDDL and DLPDL achieved very similar recognition performance, slightly better than other methods. However, it is better than other methods.

4.3   CMU PIE Database

The CMU PIE database contains 11560 images of 68 subjects. For each subject, there are about 170 images from five near frontal poses (C05, C07, C09, C27, C29). The size of each image is 32×32 pixels with 256 gray levels. In the experiment, we randomly selected 34 subjects, each of which consists of 60 images to form a new sub-database. For each subject, we randomly selected 20 images for training, and the remaining

Table 2. The recognition rate and dimensionality on the Extended Yale B database

Dimension	NN	SRC	SVM	FDDL	DLPDL
200	68.02	94.68	90.51	93.29	95.59
300	68.74	96.01	93.29	96.01	96.98
400	69.17	96.61	93.95	96.74	97.40
500	69.11	97.10	93.17	97.04	97.58

images are used for testing. The parameters of DLPDL are set as follows: $\lambda_1 = 0.005$, $\lambda_2 = 0.005$, $\gamma = 0.001$, $w = 0.05$. From the Table 4, we can see that the DLPDL algorithm achieves the highest recognition results in 500 dimensions, and the recognition performance is better than other algorithms in any dimension.

Table 3. The recognition rate and dimensionality on the AR database

Dimension	NN	SRC	SVM	FDDL	DLPDL
200	70.00	86.43	87.86	88.71	89.00
300	70.29	88.86	88.86	89.86	90.14
400	70.71	88.71	89.29	90.57	90.43
500	70.57	89.43	89.29	90.57	90.71

Table 4. The recognition rate and dimensionality on the CMU PIE database

Dimension	NN	SRC	SVM	FDDL	DLPDL
200	76.25	87.13	86.84	86.91	87.94
300	76.69	86.10	86.54	88.01	88.16
400	76.69	84.78	84.04	87.65	88.09
500	76.76	84.78	81.40	88.01	88.82

In all the experiments, DLPDL achieves the achieved the best recognition performance. And when the dimensions is relatively small, DLPDL obtained a more significant improvement. FDDL has the second best performance, this may imply the validity of discriminant information and local structural information in face classification.

5 Conclusion

In this paper, we propose a new sparse representation dictionary learning method based on discriminant locality preserving criterion. This method introduces the discriminant local preserving constraint term in the dictionary learning model, the discriminant locality preserving criteria is imposed on the sparse coding coefficients so that the sparse coding coefficient has the ability of discrimination and the local structure of the feature is also preserved. The experimental results on three face databases show that the proposed discriminant DLPDL method not only have higher recognition rate than other method but also requires fewer feature dimensions.

Acknowledgements. This work is supported by grants by the Shandong Provincial Key R&D Program (2016ZDJS01A12), the National Natural Science Foundation of China (Grant No. 61303199).

References

1. Wright, J., Yang, A., Ganesh, A., et al.: Robust face recognition via sparse representation. IEEE Trans. Pattern Anal. Mach. Intell. **31**(2), 210–227 (2009)
2. Yang, M., Zhang, L., Yang, J., et al.: Robust sparse coding for face recognition. In: Computer Vision and Pattern Recognition, pp. 625–632. IEEE (2011)
3. Yang, M., Zhang, L.: Gabor feature based sparse representation for face recognition with gabor occlusion dictionary. In: Daniilidis, K., Maragos, P., Paragios, N. (eds.) ECCV 2010, Part VI. LNCS, vol. 6316, pp. 448–461. Springer, Heidelberg (2010). doi:10.1007/978-3-642-15567-3_33
4. Yang, M., Zhang, L., Feng, X., et al.: Fisher discrimination dictionary learning for sparse representation. In: 2011 International Conference on Computer Vision, pp. 543–550. IEEE (2011)
5. Zhu, Y., Dong, J., Li, H.: Face recognition using multiple maximum scatter difference discrimination dictionary learning. In: Applied Optics and Photonics China (AOPC 2015), pp. 96750H–96750H-7 (2015)
6. Wang, X., Gu, Y.: Cross-label suppression: a discriminative and fast dictionary learning with group regularization. IEEE Trans. Image Process. **26**(8), 3859–3873 (2017)
7. Yang, M., Chen, L.: Discriminative semi-supervised dictionary learning with entropy regularization for pattern classification. In: AAAI, pp. 1626–1632 (2017)
8. Yu, W., Teng, X., Liu, C.: Face recognition using discriminant locality preserving projections. Image Vis. Comput. **24**(3), 239–248 (2006)
9. Zhong, F., Zhang, J., Li, D.: Discriminant locality preserving projections based on L1-norm maximization. IEEE Trans. Neural Netw. Learn. Syst. **25**(11), 2065–2074 (2014)
10. Chen, X., Zhang, J., Li, D.: Direct discriminant locality preserving projection with hammerstein polynomial expansion. IEEE Trans. Image Process. **21**(12), 4858–4867 (2012)
11. Lu, G.F., Zou, J., Wang, Y.: L1-norm and maximum margin criterion based discriminant locality preserving projections via trace Lasso. Pattern Recogn. **55**, 207–214 (2016)
12. Huang, S., Zhuang, L.: Exponential discriminant locality preserving projection for face recognition. Neurocomputing **208**, 373–377 (2016)
13. Rosasco, L., Verri, A., Santoro, M., et al.: Iterative Projection Methods for Structured Sparsity Regularization. MIT Technical reports (2009)
14. Yang, M., Zhang, L., Yang, J., et al.: Metaface learning for sparse representation based face recognition. In: International Conference on Image Processing, Hong Kong, pp. 1601–1604. IEEE (2010)
15. Lee, K.C., Ho, J., Kriegman, D.J.: Acquiring linear subspaces for face recognition under variable lighting. IEEE Trans. Pattern Anal. Mach. Intell. **27**(5), 684–698 (2005)
16. Martinez, A.M., Benavente, R.: The AR Face Database. CVC Technical report 24 (1998)
17. Sim, T., Baker, S., Bsat, M.: The CMU pose, illumination, and expression (PIE) database. In: 5th IEEE International Conference on Automatic Face and Gesture Recognition, pp. 53–58. IEEE (2002)

18. Van Den Berg, E., Friedlander, M.P.: Probing the Pareto frontier for basis pursuit solutions. SIAM J. Sci. Comput. **31**(2), 890–912 (2008)
19. Chang, C.C., Lin, C.J.: LIBSVM: a library for support vector machines. ACM Trans. Intell. Syst. Technol. (TIST) **2**(3), 27 (2011)
20. Turk, M., Pentland, A.: Eigenfaces for recognition. J. Cogn. Neurosc. **3**(1), 71–86 (1991)

Cost-Sensitive Alternating Direction Method of Multipliers for Large-Scale Classification

Huihui Wang[1], Yinghuan Shi[1], Xingguo Chen[1,2], and Yang Gao[1(✉)]

[1] State Key Laboratory for Novel Software Technology,
Nanjing University, Nanjing, China
gaoy@nju.edu.cn
[2] School of Computer Science and Technology, School of Software,
Nanjing University of Posts and Telecommunications, Nanjing, China

Abstract. Large-scale classification is one of the most significant topics in machine learning. However, previous classification methods usually require the assumption that the data has a balanced class distribution. Thus, when dealing with imbalanced data, these methods often present performance degradation. In order to seek the better performance in large-scale classification, we propose a novel Cost-Sensitive Alternating Direction Method of Multipliers method (CSADMM) to deal with imbalanced data in this paper. CSADMM derives the problem into a series of subproblems efficiently solved by a dual coordinate descent method in parallel. In particular, CSADMM incorporates different classification costs for large-scale imbalanced classification by cost-sensitive learning. Experimental results on several large-scale imbalanced datasets show that compared with distributed random forest and fuzzy rule based classification system, CSADMM obtains better classification performance, with the training time is significantly reduced. Moreover, compared with single-machine methods, CSADMM also shows promising results.

Keywords: Large-scale classification · Cost-sensitive learning · Alternating direction method of multipliers · Imbalanced data

1 Introduction

Large-scale classification has attracted a growing attention from industry and academia. There are many attempts to study distributed methods in a parallel/distributed fashion, aiming to face the explosive growth of the data [1]. Recently, distributed algorithms have been developed for dealing with large-scale data classification. In particular, the Alternating Direction Method of Multipliers (ADMM) is particularly suitable to solve the optimization problem in a distributed manner [2]. However, one of the challenges that deteriorate classification performance is the class imbalance problem in data [3,4]. Numerous real-world applications such as card fraud detection, health care and financial businesses prevalently suffer from the class imbalance problem [5]. Therefore, the interest of the researchers is focused towards the minority class in recent years.

© Springer International Publishing AG 2017
H. Yin et al. (Eds.): IDEAL 2017, LNCS 10585, pp. 315–325, 2017.
https://doi.org/10.1007/978-3-319-68935-7_35

Conventional classification algorithms tend to emphasis the majority class and ignore minority classes, and thus may present a performance degradation when handling with imbalanced data [4]. Therefore, it is necessary to consider the imbalance ratio (IR) and then solve it accordingly. Many methods have been developed to solve this problem [5,6]. In particular, cost-sensitive learning [7] is one of the most widely-used methods. Conventional imbalanced classification methods are usually proposed to deal with small-scale imbalanced data which runs on a single machine. Recently, only a few attempts has been studied for large-scale imbalanced classification with cost-sensitive learning [3,4]. However, these works suffer from the high time cost. Therefore, the important challenge is to design new models in scalable and parallel implementations.

To this end, we propose a novel distributed method, namely Cost-Sensitive Alternating Direction Method of Multipliers (CSADMM) for imbalanced classification in this paper. In particular, CSADMM decomposes the classification problem as a series of subproblems with class imbalance, and then solves subproblems efficiently by a dual coordinate descent method with cost-sensitive learning. All local variables of subproblems are required to reach the consensus until the final solution is found. Experiments results demonstrate that our method can achieve promising results against with state-of-the-arts and variations of single-machine methods on several imbalanced datasets.

2   Related Work

2.1   Cost-Sensitive Learning

Cost-sensitive learning [4,6] is a procedure of learning models from data with the imbalance class distribution, has attracted extensive attentions. Cost-sensitive learning takes different misclassification costs into consideration, and minimizes the total misclassification costs. Many effective classification algorithms combined with cost-sensitive learning have been developed [3,6]. In particular, Batuwita *et al.* [7] developed a fuzzy support vector machines for class imbalance learning. Also, a novel wrapper for training a cost-sensitive classifier was proposed by incorporating the evaluation measure into the objective in [8]. However, these aforementioned methods are specifically designed for the small-scale problem, which cannot be directly borrowed to large-scale classification. Recently, there only be a few attempts for studying large-scale imbalanced classification in the distributed environment. Del Río *et al.* [4] proposed a parallel Random Forest algorithm with Cost-Sensitive learning (RF-BigDataCS) to deal with large-scale imbalanced datasets. López *et al.* [3] also developed fuzzy rule based classification systems with cost-sensitive learning (Chi-FRBCS-BigDataCS) for imbalanced classification. Specifically, the penalized cost-sensitive certainty factor was applied to the misclassification costs. These algorithms are implemented in MapReduce, and their training times are expected to be further reduced.

2.2 Distributed ADMM

Alternating Direction Method of Multipliers (ADMM) has superior convergence and performance which attracts many research interests [9]. By formulating the classification problem as a global consensus problem, ADMM can decompose it as a series of subproblems and then solve them in a distributed manner. In distributed ADMM, the nodes solve subproblems by processing its individual data locally in parallel, and then aggregate local results to obtain a global result [2]. Recently, a number of distributed classification methods via distributed ADMM have been proposed because of its property of high decomposability [2]. In particular, Zhang et al. [10] developed distributed linear classification methods via ADMM, in which the subproblems can be solved in parallel by conventional optimization algorithms. Also, ADMM had been extended to SVM with online or gradient versions for general convex problems [11]. Therefore, we focus on ADMM for imbalanced classification.

3 Our Method

3.1 Framework of CSADMM

Given the data points $\{(\mathbf{x}_i, y_i)\}_{i=1}^{l}$ ($\mathbf{x}_i \in \mathbb{R}^n, y_i \in \{-1, +1\}$), we take cost-sensitive SVM (CSSVM) as the linear classification model. Let $D^+ = \{\mathbf{x}_j | y_j = 1\}$, and $D^- = \{\mathbf{x}_j | y_j = -1\}$, CSSVM solves the optimization problem as follow:

$$\min_{\mathbf{w},\xi} \quad \frac{1}{2}\|\mathbf{w}\|^2 + C(C_+ \sum_{\mathbf{x}_j \in D^+} \xi_j + C_- \sum_{\mathbf{x}_j \in D^-} \xi_j) \tag{1}$$

where $\xi_j = \max\{0, 1 - y_j \mathbf{w}^T \mathbf{x}_j\}$, $C \geq 0$ is a hyperparameter. C_+ and C_- are the misclassification cost parameters for positive and negative classes. Supposing the data is split into n datasets $(D_1, D_2, \ldots, D_n)$, the problem is reformed as:

$$\min_{\mathbf{w}_i,\ldots,\mathbf{w}_n,\mathbf{z}} \quad \frac{1}{2}\|\mathbf{z}\|^2 + C \sum_{i=1}^{n}(C_+ \sum_{\mathbf{x}_j \in D_i^+} \xi_j + C_- \sum_{\mathbf{x}_j \in D_i^-} \xi_j), \tag{2}$$

$$s.t. \quad \mathbf{w}_i = \mathbf{z}, i = 1, 2, \ldots, n,$$

This problem in (2) is a global consensus problem [1]. For simplicity, we denote $\mathbf{w} = \{\mathbf{w}_1, \ldots, \mathbf{w}_n\}$, $\boldsymbol{\lambda} = \{\boldsymbol{\lambda}_1, \ldots, \boldsymbol{\lambda}_n\}$. Its augmented Lagrangian function is:

$$L_\rho(\mathbf{w}, \mathbf{z}, \boldsymbol{\lambda}) = \frac{1}{2}\|\mathbf{z}\|^2 + C \sum_{i=1}^{n}(C_+ \sum_{\mathbf{x}_j \in D_i^+} \xi_j + C_- \sum_{\mathbf{x}_j \in D_i^-} \xi_j)$$
$$+ \sum_{i=1}^{n} \boldsymbol{\lambda}_i(\mathbf{w}_i - \mathbf{z}) + \sum_{i=1}^{n} \frac{\rho}{2}\|\mathbf{w}_i - \mathbf{z}\|^2. \tag{3}$$

Given $\mathbf{u}_i = \frac{1}{\rho}\boldsymbol{\lambda}_i$, the augmented Lagrangian can be written equally as follows:

$$L_\rho(\mathbf{w}, \mathbf{z}, \mathbf{u}) = \frac{1}{2}\|\mathbf{z}\|^2 + C\sum_{i=1}^{n}(C_+ \sum_{\mathbf{x}_j \in D_i^+} \xi_j + C_- \sum_{\mathbf{x}_j \in D_i^-} \xi_j)$$
$$+ \frac{\rho}{2}\sum_{i=1}^{n}\|\mathbf{w}_i - \mathbf{z} + \mathbf{u}_i\|^2, \tag{4}$$

3.2 Optimization Procedures of CSADMM

Since $L_\rho(\mathbf{w}, \mathbf{z}, \mathbf{u})$ is separable in terms of $\mathbf{w}_i$, the optimization for $\mathbf{w}$ can be split into n separated subproblems solved in parallel. For node i, its optimization problem can be reformed as:

$$\mathbf{w}_i^{k+1} = \arg\min_{\mathbf{w}_i} f_i(\mathbf{w}_i) + \frac{\rho}{2}\|\mathbf{w}_i - \mathbf{z}^k + \mathbf{u}_i^k\|^2, \tag{5}$$

where $f_i(\mathbf{w}_i) = C\sum_{i=1}^{n}(C_+ \sum_{\mathbf{x}_j \in D_i^+} \xi_j + C_- \sum_{\mathbf{x}_j \in D_i^-} \xi_j)$. Moreover, local variable $\mathbf{w}_i^{k+1}$ can be replaced with its relaxation form to improve the convergence speed [10] before the update for global and dual variables:

$$\mathbf{w}_i^{k+1} \leftarrow \alpha\mathbf{w}_i^{k+1} + (1-\alpha)\mathbf{z}^k.$$

Then, global variable $\mathbf{z}^{k+1}$ has a closed form solution because of its differentiable in (4), and can be updated by the average of $\mathbf{w}_i^{k+1}$ and $\mathbf{u}_i^k$ as follows:

$$\mathbf{z}^{k+1} = \frac{\rho\sum_{i=1}^{n}(\mathbf{w}_i^{k+1} + \mathbf{u}_i^k)}{1 + n\rho}. \tag{6}$$

Finally, $\mathbf{u}_i^{k+1}$ of the subproblem with dataset D_i can be updated as:

$$\mathbf{u}_i^{k+1} = \mathbf{u}_i^k + \mathbf{w}_i^{k+1} - \mathbf{z}^{k+1}. \tag{7}$$

Since in CSADMM, the objectives $\frac{1}{2}\|\mathbf{z}\|^2$ and $f_i(\mathbf{w}_i)$ are both proper convex functions, the classification problem in (2) meets the Karush-Kuhn-Tucker (KKT) conditions to guarantee the convergence results.

3.3 Subproblem Optimization

We here adopt the dual coordinate descent method similar as [12] to solve subproblems, which has the linear run-time complexity in terms of the number of samples and good performance. Formally, the lagrange equation of the primal subproblem in (5) is defined as follows:

$$L_\rho(\mathbf{w}_i^k, \boldsymbol{\alpha}) = \frac{\rho}{2}\|\mathbf{w}_i^k - \mathbf{v}_i\|^2 - \sum_{j=1}^{s}\alpha_j y_j \mathbf{w}_i^k \mathbf{x}_j + \sum_{j=1}^{s}\alpha_j,$$

where $\mathbf{v}_i = (\mathbf{z}^k - \mathbf{u}_i^k)$, and s is the number of samples in dataset D_i. For simplicity, we denote C_j as the misclassification cost of classifying a sample $\mathbf{x}_j$ into incorrect classes. If $\mathbf{x}_j$ is a positive sample, $C_j = C * C_+$, otherwise $C_j = C * C_-$. The dual optimization problem can be obtained as:

$$\min_{\boldsymbol{\alpha}} f(\boldsymbol{\alpha}) = \frac{1}{2\rho} \boldsymbol{\alpha}^T \mathbf{Q} \boldsymbol{\alpha} - \mathbf{b}_i^T \boldsymbol{\alpha},$$

$$s.t. \quad 0 \le \alpha_i \le C_j, j = 1, 2, \dots, s,$$

(8)

where $Q_{jt} = y_j y_t \mathbf{x}_j^T \mathbf{x}_t$, $\mathbf{b}_i = [1 - y_1 \mathbf{v}_i^T \mathbf{x}_1, \dots, 1 - y_s \mathbf{v}_i^T \mathbf{x}_s]^T$. Since $L_\rho(\mathbf{w}_i^k, \boldsymbol{\alpha})$ is differentiable in terms of $\mathbf{w}_i^k$, it can be reformed as $\mathbf{w}_i^k = \frac{\sum_{j=1}^{s} y_j \alpha_j \mathbf{x}_j + \rho \mathbf{v}_i}{\rho}$. The partial derivative of $f(\boldsymbol{\alpha})$, $\triangledown_j f(\boldsymbol{\alpha}^k)$ with respect to α_j can be defined as:

$$\triangledown_j f(\boldsymbol{\alpha}^k) = \frac{1}{\rho} \sum_{t=1}^{s} \alpha_t^k Q_{jt} - b_{ij} = y_j \mathbf{w}_j^k \mathbf{x}_j - 1.$$

(9)

The dual coordinate descent method updates α_j^k to α_j^{k+1} while fixing all the other variables. Therefore, in each inner iteration, the problem in (8) can be regarded as one-variable subproblem has an optimum solution when the projected partial derivative $\triangledown_j^p f(\boldsymbol{\alpha}^k) = 0$, which is defined as:

$$\triangledown_j^p f(\boldsymbol{\alpha}^k) = \begin{cases} \min(0, \triangledown_j f(\boldsymbol{\alpha}^k)) & \text{if } \alpha_j^k = 0, \\ \max(0, \triangledown_j f(\boldsymbol{\alpha}^k)) & \text{if } \alpha_j^k = C_j, \\ \triangledown_j f(\boldsymbol{\alpha}^k) & \text{if } 0 \le \alpha_j \le C_j. \end{cases}$$

(10)

If $\triangledown_j^p f(\boldsymbol{\alpha}^k) = 0$, $\alpha_j^{k+1} = \alpha_j^k$. Otherwise, α_j can be updated as follows:

$$\alpha_j^{k+1} = \min(\max(\alpha_j^k - \frac{\triangledown_j^p f(\boldsymbol{\alpha}^k)}{Q_{jj}}, 0), C_j).$$

Finally, local variable $\mathbf{w}_i$ of the primal subproblem can be updated as follows:

$$\mathbf{w}_i^{k+1} = \mathbf{w}_i^k + (\alpha_j^{k+1} - \alpha_j^k) y_j \mathbf{x}_j.$$

(11)

The dual problem (8) is similar to that in [12] which has been demonstrated to converge into an ϵ-accurate solution (i.e., $f(\boldsymbol{\alpha}) \le f(\boldsymbol{\alpha}^*) + \epsilon$) in $O(log(1/\epsilon))$ iterations [10,12]. We can easily apply the results to subproblem optimization. In summary, the overall procedure is presented in Algorithm 1.

4   Experimental Results

4.1   Datasets and Experimental Settings

The public dataset KDD Cup 1999 available in the UCI Machine Learning Repository[1] is adopted to validate the performance of CSADMM. The KDD

---

[1] http://archive.ics.uci.edu/ml.

Algorithm 1. CSADMM

Input: C, ρ, α, $\mathbf{r}^0$, $\mathbf{s}^0$, ϵ^{pri} and ϵ^{dual}.

1: $k \leftarrow 0$
2: **while** $\|\mathbf{r}^k\|_2 > \epsilon^{\mathrm{pri}}$ or $\|\mathbf{s}^k\|_2 > \epsilon^{\mathrm{dual}}$ **do**
3:     $\mathbf{w}_i^{k+1} \leftarrow \arg\min_{\mathbf{w}_i} f_i(\mathbf{w}_i) + \frac{\rho}{2}\|\mathbf{w}_i - \mathbf{z}^k + \mathbf{u}_i^k\|^2$
4:     $\mathbf{w}_i^{k+1} \leftarrow \alpha \mathbf{w}_i^{k+1} + (1-\alpha)\mathbf{z}_i^k$
5:     $\mathbf{z}^{k+1} \leftarrow \frac{\rho \sum_{i=1}^n (\mathbf{w}_i^{k+1} + \mathbf{u}_i^k)}{1+n\rho}$
6:     $\mathbf{u}_i^{k+1} \leftarrow \mathbf{u}_i^k + \mathbf{w}_i^{k+1} - \mathbf{z}^{k+1}$
7:     $\mathbf{r}^{k+1} \leftarrow \sum_{i=1}^n \mathbf{w}_i^{k+1} - \mathbf{z}^{k+1}$
8:     $\mathbf{s}^{k+1} = \rho(\mathbf{z}^{k+1} - \mathbf{z}^k)$
9:     $k \leftarrow k+1$
10: **end while**

Output: z

Cup 1999 dataset contains multiple classes, which has been derived new binary datasets by treating the classes that contains larger number of samples as majority classes, while other classes as minority classes with different degree of IR. The details of seven large-scale imbalanced datasets are tabulated in Table 1.

Table 1. Summary of imbalanced datasets. The number of samples and attributes are denoted as (#Ex.) and (#Atts.) with imbalance ratio (#IR.).

Datasets	#Ex.	#Atts.	#IR.
dosvsnormal	4,856,151	41	3.99
dosvsprb	3,924,472	41	94.58
dosvsr2l	3,884,496	41	3448.82
dosvsu2r	3,883,422	41	74680.19
normalvsprb	1,013,883	41	23.67
normalvsr2l	973,907	41	863.93
noramlvsu2r	972,833	41	18707.33

To evaluate the performance of CSADMM, the geometric mean (GM), β-F-measure (FM), the area under the ROC curve (AUC) and training time are used as evaluation metrics. In particular, We define the training time starting from the data loading, and set β to be the value of IR. For the parameters (ρ, C_+, C_-, α) which can be empirically well chosen, we can set them as predefined constants (1,IR,1,1.6) which are consistent with these in [3,10]. The hyperparameter C was chosed from $10^i (i = -2, \ldots, 4)$ by 10-fold cross-validation. For the implementation, 12 machines are used under the framework of Message Passing Interface (MPI), each of them with an Intel(R) Xeon(R) CPU E5-2650 (2.6 GHz/30 M Cache) processor and 64 GB RAM.

4.2   Comparison with Other Distributed Methods

To validate the superiority of CSADMM for large-scale imbalanced data classification, we compare it with two distributed classification methods implemented in MapReduce. The first method is distributed Random Forest (RF) [4] combined with sampling and cost-sensitive learning, respectively. The second method proposes the penalized cost-sensitive certainty factor in a Fuzzy Rule Based Classification System (Chi-FRBCS) by Chi *et al.*'s method in [3]. The details of compared methods with different strategies are listed as follows:

- **DADMM:** Classification method via distributed ADMM.
- **CSADMM:** Our method for large-scale imbalanced data classification.
- **RF-BigData:** Distributed random forest classification for large-scale data classification.
- **RF-BigDataCS:** Cost-sensitive learning is adopted in the framework of distributed random forest to deal with imbalanced data classification.
- **RUS+RF-BigData:** Random undersampling has been applied in the distributed random forest.
- **ROS+RF-BigData:** Random oversampling is used to prevent the class imbalance problem in distributed random forest.
- **SMOTE+RF-BigData:** SMOTE is adopted to generate the synthetic minority class samples in distributed random forest.
- **Chi-FRBCS-BigData:** The fuzzy rule based classification system, which can deal with large-scale data classification in the distributed environment.
- **Chi-FRBCS-BigDataCS:** Distributed Chi-FRBCS, in which different misclassification costs of rule weights are used by cost-sensitive learning.

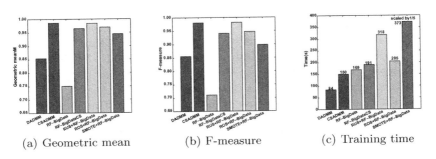

(a) Geometric mean   (b) F-measure   (c) Training time

Fig. 1. Figures (a–c) show classification performance and training time of these comparative methods.

The classification results are generated, and evaluated by their averages. Firstly, we compare the performance of CSADMM with various versions of distributed random forest. The average results are shown in Fig. 1. We can observe that our CSADMM can achieve superior performance but with much less training time, which can effectively solve the class imbalance problem. In particular, the geometric mean and F-measure results of CSADMM perform better than those of distributed random forest with the exception of RUS+RF-BigDta on KDD Cup 1999 dataset. Also, compared with distributed random forest classifiers, CSADMM can speed up the training time to obtain better classification performance. It is because distributed framework of ADMM has superior convergence and performance by the alternating minimization over the local and global variables, and has been demonstrated its effectiveness and scalability [2].

Table 2. The classification results of these comparative methods.

Dataset	DADMM		CSADMM		Chi-FRBCS-BigData		Chi-FRBCS-BigDataCS	
	AUC (%)	Time (s)	AUC (%)	Time (s)	AUC (%)	Time (s)	AUC (%)	Time (s)
dosvsnormal	99.94	**168**	**100.00**	279	99.92	4060	99.93	3408
dosvsprb	99.84	**101**	**100.00**	217	86.33	5310	95.45	5135
dosvsr2l	94.91	**65**	**100.00**	131	99.18	1789	99.99	2047
dosvsu2r	**100.00**	**140**	**100.00**	201	84.64	1266	88.80	1282
normalvsprb	97.88	**23**	**98.20**	74	86.90	262	96.59	301
normalvsr2l	50.00	**28**	**94.34**	92	52.23	279	82.29	329
normalvsu2r	50.00	61	**90.48**	**54**	50.00	350	50.00	327

Figure 1 also shows that distributed solvers combined with data sampling and cost-sensitive learning methods can improve geometric mean and F-measure on imbalanced datasets compared with DADMM and RF-BigData classifiers. We can find cost-sensitive methods modify the classification hyperplane away from the minority class, can obtain the competitive classification performance, and take less training time compared with data sampling methods (undersampling and oversampling). Although the oversampling strategy can improve classification performance in geometric mean and F-measure, the training time is much more than those of undersampling and cost-sensitive classification. In particular, SMOTE+RF-BigData performs worst and takes too much time to balance the data class distribution compared others methods. SMOTE+RF-BigData randomly chooses the k nearest neighbors from the minority class samples, and then generates the synthetic samples in a certain way, and thus it may be not suitable to deal with large-scale data because of its high computational complexity.

Secondly, we compare our method with distributed Chi-FRBCS-BigData combined with cost-sensitive learning. The comparative results between ADMM and Chi-FRBCS are tabulated in Table 2. We found that DADMM and CSADMM can significantly speed up the training times on all datasets compared with Chi-FRBCS-BigData and Chi-FRBCS-BigDataCS. Since Chi-FRBCS builds a fuzzy rule based classification system and uses the penalized cost-sensitive certainty factor to compute the rule weight, these methods take much more time to train the samples. All the AUC and training time of CSADMM perform better than these of Chi-FRBCS-BigData and Chi-FRBCS-BigDataCS, which reveals that CSADMM is an effective method for distributed classification on imbalanced data. Therefore, compared with these methods [3,4], our CSADMM obtains promising results with much less training times.

4.3 Compare with Single-Machine Methods

Now, we compare our CSADMM against several single-machine baseline methods, i.e. Support Vector Machine (SVM) and Cost-Sensitive SVM (CSSVM) by modifying LIBLINEAR [13], Cost-Sensitive Neural Networks (CSNN) [14] and AdaBoost [15]. LIBLINEAR is an open source library widely used for classification, and can adjust different misclassification costs to deal with different class samples. CSNN conveys the costs of samples to resample a new dataset for training a neural network. AdaBoost is a well-known ensemble algorithm which can be utilized to classify by constraining each weak classifier (decision tree used in this paper). The maximum numbers of iteration in CSNN, Adaboost and CSADMM are 200. Table 3 shows that CSADMM can obviously improve classification performance, and achieves significant speedup against these methods. Also, we can find AdaBoost can obtain good classification performance, but it is very time-consuming, which makes it difficult to apply for large-scale imbalanced classification. In summary, CSADMM can obtain competitive performance and significantly speed up the training process.

Table 3. The experimental results of CSADMM and single-machine methods on large-scale datasets (NaN indicates that all the samples are classified as negative samples).

Dataset	CSADMM	SVM	CSSVM	CSNN	AdaBoost	CSADMM	SVM	CSSVM	CSNN	AdaBoost
FM (%): F-measure						GM (%): geometric mean				
dosvsnormal	99.94	98.48	97.19	99.98	**99.99**	99.93	98.50	98.50	99.97	**99.98**
dosvsprb	99.92	99.82	99.66	99.95	**99.96**	99.63	99.66	99.83	99.18	**99.76**
dosvsr2l	95.13	79.47	65.93	74.02	**99.34**	97.54	81.20	81.30	79.56	**99.45**
dosvsu2r	99.98	32.56	70.00	66.67	**99.98**	99.98	83.66	84.54	70.71	**99.98**
normalvsprb	96.58	80.64	70.09	90.39	**99.93**	98.14	83.63	83.63	91.72	**99.96**
normalvsr2l	94.47	41.74	41.78	92.34	**95.56**	94.44	64.27	68.34	94.04	**97.75**
normalvsu2r	**76.77**	NaN	NaN	NaN	72.73	**87.60**	0	0	0	85.28
Average	94.69	72.12	74.11	87.23	**95.36**	96.75	72.99	73.73	76.45	**97.45**
AUC (%): the area under the ROC curve						Time (s): training time				
dosvsnormal	99.93	98.51	98.50	99.96	**99.98**	**64**	1099	1219	549	15469
dosvsprb	99.96	99.83	99.84	99.98	**99.94**	**48**	332	318	224	13741
dosvsr2l	97.57	82.97	82.23	82.19	**99.56**	**34**	856	824	192	13034
dosvsu2r	99.97	85.00	85.00	75.00	**99.98**	53	**46**	47	59	10308
normalvsprb	98.16	84.92	84.94	91.96	**99.97**	**20**	150	159	143	3418
normalvsr2l	89.32	70.33	70.33	94.22	**97.78**	**31**	209	217	129	2974
normalvsu2r	**88.33**	50.00	50.00	50.00	86.37	**21**	200	215	17	3032
Average	96.18	81.58	81.55	84.76	**97.65**	**39**	413	428	187	8854

5 Conclusion

In this paper, we proposed a novel distributed classification method namely Cost-Sensitive Alternating Direction Method of Multipliers (CSADMM) for large-scale imbalanced data classification. In particular, the classification problem is derived into a series of subproblems in CSADMM. An efficient dual coordinate descent method is adopted for subproblem optimization in parallel. Meanwhile, cost-sensitive SVM is employed to solve the class imbalance problem. The experimental results show that CSADMM performs better in classification performance, and speeds up the training time compared with RF-BigDataCS and Chi-FRBCS-BigDataCS. Moreover, compared with single-machine methods, CSADMM also achieves promising results.

Acknowledgements. The work was supported by National Natural Science Foundation of China (61403208, 61673203), and Young Elite Scientists Sponsorship Program by CAST (YESS 20160035).

References

1. Forero, P.A., Cano, A., Giannakis, G.B.: Consensus-based distributed support vector machines. J. Mach. Learn. Res. **11**, 1663–1707 (2010)
2. Boyd, S., Parikh, N., Chu, E., Peleato, B., Eckstein, J.: Distributed optimization and statistical learning via the alternating direction method of multipliers. Found. Trends® Mach. Learn. **3**(1), 1–122 (2011)
3. López, V., del Río, S., Benítez, J.M., Herrera, F.: Cost-sensitive linguistic fuzzy rule based classification systems under the mapreduce framework for imbalanced big data. Fuzzy Sets Syst. **258**, 5–38 (2015)
4. del Río, S., López, V., Benítez, J.M., Herrera, F.: On the use of mapreduce for imbalanced big data using random forest. Inf. Sci. **285**, 112–137 (2014)
5. Kumar, N.S., Rao, K.N., Govardhan, A., Reddy, K.S., Mahmood, A.M.: Undersampled k-means approach for handling imbalanced distributed data. Progress Artif. Intell. **3**(1), 29–38 (2014)
6. Veropoulos, K., Campbell, C., Cristianini, N., et al.: Controlling the sensitivity of support vector machines. In: Proceedings of the International Joint Conference on AI, pp. 55–60 (1999)
7. Batuwita, R., Palade, V.: FSVM-CIL: fuzzy support vector machines for class imbalance learning. IEEE Trans. Fuzzy Syst. **18**(3), 558–571 (2010)
8. Cao, P., Zhao, D., Zaiane, O.: An optimized cost-sensitive SVM for imbalanced data learning. In: Pei, J., Tseng, V.S., Cao, L., Motoda, H., Xu, G. (eds.) PAKDD 2013. LNCS, vol. 7819, pp. 280–292. Springer, Heidelberg (2013). doi:10.1007/978-3-642-37456-2_24
9. Goldfarb, D., Ma, S., Scheinberg, K.: Fast alternating linearization methods for minimizing the sum of two convex functions. Math. Program. **141**(1–2), 349–382 (2013)
10. Zhang, C., Lee, H., Shin, K.G.: Efficient distributed linear classification algorithms via the alternating direction method of multipliers. In: International Conference on Artificial Intelligence and Statistics, pp. 1398–1406 (2012)
11. Tao, Q., Gao, Q.K., Chu, D.J., Wu, G.W.: Stochastic learning via optimizing the variational inequalities. IEEE Trans. Neural Netw. Learn. Syst. **25**(10), 1769–1778 (2014)
12. Hsieh, C.J., Chang, K.W., Lin, C.J., Keerthi, S.S., Sundararajan, S.: A dual coordinate descent method for large-scale linear SVM. In: Proceedings of the 25th International Conference on Machine Learning, pp. 408–415. ACM (2008)

13. Fan, R.E., Chang, K.W., Hsieh, C.J., Wang, X.R., Lin, C.J.: Liblinear: a library for large linear classification. J. Mach. Learn. Res. **9**, 1871–1874 (2008)
14. Zhou, Z.H., Liu, X.Y.: Training cost-sensitive neural networks with methods addressing the class imbalance problem. IEEE Trans. Knowl. Data Eng. **18**(1), 63–77 (2006)
15. Owusu, E., Zhan, Y., Mao, Q.R.: An SVM-AdaBoost facial expression recognition system. Appl. Intell. **40**(3), 536–545 (2014)

Fuzzy 2D-LDA Face Recognition Based on Sub-image

Xingrui Zhang[1], Yulian Zhu[1(✉)], and Xiaohong Chen[2]

[1] College of Computer Science and Technology,
Nanjing University of Aeronautics and Astronautics,
Nanjing 210016, People's Republic of China
{xrzhang,lianyi_1999}@nuaa.edu.cn
[2] College of Science, Nanjing University of Aeronautics and Astronautics,
Nanjing 210016, People's Republic of China
lyandcxh@nuaa.edu.cn

Abstract. This paper proposes a novel method for face recognition, called, sub-image-based fuzzy 2D Linear Discriminant Analysis (subimage-F2DLDA) based on sub-image method, 2-Dimensional Linear Discriminant Analysis (2D-LDA) and fuzzy set theory. We first partition the whole training image set into several different sub-image sets by dividing each image into sub-images and collecting the same location together, and then redefine the within-class matrix and between-class matrix for each sub-image set, which can be computed by incorporating the membership degree matrix using fuzzy k-nearest neighbor(FKNN). Finally, we construct a nearest classifier based on fuzzy 2DLDA for each sub-image set. We construct experiments on Yale A, Extended Yale B and ORL face databases, and the results show that the proposed approach achieves better performance than compared methods on face recognition.

Keywords: Face recognition · Sub-image · Fuzzy theory · 2D-LDA

1 Introduction

Face recognition is a research hot in the area of computer vision and pattern recognition. Linear Discriminant Analysis (LDA) [1] is a popular tool for face feature extraction and face recognition. Based on LDA, a large number of its variants have been proposed and widely used in face recognition. The aim of LDA is to find a projection that minimizes the within-class scatter and maximizes the between-class scatter at the same time [2]. Since LDA, as a supervised method, makes good use of class information of samples, it usually outperforms the unsupervised method. However, LDA cannot be directly applied to the Small Sample Size (SSS) problem, since SSS leads to singularity of the within-class scatter matrix in LDA.

In order to overcome the SSS problem, researchers paid more attention on 2D-based approaches by directly extracting image features from 2D image matrices and extended LDA into 2-dimensional LDA. For example, Li *et al.* extracted

© Springer International Publishing AG 2017
H. Yin et al. (Eds.): IDEAL 2017, LNCS 10585, pp. 326–334, 2017.
https://doi.org/10.1007/978-3-319-68935-7_36

features along the row (or the column) direction of image matrices [3]; while Ye *et al.* extracted features along both the row and the column directions of image matrices [4]. Since 2D-based approaches represent the face images with original images representation, they not only effectively reduce the time complexity, but also make good use of the spatial relationship between pixels.

It is worth noting that LDA and most of its variants are based on "hard class label", that is, a sample is (or not) fully assigned to a given class. Actually, each sample contains some uncertain class information of other samples. Cover and Hart pointed out that about half of the class information for one sample is hidden in its neighbors [5]. So it is reasonable to introduce the idea of "soft class label" into feature extraction in order to make good use of the class information hidden in neighbors. Then, fuzzy set theory could show its mettle here.

Fuzzy set theory has attracted increasing attention in the fields of pattern recognition and image processing in recent years [6–9] since it could be used to classify a sample into multi-classes with the fuzzy membership degree, which plays an important role in processing the uncertainty [7]. For example, Kwak *et al.* proposed the fuzzy fisherface by adding the membership degree on feature vectors resulting from PCA [9]; Yang *et al.* combined the complete LDA and fuzzy set theory to form complete fuzzy LDA (CFLDA) [8]; and Yang *et al.* proposed fuzzy 2D-LDA that extended fuzzy fisherface to image matrices [7].

However, these methods are all based on global features and they are not robust to occlusions and illumination. While local features-based methods can extract local features effectively, and reduce the sensitivity on the variance of environment. Sub-image method is a simple and effective local features-based method which has been widely used to improve the performance of face recognition [10]. The main common point of sub-image methods is to partition a whole face image into several smaller sub-images and then to extract local features from each sub-image respectively. Extensive experimental results have shown that the sub-image method has better performance than the global approach.

Inspired by the successes of the sub-image method, we combine the sub-image method and fuzzy set theory, and propose a novel method called subimage-F2DLDA to improve the performance of face recognition. Specifically, we first partition each image into some sub-images and construct several different sub-image sets by collecting all sub-images sharing the same location. Then, we redefine the within-class matrix and between-class matrix for 2D-LDA by adding membership degree matrix which is computed by using FKNN. Finally, we make a final decision with majority vote rule. The experiments are constructed on Yale A, Extended Yale B and ORL face databases, and the results show that the proposed approach outperforms compared methods on face recognition.

The rest of the paper is organized as follows. In Sect. 2 the proposed method for face recognition is formulated. In Sect. 3 experimental comparisons carried out on three face databases are reported. Finally, the conclusions are included in the Sect. 4.

2 The Proposed Approach

The proposed approach can be summarized as follows: (1) partition the images into patches; (2) compute the fuzzy membership degree using FKNN; (3) construct Fuzzy 2D-LDA by redefining the within-class scatter matrix and between-class scatter matrix; (4) classify each patch of an unknown test sample and make a final decision. Next, we will describe the proposed method in detail.

2.1 Partitioning Sub-image

Currently, there are two popular techniques to implement the image partition: local components and local regions. Since local regions methods can usually obtain better performance than local components ones, we adopt the simplest rectangular region to partition images in this paper, with the detailed process is shown in Fig. 1. Specifically, we first divide each face image of the training set into L equally-sized sub-images in a non-overlapping way. Then we collect all sub-images with the common attributes of all training images to form a specific sub-image's training set. In this way, we can obtain L different sub-image's training sets.

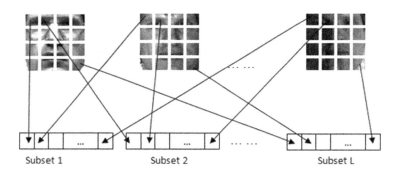

Fig. 1. The process of partitioning sub-images in a non-overlapping way. L represents the number of patches of an image.

2.2 Computing Class Membership Degree by Using Fuzzy K-Nearest Neighbor(FKNN)

Suppose that there are N training images A_{ij} $(i = 1, 2, \cdots, C; j = 1, 2, \cdots, N_i)$ belonging to C classes and these classes have $N_1, N_2, \cdots, N_C$ $\left(\sum_{i=1}^{C} N_i = N \right)$ face images respectively. Define the membership grade of j-th sample in i-th class as u_{ij} $(i = 1, 2, \cdots, C; j = 1, 2, \cdots, N)$. It satisfies three obvious properties: $\sum_{i=1}^{C} u_{ij} = 1, 0 < \sum_{j=1}^{N} u_{ij} < N, 0 \leq u_{ij} \leq 1$. u_{ij} can be realized through the following steps [6].

Step 1: Compute the Euclidean distance matrix between pairs of the feature vectors in the training set.

Step 2: Set the diagonal elements of the distance matrix to infinity (practically place large numeric values there).

Step 3: Sort the distance matrix in an ascending order by columns. Collect the class labels of k-nearest neighbors for next step.

Step 4: Compute the membership grade to class "i" for j-th pattern according to the expression proposed in the literature [11]:

$$u_{ij} = \begin{cases} a + b\,(n_{ij}/k)\,, \text{if } i = \text{ the label of the } j \text{ -th pattern} \\ b\,(n_{ij}/k)\,, i \neq \text{ the label of the } j \text{ -th pattern} \end{cases} \quad s.t. \begin{cases} a + b = 1 \\ a > b \\ 0 \leq a \leq 1 \\ 0 \leq b \leq 1. \end{cases} \quad (1)$$

where n_{ij} stands for the number of neighbors of the j-th data (pattern) that belong to the i-th class; a and b are two constants. The design of u_{ij} can ensure that the membership degree of the j-th sample to the i-th class is larger than other classes if the j-th sample belongs to the i-th class.

2.3   Constructing both Fuzzy Within-Class and Between-Class Matrices of Fuzzy 2D-LDA

After computing the class membership degree, we can construct fuzzy 2D-LDA by incorporating fuzzy class membership degree into within-class and between-class matrices. They can be redefined as:

$$S_{fw} = \sum_{i=1}^{C} \sum_{j=1}^{\widetilde{N_i}} u_{ij}^p \left(A_{ij} - m_i\right) \left(A_{ij} - m_i\right)^T. \tag{2}$$

$$S_{fb} = \sum_{i=1}^{C} \widetilde{N_i} u_{ij}^p \left(m_i - m\right) \left(m_i - m\right)^T. \tag{3}$$

where m, m_i and $\widetilde{N_i}$ is the mean of total data, the mean vectors of class i and fuzzy number of elements belonging to class i respectively:

$$m = \frac{1}{N_F} \sum_{i=1}^{C} \sum_{j=1}^{N} u_{ij}^p A_j = \frac{1}{N_F} \sum_{i=1}^{C} \widetilde{N_i} m_i. \tag{4}$$

$$m_i = \frac{1}{\widetilde{N_i}} \sum_{j=1}^{N} u_{ij}^p A_j. \tag{5}$$

$$\widetilde{N_i} = \sum_{j=1}^{N} u_{ij}^p. \tag{6}$$

N_F is the fuzzy number of elements in the total data set and p is a constant which controls the influence of fuzzy membership degree [11].

$$N_F = \sum_{i=1}^{C} \sum_{j=1}^{N} u_{ij}^p = \sum_{i=1}^{C} \widetilde{N_i}. \tag{7}$$

Similar with LDA, the aim of our fuzzy 2D-LDA is also to maximize the ratio of between-class scatter matrix and within-class scatter matrix. So the objective function can be formulated:

$$J(W) = \arg\max_W \frac{W^T S_{fb} W}{W^T S_{fw} W}. \tag{8}$$

The optimal solution W can be easily obtained by solving a generalized eigen-equation.

2.4 Classification and Final Decision

After obtaining the projection vectors by using fuzzy 2D-LDA for each sub-image training set, we can construct a nearest neighbor classifier for each subset as base classifier $C_i\,(i = 1, 2, \cdots, L)$. For an unknown image P, we first divide the image into L sub-images $P_j\,(j = 1, 2, \cdots, L)$ and project each sub-image into corresponding projection vectors. Then we recognize each sub-image using corresponding nearest neighbor classifier which is constructed in Sect. 2.3. There are L sub-images for the unknown image P, therefore L classification results can be obtained. Finally, we use majority voting rule as combination rule to make the final decision. Let the base classifier $C_i\,(P_j) \in \{0, 1\}$, ($C_i\,(P_j) = 1$ means P_j belongs to i-th class; otherwise, it not belongs to i-th class), we can represent the class label of P as follows:

$$identify\,(P) = \arg\max_{1 \leq i \leq C} \left(\sum_{j=1}^{L} C_i\,(P_j) \right). \tag{9}$$

3 Experimental Analysis

3.1 The Used Databases

In this paper, we conduct experiments on Yale A, Extended Yale B and ORL databases which are three popular face databases. We get the three databases from Deng Cai's website: http://www.cad.zju.edu.cn/home/dengcai/. Some of their samples are shown in Fig. 2. Then we would introduce the three databases.

The Yale A database has 15 subjects with 11 images per subject, including the illumination variation, expression variation, occlusion (glasses/no glasses) as well as scale changes. Some samples from Yale A database are shown in Fig. 2 (a).

Figure 2 (b) shows 32 images of one person from Extended Yale B database. Extended Yale B database comprises about 64 images per subject under the large illumination variation which is mainly used to test the robustness of the algorithm to light changes.

The ORL database includes 400 face images coming from 40 subjects and each subject has 10 images. These pictures were taken in different environments with slight vary in position, rotation, scale, and facial expression. For some subjects, the images were taken at different times, the lighting, facial expressions and facial details (glasses/no glasses). Figure 2 (c) shows 10 images of one person in ORL database.

Besides, some other information of the three databases is shown in Table 1. The "Train set size" in Table 1 means the number of images of per subject that are taken to form the training set in random way, and the rest are considered to be the testing set [11].

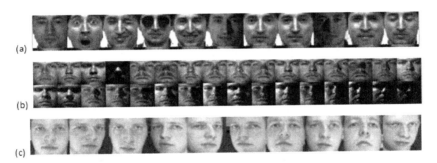

Fig. 2. Some samples of three face databases: (a) Yale A database, (b) Extended Yale B database, (c) ORL database.

Table 1. The Related Parameters of Used Face Databases

Data set	Image size	Number of subjects	Images per subject	Train set size
Yale A	64×64	15	11	3 4 5 6 7 8
Ext.Yale B	32×32	38	64	5 10 20 30 40
ORL	64×64	40	10	3 4 5 6 7

3.2 Experimental Settings

In this paper, we set $a = 0.51, b = 0.49, p = 1, k = 5$ in FKNN and set the patch size to 8×8 pixels for all databases following settings in paper [10], [12]. Besides, according to the parameters settings in paper [12], the dimension reduced using fuzzy 2D-LDA is set to 2 for Yale A and ORL databases; while Extended Yale B database's is set to 1. Each experiment was repeated 20 times for random choices of the training set in a fixed training set size. And the average recognition error rates and standard deviations are reported.

3.3 Experimental Results

We evaluate the performance of the proposed method through the average recognition error rate and the standard deviation. The results on Yale A, Extended Yale B and ORL databases are shown in Tables 2, 3 and 4 respectively (The "G" means "Train set size"). The better results are highlighted in bold.

Table 2 shows experimental results on Yale A database. We could find that the average error rates of the proposed method on Yale A database are the lowest. Compared to fuzzy-2DLDA which is based on global features, our method reduces at least 10%. Therefore, we believe that performing fuzzy-2DLDA in sub-image set is effective to reduce the error rate.

Table 3 also shows that our proposed subimage-F2DLDA achieves the lowest average error rates. Compared to subimage-2DLDA which uses "hard class label", our method reduces about 3% in most of experiments. Also, with the increase of the train set size, the advantage of the proposed method is more significant compared to subimage-2DLDA. The results indicate that introducing the "soft class label" is necessary.

From Table 4 we can also find the proposed method has a good performance on ORL database. The ORL database includes some images with slight pose variation which means the proposed method works well on slight pose variation.

In addition, we can also observe that the approaches based on sub-image outperform global methods, that is, the subimage-2DLDA and the subimage-F2DLDA outperform 2D-LDA and fuzzy 2D-LDA, respectively. The essence is that the local features are more important than global features. It is worth noting that sub-image methods maybe fail to work well when images involve more pose variation because local feature extracted from each local region is less helpful to correct recognition while the global feature is more important than the local feature in this case.

Selection of parameter is a key factor to any algorithm. Here, we test the effect of changing k-value (the number of nearest neighbors in FKNN) on the performance of our proposed method. We test on Yale A database and the train set sizes are set to 3, 4, 5, 6 respectively. The results are shown on Fig. 3, from which we can find that the average error rates have just slight fluctuation with the change of k-value. It means that our method has a good stability to the change of k-value.

Table 2. The Performance on Yale A Database (Average Error Rates/Standard Deviations%).

	G=3	G=4	G=5	G=6	G=7	G=8
2D-LDA	40.0/4.7	33.6/4.0	29.9/4.0	27.2/3.6	27.8/3.6	25.0/6.0
Fuzzy 2D-LDA	44.5/4.9	36.9/4.2	31.3/4.1	30.7/4.2	28.8/3.3	27.1/6.3
Subimage-2DLDA	31.8/3.8	25.8/4.5	21.7/5.2	18.8/4.7	19.3/4.0	16.7/6.2
Subimage-F2DLDA	**29.5/3.9**	**25.1/3.6**	**19.6/3.7**	**18.5/3.8**	**18.6/4.0**	**15.8/5.0**

Table 3. The Performance on Extended Yale B Database (Average Error Rates/Standard Deviations%).

	G=5	G=10	G=20	G=30	G=40	G=50
2D-LDA	51.3/2.8	36.3/2.0	25.8/2.2	20.5/1.8	16.8/0.9	14.6/1.4
Fuzzy 2D-LDA	51.8/2.3	36.5/1.4	25.0/1.3	19.8/1.2	16.8/0.9	14.9/1.1
Subimage-2DLDA	40.7/2.2	27.8/1.6	19.3/0.9	15.0/0.9	13.0/1.0	11.9/1.2
Subimage-F2DLDA	**40.4/2.4**	**27.5/1.9**	**16.9/1.4**	**13.0/1.0**	**10.0/1.2**	**8.6/1.1**

Table 4. The Performance on ORL Database (Average Error Rates/Standard Deviations%).

	G=3	G=4	G=5	G=6	G=7
2D-LDA	20.8/3.4	14.5/2.7	11.1/2.6	9.5/2.1	8.4/3.0
Fuzzy 2D-LDA	22.7/4.7	16.3/3.8	12.4/3.3	9.6/2.1	8.7/2.4
Subimage-2DLDA	18.6/1.8	11.6/2.3	8.4/2.0	6.2/2.0	5.2/2.0
Subimage-F2DLDA	**17.7/1.5**	**11.0/2.1**	**7.6/2.2**	**5.5/1.8**	**4.8/2.0**

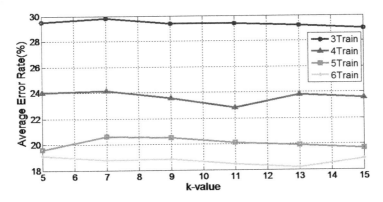

Fig. 3. The average error rates of proposed method with the change of k-value on Yale A database.

4   Conclusions

In this paper, we propose a subimage-F2DLDA method for face recognition based on sub-image. We divide each training image into some sub-images and redefine fuzzy within-class and between-class scatter matrices by incorporating fuzzy membership degrees. Experimental results on Yale A, Extended Yale B and ORL databases showed the new method is very effective to the variation caused by varying illumination, viewing conditions and facial expression etc.

Acknowledgments. This work is supported by the Fundamental Research Funds for the Central Universities, no.NS2017071 and NP2014081, National Natural Science Foundations of China (NSFC) under Grant no.61703206 and 61403193.

References

1. Martínez, A.M., Kak, A.C.: Pca versus lda. IEEE Trans. Pattern Anal. Mach. Intell. **23**(2), 228–233 (2001)
2. Xu, J., Gu, Z., Xie, K.: Fuzzy local mean discriminant analysis for dimensionality reduction. Neural Process. Lett. **44**(3), 701–718 (2016)
3. Li, M., Yuan, B.: 2d-lda: A statistical linear discriminant analysis for image matrix. Pattern Recogn. Lett. **26**(5), 527–532 (2005)
4. Ye, J., Janardan, R., Li, Q.: Two-dimensional linear discriminant analysis. In: Advances in Neural Information Processing Systems, pp. 1569–1576 (2005)
5. Cover, T., Hart, P.: Nearest neighbor pattern classification. IEEE Trans. Inf. Theory **13**(1), 21–27 (1967)
6. Li, X., Song, A.: Fuzzy msd based feature extraction method for face recognition. Neurocomputing **122**, 266–271 (2013)
7. Yang, W., Yan, X., Zhang, L., Sun, C.: Feature extraction based on fuzzy 2dlda. Neurocomputing **73**(10), 1556–1561 (2010)
8. Yang, W., Yan, H., Wang, J., Yang, J.: Face recognition using complete fuzzy lda. In: 19th International Conference on Pattern Recognition, 2008, ICPR 2008, pp. 1–4. IEEE (2008)
9. Kwak, K.-C., Pedrycz, W.: Face recognition using a fuzzy fisherface classifier. Pattern Recogn. **38**(10), 1717–1732 (2005)
10. Chakrabarty, A., Jain, H., Chatterjee, A.: Volterra kernel based face recognition using artificial bee colonyoptimization. Eng. Appl. Artif. Intell. **26**(3), 1107–1114 (2013)
11. Yang, J., Sun, Q.-S.: A novel generalized fuzzy canonical correlation analysis framework for feature fusion and recognition. Neural Process. Lett. 1–16 (2017). doi:10.1007/s11063-017-9600-z
12. Kumar, R., Banerjee, A., Vemuri, B.C.: Volterrafaces: discriminant analysis using volterra kernels. In: IEEE Conference on Computer Vision and Pattern Recognition, 2009, CVPR 2009, pp. 150–155. IEEE (2009)

Evolving Technical Trading Strategies Using Genetic Algorithms: A Case About Pakistan Stock Exchange

Basit Tanvir Khan, Noman Javed[✉], Ambreen Hanif[✉],
and Muhammad Adil Raja[✉]

Namal College, Mianwali, Pakistan
{basit2013,noman.javed,ambreen.hanif,adil.raja}@namal.edu.pk

Abstract. Finding optimum trading strategies that maximize profit has been a human desire since the inception of the first stock market. Many techniques have been employed ever since to accomplish this goal without sacrificing much computational power and time. In this paper, Genetic Algorithms (GAs) are used to achieve the aforementioned objectives. The performances of trading strategies devised by the GA are compared with the performance of the infamous Buy and Hold (B&H) Strategy. The stocks on which the performances are compared belong to Pakistan Stock Exchange (PSX). The strategies generated by GA outperform the B&H strategies on these stocks.

1 Introduction

Ever since the inception of stock markets, there has been a consistent pursuit of devising profit maximizing strategies. Two schools of thoughts have emerged after decades of research in the domain: fundamentalists and technicians. The former believe in long term investment and focus more on the companies' fundamentals whereas the latter believe in digging the current trends for predicting short term stock movements. Efficient Market Hypothesis (EMH) [1,2] and Random Walk Theory (RWT) [3] consider current price as the reflective indicator of every impact on stock and think of it as unpredictable. However, a lot of approaches suggest the contrary by presenting empirical evidence.

A number of technical indicators (TIs) have been developed by financial and behavioral economists to gain insights about the price, volume, momentum, and trend of the shares. Short term investors rely heavily on these indicators to make better investment decisions. These indicators are also used to determine the past performance of a security, a phenomenon called back-testing. On the basis of information gathered from past performance, these indicators can be used to construct sophisticated trading rules which vary from individual to individual. These indicators can also be employed in various combinations to devise trading strategies.

© Springer International Publishing AG 2017
H. Yin et al. (Eds.): IDEAL 2017, LNCS 10585, pp. 335–344, 2017.
https://doi.org/10.1007/978-3-319-68935-7_37

With the emergence of the discipline of data science, graceful advancements in machine learning (ML) and easy access to high-performance computing resources, stock forecasting is regaining traction of research and industry. Approaches based on statistics, artificial neural networks (ANNs), genetic algorithms (GAs) and natural language processing have been developed to attack the problem. Objectives of these proposed techniques vary from individual stock price prediction to market index predictions. Problems that are addressed with such techniques range from building an optimum portfolio to finding fraudulent behaviors.

Evolutionary algorithms (EAs) in general and GAs, in particular, have been used extensively to solve optimization problems. Selecting optimal trading strategies based on the past performance is an optimization problem. Pakistan Stock Exchange (PSX) often enjoys the status of being a top performer in Asia as well as the other economically active regions [4]. No such study exists for PSX to the best of our knowledge that exploits ML approaches and TIs to study the behavior of stocks. Our aim is to evolve stock specific trading strategies through GAs. As stated earlier, these trading strategies are combinations of TIs. To this end, this means the problem is to find the optimal combination of TIs that outperforms other combinations.

The remainder of this paper is organized as follows: Sect. 2 sheds some light on the existing literature. In Sect. 3 we introduce GAs. In Sect. 4 the adaptation of GA to this project is discussed. The computational results are discussed in Sect. 5. Finally, Sect. 6 concludes this paper.

2   Related Work

Stock market prediction is a cumbersome and an intricate task. Stocks have different moods, behaviors and there are multiple other factors which affect the stock market prediction. Methods for stock market forecasting are normally divided into three main categories: conventional, market and computational intelligence based methods [5].

Conventional methods refer to the techniques designed and developed by the statisticians and economists to predict stocks. Market-based methods are further divided into two categories: fundamental analysis and technical analysis [6]. Studies have also been conducted to examine the performance of the TTRs for intra-day trading [7], foreign exchange trade [8] as well as portfolio management [6].

As far as the computational intelligence techniques are concerned, the stock prediction systems are mostly based on ANNs or fuzzy logic techniques [9]. It is also stated that in many works where ML techniques were employed, better results were obtained than the conventional methods. ANNs are usually provided with stock's price, volume and/or some other features like interest rates etc. [9]. Deep neural networks have also been explored for predicting the behavior of stock market [10].

Some computational intelligence techniques are based on the use GAs. Applications of GAs range from parameter selection of the TTRs [11,12] to the development of investment strategies [13–16]. In [11], an approach has been proposed that employs a GA for optimizing the parameters of a single TTR, MACD which has 3 parameters with an additional parameter to be used as the window size of history. Another study that is based on the optimization of the parameter of TTRs is conducted where the genome used in the GA encodes the window sizes of MACD and RSI [17]. Genetic programming (GP) has also been used for prediction [18,19].

Our study mainly focuses on the use of GA on PSX to generate a trading strategy for maximizing the capital gain and we will compare the results with the buy and hold (B&H) strategy.

3   Genetic Algorithms (GAs)

A GA is a coarse emulation of Darwinian evolution [20]. While Darwinian evolution informs us about the evolution and improvement of living organisms over multiple generations, GAs are used to evolve improved solutions to a user-specified computational problem.

A GA starts by generating a random population of candidate solutions for the problem at hand. Individuals from this population are chosen randomly to create offspring for the subsequent child population. Offspring are created by recombining the selected individuals using genetic operators of crossover and mutation. Crossover recombines chromosomes of two parents to create two new children. Mutation slightly perturbs a part of an individual chromosome to bring about a change in its behavior.

Once enough children have been created, their fitness is evaluated. Fitness criteria can be problem dependent. Normally, individuals that tend to meet a certain user specified objective are considered more fit. As a last step replacement of the less fit genetic material is performed. Less fit individuals of a population are simply removed. Individuals having higher fitness are retained for creating the next generation of individuals. To this end, a GA is an iterative algorithm. It runs over multiple generations in pursuit of creating an individual with desired fitness.

A GA can be applied to problem-solving in a variety of contexts. Mathematical optimization is one of the most common application niches of GAs. In our work, we have applied GAs for feature selection. A comprehensive and vivid description of the application of GAs for feature selection is given in [21].

4   Evolving Technical Trading Strategies Using GA

In order to derive the Technical Trading Rules (TTRs), we implemented a simple GA in R. Here we reflect on the various components of our GA system and its implementation.

4.1   Generating Initial Population

We created an initial population of randomly generated individuals as in any GA. Since we employed the GA for feature selection, we leveraged from bit-vectors for solution representation. To this end, every individual's chromosome was a bit-vector with a value of "1" representing that the respective feature is selected and a value of "2" representing otherwise.

So to build a basic chromosome which has all the features i.e. all the TIs selected, a sequence of contiguous 1s is generated. This sequence selects all the features in the order shown in Fig. 1.

EMA	MACD	RSI	CCI	Bollinger Bands	Stochastics	Parabolic SAR	TRIX	Aroon	Chaikin Oscillator	OBV	ADX

Fig. 1. A chromosome with a sequence of contiguous 1s selects all the features in this order. Setting any bit to 0 will reject that feature.

These are twelve TIs so a basic chromosome will have 12 components: [1, 1, 1, 1,   1, 1, 1, 1,   1, 1, 1, 1] Each chromosome is a trading strategy. In any trading strategy, at least one TI must be present thus there will be at least a single "1" in every chromosome. A comprehensive elaboration of these TIs is presented in Table 1.

4.2   Fitness Function

In order to evaluate the viability of a solution produced by a GA, a fitness function plays a very crucial role in evolutionary computation. It is used to assign a fitness value to every individual candidate solution produced by a GA. In our case, every solution of a GA represents a trading strategy. To determine the fitness of a candidate trading strategy, we calculate the capital gain that is generated using this strategy for a particular stock.

The implementation of the function is such that it can calculate the capital gain by using any combination of the TIs presented to it. A buy order is executed whenever the first *buy signal* is generated by the indicators present in the trading strategy. When a sell order is executed the buying price is subtracted from the selling price thus generating the profit or loss which is recorded as capital gain or loss. Trades are executed during the whole training period and their respective profits are recorded. The cumulative value, representing the capital gain, of these profits represents the fitness value of the trading strategy. Thus, the strategy yielding the maximum capital gain in a specific period is considered to be the fittest in that time frame. This phenomenon is termed as back-testing in the stock community.

To calculate TIs, the *TTR* package of R developed by Joshua Ulrich is used. This package has the capability to calculate more than 50 TIs including the ones used in our project.

Table 1. An explanation of TIs

TI	Details
EMA	Exponential Moving Average is weighted Moving Average that applies more weight to the more recent changes in price while generating a trading signal
MACD	Moving Average Convergence Divergence uses a combination of EMAs in order to generate trading signals
RSI	Relative Strength Index is a momentum indicator that compares the recent gains and losses over a certain time period to measure the speed and momentum of the price movements of a stock
CCI	The Commodity Channel Index is a momentum indicator and like other momentum indicators, it oscillates about a certain value which is the zero line
BBands	Bollinger Bands are a volatility indicator and are very versatile. This indicator is made up of 3 components which are calculated using the price of a stock: the Upper Band, the Lower Band and the MA line
SMI	SMI indicator is also referred to as the Stochastics Indicator. It is a momentum indicator that compares the closing price of a stock with the range of its price i.e. the difference between the high and low prices over a certain period of time, which is usually of 14 days
PSar	Parabolic SAR combines both the price and the date i.e. the time of the respective prices in order to generate potential trading signals
TRIX	The Triple Exponential Average (TRIX) is an indicator that illustrates the percent rate of change of the moving average that is triple exponentially smoothed
Aroon	The Aroon indicator is used to find the condition of a market whether it is up-trending, down-trending or is in a range-bound i.e. trendless market. The indicator is made up of 2 lines namely Aroon Up - which measures the strength of an uptrend and Aroon Down - which measures the strength of downtrend
CHO	Chaikin Oscillator, also known as the Volume Accumulation Indicator is a volume indicator that consists of the 2 EMA (of usually a 3-day and a 10-day period) of the Accumulation Distribution Line (ADL) indicator. The ADL indicator uses volume of a stocks trade to confirm the price trends and warn about the weak movements that may result in a price reversal
OBV	On Balance Volume (OBV) is yet another volume indicator that combines both the price and the volume of a stock in order to determine whether the price movements are strong or weak
ADX	The Average Directional Index (ADX) is a TI that is used to verify the markets trend. It is a combination of 2 other indicators namely Average True Range and Directional Movement index which are combined to generate the trading signals and other interpretations that are derived from ADX

4.3 Genetic Operators

Here we reflect on how we implemented the genetic operators of our GA. Selection strategy is presented first. There are several methods to perform selection. The main goal of this method is to select individuals from a parent generation to create a new offspring generation. A better strategy is one which respects randomness albeit with a bent for choosing more fit individuals with a higher probability. So selection is biased towards chromosomes that are more fit.

In this project, we employed tournament selection. It has proven to be a better selection strategy as reported in literature. Tournament selection is conducted as follows.

- Seven chromosomes from the initial population of sixty chromosomes are selected purely at random.
- From these *seven*, the *two* chromosomes that are the fittest are selected to be the parents of the new generation.

- The selected individuals are treated with crossover and mutation operators to create two new offspring.
- The first two steps mentioned above are repeated till the size of the new generation equals the size of the previous generation.

As discussed before, the crossover operator is used to recombine two parent individuals to create offspring. Our GA performs crossover in the following manner.

- It selects two individuals purely at random.
- It chooses a random split point for both the chromosomes.
- Swaps the parts after the split points of two chromosomes with each other to create two new children.

Mutation induces random perturbations in the gene code of individuals of an offspring population. By inducing random perturbations, the hope is to increase diversity in the gene code. Mutation is conducted on every individual of a generation by these simple steps:

- Select a chromosome;
- Choose a totally random point in the chromosome.
- Flip the bit at the chosen position with a probability of 0.5.

Replacement is the final step of a single cycle of a GA. In our scheme, both the parent and the offspring generations are combined to make a population. In this population, the fitness of all the individuals is calculated and assigned to them. Then, sorting of the individuals is done in the descending order based on their fitness. The first half of this population is the fittest and, thus, it is kept while the second half is discarded.

This approach is known as Elitism where the elite individuals – in terms of their fitness – are favored upon the individuals that are less fit regardless of whether they belong to the parent population or the child population.

4.4   Accumulation of Trading Strategies

The steps discussed in Sect. 4 are repeated, in a sequence, over a certain number of iterations. Each iteration contains a generation of the GA. The repetition is done until a stopping criterion has been met which can be any condition i.e. when a specific number of iterations has been exhausted or when the results being produced by the simulated GA come out to be the same. We have used the former approach by keeping the number of generations limited to 50.

A new population that is generated at the end of the last iteration contains the fittest chromosomes. After applying elitist replacement, the population is arranged in descending order to place the fittest chromosome at the very top. After the GA simulation has stopped, the fittest three chromosomes from the latest generation are selected to be the outcome of the GA. These chromosomes have the highest fitness which means that the trading strategies translated from them yield the highest capital gains. Since GAs are prone to over-fitting, the trading strategy derived from the GA can perform optimally in the training period but might not perform well for unseen data. So, in order to verify the practical viability of the results, we measure the performance of the top three trading strategies in testing period and compare them with "B&H" strategy.

5   Experiments and Results

Pakistan Stock Exchange (PSX) often enjoys the status of being a top performer in Asia as well as other economically active regions [4]. Its status has recently been reclassified to an emerging market by MSCI. This makes it an attractive venue for investors. In order to find suitable stock specific trading strategies, GAs are supplied with data of sixteen years from 2000 to 2015, including both years. The dataset contains opening price, high, low, closing price and volume. Each row of the dataset represents a trading day. Data from 2000 to 2011 is considered as in-sample data. To evaluate the performance strategies evolved by GA, they are supplied with out-sample data for back-testing. Data of eight companies are picked for this purpose. These companies are selected from different sectors and on the basis of availability of data.

Top three strategies evolved with GA are compared against the well known B&H strategy which is a yardstick that is usually recommended by the fundamentalists. The capital gain or loss made from the transactions of GA devised strategies is accumulated over the period of four years and compared with the difference in the value of B&H as shown in Fig. 2.

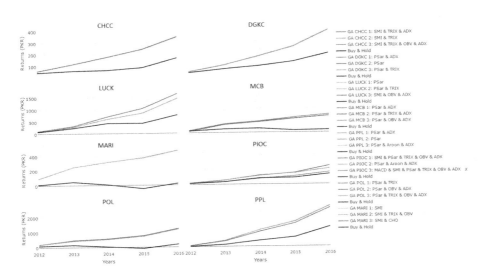

Fig. 2. Accumulative gain for testing period.

It is evident from Fig. 2 that GA devised strategies outperform B&H in all cases. In all of these cases, difference with B&H is significant enough to ignore transaction costs or other factors.

To gain further insights, Year on Year (YoY) performance of these strategies is plotted against the B&H in Fig. 3.

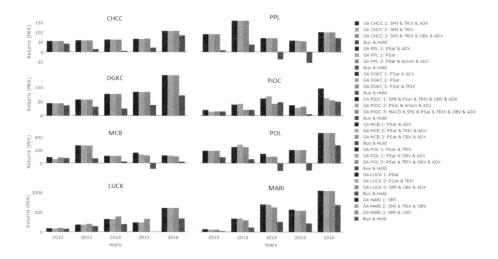

Fig. 3. Annual capital gain for testing period.

The YoY performance of GA devised strategies appears to be consistent and significantly better than B&H. The results of the stocks PIOC and LUCK reveal that the strategy ranked as best by the GA is beaten by another one. This is quite possible as the test data is unseen and chances of one strategy performing better than the other exist in that period.

The performance of all these strategies is also measured on quarterly basis. The results from respective quarters of the year are averaged for four years period as shown in Fig. 4.

The Same pattern emerges in average quarterly performance reflecting the consistency of GA generated strategies. These results reveal that GAs have the potential to find better trading strategies which can perform consistently on unseen data.

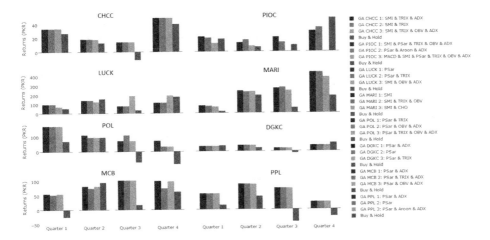

Fig. 4. Average quarterly returns.

6 Conclusion

The trading strategies devised by the GA clearly and significantly outperform the B&H strategy which has been a popular yardstick in the research conducted in the past as well as the research that is currently being conducted. These strategies yield better results in the complete testing period, on the yearly and quarterly basis. The results can be further refined by adding more TIs to the project and measuring the performance of them on a greater number of stocks. The next logical extension of the project is to discover the stock specific optimal trading rules using GAs.

References

1. Fama, E.F.: Efficient capital markets: a review of theory and empirical work. J. Finan. **25**, 383–417 (1970)
2. Fama, E.F.: Efficient capital markets: Ii. J. Finan. **46**, 1575–1617 (1991)
3. Fama, E.F.: Random walks in stock-market prices. Finan. Anal. J. **21**, 55–59 (1965)
4. Kim, N., Mangi, F.: What's next for Asia's best-performing stock market? (2016)
5. Pimenta, A., Guimarães, F.G., Carrano, E.G., Nametala, C.A.L., Takahashi, R.H.: Goldminer: a genetic programming based algorithm applied to Brazilian stock market. In: 2014 IEEE Symposium on Computational Intelligence and Data Mining (CIDM), pp. 397–402. IEEE (2014)
6. Murphy, J.J.: Technical Analysis of the Financial Markets: A Comprehensive Guide to Trading Methods and Applications. Penguin, New York (1999)
7. Neely, C.J., Weller, P.A.: Intraday technical trading in the foreign exchange market. J. Int. Money Finan. **22**, 223–237 (2003)
8. Cohen, J.B., Zinbarg, E.D., Zeikel, A.: Investment analysis and portfolio management. homewood, ill.: Richard d. irwin (1967)

9. Leigh, W., Frohlich, C.J., Hornik, S., Purvis, R.L., Roberts, T.L.: Trading with a stock chart heuristic. IEEE Trans. Syst. Man Cybern. Part A Syst. Hum. **38**, 93–104 (2008)
10. Kraus, O.Z., Grys, B.T., Ba, J., Chong, Y., Frey, B.J., Boone, C., Andrews, B.J.: Automated analysis of high-content microscopy data with deep learning. Mol. Syst. Biol. **13**, 924 (2017)
11. Fernández-Blanco, P., Bodas-Sagi, D.J., Soltero, F.J., Hidalgo, J.I.: Technical market indicators optimization using evolutionary algorithms. In: Proceedings of the 10th Annual Conference Companion on Genetic and Evolutionary Computation, pp. 1851–1858. ACM (2008)
12. Esfahanipour, A., Mousavi, S.: A genetic programming model to generate risk-adjusted technical trading rules in stock markets. Expert Syst. Appl. **38**, 8438–8445 (2011)
13. Gorgulho, A., Neves, R., Horta, N.: Applying a GA kernel on optimizing technical analysis rules for stock picking and portfolio composition. Expert Syst. Appl. **38**, 14072–14085 (2011)
14. Gorgulho, A., Neves, R., Horta, N.: Using gas to balance technical indicators on stock picking for financial portfolio composition. In: Proceedings of the 11th Annual Conference Companion on Genetic and Evolutionary Computation Conference: Late Breaking Papers, pp. 2041–2046. ACM (2009)
15. Wagman, L.: Stock portfolio evaluation: an application of genetic-programming-based technical analysis. Genet. Algorithms Genet. Program. Stanford **2003**, 213–220 (2003)
16. Yan, W., Clack, C.D.: Evolving robust GP solutions for hedge fund stock selection in emerging markets. Soft Comput. **15**, 37–50 (2011)
17. Kimoto, T., Asakawa, K., Yoda, M., Takeoka, M.: Stock market prediction system with modular neural networks. In: 1990 IJCNN International Joint Conference on Neural Networks, pp. 1–6. IEEE (1990)
18. Iba, H., Nikolaev, N.: Genetic programming polynomial models of financial data series. In: Proceedings of the 2000 Congress on Evolutionary Computation, vol. 2, pp. 1459–1466. IEEE (2000)
19. Kaboudan, M.A.: Genetic programming prediction of stock prices. Comput. Econ. **16**, 207–236 (2000)
20. Goldberg, D.E.: Genetic Algorithms in Search, Optimization and Machine Learning, 1st edn. Addison-Wesley Longman Publishing Co. Inc., Boston (1989)
21. Xue, B., Zhang, M., Browne, W.N., Yao, X.: A survey on evolutionary computation approaches to feature selection. IEEE Trans. Evol. Comput. **20**, 606–626 (2016)

A Hybrid Evolutionary Algorithm based on Adaptive Mutation and Crossover for Collaborative Learning Team Formation in Higher Education

Virginia Yannibelli[1,2(✉)] and Analía Amandi[1,2]

[1] ISISTAN Research Institute, UNCPBA University, Campus Universitario,
Paraje Arroyo Seco, 7000 Tandil, Argentina
{vyannibe,amandi}@exa.unicen.edu.ar
[2] CONICET, National Council of Scientific and Technological Research,
Buenos Aires, Argentina

Abstract. In this paper, we address a collaborative learning team formation problem in higher education environments. This problem considers a grouping criterion successfully evaluated in a wide variety of higher education courses and training programs. To solve the problem, we propose a hybrid evolutionary algorithm based on adaptive mutation and crossover processes. The behavior of these processes is adaptive according to the diversity of the evolutionary algorithm population. These processes are meant to enhance the evolutionary search. The performance of the hybrid evolutionary algorithm is evaluated on ten different data sets, and then, is compared with that of the best algorithm previously proposed in the literature for the addressed problem. The obtained results indicate that the hybrid evolutionary algorithm considerably outperforms the previous algorithm.

Keywords: Collaborative learning · Collaborative learning team formation · Team roles · Evolutionary algorithms · Hybrid evolutionary algorithms · Adaptive evolutionary algorithms · Simulated annealing algorithms

1 Introduction

In higher education environments, collaborative learning is a pedagogical approach usually used to both complement and enrich the individual learning of students. This approach requires organizing students into collaborative learning teams. The students of each collaborative learning team must work together to achieve shared learning goals. The collaborative learning teams must be formed so that students can acquire new knowledge and skills through the interaction with their peers, improving their individual learning. In this context, the grouping criterion (i.e., the criterion to form collaborative learning teams) is highly relevant because of the composition of each collaborative learning team affects the learning level and the social behavior of their student members as well as the performance of the team [1, 2]. Moreover, the way in which the grouping criterion is applied (i.e., either manually or automatically) is very

© Springer International Publishing AG 2017
H. Yin et al. (Eds.): IDEAL 2017, LNCS 10585, pp. 345–354, 2017.
https://doi.org/10.1007/978-3-319-68935-7_38

relevant because of many known grouping criteria require a considerable amount of knowledge, time and effort to be manually applied [10]. In such cases, it is possible to reduce considerably the workload of professors and also optimize the collaborative learning team formation through automation.

Different works in the literature have described and addressed the problem of forming collaborative learning teams automatically from the students [4, 10]. These works differ in relation to several aspects including the grouping criteria considered, and the algorithms utilized. In this regards, to the best of our knowledge, only few of these works consider grouping criteria that have been both successfully and widely evaluated in higher education environments.

In [5], the authors describe the problem of forming collaborative learning teams automatically from the students enrolled in a given course. As part of the problem, the authors consider a grouping criterion successfully evaluated in a wide variety of higher education courses and training programs. Such grouping criterion corresponds to the criterion defined by Belbin's team role model [3]. This criterion considers the team roles of students, and implies forming well-balanced teams regarding the team roles of their members. A team role is the way in which a person tends to behave, contribute and interrelate with others throughout a collaborative task. In this respect, the Belbin's model [3] defines nine team roles and balance conditions. Many different studies in the literature indicate that collaborative learning team formation in higher education environments according to the Belbin's criterion leads to good interactions and discussions during the learning process, improves the social behavior of the students, enhances the learning process of the students, and impacts positively on the learning level of the students as well as on the performance of teams [4]. Thus, it is considered that the collaborative learning team formation problem described in [5] is really valuable in the context of higher education environments.

In this paper, we present a hybrid evolutionary algorithm to solve the collaborative learning team formation problem described in [5]. This algorithm utilizes adaptive mutation and crossover processes that adapt their behavior according to the diversity of the evolutionary algorithm population. The utilization of these adaptive processes is meant to improve the evolutionary search performance [6, 12, 13].

We present this hybrid evolutionary algorithm because of the following reasons. The collaborative learning team formation problem described in [5] is an NP-Hard optimization problem. In this respect, evolutionary algorithms with adaptive mutation and crossover processes have been proven to be more effective than evolutionary algorithms with non-adaptive mutation and crossover processes in the resolution of a wide variety of NP-Hard optimization problems [6, 12, 13]. Therefore, we consider that the hybrid evolutionary algorithm presented could outperform the best algorithm previously presented in the literature for the addressed problem. We refer to the hybrid evolutionary algorithm presented in [8].

The remainder of the paper is organized as follows. In Sect. 2, we describe the problem addressed. In Sect. 3, we present the hybrid evolutionary algorithm. In Sect. 4, we present the computational experiments carried out to evaluate the performance of the hybrid evolutionary algorithm and an analysis of the results obtained. In Sect. 5, we present related works. Finally, in Sect. 6 we present the conclusions of the present work.

2 Problem Description

In this paper, we address the collaborative learning team formation problem described in [5]. We present below a description of this problem.

Suppose a course S has n students enrolled, $S = \{s_1, s_2, ..., s_n\}$, and the professor must organize the n students into g teams, $G = \{G_1, G_2, ..., G_g\}$. Each G_i team is composed of a z_i number of students, and each student can only belong to one team. Regarding team size, students must be organized so that the g teams have a similar number of students each. Specifically, the difference among the sizes of the teams must not exceed one. The values of the terms S, n and g are known.

As regards the students, it is considered that they naturally play different team roles when participating in a collaborative task. Regarding the team roles that can be played by the students, the nine team roles defined in Belbin's model [3] are considered. Table 1 presents these nine roles and a brief description of the features of each.

According to Belbin's model [3], it is considered that each student naturally plays one or several of the nine roles presented in Table 1. In this sense, the roles naturally played by each student are known data. These roles may be obtained through the Belbin Team-Role Self-Perception Inventory (BTRSPI) developed by Belbin [3].

As part of the problem, teams must be composed so that the balance among the team roles of their members is maximized. This grouping criterion requires analyzing the balance level of the formed teams. To analyze such level, the balance conditions established by Belbin are considered [3]. Regarding these conditions, Belbin [3] states that a team is balanced if each role specified in his model is played naturally by at least one team member. In other words, in a balanced team, all team roles are naturally played. Further, Belbin states that each role should be naturally played by only one team member [3]. Belbin states that a team is unbalanced if some roles are not played naturally or if several of its members play the same role naturally (i.e., duplicate role) [3].

Table 1. Belbin's role characteristics.

Role	Characteristics
Plant (PL)	Creative, imaginative, unorthodox. Solves difficult problems.
Resource Investigator (RI)	Extrovert, enthusiastic, communicative. Explore opportunities. Develops contacts.
Co-ordinator (CO)	Mature, confident, a good chairperson. Clarifies goals, promotes decision-making, delegates well.
Shaper (SH)	Challenging, dynamic, thrives on pressure. Has the drive and courage to overcome obstacles.
Monitor Evaluator (ME)	Sober, strategic and discerning. Sees all options. Judges accurately.
Teamworker (TW)	Co-operative, mild, perceptive and diplomatic. Listens, builds, averts friction.
Implementer (IM)	Disciplined, reliable, conservative and efficient. Turns ideas into practical actions.
Completer/Finisher (CF)	Painstaking, conscientious, anxious. Searches out errors and omissions. Polishes and perfects.
Specialist (SP)	Single-minded, self-starting, dedicated. Provides skills in key areas.

The grouping criterion considered as part of the problem is modeled by Formulas (1), (2) and (3). Formulas (1) and (2) model the balance conditions defined by Belbin [3]. Formula (1) analyzes the way in which a given r role is played within a given G_i team and then gives a score accordingly. When r is naturally played by only one member of G_i team, then 1 point is awarded to G_i. Conversely, when r is not naturally played by any member of G_i, or otherwise r is naturally played by several members of G_i, then 2 points and p points are taken off respectively.

Formula (2) defines the balance level of a given G_i team. This level is established based on the scores obtained by G_i, through Formula (1), regarding the nine roles. Thus, the greater the number of non-duplicate roles (i.e., roles played naturally by only one member of G_i), the greater the balance level assigned to G_i. Conversely, the fewer the number of roles played naturally, or the more duplicate roles, the lower the balance level assigned to G_i. The balance conditions defined by Belbin [3] can be seen in Formula (2). By using this formula, a perfectly balanced team (i.e., a team in which each of the nine roles is played naturally by only one team member) will obtain a level equal to 9.

Formula (3) maximizes the average balance level of g teams defined from the n students of the course. In other words, this formula aims to find a solution (i.e., set of g teams) that maximizes the average balance level of g teams. This is the optimal solution to the addressed problem. In Formula (3), set C contains all the sets of g teams that may be defined from the n students. The term G represents a set of g teams belonging to C. The term $b(G)$ represents the average balance level of the g teams belonging to set G. Then, Formula (3) utilizes Formula (2) to define the balance level of each G_i team belonging to the G set. Note that in the case of a G set of perfectly balanced g teams, the value of the term $b(G)$ is equal to 9.

For a more detailed discussion of Formulas (1), (2) and (3), we refer to [5].

$$
nr(G_i,\ r) = \begin{cases} 1 & \text{if } r \text{ is naturally played by only one member of } G_i \\ -2 & \text{if } r \text{ is not naturally played in } G_i \\ -p & \text{if } r \text{ is naturally played by } p \text{ members of } G_i \end{cases} \tag{1}
$$

$$
nb(G_i) = \sum_{r=1}^{9} nr(G_i, r) \tag{2}
$$

$$
\max_{\forall G \in C} \left(b(G) = \frac{\sum_{i=1}^{g} nb(G_i)}{g} \right) \tag{3}
$$

3   Hybrid Evolutionary Algorithm

The general behavior of the hybrid evolutionary algorithm proposed for the addressed problem is described as follows. Considering a course with n students who shall be organized into g teams, the algorithm starts creating a random initial population of feasible encoded solutions. In this population, each solution codifies a feasible set of g teams which may be defined from the n students. To encode these solutions, the representation proposed in [5] is used. Then, the algorithm decodes and evaluates each solution of the population by a fitness function. The set of g teams inherent to each solution is built by the decoding process proposed in [5], and then evaluated regarding the optimization objective of the problem. As mentioned in Sect. 2, this objective is maximizing the balance level of the g teams formed from n students. Thus, the fitness function evaluates the balance level of the g teams represented by each solution and defines a fitness level for each solution (i.e., the fitness function calculates the value of the term $b(G)$ corresponding to each solution by Formulas (3), (2) and (1)). To such evaluation, the function uses knowledge of the students' roles.

Once each solution of the population is evaluated, a well-known parent selection process named roulette wheel selection process [6] is used to decide which solutions of the population will compose the mating pool. By this process, the highest fitness solutions will have more probability of being selected for the mating pool. After the mating pool is complete, the solutions in the mating pool are paired. Then, a crossover process named partially mapped crossover [6] is applied to each of these pairs of solutions with an adaptive probability AP_c, to generate new feasible ones. Then, a mutation process named insert mutation [6] is applied to each solution obtained by the crossover process, with an adaptive probability AP_m. Then, the traditional fitness-based steady-state selection process [6] is applied in order to define which solutions from the solutions in the population and the solutions generated from the mating pool will integrate the new population. This survival selection process preserves the best solutions found by the hybrid evolutionary algorithm [6]. Finally, an adaptive simulated annealing algorithm is applied to each solution of the new population, excepting the best solution which is preserved.

This process is repeated until a predefined number of iterations is reached.

3.1   Adaptive Mutation and Adaptive Crossover

The above-mentioned crossover and mutation processes are applied with adaptive crossover and mutation probabilities, respectively. In this respect, we defined the adaptive crossover probability AP_c and the adaptive mutation probability AP_m. These probabilities are defined by Formulas (4) and (5), where PD refers to the population diversity, and PD_{MAX} refers to the maximum PD attainable. In Formula (4), the terms C^H and C^L represent to the upper and lower bounds for the crossover probability, respectively. In Formula (5), the terms M^H and M^L represent to the upper and lower bounds for the mutation probability, respectively. Then, f_{max} is the maximal fitness of the population, f_{min} is the minimal fitness of the population, and f is the fitness of the solution to be mutated.

The term *PD* is defined by Formula (6), where f_{max} is the maximal fitness of the population, f_{avg} is the average fitness of the population, and $(f_{max} - f_{avg})$ is a measure of the population diversity. This measure has been proposed by Srinivas and Patnaik [7], and is one of the population diversity measures most well-known in the literature [6].

The term PD_{MAX} is defined by Formula (7), where f_{MAX} and f_{MIN} correspond to the upper and lower bounds for the fitness function, respectively.

By Formulas (4)–(7), AP_c and AP_m are adaptive regarding the population diversity. When the population diversity decreases, AP_c and AP_m are increased, promoting the exploration of unvisited regions of the search space. This is important to prevent the premature convergence of the evolutionary search. When the population is diverse, AP_c and AP_m are decreased, promoting the exploitation of visited regions of the search space.

By Formula (5), AP_m is also adaptive according to the fitness of the solution to be mutated. In this regards, lower values of AP_m are defined for high-fitness solutions, and higher values of AP_m are defined for low-fitness solutions. This is meant in order to preserve high-fitness solutions, while disrupting low-fitness solutions to promote the exploration of the search space.

$$AP_c = \left(\frac{PD_{MAX} - PD}{PD_{MAX}}\right) * \left(C^H - C^L\right) + C^L \tag{4}$$

$$AP_m = \left(\frac{f_{max} - f}{f_{max} - f_{min}}\right) * \left(\frac{PD_{MAX} - PD}{PD_{MAX}}\right) * \left(M^H - M^L\right) + M^L \tag{5}$$

$$PD = (f_{max} - f_{avg}) \tag{6}$$

$$PD_{MAX} = (f_{MAX} - f_{MIN}) \tag{7}$$

3.2 Adaptive Simulated Annealing Algorithm

The adaptive simulated annealing algorithm applied is a variant of the one proposed in [8], and is described below.

The adaptive simulated annealing algorithm is an iterative process. This process starts from a given encoded solution s and a given initial value T_0 for the temperature parameter. In each iteration, the algorithm generates a new encoded solution s' from the current encoded solution s by a move operator, and then decides if s should be replaced by s'. If the fitness value of s' is higher than that of s, the algorithm replaces to s by s'. Otherwise, if the fitness value of s' is not higher than that of s, the algorithm replaces to s by s' with an acceptance probability equal to $exp(-\Delta/T)$, where T is the temperature current value, and Δ is the difference between the fitness values of s and s'. This probability mainly depends on T. When T is high, the probability is also high, and vice versa. T is reduced by a given cooling factor at the end of each iteration. This process is repeated until a predefined number of iterations is reached.

Regarding the initial value T_0 for the temperature parameter, we defined this value according to the population diversity. Specifically, T_0 is inversely proportional to the population diversity, and is calculated by the next formula: $T_0 = 1/PD$, where PD refers to the population diversity, as mentioned in Sect. 3.1. By this formula, when the population is diverse, T_0 is low, and therefore, the acceptance probability of the simulated annealing algorithm is also low. As consequence of this, the algorithm fine-tunes the solutions of the population, promoting the exploitation of visited regions of the search space. When the population diversity decreases, T_0 increases, and therefore, the acceptance probability of the simulated annealing algorithm also increases. As consequence of this, the algorithm introduces diversity into the population, promoting the exploration of unvisited regions of the search space. Thus, the algorithm is adaptive to promote either the exploitation or exploration of the search space.

Regarding the move operator of the simulated annealing algorithm, we applied a well-known operator named swap mutation [6].

4 Computational Experiments

We used the ten data sets introduced in [5] to evaluate the performance of the hybrid evolutionary algorithm. Table 2 presents the main characteristics of these data sets. For a detailed description of the team roles of the students in each data set, we refer to [5]. Each data set has a known optimal solution with a fitness level equal to 9. These optimal solutions are considered here as references.

Table 2. Main characteristics of the data sets.

Data set	Number of students (n)	Number of teams to be built (g)
1	18	3
2	24	4
3	60	10
4	120	20
5	360	60
6	600	100
7	1200	200
8	1800	300
9	2400	400
10	3000	500

We run the algorithm 30 times on each data set. After each run, this algorithm provided the best solution achieved. To develop these runs, we set the algorithm parameters as follows: population size = 80; number of iterations = 200; crossover process: $C^H = 0.9$ and $C^L = 0.5$; mutation process: $M^H = 0.2$ and $M^L = 0.01$; survival selection process: replacement factor = 40; simulated annealing algorithm: number of iterations = 20 and cooling factor = 0.9. The algorithm parameters were set based on preliminary experiments that showed that these values led to the best and most stable results.

We analyzed the results obtained by the hybrid evolutionary algorithm for each of the ten data sets. Specifically, we analyzed the average fitness value of the solutions reached for each data set, and the average computation time of the runs performed on each data set. The experiments were performed on a personal computer Intel Core 2 Duo at 3.00 GHz and 3 GB RAM under Windows XP Professional Version 2002. The algorithm was implemented in Java.

For the data sets 1–7 (i.e., the seven less complex data sets), the algorithm reached an average fitness value equal to 9. This means that the algorithm achieved an optimal solution in each run. For the data sets 8–10 (i.e., the three more complex data sets), the algorithm reached average fitness values equal to 8.87, 8.81 and 8.76, respectively. This means that the algorithm reached very near-optimal solutions for each data set.

Regarding the time required by the algorithm, we may mention the following. For the data sets 1–6 (i.e., the six less complex data sets), the average time required was 0.18, 0.46, 3.78, 6.01, 14.9 and 19.03 seconds, respectively. For the data sets 7–10 (i.e., the four more complex data sets), the average time required was 72.4, 133.07, 211.18 and 303.84 seconds, respectively. Considering the complexity of the problem instances represented by the data sets, the average computation times required by the algorithm are considered acceptable.

4.1   Comparison with a Competing Algorithm

In this section, we compare the performance of the hybrid evolutionary algorithm with that of the best algorithm previously presented in the literature for the addressed problem. We refer to the hybrid evolutionary algorithm presented in [8].

For simplicity, we will refer to the hybrid evolutionary algorithm presented in [8] as algorithm H. Like the hybrid evolutionary algorithm presented here, the algorithm H incorporates an adaptive simulated annealing algorithm within the framework of an evolutionary algorithm. Unlike the hybrid evolutionary algorithm presented here, the algorithm H uses non-adaptive crossover and mutation processes. These processes do not consider the population diversity.

In the experiments reported in [8], the algorithm H has been evaluated on the ten data sets presented in Table 2, and has obtained the results that are mentioned below. These experiments were performed on a personal computer Intel Core 2 Duo at 3.00 GHz and 3 GB RAM under Windows XP Professional Version 2002. The algorithm was implemented in Java.

For the data sets 1–5, the algorithm H obtained an average fitness value equal to 9. For the data sets 6–10, the algorithm H obtained average fitness values equal to 8.97, 8.86, 8.77, 8.74 and 8.7, respectively. In relation to the time required by the algorithm H, for the data sets 1–6, the average time required was 0.29, 0.721, 5.81, 9.24, 21.46 and 29.27 seconds, respectively. For the data sets 7–10, the average time required was 103.43, 190.1, 301.69 and 405.118 seconds, respectively.

Comparing the results obtained by the algorithm H and the hybrid evolutionary algorithm proposed here, we can mention the following points. Both algorithms have obtained an optimal average fitness value for the data sets 1–5 (i.e., the less complex data sets). However, the average fitness value obtained by the hybrid evolutionary algorithm for the data sets 6–10 (i.e., the five more complex data sets) is significantly higher than that obtained by the algorithm H. Besides, the average time required by the hybrid evolutionary algorithm for each one of the data sets is much lower than that required by the algorithm H.

Therefore, the hybrid evolutionary algorithm outperforms the algorithm H on the more complex data sets. The main reason for this is that, unlike the algorithm H, the hybrid evolutionary algorithm utilizes adaptive mutation and crossover processes. These processes adapt their behavior according to the population diversity, to promote either the exploration or exploitation of the search space, and therefore, enhance the performance of the evolutionary search. Thus, the hybrid evolutionary algorithm can reach better solutions in less computation time than algorithm H on the more complex data sets.

5   Related Works

To the best of our knowledge, only few works in the literature address the problem of automatically forming collaborative learning teams based on the Belbin's model [4, 10]. These works differ mainly in relation to the modeling of this problem, and the algorithms proposed to solve it.

In the framework proposed in [11], this team formation problem is modeled as a constraint satisfaction problem, and is solved by a DLV constraint satisfaction solver. In the tool proposed in [4], this team formation problem is modeled as a coalition structure generation problem, and is solved by means a linear programming method.

The two above-mentioned works propose exhaustive search algorithms to solve the problem. However, this kind of algorithms only can solve very small instances of the problem in a reasonable period of time.

In [5, 8, 9], different evolutionary algorithms are presented with the aim of solving problem instances with very different complexity levels. Such algorithms, particularly the algorithm presented in [8], achieved promising results. However, these algorithms use non-adaptive mutation and crossover processes for developing the evolutionary search.

6   Conclusions

In this paper, we proposed a hybrid evolutionary algorithm to solve the collaborative learning team formation problem described in [5]. This algorithm utilizes adaptive mutation and crossover processes, to improve the evolutionary search performance. The behavior of such processes is adaptive in order to promote either exploration or exploitation of the search space, according to the population diversity. The presented computational experiments show that the hybrid evolutionary algorithm significantly outperforms the best algorithm previously proposed in the literature for solving the addressed problem.

In future works, we will evaluate other adaptive mutation and crossover processes. Besides, we will evaluate adaptive parent selection and survival selection processes. Moreover, we will evaluate the integration of other adaptive search and optimization techniques into the framework of the evolutionary algorithm.

References

1. Barkley, E.F., Cross, K.P., Howell Major, C.: Collaborative Learning Techniques. Wiley, New York (2005)
2. Michaelsen, L.K., Knight, A.B., Fink, L.D.: Team-Based Learning: A Transformative Use of Small Groups in College Teaching. Stylus Publishing, Sterling (2004)
3. Belbin, R.M.: Team Roles at Work, 2nd edn. Taylor & Francis, London (2011)
4. Alberola, J., Del Val, E., Sanchez-Anguix, V., Palomares, A., Teruel, M.: An artificial intelligence tool for heterogeneous team formation in the classroom. Knowl.-Based Syst. **101**(1), 1–14 (2016)
5. Yannibelli, V., Amandi, A.: A deterministic crowding evolutionary algorithm to form learning teams in a collaborative learning context. Expert Syst. Appl. **39**(10), 8584–8592 (2012)
6. Eiben, A.E., Smith, J.E.: Introduction to Evolutionary Computing, 2nd edn. Springer, Heidelberg (2015)
7. Srinivas, M., Patnaik, L.M.: Adaptive probabilities of crossover and mutation in genetic algorithms. IEEE Trans. Syst. Man Cybern. **24**(4), 656–667 (1994)
8. Yannibelli, V., Amandi, A.: A Hybrid Algorithm Combining an Evolutionary Algorithm and a Simulated Annealing Algorithm to Solve a Collaborative Learning Team Building Problem. In: Pan, J.-S., Polycarpou, Marios M., Woźniak, M., de Carvalho, A.C.P.L.F., Quintián, H., Corchado, E. (eds.) HAIS 2013. LNCS, vol. 8073, pp. 376–389. Springer, Heidelberg (2013). doi:10.1007/978-3-642-40846-5_38
9. Yannibelli, V., Amandi, A.: A Memetic Algorithm for Collaborative Learning Team Formation in the Context of Software Engineering Courses. In: Cipolla-Ficarra, F., Veltman, K., Verber, D., Cipolla-Ficarra, M., Kammüller, F. (eds.) ADNTIIC 2011. LNCS, vol. 7547, pp. 92–103. Springer, Heidelberg (2012). doi:10.1007/978-3-642-34010-9_9
10. Cruz, W.M., Isotani, S.: Group Formation Algorithms in Collaborative Learning Contexts: A Systematic Mapping of the Literature. In: Baloian, N., Burstein, F., Ogata, H., Santoro, F., Zurita, G. (eds.) CRIWG 2014. LNCS, vol. 8658, pp. 199–214. Springer, Cham (2014). doi:10.1007/978-3-319-10166-8_18
11. Ounnas, A., Davis, H.C., Millard, D.E.: A framework for semantic group formation in education. Educational Tech. Soc. **12**(4), 43–55 (2009)
12. Rodriguez, F.J., García-Martínez, C., Lozano, M.: Hybrid metaheuristics based on evolutionary algorithms and simulated annealing: taxonomy, comparison, and synergy test. IEEE Trans. Evol. Comput. **16**(6), 787–800 (2012)
13. Talbi, E. (ed.): Hybrid Metaheuristics. SCI, vol. 434. Springer, Heidelberg (2013)

Object Recognition Based on Dynamic Random Forests and SURF Descriptor

Khaoula Jayech[✉] and Mohamed Ali Mahjoub

LATIS Research Lab, National Engineering School of Sousse,
University of Sousse, Sousse, Tunisia
jayech_k@yahoo.fr, mohamedali.mahjoub@eniso.rnu.tn

Abstract. Visual object recognition is an extremely difficult computational problem. It is still a challenging task for computer vision systems related especially to the high variability of the image of objects that may vary somewhat in different viewpoints, in many different sizes and scales or even when they are translated or rotated. In this study, we investigate the combination of a new dynamic random forests and SURF descriptor for object recognition. We have carried out experiments on two benchmark object recognition datasets: CIFAR-10 and STL-10. The experimental results show the superior ability of our proposed approach, compared to the standard RF in terms of recognition rate and execution time.

Keywords: Dynamic RFs · SURF descriptor · Object recognition

1 Introduction

Object recognition is a central research topic and a well-studied problem in computer vision. It is useful in a lot of applications such as video stabilization, automated vehicle parking systems and cell counting in bio-imaging. Object recognition consists in finding and identifying objects in an image in order to classify it. Several studies have been proposed [1–5]; however, this problem is still a challenging task. The main problem is that each object in the world can have an infinite number of different 2-D images onto the retina as the lighting, pose, position and backgrounds of an object vary relative to the viewer [6]. Nevertheless, the success of any image classification depends on various factors such as the choice of a suitable classification procedure. Several supervised classifiers have been suggested in the literature like the Recurrent Convolutional Neural Networks (RCNNs) [5], the Support Vector Machine (SVM) [7], the Deep Convolutional Neural Networks (DCNNs) combined with the Random Forests (RF) [1] and others. On the other hand, multiple classifier systems prove to be efficient in object recognition, mainly when using bagging, random-subspace and boosting classifiers. In this paper, we present an object recognition system based on the RFs and a new version of Dynamic RFs (DRFs). RFs are a nonparametric statistical method, which demonstrates its high efficiency in a lot of applications, both for regression and supervised-classification problems. Indeed, the RFs involve using a large number ($\approx$400) of randomly constructed decision trees in parallel before they are averaged. Therefore, the

© Springer International Publishing AG 2017
H. Yin et al. (Eds.): IDEAL 2017, LNCS 10585, pp. 355–364, 2017.
https://doi.org/10.1007/978-3-319-68935-7_39

RFs can lead to overcome several problems with decision trees, including: (i) running efficiently on large data bases, (ii) providing effective methods for estimating missing data, (iii) handling thousands of input variables without variable deletion, (iv) giving estimates of which variables are important in the classification, and (v) reduction in overfitting. Actually, by averaging several trees, there is a significantly lower risk of overfitting. Yet, the RFs are more computationally expensive. To enhance the performance of RFs, several extensions have been put forward. The current study aims to develop a new RF extension known as the DRFs which are able to improve the object recognition accuracy of the proposed system. The main contributions of this work are: (i) modifying the DRF algorithm by adding a stopping criterion to the DRF classifier training so that the number of trees need not grow, and (ii) changing the way in which samples are randomly chosen for training each newly added tree by considering the sample weights and using only 50% of them, randomly chosen according to their weights.

The outline of the paper is as follows. In Sect. 2, we discuss some related work. Section 3 introduces and details the different steps of our developed system. Section 4 runs the experiments and the analysis. Finally, Sect. 5 concludes the paper and draws future work directions.

2 Related Work

The literature review has shown limited research work, which has investigated the use of the RF classifier for object recognition. However, those classifiers have presented good results in different domains of pattern recognition. As defined by Breiman in [9], the RF is a well-known and an excellent ensemble method that analyzes feature selection and classification. In fact, ZHAO in [2] employed the RF to automatically select features and classify landforms based on their topographical characteristics. The experiment demonstrated that the RF method could effectively extract and recognize the feature of landforms. A handwritten digit recognition system based on the RFs and the CNNs was developed by Zamani in [10]. The authors performed some experiments with different preprocessing steps, feature types, and baselines. The results proved that the RFs performed better than the CNNs. BAI in [1] suggested exploring more information of the image contents in order to determine their categories by extracting multiple layers of deep neural networks. In fact, the authors put forward a novel feature selection method that could be used to reduce redundancy in features obtained by deep neural networks for an RF-based classification. The proposed method provided a potential way for improving the performances of other experts and intelligent systems. A review of the different applications of the RF classifier was presented by BELGIU in [11]. This review revealed that the RF classifier could successfully handle high data dimensionality and multi-collinearity, being both fast and insensitive to over-fitting. This brief state-of-the-art shows the efficiency of using the RFs in object recognition compared with other classifiers like the CNN, the SVM and others.

3   System Overview

In this section, the architecture of the proposed object recognition system based on the new DRFs is presented in Fig. 1a. It includes the following steps: (1) descriptors extraction based on Speeded Up Robust Features (SURF), (2) quantification step, (3) modeling and training.

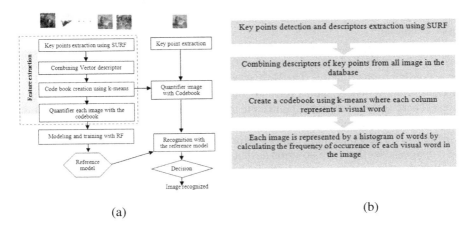

(a)   (b)

Fig. 1. (a) Architecture of recognition system based on RF, (b) Descriptors extraction and quantification steps

3.1   Descriptors Extraction Based on SURF

For the representation of objects, we use the SURF descriptors. SURF is a patented local feature detector and descriptor. It is inspired from the SIFT descriptor but it is several times faster and is more robust against different image transformations such as invariance to translation, rotation, scaling and robustness to occlusions. This method involves three steps like the SIFT, but the details in each step are different: (i) point of interest detection and extraction, (ii) descriptors computing, and (iii) point of interest matching.

The SURF method uses the Fast-Hessian for the point of interest detection and an approximation of the Haar wavelets to compute the descriptors. The fast Hessian is based on the study of the Hessian matrix as indicated in Eq. 1:

$$H(s, y, \sigma) = \begin{bmatrix} L_{xx}(x, y, \sigma) & L_{xy}(x, y, \sigma) \\ L_{xy}(x, y, \sigma) & L_{yy}(x, y, \sigma) \end{bmatrix} \tag{1}$$

where Lij (x, y, σ) is the second derivative along the directions i and j of L with:

$$L(x, y, \sigma) = G(x, y, \sigma) * I(x, y) \tag{2}$$

$$\text{where} \qquad G(x, y, \sigma) = \frac{1}{\sigma\sqrt{2\pi}} e^{\frac{\left(x^2 + y^2\right)}{2\sigma^2}} \qquad (3)$$

and I is the starting image.

The determinant maximization of this matrix makes it possible to obtain the point of interest coordinates on a given scale. This step brings a point of interest invariance of according to the scale. This step, therefore, makes it possible to detect the point of interest candidate.

The computation of the descriptors is done using Haar wavelets. They make it possible to estimate the local orientation of the gradient and thus to provide the invariance according to the rotation. The Haar wavelet responses are computed in x and y in a circular window whose radius depends on the scale factor of the detected point of interest. These specific responses contribute to the formation of the feature vector corresponding to the key point.

In point of interest matching, the search for the best similarity between the descriptors of two images, uses the same criterion as that used in the SIFT algorithm, i.e. the Euclidean distance.

3.2 Quantification Step: Creating Vocabulary of Visual Words

The representation of the bag of words becomes one of the most popular methods to represent the content of an image. It has been applied successfully to the categorization of objects. In a typical representation of the bag-of-words model, points of interest must first be identified from the image. These points of interest are represented by vectors in a large space. To effectively manage the points of interest, the key idea is to quantify each extracted point in one of the visual words. Then each image is represented by a histogram of visual words, which is often called the bag-of-words representation, and consequently converts the problem of categorizing objects into a text categorization problem. A grouping procedure in a large number of groups (e.g. K_means) is often applied to the points of interest of all images in the database, with the center of each group corresponding to a visual word.

3.3 Modeling and Training

The suggested approach is based on a new DRF considered as an RF extension. In the next section, we will recall the definition of the RFs and describe the algorithm of DRF construction.

Random forests

The RFs are composed of a set of decision trees. These trees are distinguished from one to another by data sub-sample on which they are trained. These sub-samples are drawn at random in the initial dataset. The RFs are used to calculate a vote for the most popular class (classification). Otherwise, their responses are combined (averaged) to obtain an estimate of dependent variables (regression). The different steps of RF construction were introduced by Breiman in [9]. The complexity of the Forest-RI algorithm is as follows:

$$C\left(\text{forest}_{RI}\right) = \Theta(L) \times C(\text{RndTree})$$
$$C\left(\text{forest\_RI}\right) = \Theta((L \times N \times k \times \log(N)) \tag{4}$$

where L is the number of trees, N is the size of data, and K is the number of randomly selected characteristics.

Several RF extensions known as the DRFs, have been developed in literature so as to enhance the performance of the RFs.

Dynamic random forest

The DRF algorithm was the first solution proposed by Simon Bernard, in [12], for the dynamic induction of an RF. Its objective was to construct a set of trees that could first of all give the best predictions and; secondly avoid as much as possible incorrect predictions using a data weighting. For the latter, a function is used, which is proportionally inverse to the confidence score and which represents an estimate of the ability of the current forest to predict the best class. Bernard put forward multiple variants of the DRF algorithm, each using a different weighting function defined on the interval [0, 1] and inversely proportional to the proposed confidence score. Three weighting functions were defined: polynomial, exponential and inverse. The results of the experiments made by Simon Bernard on the performance evolution in generalization utilizing the DRF algorithm illustrate that the adaptive weighting acts rapidly on these RF performances. It shows that with a certain number of trees, we can reach a maximum performance gain; whereas, adding more trees to the forest does not give better performances in generalization. These results highlight the possibility of setting up a stop criterion. The complexity of the DRF is as follows:

$$C(\text{DRF}) = \Theta(L \times (2N + C(\text{Rtree})))$$
$$C(\text{DRF}) = \Theta(L \times (2N + (N \times k \times \log(N)))) \tag{5}$$

where L is the number of trees, N is the size of data, and K is the number of randomly selected characteristics.

Algorithm 1. Developed algorithm: New DRFs

New DRF

Input
A: training set,
M: number of randomly selected variables,
L: number of trees,
$W_\alpha(c(x, y))$: weighting function (c: ratio of trees in the ensemble of decision trees that predict the true class.)
Output: RF
Begin
1: **for** $x_i \in A$ **do**
2: $D_1(x_i) = 1/N$/// weighting vector
3: t=false
4: l=1
5: **while (t=false) & (l<=L) do**
6 :   tree ←empty tree
7:    Z ←0
8:    A_l ← weighted bootstrap set
9:    tree. Root← RndTree (tree.root, A_l)
10:    Random Forest← RandomForest ∪ tree
11:    Models{l}← RandomForest
12:    **for** $x_i \in A$ **do**
13:     $D_{l+1}(x_i)$ ← $W_\alpha(c(x_i, y_i))$
14 :   Z← Z + $D_{l+1}(x_i)$
15 :   **for** $x_i \in A$ **do**
16 : $D_{l+1}(x_i) \leftarrow \frac{D_{l+1}(x_i)}{Z}$/// normalizing weights
17: if (l>50)
18: if (rate(Models{l}, S)-rate(Models{l-50}, S)) <0.001
19:     t=true
20: l=l+1
21: **return** RF
End

Training the DRF is different from training the RF on the process of a partition-rule selection. Each rule divides a group of training data into two groups based on certain evaluation criteria. These criteria in using the RF are based on the size of each class of problems in each group. In utilizing the DRF, those sizes are replaced by the sum of weights of data contained in the sub-groups.

In this study, the main contribution of our algorithm, as shown in Algorithm 1, is to put forward a stop criterion so as to stop the process when the maximum performance gain is obtained. Moreover, based on the basic idea of the DRF algorithm, which consists in focusing the induction of the next tree on the data misclassified by the current forest using an adaptive weighting, we decide instead of randomly choosing the next tree learning data to choose the data misclassified by the majority of trees constituting the current forest. In addition, by studying the out-of-bag measures of the bagging method, we find that almost 40% of the data is unknown to the tree. Hence, every time a tree is added to the forest, a 40% portion of the data represents the data misaligned by the majority of trees. As a result, in the proposed method, we decide to train each added tree on a set of data of size N/2, where N represents the size of the initial learning database. The data belonging to this set are those which have the maximum weighting values. The aim of this method is to reduce the time execution and focus the induction of the trees on the misclassified data. The function rate (x, y) is a function that returns the recognition rate obtained by the model x and the testing set y. The complexity of the new DRF algorithm is as follows:

$$C(\text{DRF}) = \theta(L \times (2N + \tau(\text{Rtree})))$$
$$C(\text{DRF}) = \theta\left(L \times \left(2N + \left(\frac{N}{2} \times k \times log\left(\frac{N}{2}\right)\right)\right)\right) \tag{6}$$

where L is the number of trees, N is the size of data, and K is the number of characteristics.

4   Experiments, Results and Discussion

4.1   CIFAR and STL Datasets

In order to evaluate the performance of the suggested system, the experiments are conducted on the benchmark object recognition datasets: CIFAR-10 and STL-10. The CIFAR-10 dataset consists of 60,000 32 × 32 color images in 10 classes, with 6,000 images per class. There are 50,000 training images and 10,000 test images. The CIFAR-10 classification is a common benchmark problem in machine learning. The problem is to classify RGB 32 × 32 pixel images across 10 categories: airplanes, automobiles, birds, cats, deers, dogs, frogs, horses, ships, and trucks. We use also the STL-10 dataset that contains 96 × 96 RGB images in 10 categories. This dataset has 500 images per class for training and 800 for testing.

4.2   Feature Extraction

Features play a very important role in the area of visual object recognition. Consequently, before getting features, various image preprocessing techniques, like resizing and normalization, are applied on a sampled image. After that, feature extraction techniques are applied based on the SURF descriptor to get features that will be useful in object recognition. Figure 2 represents the different steps which we follow to extract the features.

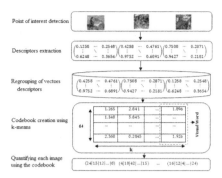

Fig. 2. Feature extraction steps

4.3 Experimental Protocol

To test the effectiveness of the new DRF model, we try to vary the input arguments which are the number of trees, the number of randomly selected variables to consider at each node, and the weighting function in order to optimize our model. We represent the results of experiments in the Table 1, Fig. 3a and b. Simon Bernard in [12] put forward three types of weighting functions (polynomial, exponential and inverse). Each function has as an input the parameter α. This parameter allows accentuating or weakening the differences in weighting between two data, which have two different values of C (x, y).

$$W_\alpha(c(x, y)) = 1 - c(x, y)^\alpha$$
$$W_\alpha(c(x, y)) = \exp(-\alpha \times c(x, y))$$
$$W_\alpha(c(x, y)) = \frac{1}{c(x, y)^\alpha}$$

(7)

Table 1. Recognition rate obtained with different functions used by DBF

Test	Function								
	Polynomial (%)			Exponential (%)			Inverse (%)		
	1	2	5	2	10	50	1	2	5
CIFAR-10	97.33%	96.0%	95.0%	95.3%	94.67%	90.96%	94.33%	93.67%	91.67%
STL-10	76.33%	75.97%	75.06%	75.06%	74.33%	73.33%	74.02%	72.67%	72.15%

We provide in Table 1 the recognition rate obtained with each function and each value of α so as to optimize our model and to find the function and the value that returns the best recognition rate. As it can be noted, the best results are obtained when using the polynomial function; and there will be no difference in the recognition rate when we change the value of α. Accordingly, we use the value 1.

As it is depicted in Fig. 3a, the recognition rate increases with the number of trees to obtain a sort of stability and the time execution grows with the size of bootstrap samples. We try with those experiments to find a compromise between the recognition

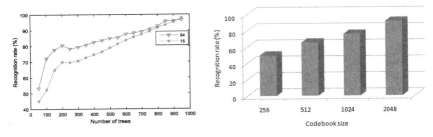

Fig. 3. (a) Recognition rate vs number of trees and number of randomly selected characteristics, (b) Recognition rate vs. codebook size

rate and the time execution. We find that the best performance is obtained with 900 trees and 15 variables.

Figure 3b shows that the recognition rate goes up with the codebook size, and we find that the best rate is obtained with a codebook size of 2,048. Figure 4a indicates that the DRF returns a better recognition rate than the static RF. A dynamic induction procedure satisfies its objective by significantly reducing, in a large majority of cases, the error, while comparing it with the static induction processes. Added to that, the adaptive weighting improves efficiently and rapidly the forest performance in the first iterations of the process. The static RF has no guarantee that each tree added to the current forest ameliorates its performance, compared with the DRF. Also, we notice that the new DRF enhances the performance of the DRF and reduces the time execution to 50%, as it is shown in Fig. 4b.

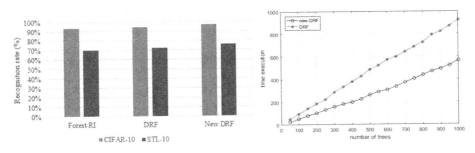

Fig. 4. (a) Obtained recognition rate, (b) Time execution vs. number of trees

5   Conclusion

An efficient object recognition system based on new DRFs has been presented. Extensive experiments with various numbers of trees, characteristics and codebook size have been performed on the CIFAR-10 and STL-10 object recognition datasets. The developed model of new DRFs has performed better to the state-of-the-art models. It has been also faster than the DRF, proposed by Bernard. In the future, we will investigate other feature types and training methods that will close the gap on the DRFs and the state-of-the-art methods on those datasets.

References

1. Bai, S.: Growing random forest on deep convolutional neural networks for scene categorization. Expert Syst. Appl. **71**, 279–287 (2017)
2. Zhao, W., Xiong, L., Ding, H.: Automatic recognition of loess landforms using Random Forest method. J. Mt. Sci. **14**(5), 885–897 (2017)
3. Krizhevsky, A., Hinton, G.: Learning multiple layers of features from tiny images (2009)
4. Gecer, B., Azzopardi, G., Petkov, N.: Color-blob-based COSFIRE filters for object recognition. Image Vis. Comput. **57**, 165–174 (2017)
5. Liang, M., Hu, X.: Recurrent convolutional neural network for object recognition. In: Proceedings of the IEEE Conference on Computer Vision and Pattern Recognition, pp. 3367–3375 (2015)
6. Dicarlo, J., Cox, D.: Untangling invariant object recognition. Trends Cogn. Sci. **11**(8), 333–341 (2007)
7. Zhang, L., He, Z., Liu, Y.: Deep object recognition across domains based on adaptive extreme learning machine. Neurocomputing **239**, 194–203 (2017)
8. Bernard, S., Adam, L.: Using random forests for handwritten digit recognition. In: Ninth International Conference on Document Analysis and Recognition, vol. 2, pp. 1043–1047 (2007)
9. Breiman, L.: Random forests statistics. Mach. Learn. **45**, 5–32 (2001). Department, University of California, Berkeley
10. Zamani, Y., Souri, Y., Rashidi, H., Kasaei, S.: Persian handwritten digit recognition by random forest and convolutional neural networks. In: Conference on Machine Vision and Image Processing (2015)
11. Belgiu, M., Dragut, L.: Random forest in remote sensing: A review of applications and future directions. ISPRS J. Photogrammetry Remote Sens. **114**, 24–31 (2016)
12. Bernard, S., Adam, S., Heutte, L.: Dynamic random forests. Pattern Recogn. Lett. **33**(12), 1580–1586 (2012)

Reducing Subjectivity in the System Dynamics Modeling Process: An Interdisciplinary Approach

Jae Un Jung[✉]

Department of MIS, Dong-A University, Busan, Korea
imhere@dau.ac.kr

Abstract. Over the last six decades, the system dynamics (SD) methodology has been used to address a variety of issues that are inherent to complex systems. In the SD methodology, circular-causality-centric problem definitions and computer simulations are uniquely applied to narrow the scope of a problem and facilitate an understanding of the diverse phenomena that arise from the underlying problem structure. With unclear but conceivable causality between the variables that constitute a problem, intentionally choosing to focus on the circular causality (causal loop) may result in erroneous models. The SD modeling tools, causal loop diagram for mental modeling, and stocks-and-flows diagram for physical simulation modeling are all interconnected, and it is therefore important to maintain consistency between their outputs to ensure the procedural validity. However, the modeling activities are normally performed individually, which introduces ambiguity and subjectivity into the SD modeling process. To address this issue, this research presents an integrated SD modeling support system that employs graph theory, text mining, and social network analysis approaches, and can be used as an extended SD modeling framework. This system is expected to facilitate SD modeling and to reduce errors in the SD modeling process. In the future, it will be utilized to implement an advanced computer-aided SD modeling toolchain and methodology.

Keywords: System dynamics · Modeling process improvement · SD modeling support system · Interdisciplinary approach · Computer aided modeling tool

1 Introduction

System dynamics (SD) is a methodology used in complex systems research that focuses on the circular causality that represents a problem [1–3]. This unique strategy simplifies a dynamic problem by way of an analysis to identify the underlying cause of the dynamic phenomena using a causal loop. It should be noted that this approach results in a loss of information because it prunes the fragmented causality outside of the circular causality [4, 5].

This methodology defines a dynamic problem by developing a mental model using causal-loop-diagramming and a simulation model using stocks-and-flows-diagramming. The order in which these two methods are applied in the SD modeling process can be changed, although a stocks-and-flows diagram (SFD) is normally

© Springer International Publishing AG 2017
H. Yin et al. (Eds.): IDEAL 2017, LNCS 10585, pp. 365–375, 2017.
https://doi.org/10.1007/978-3-319-68935-7_40

developed based on a causal loop diagram (CLD). In some cases, only one of the two is employed.

Note that in the SD modeling process, there is a strong probability of subjective bias due to the mental representation held by the researcher [6]. For example, the desire to define a problem as a closed system (circular causality) with incomplete information may distort or invalidate the corresponding model [7]. For this reason, arguments are consistently raised regarding the methodological reliability and validity of the SD approach, which prevents it from being widely adopted in scientific research [8, 9].

To eliminate the judgement of the researcher (subjectivity) and to reduce errors in the SD modeling process, Kim and Anderson [10] suggested a method of generating a CLD from purposive text data, and Burns [11] presented an algorithm for converting a CLD to an SFD.

Even though conventional SD modeling theory does not explain the structural characteristics and principles of CLDs, from graph theory, CLD can be classified as labeled or signed directed graphs (digraphs) [12–14]. As such, text mining and social network analysis (SNA) techniques can be used to generate a signed digraph by extracting and connecting keywords (variables) [15–19].

In addition, if a CLD shares a particular topic (i.e., information) with an SFD, they are conceptually convertible from one to the other, as seen in Burns [11]. By discovering and normalizing their relationship, a programming approach can be used to develop a conversion algorithm (i.e., a sequence of logic and rules) between the two diagrams.

On the basis of these results, this research proposes: (1) an integrated SD modeling support system (extended SD modeling framework) as a way to improve the existing SD modeling process, (2) with the aim to promote the use of the SD methodology.

This system, which will be implemented during the follow-up research, provides insight into the advancement of the SD modeling process.

2   Conventional and New Approaches to SD Modeling

Casual-loop and stocks-and-flows diagramming constitute the two main axes of the SD modeling process.

The first axis is the use of CLDs. The primary advantage of CLDs is that they provide an easy way to visualize, explain, and understand dynamic problems using simple notation; however, CLDs are sometimes confused with causal and/or cognitive maps. In terms of circular causality (direction and circularity), all of these have similar, but not identical, forms. Causal and cognitive maps show causality or the directions of variables on their maps, but are not intended to present circularity unlike the CLD [20].

A CLD can be fundamentally classified as a signed or labeled digraph according to graph theory, although the CLD notation does include unique notation that is not explained in graph theory to refer to the time delay and the polarity of the dominant loop (see Fig. 1). On the topic of CLD notation, Richardson [21] and Jung and Kim [22] raised questions about ambiguity, and proposed alternative symbol systems to provide more exact and concrete expressions of dynamic problems.

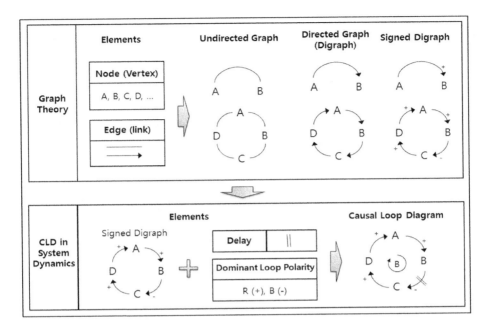

Fig. 1. Graph types and CLD

The second axis is stocks-and-flows diagramming, which is applied to simulation modeling. This tool shares topics or information with the causal loop diagramming process. It is important to maintain consistency between the CLD and SFD to ensure the procedural validity of SD modeling. In this regard, Jung and Kim [23] studied the coherence between the two different models using the population dynamics model, which is well-known but defective.

On the other hand, to objectify the existing SD modeling process, Burns [11] and Binder et al. [24] formulated rules to convert SFDs into CLDs and vice versa using a mathematical approach. Kim and Anderson [10] introduced a conceptual method to generate a CLD from the purposive text data of the Federal Open Market Committee.

This research proposes an integrated SD modeling support system based on an interdisciplinary approach that includes graph theory (i.e., text mining and SNA techniques) and a software (S/W) engineering approach (i.e., algorithms and programming).

3   SD Modeling Algorithms for CLDs and SFDs

3.1   Causal Loop Diagramming Algorithm

In order to eliminate the subjectivity of researchers during causal loop diagramming, this section presents an advanced causal loop diagramming process that combines text mining and SNA approaches, as shown in Fig. 2.

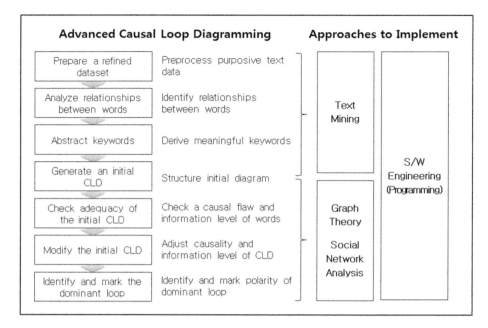

Fig. 2. Advanced causal loop diagramming process and implementation approaches

The following pseudocode, which is programming language agnostic, shows an algorithmic approach to causal loop diagramming.

PROGRAM CausalLoopDiagramming:
    Step1: Prepare a refined dataset
        # Input a text file (dataset) to the variable data
        data = Load(textFile);
        # Convert the dataset to the corpus type consisting only of words
        data.cor = Corpus(data);
        data.cor = Transform(data.cor, tolower, removePunctuation,…);
        i=Length(data.cor);
        FOR (i , i>0, i--) { IF (Type(data.cor[i]!= text)
          THEN Delete(data.cor[i])};
        # Remove insignificant words and filter out duplicated words
        FOR (j , j>0, j--) {
        Print("Is this word insignificant or duplicated "+ data.cor[j]+"?");
        input = Scan.nextLine().charAt(0);
        IF(input =='y') THEN Delete(data.cor[j])};
        REPEAT{data.cor = Synonym(data.cor, syn_a, syn_b) };

Step2: *Analyze relationships between words*
   # Make a relation matrix with refined words
   data.matrix = RelationMatrix(data.cor);
   *data.matrix = data.matrix * Inverse(data.matrix);*
   data.cld = data.matrix[matrix.data>0];

Step3: *Abstract keywords*
   # Derive words that have relationships with the others more than once
   i=LinkedNodes(data.cld);
   FOR(i, i>1,i--) {IF(CheckSignificance(data.cld[i] != "y")
   THEN Delete(data.cld[i])};

Step4: *Generate an initial CLD*
   # Structure initial diagram on a screen
   Plotting{Graph.adjacency(data.cld, mode = directed)};

Step5: *Check adequacy of the initial CLD*
   # Check a pointless direction or causal flaw (conflict) between words
   FOR(i, i>0,i--) { FOR(j, j>0, j--) {
   IF(data.cld[i,j] == 1) THEN PlotSign(data.cld[i,j],"+");
                    ELSE PlotSign(data.cld[i,j],"-") } };
   REPEAT{IF(FirstPoint(data.cld[i,...])==1 &&
   LastPoint(data.cld[...,i])!=1) THEN Delete(Pointless)};

Step6: *Modify the initial CLD*
   # Adjust causality and information level of CLD
   REPEAT{ Replace(data.cld[i,i+1], data.cld[i+1,i]);
   Replace(data.cld[j,n], 0);
   Replace(data.cld[k,l], 1) };

Step7: *Identify and mark the dominant loop*
   # Judge polarity of dominant loop
   REPEAT{ IF(Point(data.cld[i,i+1]!=1) THEN k=k+1};
   IF(k%2 != 0) THEN loopSign="B" ELSE loopSign="R";
   # Mark a dominant sign in the center of the loop
   DominantLoopSign(Plot(data.cld), centered, loopSign);
END.

This algorithm is intended to automatically generate a CLD; however, decoding the meanings of words in text data requires a qualitative interpretation. To systematically address this weakness, supplementary modules for SD modeling are suggested in Sect. 4.

3.2 Stocks-and-Flows Diagramming Algorithm

While causal loop diagramming aids in the understanding of a dynamic problem through a logical representation, it is difficult to identify concrete dynamics over time by numerical interpretation. For this reason, the SD employs stocks-and-flows diagramming to provide a deeper understanding of a dynamic problem by way of a physical simulation.

Typically, both diagramming methods are interactive and their respective artifacts are consistent if they define the same problem. However, sometimes the models are developed separately, which makes validation of the procedural consistency difficult. This is a critical issue from a scientific (methodological) point of view.

To maintain consistency, Burns [11] and Binder et al. [24] presented algorithmic approaches for converting a CLD into an SFD, which require the elements of a CLD to be decomposed and overlapped with those of an SFD. As this is challenging, the referenced study assumed that a CLD had been fully developed before generating an SFD. The elements and notations as well as illustrations of CLDs and SFDs are shown in Fig. 3.

	Element & Notation	Illustration
CLD	• variable(node/vertex): noun • causal direction(directed edge/arc): → • link polarity: +(positive), -(negative) • dominant polarity of circular causality: R(reinforcing), B(balancing) • delay: ‖	(illustration)
SFD	• stock: □ • flow: ⇒ • variable(auxiliary, rate): ○ • system boundary: △	(illustration)

Fig. 3. Elements of CLD and SFD

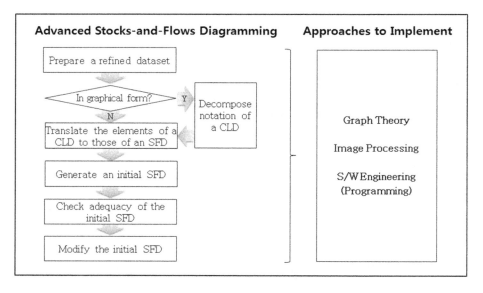

Fig. 4. Advanced stocks-and-flows diagramming process and implementation approaches

If the dataset that makes up a CLD that will be used to construct an SFD is not in the graphical form, the step containing the graphical decomposition of the notation is not required. In other words, the data describing the CLD just before it is plotted on a screen logically represents a CLD, but is not yet a graphical CLD. Even so, this data can be used as a dataset for constructing an SFD.

With this in mind, this section introduces an alternative algorithmic approach to stocks-and-flows diagramming as shown in Fig. 4.

The following pseudocode presents an algorithmic approach to CLD-based stocks-and-flows diagramming.

```
PROGRAM StocksAndFlowsDiagramming:
    Step1: Prepare a refined dataset
        # Input a dataset and check data types
        data=Load(dataset);
        # If the prepared data is the graphical type, decompose elements of a
        CLD through the image preprocessing, otherwise go to Step 2
        IF(data.type == image){
            THEN data.cld=Decompose(ImageProcessing(dataset));
            ELSE IF(data.type == matrix){ Print("Well prepared");
                ELSE Print("Please check data type")}};
    Step2: Translate the elements of a CLD to those of an SFD
        # Convert the elements
        REPEAT{ IF(TransRule(data.cld[i],variable)==1) THEN
                data.variable[i]="stock" ELSE data.variable[i]="auxiliary";
            IF( TransRule(data.cld[j],link)==1) THEN data.link[j]="flow";
                ELSE data.link[j]="arrow"};
        data.sfd= compose(data.variable, data.link);
    Step3: Generate an initial SFD
        # Structure an initial SFD on a screen
        Plotting(Graph(data.sfd, mode=sfd);
    Step4: Check adequacy of the initial SFD
        # Check incorrect translation
        REPEAT{ IF(data.variable[i,i+1]=="stock" && data. variable
        [i,i+1]== data. variable [i+1,i+2])
        THEN errorCheck[i,i+1]="f";
        Print("Need to modify stock and auxiliary variables: "+i)};
    Step5: Modify the initial SFD
        # Modify the feature of the translated elements
        REPEAT{ IF(errorCheck=="f")
            THEN Replace(data.link[j,j+1], reverse(data.variable[j,j+1])};
    END.
```

This algorithm can structurally generate an SFD, but cannot define a numerical or logical function in each variable. For practical simulation purposes, a mathematical definition of each variable should be provided; however, this is not considered here.

4   Integrated SD Modeling Support System

The algorithms proposed in Sect. 3 are anticipated to be fast and to develop perfect SD models; however, further verification and refinement is required to ensure their suitability for robust modeling over the long term.

From a practical perspective, the algorithmic approach is significant for quickly and systematically generating an initial SD model. This advantage is clarified in this section, and the performance of the algorithms is validated in terms of an initial solution. In summary, this section presents an integrated SD modeling support system that consists of SD modeling algorithms and their associated support modules, as shown in the left side of Fig. 5.

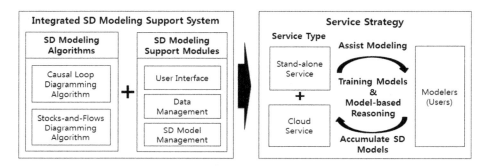

Fig. 5. Integrated SD modeling support system and service strategy

The SD modeling support modules consist of user interface, data management, and SD model management functions. The user interface module facilitates the overall usage of this system on a screen, the data management module assists with the input, deletion, and modification of the data used for SD modeling, and the SD model management module stores, deletes, and modifies the SD models (i.e., CLD and SFD models) generated by the SD modeling algorithms.

Similarly, SD simulation S/W packages, such as Powersim [25], Stella [26], and Vensim [27], support both diagramming and simulation, but depend solely on the work of researchers. As a new entrant in the field of SD project management, DynamicVu [28] focuses on managing data for SD modeling and does not provide a modeling function. Thus, the system proposed in this research is functionally different from existing systems, although it has not yet been implemented.

In general, there are two methods typically used to deliver these types of S/W packages to SD modelers. The first is as a standalone S/W package, and the second is a cloud-based option. Conventional SD standalone S/W packages are installed locally onto the computer of a modeler. In contrast, a cloud-based S/W package is not installed locally, but is run from the cloud. In this model, users pay for access in terms of the simulation length or data bandwidth required. The service types depend on the environment (i.e., device, location, version, etc.) and the required modeling performance. Another SD simulation S/W package, the Sysdea [29] is a cloud-based offering that provides simulation S/W online.

Considering that in the long run, the integrated SD support system intends to quickly, accurately, and automatically generate SD models, the reliability of this system and the validity of the generated artifacts (i.e., CLDs and SFDs) are dependent on the sophistication of the modeling algorithms. Therefore, the suggested system requires a method to train the modeling rules (i.e., logic) and data (i.e., diverse modeling cases) provided by SD modelers. When combined with the S/W service strategy, diverse and rich SD models can be efficiently constructed. In addition, the accumulated models (modeling cases) can be used for model-based reasoning to improve the modeling performance [30]. The right hand side of Fig. 5 illustrates an additional alternative to the SD modeling service, which is to obtain feedback from modelers during advanced SD modeling.

5   Discussion

The impetus for this research was the need to reduce imprecise and subjective interventions in the SD modeling process, and thereby improve its objectivity and robustness. To this end, this research presented an integrated SD modeling support system that consisted of SD modeling algorithms for generating the initial diagrams and SD modeling support modules for improving the initial solutions.

The algorithmic approach to causal loop diagramming was developed in reference to graph theory and relevant technical principles, including text mining and SNA. The stocks-and-flows diagramming algorithm incorporated a programming approach to data conversion. Finally, the SD modeling support modules leveraged S/W engineering principles.

The technical principles employed in the suggested system are applicable to SD modeling, but are not considered in classical SD theory. In this context, the proposed interdisciplinary approach is noteworthy because it demonstrates the extendibility of existing SD theory by successfully integrating it with other relevant theories (techniques).

While the suggested modeling support system has not yet been implemented, the results of this research will be useful as a blueprint for further development of an advanced computer-aided SD modeling tool and method.

As time passes, more complicated problems will arise in the real world. It is therefore critical that the SD approach be hardened by conducting diverse interdisciplinary experiments or challenges to objectify the SD modeling process. In this sense, system dynamists also need to explore statistical or probabilistic principles, instead of simply being opposed to them on principle. It is possible that SD simulation will also require stochastic or probabilistic modeling, which is a theoretical root of simulation, although these were not considered in this research.

As part of the future work, the follow-up research will implement the suggested modeling algorithms and modeling support modules, based on concrete definitions prepared at that time.

Acknowledgments. This work was supported by the Ministry of Education of the Republic of Korea and the National Research Foundation of Korea (NRF-2017S1A5A8018867).

References

1. Forrester, J.W.: Industrial Dynamics. MIT Press, Cambridge, Reprinted by Pegasus (1961)
2. Forrester, J.W.: Urban Dynamics. MIT Press, Cambridge, Reprinted by Pegasus (1969)
3. Sterman, J.D.: Business Dynamics - System Thinking and a Complex World. McGraw-Hill, Boston (2000)
4. Morecroft, J.D.W.: A critical review of diagramming tools for conceptualizing feedback system models. Dynamica **8**(1), 20–29 (1982)
5. Illari, P.M., Russo, F., Williamson, J.: Causality in the Sciences. Oxford University Press, Oxford (2011)
6. Sterman, J.D.: All models are wrong: reflections on becoming a systems scientist. Syst. Dyn. Rev. **18**(4), 501–531 (2002)
7. Featherston, C.R., Doolan, M.: A critical review of the criticisms of system dynamics. In: 30th International Conference of the System Dynamics Society, St. Gallen, Switzerland, pp. 1–13 (2012)
8. Barlas, Y.: Formal aspects of model validity and validation in system dynamics. Syst. Dyn. Rev. **12**(3), 183–210 (1996)
9. Jung, J.U., Kim, H.S.: A study on theoretical improvement of causal mapping for dynamic analysis and design. Korean Syst. Dyn. Rev. **10**(1), 33–60 (2009)
10. Kim, H., Anderson, D.F.: Building confidence in causal maps generated from purposive text data: mapping transcripts of the federal reserve. Syst. Dyn. Rev. **28**(4), 311–328 (2012)
11. Burns, J.R.: Simplified translation of CLD's into SFD's. In: 19th International Conference of the System Dynamics Society, Atlanta, Georgia, pp. 1–28 (2001)
12. Ruohonen, K.: Graph Theory. Tampere University of Technology, Finland (2013)
13. Bang-Jensen, J., Gutin, G.: Digraphs Theory, Algorithms and Applications. Springer, Heidelberg (2007)
14. Teng, T.H.: Signed digraphs and their applications. Soochow J. Math. **29**(1), 69–81 (2003)
15. Silge, J., Robinson, D.: Text Mining with R. O'Reilly Media, Sebastopol (2017)
16. Sonawane, S.S., Kulkarni, P.A.: Graph based representation and analysis of text document: a survey of techniques. Int. J. Comput. Appl. **96**(19), 1–8 (2014)
17. Felman, R., Sanger, J.: The Text Mining Handbook. Cambridge University Press, Cambridge (2007)
18. Sorgente, A., Vettigli, G., Mele, F.: Automatic extraction of cause-effect relations in natural language text. In: Proceedings of the 7th International Workshop on Information Filtering and Retrieval, Turin, Italy, pp. 37–48 (2013)
19. The R Project for Statistical Computing. www.r-project.org
20. Decision Explorer. https://banxia.com/dexplore/resources/whats-in-a-name
21. Richardson, G.P.: Problem in causal loop diagrams revisited. Syst. Dyn. Rev. **13**(3), 247–252 (1997)
22. Jung, J.U., Kim, H.S.: A study on ensuring validity and increasing power of expression on causal maps. Korean Syst. Dyn. Rev. **8**(1), 97–115 (2007)
23. Jung, J.U., Kim, H.S.: An improvement of coherence and validity between CLD and SFD of system dynamics. J. Digit. Convergence **12**(6), 66–77 (2014)

24. Binder, T., Vox, A., Belyazid, S., Haraldsson, H., Svensson, M.: Developing system dynamics models from causal loop diagram. In: 22nd International Conference of the System Dynamics Society, Oxford, Great Britain, pp. 1–21 (2004)
25. Powersim. www.powersim.com
26. Stella. www.iseesystems.com
27. Vensim. www.vensim.com
28. DynamicVu. www.dynamicvu.com
29. Sysdea. www.sysdea.com
30. Aamdt, A., Plaza, E.: Case-based reasoning: foundational issues, methodological variations, and system approaches. AI Commun. 7(1), 39–59, IOS Press (1994)

The Theory of Modified Rings Game

Yushuang Wu[1], Yuhao Lin[1], Xiaoyu Chen[1], and Xingguo Chen[1,2(✉)]

[1] School of Computer Science and Technology,
Nanjing University of Posts and Telecommunications, Nanjing 210023, China
chenxg@njupt.edu.cn
[2] State Key Laboratory for Novel Software Technology, Nanjing University,
Nanjing 210023, China

Abstract. The highly addictive stochastic puzzle game Rings has recently invaded the mobile devices. In this paper we propose the theory of a modified Rings in terms of NP-completeness and decidability. We show the NP-completeness by reduction from the 3-PARTITION problem, and the decidability by reduction from the Post Correspondence Problem.

Keywords: NP-completeness · Decidability · Rings game

1 Introduction

The puzzle game Rings is an addictive single-player, nondeterministic video game, which was created by Gamezaur from Kraków, Poland in Sept. 2016. It is played on a 3×3 board, shown in Fig. 1. Each square of the board can hold at most 3 rings of different sizes, and each ring can be given of 3 sizes and in 8 colors. A new game piece is generated randomly with one or two rings below the game board, and the color of each ring is selected randomly in 8 colors. Three pieces appear in each round, and the player can move them, one by one, onto the board in any order. While pieces can only be moved to alternative squares on the board. (Call a square is alternative, if it can hold the game piece). Rings will be cleared if a square is fully filled by rings in the same color, or in a row, column, or diagonal, each square holds at least one ring in the same color. The reward is the number of the cleared rings, and the game is over when there is no alternative squares on the board. The objective of the player is to maximize the total rewards.

The theoretical analysis is crucial to designing an algorithm, with issues like complexity and decidability in special cases. The complexity of several games (or their generalized versions) and puzzles has been studied. Othello game on an $n \times n$ board proves to be PSPACE-complete by reduction from generalized geography played on bipartite graphs with maximum degree 3 [1]. Tetris game proves to be NP-complete by reduction from the 3-PARTITION [2–4], and undecidable in some special cases [5]. The game of 2048 proves to be NP-complete by reduction from the 3-SAT problem [6], and PSPACE-complete by reduction from Nondeterministic Constraint Logic [7].

© Springer International Publishing AG 2017
H. Yin et al. (Eds.): IDEAL 2017, LNCS 10585, pp. 376–386, 2017.
https://doi.org/10.1007/978-3-319-68935-7_41

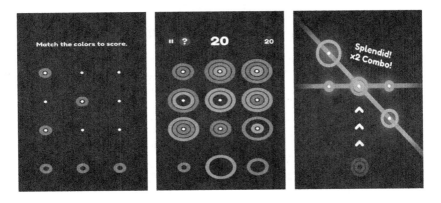

Fig. 1. The game of rings.

In this paper we propose the theory of a modified Rings in terms of NP-completeness and decidability. We show the NP-completeness by reduction from the 3-PARTITION problem, and analyse the decidability by reduction from the Post Correspondence Problem.

2   The Modified Rings Game

2.1   Generalization and Modification

It is natural to generalize the Rings board to $n \times n$, where $n \geq 3$, and it is an odd number, and generalized and modified rules are as follows:

- Rings still have 3 sizes and 8 colors.
- Each square can only hold $\lceil n/3 \rceil$ rings of each size.
- One game piece appears in one round.
- One new rule is added that for every clearing, equal number of rings in the same color will be cleared until a square is short of rings in this color. Besides, rings with smaller size will be cleared in priority.

Call this generalized game as the modified Rings game, and the spirit of the original game is preserved.

2.2   Rules

Then we give some definitions and variants of the game of modified Rings, to formalize rules of this game.

The board. The board is indexed from up-to-bottom and left-to-right, and each square (i, j), can hold $\lceil n/3 \rceil$ rings of each size. If we view the board from 3-dimension angle, then the number of rings a square holds can be viewed as the height above this square. Thus the maximum height is n, and if the height above a square reaches n, we define the square as *full*.

Game pieces. The game piece is a group of rings, consisting of not more than $n - 1$ rings, of two sizes at most. Colors of rings are labelled with clr_i, with $i = 1, 2, \ldots, 8$, and 3 sizes are labelled with S, M, and, L.

Thus the state of a piece $P = {<}(i, j), sz, clr{>}$ is a triple, consisting of:

- the *position* $(i, j) \in \{1, \ldots, n\} \times \{1, \ldots, n\}$ of a square on the board.
- the *size information* $sz = \{a_s, a_m, a_l\}$, with non-negative integers $a_s, a_m, a_l \in (1, \lceil n/3 \rceil)$ indicating the number of rings of size S, M and L. For each square, when the number of rings of one size reaches $\lceil n/3 \rceil$, we call the square as *occupied* by this size.
- the *color information* $clr = \{b_1, b_2, \ldots, b_8\}$, with the non-negative integer b_i indicating the number of rings in the i_{th} color.

Note that $h = a_s + a_m + a_l = \Sigma_{i=1}^{8} b_i$. After being placed onto a square, a piece is *fixed*, which means that no moves are legal for this piece. A new game piece appears when the present one is fixed.

Clearing. The player can continuously place a new piece onto a square until all the squares are *occupied* by all three sizes, which means the loss of the game. Thus the player aims to create more *clearings*, which occur when there are same-color rings on each square in a line (called as a *line-clearing*), or a full square is concolorous (called as a *self-clearing*). In the first case, for each square in the line, equal number of same-color rings will be cleared until there is one square short of rings in this color. (Rings of smaller size will be first cleared, and if with the same color and size, rings that are first to be placed will be cleared at first). That more than one clearings occur at a time is legal. The score increases with clearings, and more reward will be given with more rings cleared at a time.

3   NP-Completeness of Rings

In this section, we will analyse the complexity of some decision problems related to modified Rings game. We start from introducing the 3-PARTITION problem, and then employ the method of reduction to map an instance of Rings to the instance of 3-PARTITION, with the proof of its completeness and soundness in the end.

3.1   The Decision Problem

We first consider the following decision problem, called RINGS CLEANING:

Instance. Given an initial board with rings placed partially, and an ordered sequence of game pieces.

Question. Is it possible to clear all the rings in the sequence and leave the board empty in the end?

It is not difficult to classify RINGS CLEANING as NP problem. We now prove its NP-hardness by reduction, so that its NP-completeness can be proved. Here we use the 3-PARTITION Problem:

Instance. Given a sequence of positive integers $t_1, t_2, \ldots, t_{3s}$ and a positive integer T. For all i's with $1 \leq i \leq 3s$, $T/4 < t_i < T/2$ and $\Sigma_{i=1}^{3s} t_i = sT$.

Question. Can the set $\{t_1, t_2, \ldots, t_{3s}\}$ be partitioned disjointly into s subsets $A_1, A_2, \ldots, A_s$, with 3 integers in each subset, so that for all j with $1 \leq j \leq 3s$, we have $\Sigma_{t_i \in A_j} t_i = T$? (Call it a "yes" instance if this is the case and a "no" instance otherwise).

We use the following result:

Theorem 1. *3-PARTITION is NP-complete in the strong sense* [8].

We might as well limit our attention to 3-PARTITION instances for which s is a square number, and $T + 3 \leq \lceil n/3 \rceil$, because the number n will be big enough when considering the complexity. For technical reasons these will make the reduction easier without affecting the complexity of instances.

3.2   Reduction

The initial Rings board. As shown in Fig. 2, we first construct the initial board as follows:

- squares in the first row and the first column will be occupied in advance, called *locked squares*. They are occupied by M and L rings, and the total height is $n - 1$. Rings on squares (1 to $n - 1, 1$) are in color clr_8, on squares $(1, 1$ to $n-1)$ are in color clr_7, and on the crossover point $(1, 1)$ is in color clr_6.
- Rings on a locked square are in a single color. One ring of size S, with the same color as the locked square is called a *key*, which can trigger a self-clearing on the locked square.

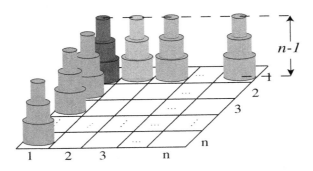

Fig. 2. An initial board example. In this example, clr_8 is green, clr_7 is blue, and clr_6 is red. (Color figure online)

It is noticed that the initial gameboard is constructable in polynomial time. Some other necessary explanations for the construction are as follows:

- For any locked square, any piece in the sequence is prevented to be placed onto locks except a key.
- Line-clearing will not occur, because there are no same-color rings in the first row or column.
- Self-clearing will not occur, too, because no squares are full.

The sequence of Rings pieces. In the first phase, $(n-1)^2$ code-pieces are given successively. A code-piece is coded by $t_i + 1$ rings of the same size, with one ring in color clr_1, and t_i rings in color clr_2. t_i corresponds to one of the three integers in a triple subset in the instance of 3-PARTITION problem, shown in Fig. 3. Assume $s = (n-1)^2$, then the number of subsets s can be mapped to the number of squares $(n-1)^2$. The sum of t_i is equal to $s(T+3)$, with $i = 1, 2, \ldots, 3s$ denoting $3s$ code-pieces, so that $3s$ code-pieces can be placed onto s squares with a same total height.

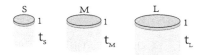

Fig. 3. Codepieces of size S, M, and L.

In the second phase, there will be $2n+1$ keys for opening the locks and $n-1$ pieces for clearing the entire board, which are called clearing-pieces. There are three types of keys: key_1: one ring of size S in the 8_{th} color, for locks in the first column except $(1,1)$; key_2: one ring of size S in the 7_{th} color, for locks in the first row except $(1,1)$; key_3: one ring of size S in the 6_{th} color, for the locks at $(1,1)$.

There will be one key_1 and one clearing-piece given in turn, which circulates for $n-1$ times. Then $n-1$ key_2's and one key_3 will be provided. The process will be precisely shown in the next part.

3.3   Completeness

Given an "yes" instance of 3-PARTITION $\mathcal{P} = <t_1, t_2, \ldots, t_{3s}, T>$ with T divisible by three, s is a square number, and $T + 3 \leq (\sqrt{s} + 1)/3$, we can reduce it to an instance of the modified Rings game whose board can be cleared, given an initial board and an ordered sequence. We divided the process of clearing into two phases, and the first one can be reduced to the process of partitioning in 3-PARTITION problem.

The first phase: Place code-pieces. The first phase is placing code-pieces until locks are opened. In this phase, code-pieces are prevented to be placed on locked squares, and no line-clearing will be triggered, because colors of rings in code-pieces are different from those in the first row and column.

Pieces are placed as below: In order to clear all the rings on bucket-squares, we have to make sure that each square shares equal total height at last. Further, each square will only hold three pieces, and then the total height will be $t_1+t_2+t_3+3 = T + 3 \leq \lceil n/3 \rceil$, where t_i is the "code" of i_{th} piece placed on the square. Placing pieces reasonably is equally hard to dividing intergers into triples. Therefore, it is a natural reduction from the instance of 3-PARTITION.

The second phase: Unlock and clear. Not until all the code-pieces are fixed, the second phase begins. As shown in Fig. 4, the player will be circularly given one key_1 and one *clearing − pieces*, to open a lock in the first column, and to clear the entire row. At last, n key_2's and key_3 will be given to clear rings on squares in the first row. Thus, the entire board will be left empty.

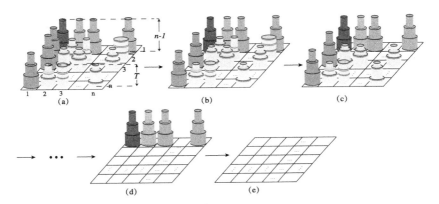

Fig. 4. The process of opening locks and clearing the board in the second phase: a key_1 unlocks the square at $(2,1)$, change the board from (a) to (b), and in (c) a clearing-piece clears the second row; such clearings circulate for $n-1$ times until (d), and then $n-1$ key_2's and one key_3 clear locks in the first row.

3.4   Soundness

Call valid a *trajectory sequence* that, adhering to the rules, clears all the rings to ensure a "yes" instance. A move is also valid if it is in a valid trajectory sequence.

Proposition 1. *In the first phase, it is invalid that the total height above a square is unequal to $T + 3$.*

Proof. The total number of rings in the code-pieces is $s(T + 3)$, thus after all the code-pieces are fixed, if any square have a stick-out height, there must be a square with a caved-in height, which will cause some rings unable to be cleared at last, shown in Fig. 5.

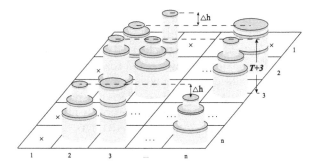

Fig. 5. Locked squares are labelled with ×. If the height above a square (in the second row) is higher than T + 3 by Δh, then there must be a square, whose height is lower than T + 3 by Δh(in the last row).

Proposition 2. *In the first phase, each square can hold 3 pieces and 3 pieces only, invalid otherwise.*

Proof. There are three factors as guarantees: (1) $T + 3 \leq \lceil n/3 \rceil$. (2) Proposition (1). (3) $T/4 < t_i < T/2$.

4 Decidability

In this section, we discuss the issue of decidability in Rings games, and the result is valid for the board of size 5×5. In the Rings game, it is a existential question to decide whether there is a sequence, which consists of certain pieces we give, to clear rings in certain colors. Inspired by [5], we construct a special initial board and give a set of pieces, then by reduction, an instance of the modified Rings game can be mapped to the Post Correspondence Problem.

4.1 Reduction

First, recall the instance of the Post Correspondence Problem (PCP).

Instance. Given two string sequences $(u_1, u_2, \ldots, u_n)$ and $(v_1, v_2, \ldots, v_n)$ over a two-letter alphabet $\{a, b\}$.

Question. For some $k \geq 1$, and $1 < i_1, i_2, \ldots, i_k < n$, is it decidable that whether there exists words in the two sequences, with the form $u_{i_1}, u_{i_2}, \ldots,$ $u_{i_k} = v_{i_1}, v_{i_2}, \ldots, v_{i_k}$?

Instance. Given an special initial board, and a certain set of game pieces.

Question. Is it decidable that whether the set contains a sequence to clear rings with some certain colors?

To be mapped to the PCP, let two colors correspond to two letters in the alphabet $\{a, b\}$, with gray for a and purple for b, and another color yellow for x

as the occupation and division. And let two sizes S, M correspond to two stacks to hold two words, so that rings of one size cannot be placed into the other stack. In the first phase, We first alternately arrange rings of size S to form strings in the stack S, with the same number of x's in the stack M, arrange rings of size M in the stack M, with the same number of x's in the stack S. If strings are placed according to a solution of PCP, two stacks will hold the same word (a "perfect" configuration).

In the second phase, language $\mathcal{L}$ picks pieces from the set into two stacks to try to clear rings in gray, purple, and yellow. At the end of the sequence, a final piece will be placed to verify whether $\mathcal{L}$ corresponds to a solution of the PCP, and if it does, we call $\mathcal{L}$ a "yes" language.

4.2 Proof

The initial board. Considering the board of size 5, then every square can only hold 2 rings of each size at most, and we assume that the word length is 4, so that there will be 8 pieces picked by $\mathcal{L}$ and a final piece to verify. it reflects our main idea of the construction and reduction, and in cases of a longer word length or a different board size, it is not difficult to generalize our result.

In the reduction, it is important to construct a special initial board with such properties:

Property 1: In the first phase, pieces should be held in stacks in a fixed order, instead of being determined by the player, so that ordered strings and words can be formed.

Property 2: In the second phase, if pieces are placed according to a "yes" language, rings in certain colors can be cleared in the end, which guarantees the validness of the reduction.

For Property 1, we consider a method, called *"one-trigger-one"*, to restrict the alternative position of a piece. It is based on giving only one alternative square for the ready piece, and making other squares occupied (Call them as locked). Thus the piece can only be placed onto the certain square in the stack (call them *stack-squares*), and a lock will be triggered to be opened, prepared for the next piece.

We consider successive line-clearings to be triggered in different directions, so that it is easier to arrange stack squares. Furthermore, if line-clearings triggered by the final piece can clear rings on all stack-squares, then verifying the guessed solution will be straightforward. The arrangement of stack-squares in our instance and the order of triggering is shown in Fig. 6, where stack-squares are arranged on two diagonals.

For Property 2, it depends on the state of pieces in the sequence, and states of initial rings on stack-squares. We first label a letter with the name of the stack it belongs to, e.g. $a(S)$ represents the letter a belongs to the stack S. And then use $x(M)$'s to occupy stack-squares of S, use $x(S)$'s to occupy stack-squares of M, so that strings are divided by them, and every stack-square have equal number of yellow rings, which can be cleared by the final piece.

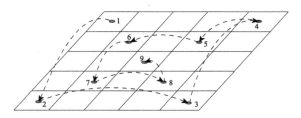

Fig. 6. The distribution of stack-squares and the order of triggering.

Then, we place purple or gray rings on each stack-square in advance to construct a "perfect" configuration. Thus, the objective can be reached only by placing correct pieces into stacks. For example, if both stack hold the word "*baab*", formed by strings "*b, a, ab*" in stack S, and strings "*ba, a, b*" in stack M, we place in turn "$a(S)$, $a(M)$, $b(M)$, $b(S)$, $b(M)$, $b(S)$, $a(S)$, $a(M)$" onto stack-squares. Note that we use a trick to place into the stack a letter which is the opposite of what it is in the word, so that if $\mathcal{L}$ is a solution of the PCP, every stack-square has both gray and purple rings.

The sequence. Pieces in the sequence are generated according to $\mathcal{L}$, which is guessing a solution of PCP. With $\mathcal{L}$ repeatedly picking an index $i, 1 \leq i \leq n$, pieces are given according to the string u_i and then v_i. For example, for $i = 2$ and $u_i = ab, v_i = b$, given pieces are $a(S), b(S), b(M)$, and after placing these pieces the next i will be picked.

For each piece, it contains a ring in gray or purple as the "letter", and a ring in some certain colors to trigger a clearing, called as *trigger-colors*, which are determined by the direction of its triggered line-clearing.

The final piece covers two yellow rings (x), a gray ring (a), and a purple ring (b).

An example. As shown in Fig. 8(a), we construct a complex while clever initial board, and we will use a "yes" $\mathcal{L}$ (a solution of the PSP), to indicate why the initial board should be constructed as so and how to verify the correctness of $\mathcal{L}$. The sequence generated by $\mathcal{L}$ is shown in Fig. 7, it represents the word as "*baab*", formed by strings "*b, a, ab*" in stack S, and strings "*ba, a, b*" in stack M. We define that trigger-colors are circularly "red, blue, orange, red", because directions of triggered line-clearings are circular. Therefore if we place pieces in this sequence one by one, in the order is definitely the same as shown in Fig. 6. After a series of clearings triggered, when the final piece is ready to be placed, the board will

Fig. 7. With 9 pieces in the sequence, the piece in the third row is the final piece.

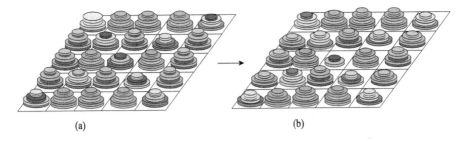

Fig. 8. Figure(a): the initial board we construct; Figure(b): the board when the final piece is ready to be placed.

be as shown in Fig. 8(b), with rings above 8 stack-squares all consists of a gray ring, a purple ring, and two yellow ones. Therefore, when the final piece is placed at $(3, 3)$, rings in color gray, purple, and yellow will be cleared, and $\mathcal{L}$ is verified to be a solution to the instance of the PCP.

5 Conclusion

In this paper, we generalize the game of Rings and modify it, and several issues are discussed that are related to the modified Rings. We first reduce the instances of 3-PARTITION to the instances of our modified Rings game, so that its NP-completeness is proved. Then, a special case of the Rings game is discussed, which can be reduced from the Post Correspondence Problem, thus it is of undecidability.

Our future work is to consider about variants of the game rules and different generalization, so that issues of NP-completeness, PSPACE-completeness, and (un)decidability will be researched further.

Acknowledgements. The work was supported by National Natural Science Foundation of China (61403208), Natural Science Foundation of Jiangsu Province (BK20161516), Scientific and Technological Support Project (Society) of Jiangsu Province (BE2016776), and Science Foundation of Nanjing University of Posts and Telecommunications (NY214014).

References

1. Iwata, S., Kasai, T.: The othello game on an n × n board is pspace-complete. Theor. Comput. Sci. **123**(2), 329–340 (1994)
2. Hoogeboom, H.J., Kosters, W.A.: The theory of tetris. Nieuwsbrief van de Nederlandse Vereniging voor Theoretische Informatica **9**, 14–21 (2005)
3. Breukelaar, R., Hoogeboom, H.J., Kosters, W.A.: Tetris is hard, made easy. In: Leiden Institute of Advanced Computer Science, Universiteit Leiden (2003)
4. Demaine, E.D., Hohenberger, S., Liben-Nowell, D.: Tetris is hard, even to approximate. In: Warnow, T., Zhu, B. (eds.) COCOON 2003. LNCS, vol. 2697, pp. 351–363. Springer, Heidelberg (2003). doi:10.1007/3-540-45071-8_36

5. Hoogeboom, H.J., Kosters, W.A.: Tetris and decidability. Inf. Process. Lett. **89**(6), 267–272 (2004)
6. Abdelkader, A., Acharya, A., Dasler, P.: 2048 without new tiles is still hard. In: 8th International Conference on Fun with Algorithms, vol. 49, pp. 1–14 (2016)
7. Mehta, R.: 2048 is (PSPACE) hard, but sometimes easy. arXiv preprint (2014). arXiv:1408.6315
8. Gary, M.R., Johnson, D.S.: Computers and Intractability: A Guide to the Theory of NP-Completeness (1979)

Markov Random Field Based Convolutional Neural Networks for Image Classification

Yao Peng[(✉)] and Hujun Yin

School of Electrical and Electronic Engineering, The University of Manchester,
Manchester M13 9PL, UK
yao.peng@postgrad.manchester.ac.uk, hujun.yin@manchester.ac.uk

Abstract. In image classification, deriving efficient image representations from raw data is a key focus as it can largely determine the performance of a vision system. Conventional methods extract low-level features based on experiments or certain theories, whilst deep learning approaches learn image representations hierarchically with multiple layers of abstraction from vast number of sample images. Markov random fields are generative, flexible and stochastic image texture models, in which global image representations can be obtained by means of local conditional probabilities. Texture has been strongly linked to human visual perception. The ability of deriving global description from local structure shares compatibility with convolutional neural networks. Inspired by this property, we investigate the combination of Markov random field models with deep convolutional neural networks for image classification. Various filters from Markov random field models are first derived to form the features maps. Then convolutional neural networks are trained with prefixed filter banks. Comprehensive experiments conducted on the MNIST dataset, EMNIST database and CIFAR-10 object database are reported.

Keywords: Image classification · Image representations · Markov random field · Convolutional neural networks

1 Introduction

Image classification is a problem of identifying objects or region of interest by learning the feature vectors and the contextual constraints of pixels of the objects in an image. A major difficulty comes from the large amount of intra-class variability arising from different illumination condition, viewpoint, misalignment and deformations. To minimise intra-class variability, numerous efforts have been devoted to building up robust image representations for subsequent classification. Conventional feature extraction methods handcraft low-level features to describe image properties. Representative examples include Gabor features and scale invariant feature transform (SIFT). Although the low-level features can be carefully handcrafted with great success for certain data and tasks, designing effective features for new data and tasks usually requires prior knowledge and

H. Yin et al. (Eds.): IDEAL 2017, LNCS 10585, pp. 387–396, 2017.
https://doi.org/10.1007/978-3-319-68935-7_42

careful engineering, which has motivated the development of learning robust and effective representations directly from raw data [1, 2].

Recently deep learning provides a plausible way of automatically learning multiple level features, by using multiple processing layers to learn image representations with multiple levels of abstraction [3]. As one key ingredient of the success to deep learning, convolutional neural networks (CNNs) possess the ability of learning large-scale invariances, which are stable to deformations [4]. Recent advances have witnessed increasing number of applications of deep networks to a wide range of image classification tasks.

Unlike these feature-based approaches, Markov random fields (MRFs) provide a way of modelling image properties and encoding contextual constraints which are ultimately indispensable in visual interpretation and image understanding [5–8]. As a stochastic process approach, MRFs take pixel intensity levels in an image as random variables and establish contextual dependencies among pixels using conditional probabilities [9]. Specifically, the clique-based structure makes them particularly well suited for capturing local relations and neighbourhood dependencies. MRF models derive global description by specifying local properties of images, which are consistent with weight sharing and local connectivity properties in CNNs.

In this paper, we explore an image classification framework by combining the respective strengths of CNNs and MRFs. We introduce a Markov random field based convolutional neural networks (MRF-CNN) for image classification. By means of modelling different MRFs, various filters can be derived to form feature maps. CNNs are then trained with prefixed filter banks. The effectiveness of the proposed method has been verified on the MNIST [10], EMINST [11] and CIFAR-10 [12] image databases.

The remainder of this paper is structured as follows. Section 2 provides a brief review of related work. Markov random fields and Gibbs distributions are discussed in Sect. 3. Section 4 presents experimental results, followed by conclusions in Sect. 5.

2   Related Work

MRF based image analysis: In computer vision problems, the intensity level of a single point in an image is highly dependent on the intensity levels of neighbouring pixels unless the image is simply random noise [13]. MRFs model the spatial interrelationships between related random variables and the joint probability distribution can be specified by Gibbs distributions with potential functions. The study of MRFs has had a long history, originate from the Ising model on ferromagnetism [14], with a variety of applications such as texture synthesis [13], image segmentation [15], restoration [9] and classification [16]. There is a relatively efficient optimisation algorithm, simulated annealing [9], that can be used to find the globally optimal realisation.

Image classification with deep learning: Deep learning techniques currently achieve state-of-the-art performance in a multitude of problem domains such as vision, audio, robotics, and language processing. In image classification, with the increased availability of data, attention within deep learning community has recently shifted to exploring architectures with more depth. The breakthrough of large scale image classification came in 2012 when Krizhevsky *et al.* developed the AlexNet [17] and achieved the best performance in the ImageNet large scale visual recognition competition. Since then, several researches have made significant improvements in classification accuracy by modifying network architectures. Common ways include reducing filter size [18], expanding the network depth [19,20], modifying the convolution [21–23] or pooling [24,25] operations.

Although deep CNNs have been successful and many variations have been proposed for various vision tasks, arguably the first instance that has led to a clear mathematical justification is the wavelet scattering networks (ScatNet) [26]. ScatNet computes translation invariant image representations and cascades wavelet transform convolutions with non-linear modulus and average operators. Such prefixed filter banks used in similar multistage architectures of CNNs have demonstrated state-of-the-art performance in handwritten digits and texture recognition. Principal component analysis networks (PCANet) [27] and discrete cosine transform networks (DCTNet) [28] have shown worked well in face recognition by processing an input image with a layer-wise convolution with PCA and DCT filters and followed by binarisation, block-wise histograming. These methods used relatively small architectures and achieved considerably good results in certain tasks but with poor generalisation abilities.

3   Markov Random Field and Gibbs Distribution

A random field, $F = \{F_1, F_2, \cdots, F_m\}$, is a finite collection of random variables defined on the set $\mathcal{S}$ and each F_i is a random variable taking a value f_i from the label set $\mathcal{L}$. Assuming $f = \{f_1, f_2, \cdots, f_m\}$ is one possible configuration or realisation of the field F and denotes $\mathbb{F}$ as the set of all possible configurations of F. F is considered as an MRF on $\mathcal{S}$ with respect to the neighbourhood system $\mathcal{N}$ if following properties are satisfied:

– Positivity: Any configuration is positive.

$$P(f) > 0, \ \forall f \in \mathbb{F} \tag{1}$$

– Markovianity: The conditional probability of any site given the others only depends upon the configuration of its neighbours.

$$P(f_i \mid f_{\mathcal{S}-\{i\}}) = P(f_i \mid f_j, \ j \in \mathcal{N}_i) = P(f_i \mid \mathcal{N}_i) \tag{2}$$

where $\mathcal{S} - \{i\}$ denotes the all sites excluding site i, $f_{\mathcal{S}-\{i\}}$ is the set of labels at the sites in $\mathcal{S} - \{i\}$.

An MRF is an undirected graph model that explicitly represents random variables and their conditional dependencies. With its conditional probability distribution, it is still difficult to specify an MRF due to several reasons [29]. Fortunately, the Hammersley-Clifford theorem [29, 30] provides a mathematically means of specifying MRFs by demonstrating the equivalence between MRFs and Gibbs distribution with regard to the same graph.

Let $P(f)$ denote a Gibbs distribution on the set $\mathcal{S}$. Then the joint probability $P(f)$ takes the form

$$P(f) = \frac{1}{Z} e^{-\frac{1}{T} U(f)} \tag{3}$$

where

$$U(f) = \sum_{c \in \mathcal{C}} V_c(f) \tag{4}$$

is the energy function that sums clique potentials $V_c(f)$ over all possible cliques $\mathcal{C}$ and stays positive for all possible configurations. A clique c is either single pixel or a collection of neighbouring pixels. T is a constant called the temperature, and Z is a normalising constant defined by

$$Z = \sum_{f \in \mathbb{F}} e^{-\frac{1}{T} U(f)} \tag{5}$$

which is also called the partition function.

While MRFs only specify the conditional dependencies, Gibbs distributions provide explicit probability function for each clique. The Hammersley-Clifford theorem states that F is an MRF on $\mathcal{S}$ with respect to a neighbourhood system $\mathcal{N}$ if and only if the corresponding $P(f)$ is a Gibbs distribution, which indicates that each MRF can be measured by the Gibbs distribution, and every Gibbs distribution defines an MRF.

Multi-level logistic model [9, 31] is a typical category of MRF models in which a clique potential depends on the type and the local configuration of that clique. An isotropic multi-level logistic model share the same potential functions defined by β and usually depicts blob-like regions, while an anisotropic model tends to generate texture-like images [9, 31, 32]. To generate MRFs, several algorithms have been developed [9, 13, 33]. Among these, simulated annealing is a practically efficient method as the global optimal solution can be achieved as long as cooling process is slow enough.

4   Proposed MRF-CNN, Experiments and Discussions

The main advantage of MRF models is that they provide flexible and natural models for the interaction between spatially related random variables in their neighbourhood systems via clique functions [34]. As a statistical model, MRF models pixels dependently together with their spatial constraints and thus avoids losing information in feature extraction. In the proposed MRF-CNN, we defined multi-level logistic MRF models (label set $\mathcal{L} = \{0, 1\}$) with parameter β specified on a fifth-order neighbourhood system as

$$\beta = (\beta_{11}, \beta_{12}; \beta_{21}, \beta_{22}; \beta_{31}, \beta_{32}; \beta_{41}, \beta_{42}; \beta_{51}, \beta_{52})$$

$$= \begin{bmatrix} \beta_{51} & \beta_{41} & \beta_{32} & \beta_{42} & \beta_{52} \\ \beta_{41} & \beta_{21} & \beta_{12} & \beta_{22} & \beta_{42} \\ \beta_{31} & \beta_{11} & 0 & \beta_{11} & \beta_{31} \\ \beta_{42} & \beta_{22} & \beta_{12} & \beta_{21} & \beta_{41} \\ \beta_{52} & \beta_{42} & \beta_{32} & \beta_{41} & \beta_{51} \end{bmatrix} \tag{6}$$

By changing the value of β, different distributions can be modelled. Figure 1(a) shows samples taken from multi-level logistic distributions for specified parameter values displaying horizontal, vertical and diagonal line textures. Figure 1(b) displays different MRF patterns (such as checkboard-like patterns and blob-like textures) generated using multi-level logistic distributions with random parameters. The images were simulated using five hundred iterations of a Metropolis sampler with simulated annealing strategy. All images were initialised randomly with a size of 100×100 from a uniform distribution, and the temperature initialised at $T = 1$ was annealed by a factor of 0.999 per iteration.

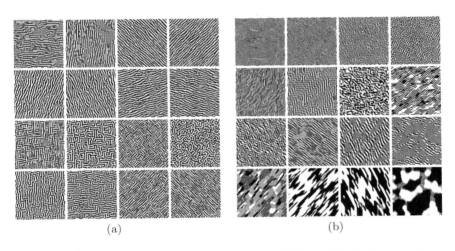

(a)   (b)

Fig. 1. Sample images from the multi-level logistic MRF models for (a) specified and (b) random parameters.

With sample images generated, we extracted filters by randomly selected $n \times n$ patches from each image and repeated the process 5000 times. The resulting filters were the average of those patches from every sample pattern. Figure 2 shows filters extracted from the samples in Fig. 1(a) and (b). Apart from these MRF filters, the DCT and Gabor filters were also employed as prefixed convolutional operators. With prefixed filter bank, we trained other layers of the CNNs using batch gradient descent with backpropagation algorithm.

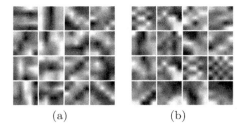
(a)   (b)

Fig. 2. 5×5 filters sampled from MRF patterns in Fig. 1(a) and (b).

4.1 Datasets

To evaluate the performance of the proposed MRF-CNN method, experiments were conducted on the MNIST [10], EMINST [11] and CIFAR-10 [12] image databases. The MNIST database of handwritten digits, formed from larger NIST database, contains 60000 training and 10000 test images, in which digits have been size-normalised and centred in a fixed size. The EMNIST dataset is an extension of MNIST to handwritten letters, which follows the same conversion process and shares the same images structures as the MNIST database. As a result, six subsets (by_class, by_merge, balanced, letters, digits and MNIST) were produced for different classification tasks. The CIFAR-10 image database is a collection of 60000 32×32 colour images split in 10 classes (airplane, automobile, bird, cat, deer, dog, frog, horse, ship and truck) with 5000 training and 1000 test images per class. Whilst each image contains one notable instance of its respective class object, there are variations in viewpoint and scale present.

4.2 MNIST Experiments

The first experiment was conducted on the MNIST digits. The only pre-processing taken was mean substraction.

The architecture used for the MNIST dataset contains 9 major layers excluding the input. The first convolutional layer consists of a prefixed filter bank and the filter parameter stays the same during training. Followed by a 2×2 max-pooling layer, the second convolutional layer has filters of size 1×1, which learns a non-linear combination of prefixed filter bank and compresses the number of feature maps to avoid large parameter space. Following another convolution-pooling block, a dropout layer is used to prevent the overfitting. Finally, fully connected layers are present with a softmax layer producing probability distribution over classes. Among this architecture, batch normalisation is inserted in each convolutional layer between the convolutions and activations excluding the first and last ones.

We initialised the convolutional weights in the model from a normal distribution with zero biases. The model was trained for 100 epochs using stochastic gradient descent with a batch size of 200 and momentum of 0.9. The learning rate was initialised with 0.01 and annealed by a factor of 0.989 per epoch. In this experiment, we evaluated the model with MRF based filter banks. Experiments were repeated with randomly selected prefixed filter banks and mean error rate was computed. Results are shown in Table 1 and compared with some state-of-the-art algorithms.

Table 1. Classification results on the MNIST dataset.

Methods	Error rate (%)
CNN [22]	0.53
Stochastic pooling [24]	0.47
Maxout [35]	0.45
DropConnect [36]	0.57
Network in network [23]	0.47
ScatNet [26]	0.43
PCANet [27]	0.62
MRF-CNN	**0.38**±0.01

4.3 EMNIST Experiments

In this experiment, the EMNIST images were normalised by subtracting their mean. The architecture stays the same as for MNIST dataset except that the number of feature maps in the last fully connected layer varies for different subsets.

The experimental protocols (initialisation, number of epochs, batch size, momentum and learning rate) were the same as for the MNIST database. Results for the six subsets are shown in Table 2.

Table 2. Summary of the results for different subsets on the EMNIST dataset.

EMNIST subsets	Error rate (%)		
	Linear classifier [11]	OPIUM classifier [11]	MRF-CNN
Balanced	49.17	21.98±0.92	**9.71**±0.13
By class	48.19	30.29±1.47	**12.23**±0.29
By merge	49.49	27.43±1.18	**9.06**±0.18
Letters	44.22	14.85±0.12	**4.56**±0.04
Digits	15.30	4.10±0.40	**0.25**±0.01
EMNIST MNIST	14.89	3.78±0.14	**0.33**±0.01
Average	36.86	17.07±0.71	**6.05**±0.11

4.4   CIFAR-10 Experiments

The third experiment was conducted on the CIFAR-10 image dataset. The original 32×32 RGB images were first normalised using data standardisation strategy to have zero mean and unit variance. Then the PCA whitening was applied to project images onto eigenvectors to remove correlations.

The architecture used starts with a prefixed filter bank as the first convolutions, a 1×1 convolutional layer as regularisations, and a dropout layer to prevent overfitting. Followed with two convolution-pooling-dropout blocks, fully connected layers with softmax loss are used for classification.

The prefixed filter banks used in three colour channel were extracted from the same MRF samples. We initialised the convolutional weights in the models from a normal distribution with zero biases. The models was trained for 300 epochs using stochastic gradient descent with 200 examples per batch and momentum of 0.9. The learning rate was initialised with 0.01 and annealed by a factor of 0.989 per epoch. Classification results compared with several state-of-the-art algorithms are shown in Table 3. The proposed MRF-CNN achieved considerably good results with relatively smaller models and fewer parameters.

Table 3. Classification results on the CIFAR-10 dataset.

Methods	Error rate (%)
MCDNN [21]	11.21
Stochastic pooling [24]	15.13
Maxout [35]	11.68
DropConnect [36]	9.32
Network in network [23]	10.41
All-CNN [25]	9.08
MRF-CNN	**8.75**±0.14

5   Conclusions

This paper presents an approach to image classification by combining Markov random field models, to extract salient information from the raw data, with the convolutional neural networks, to implicitly learn high-level features. We generate multiple patterns by the choice of different potential functions for MRF models. Numerous filters are derived from these patterns to extract various features from raw inputs. With various types of features extracted from different distributions of MRFs, we apply deep convolutional networks with these prefixed filter banks for image classification.

Results have validated the effectiveness of proposed Markov random field based convolutional neural networks. The proposed method combines respective strengths of MRFs and CNNs to bring the expressive power of deep architectures to probability formulations. Whilst it has been previously demonstrated that low-level features do not generalise well [27,28], this work has shown the opposite and confirms the powerfulness of MRFs in image modelling and feature extraction. Future work will entail applying this method to more comprehensive datasets, and building up an efficient unsupervised deep neural networks, in which feature are extracted using prefixed filter banks and only classifiers are trained in supervised mode.

References

1. Bengio, Y., Courville, A., Vincent, P.: Representation learning: a review and new perspectives. IEEE Trans. Pattern Anal. Mach. Intell. **35**(8), 1798–1828 (2013)
2. Hinton, G.E., Osindero, S., Teh, Y.W.: A fast learning algorithm for deep belief nets. Neural Comput. **18**(7), 1527–1554 (2006)
3. LeCun, Y., Bengio, Y., Hinton, G.E.: Deep learning. Nature **521**(7553), 436–444 (2015)
4. LeCun, Y., Kavukcuoglu, K., Farabet, C.: Convolutional networks and applications in vision. In: Proceedings of IEEE International Symposium on Circuits and Systems, pp. 253–256 (2010)
5. Julesz, B.: Visual pattern discrimination. IRE Trans. Inf. Theor. **8**(2), 84–92 (1962)
6. Julesz, B.: Textons, the elements of texture perception, and their interactions. Nature **290**(5802), 91–97 (1981)
7. Li, S.Z.: A Markov random field model for object matching under contextual constraints. In: Proceedings of IEEE Computer Society Conference on Computer Vision and Pattern Recognition, p. 866 (1994)
8. Li, S.Z.: Modeling image analysis problems using Markov random fields. Stochast. Processes Model. Simul. **20**(5), 1–43 (2000)
9. Geman, S., Geman, D.: Stochastic relaxation, gibbs distributions, and the Bayesian restoration of images. IEEE Trans. Pattern Anal. Mach. Intell. **6**(6), 721–741 (1984)
10. LeCun, Y., Bottou, L., Bengio, Y., Haffner, P.: Gradient-based learning applied to document recognition. Proc. IEEE **86**(11), 2278–2324 (1998)
11. Cohen, G., Afshar, S., Tapson, J., van Schaik, A.: Emnist: an extension of mnist to handwritten letters (2017). arXiv preprint: arXiv:1702.05373
12. Krizhevsky, A., Hinton, G.E.: Learning multiple layers of features from tiny images. Master Thesis, the University of Toronto (2009)
13. Cross, G.R., Jain, A.K.: Markov random field texture models. IEEE Trans. Pattern Anal. Mach. Intell. **5**(1), 25–39 (1983)
14. Ising, E.: Beitrag zur theorie des ferromagnetismus. Zeitschrift für Physik **31**(1), 253–258 (1925)
15. Yin, H., Allinson, N.M.: Unsupervised segmentation of textured images using a hierarchical neural structure. Electron. Lett. **30**(22), 1842–1843 (1994)
16. Nishii, R., Eguchi, S.: Image classification based on Markov random field models with jeffreys divergence. J. Multivar. Anal. **97**(9), 1997–2008 (2006)
17. Krizhevsky, A., Sutskever, I., Hinton, G.E.: Imagenet classification with deep convolutional neural networks. In: Advances in Neural Information Processing Systems, pp. 1097–1105 (2012)

18. Zeiler, M.D., Fergus, R.: Visualizing and understanding convolutional networks. In: Fleet, D., Pajdla, T., Schiele, B., Tuytelaars, T. (eds.) ECCV 2014. LNCS, vol. 8689, pp. 818–833. Springer, Cham (2014). doi:10.1007/978-3-319-10590-1_53

19. Simonyan, K., Zisserman, A.: Very deep convolutional networks for large-scale image recognition (2014). arXiv preprint: arXiv:1409.1556

20. Szegedy, C., Liu, W., Jia, Y., Sermanet, P., Reed, S., Anguelov, D., Erhan, D., Vanhoucke, V., Rabinovich, A.: Going deeper with convolutions. In: Proceedings of the IEEE International Conference on Computer Vision and Pattern Recognition, pp. 1–9 (2015)

21. Cireşan, D., Meier, U., Schmidhuber, J.: Multi-column deep neural networks for image classification. In: Proceedings of the IEEE International Conference on Computer Vision and Pattern Recognition, pp. 3642–3649. IEEE (2012)

22. Jarrett, K., Kavukcuoglu, K., LeCun, Y., et al.: What is the best multi-stage architecture for object recognition? In: Proceedings of the IEEE International Conference on Computer Vision, pp. 2146–2153 (2009)

23. Lin, M., Chen, Q., Yan, S.: Network in network (2013) arXiv preprint: arXiv:1312.4400

24. Zeiler, M.D., Fergus, R.: Stochastic pooling for regularization of deep convolutional neural networks (2013). arXiv preprint: arXiv:1301.3557

25. Springenberg, J.T., Dosovitskiy, A., Brox, T., Riedmiller, M.: Striving for simplicity: the all convolutional net (2014). arXiv preprint: arXiv:1412.6806

26. Bruna, J., Mallat, S.: Invariant scattering convolution networks. IEEE Trans. Pattern Anal. Mach. Intell. **35**(8), 1872–1886 (2013)

27. Chan, T.H., Jia, K., Gao, S., Lu, J., Zeng, Z., Ma, Y.: PCANet: a simple deep learning baseline for image classification? IEEE Trans. Image Process. **24**(12), 5017–5032 (2015)

28. Ng, C.J., Teoh, A.B.J.: Dctnet: a simple learning-free approach for face recognition. In: Proceedings of Asia-Pacific Conference on Signal and Information Processing Association Annual Summit, pp. 761–768 (2015)

29. Besag, J.: Spatial interaction and the statistical analysis of lattice systems. J. R. Stat. Soc. Ser. B **36**, 192–236 (1974)

30. Hammersley, J.M., Clifford, P.E.: Markov random fields on finite graphs and lattices. Unpublished manuscript (1971)

31. Elliott, H., Derin, H., Cristi, R., Geman, D.: Application of the gibbs distribution to image segmentation. In: Proceedings of IEEE International Conference on Acoustics, Speech, and Signal Processing, vol. 9, pp. 678–681 (1984)

32. Derin, H., Elliott, H.: Modeling and segmentation of noisy and textured images using gibbs random fields. IEEE Trans. Pattern Anal. Mach. Intell. **9**(1), 39–55 (1987)

33. Kashyap, R., Chellappa, R.: Estimation and choice of neighbors in spatial-interaction models of images. IEEE Trans. Inf. Theor. **29**(1), 60–72 (1983)

34. Dass, S.C.: Markov random field models for directional field and singularity extraction in fingerprint images. IEEE Trans. Image Process. **13**(10), 1358–1367 (2004)

35. Goodfellow, I.J., Warde-Farley, D., Mirza, M., Courville, A., Bengio, Y.: Maxout networks (2013). arXiv preprint: arXiv:1302.4389

36. Wan, L., Zeiler, M., Zhang, S., Cun, Y.L., Fergus, R.: Regularization of neural networks using dropconnect. In: Proceedings of the International Conference on Machine Learning, pp. 1058–1066 (2013)

Using the Multivariate Normal to Improve Random Projections

Keegan Kang[(✉)]

Singapore University of Technology and Design, Singapore 487372, Singapore
keegan_kang@sutd.edu.sg

Abstract. Random projection is a dimension reduction technique which can be used to estimate Euclidean distances, inner products, angles [9], or even l_p distances (for even p) [10] between pairs of high dimensional vectors. We extend the work of Li [9] and our prior work [7] to show how marginal information, principal components, and control variates can be used with the multivariate normal distribution to improve the accuracy of the inner product estimate of vectors. We call our method *COntrol Variates For Estimation via First Eigenvectors (COVFEFE)*. We demonstrate the results of COVFEFE on the *Arcene* and *MNIST* datasets.

1 Introduction

Random projection is one of the techniques in estimating distances between high dimensional vectors. A random projection matrix is a linear map R which maps vectors in $\mathbb{R}^p$ to $\mathbb{R}^k$, where $k \ll p$. In this smaller dimensional space, distances are approximately preserved. For vectors $\mathbf{u}_1, \ldots, \mathbf{u}_n \in \mathbb{R}^p$, we compute $\mathbf{v}_i = R\mathbf{u}_i$. Under certain distance metrics dist, we have that $\text{dist}(\mathbf{v}_s, \mathbf{v}_t) \approx \text{dist}(\mathbf{u}_s, \mathbf{u}_t)$. Therefore, analysis of the data $\mathbf{v}_1, \ldots, \mathbf{v}_n$ in this lower dimensional space can be done as a proxy to the data $\mathbf{u}_1, \ldots, \mathbf{u}_n$ in the higher dimensional space.

In the ordinary random projection algorithm, a matrix $R_{p \times k}$ with entries r_{ij} consisting of i.i.d. Gaussian random variables with mean 0 and variance 1 is constructed. For any two vectors $\mathbf{u}_i, \mathbf{u}_j$, we have $\mathbf{v}_i = R^T\mathbf{u}_i, \mathbf{v}_j = R^T\mathbf{u}_j$. Consider the sth element v_{is}, v_{js} in both $\mathbf{v}_i$ and $\mathbf{v}_j$. It can be seen that $v_{is}v_{js} = \left(\sum_{t=1}^{p} r_{ts}u_{it}\right)\left(\sum_{t=1}^{p} r_{ts}u_{jt}\right)$. The expected value of these elements yield $\mathbb{E}[v_{is}v_{js}] = \sum_{t=1}^{p} u_{it}u_{jt} = \langle \mathbf{u}_i, \mathbf{u}_j \rangle$, with variance $\text{Var}[v_{is}v_{js}] = 1 + a^2$, where a denotes the true inner product.

By the Law of Large Numbers, taking the mean of k such terms gives us the inner product, and it is up to the user to choose this value. In fact, due to the Johnson-Lindenstrauss Lemma [5], the user can choose some relative error ε and probability p, and pick k based on these values. The pairwise estimates $\hat{\theta}$ of distances in this lower dimensional space are guaranteed to lie within $(1 \pm \varepsilon)$ of the true estimate θ with probability p. This makes random projections an attractive dimension reduction method, as it costs only $O(npk + n^2k)$ to compute all pairwise estimates.

H. Yin et al. (Eds.): IDEAL 2017, LNCS 10585, pp. 397–405, 2017.
https://doi.org/10.1007/978-3-319-68935-7_43

While there are many pre-processing methods [4, 11, 15] to reduce the dimension of data, the fact that we can keep the same relative error independent of p makes random projections an attractive alternative.

Statistical theory can be used to further improve these estimates of inner products. Several instances on how such theory can be used include taking advantage of marginal information [9], computing maximum likelihood estimators [9], or variance reduction techniques in Monte Carlo methods [7]. In this paper, we explain how we use control variates with the multivariate normal to improve the inner product estimate of vectors which are close to being uncorrelated.

We assume that all vectors $\mathbf{u}_i$ in this paper are normalized to have norm of 1, for ease of notation and convenience.

2   Related Work

Suppose the random projection matrix R consists of i.i.d. entries which are Gaussian with zero mean and variance 1. Suppose we compute $\mathbf{v}_i = R\mathbf{u}_i$, for $i \in \{1, 2\}$, where $\mathbf{v}_i = (v_{i1}, v_{i2}, \ldots, v_{ik})$. The pair (v_{is}, v_{js}) can be seen as being drawn from a bivariate normal where $\begin{pmatrix} v_{1.} \\ v_{2.} \end{pmatrix} \sim N \left(\begin{pmatrix} 0 \\ 0 \end{pmatrix}, \Sigma \right)$ with $\Sigma = \begin{pmatrix} 1 & a \\ a & 1 \end{pmatrix}$ as we assume that the vectors $\mathbf{u}_i$ are normalized. Here, a is denoted to be the true inner product $\langle \mathbf{u}_1, \mathbf{u}_2 \rangle$. To avoid confusing notation in later sections, we denote the pair $\mathbf{w}_s := (v_{is}, v_{js})$.

Li [9] was one of the first to take advantage of knowing the margins (norms $\|\mathbf{u}_i\|$). By viewing the k tuples $(v_{1.}, v_{j.})$ as an independent draw from the bivariate normal, with the log likelihood given by $l(a) = -\frac{k}{2}\log(1 - a^2) - \frac{k}{2}\frac{1}{1-a^2}\sum_{s=1}^{k}\left(v_{1s}^2 - 2v_{1s}v_{2s}a + v_{2s}^2\right)$.

By differentiating the expression $l(a)$ and setting it to zero, the optimal value of $\hat{a}$ is the solution to $a^3 - a^2(\mathbf{v}_1^T\mathbf{v}_2) + a(\|\mathbf{v}_2\|_2^2 + \|\mathbf{v}_1\|_2^2 - 1) - \mathbf{v}_1^T\mathbf{v}_2 = 0$ which can be found via numerical methods. The variance of this inner product estimate becomes $\text{Var}[\hat{a}] = \frac{(1-a^2)^2}{1+a^2}$ which is lower than the variance of the ordinary inner product estimate.

The method of control variates [14] uses the same random numbers to estimate a random variable X, as well as to estimate a different variable Y of which the true mean μ_Y is known. As the errors in estimation will be correlated, knowing how far off our empirical observation of Y is from the true mean of Y will allow us to correct our observation for the random variable X. Mathematically, we can write $\mathbb{E}[X + c(Y - \mu_Y)] = \mathbb{E}[X] + c(\mathbb{E}[Y] - \mu_Y) = \mathbb{E}[X]$. Y is called the *control variate*, and c is called the *control variate correction*. The goal is to find c such that $\text{Var}[X + c(Y - \mu_Y)] = \text{Var}[X] + c^2\text{Var}[Y] + 2c\text{Cov}[X, Y]$ is minimized. From calculus, the optimal value of c is $-\frac{\text{Cov}[X,Y]}{\text{Cov}[Y]}$ and thus we have

$$\text{Var}[X + \hat{c}(Y - \mu_Y)] = \text{Var}[X] - \frac{\text{Cov}[X, Y]^2}{\text{Var}[Y]} \tag{1}$$

The amount of variance reduction depends on the correlation between X and Y. Furthermore, a variance reduction is always (theoretically) guaranteed.

Our prior work [7] used the control variate $Y = \|\mathbf{v}_1\|_2^2 + \|\mathbf{v}_2\|_2^2$ to estimate the inner product. In expectation, $\mathbb{E}[Y] = \|\mathbf{u}_1\|_2^2 + \|\mathbf{u}_2\|_2^2 = 2$ which is known. If the entries in R were Gaussian, then $\text{Cov}[X, Y]$ and $\text{Var}[X, Y]$ can be substituted from normal theory. Otherwise, the empirical covariance and variance needs to be computed from the data.

The variance of our inner product estimate θ under control variates is given by $\text{Var}[\theta] = \frac{(1-a^2)^2}{1+a^2}$.

Our method and Li's MLE [9] achieved the same (improved) variance for the inner product. We show the improved theoretical variance reduction in Fig. 1. Both methods achieve minimal variance reduction when the vector pairs are almost orthogonal to each other. We will correct this in our proposed method.

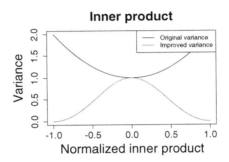

Fig. 1. Comparison of improved variance. O denotes the original variance. Recall that for normalized vectors, an inner product of 1 means the vectors are identical, an inner product of 0 means the vectors are orthogonal, and an inner product of -1 means the vectors are in opposite directions.

3   Our Contributions

We describe a method (COVFEFE) to estimate the inner product of vectors which are almost uncorrelated by looking at the multivariate normal. Our method keeps to the same order of time as ordinary random projections after the pre-processing period. We demonstrate our method on actual data.

3.1   The Multivariate Normal

We describe our method in 3 dimensions, before generalizing this to n dimensions. Suppose for the vectors $\mathbf{u}_1, \mathbf{u}_2$, we generate the known vector $\mathbf{e}_1$. We compute the inner product $a_{13} := \langle \mathbf{u}_1, \mathbf{e}_1 \rangle$ and $a_{23} := \langle \mathbf{u}_2, \mathbf{e}_1 \rangle$. With the random matrix R, we compute $\mathbf{v}_1 = R^T \mathbf{u}_1$, $\mathbf{v}_2 = R^T \mathbf{u}_2$, $\mathbf{f}_1 = R^T \mathbf{e}_1$. Then the triple (v_{1s}, v_{2s}, f_{1s}) is drawn from the multivariate normal, given by

$$\begin{pmatrix} v_{1.} \\ v_{2.} \\ f_{1.} \end{pmatrix} \sim N \left(\begin{pmatrix} 0 \\ 0 \\ 0 \end{pmatrix}, \Sigma \right) \qquad \text{where} \qquad \Sigma = \begin{pmatrix} 1 & a & a_{13} \\ a & 1 & a_{23} \\ a_{13} & a_{23} & 1 \end{pmatrix} \qquad (2)$$

All entries in Σ are known, except for a. We denote the triple $\mathbf{w}_. = (v_{1.}, v_{2.}, f_{1.})$.

We can extend this by considering vectors $\mathbf{e}_2, \mathbf{e}_3, \ldots, \mathbf{e}_s$ and computing all $\langle \mathbf{u}_i, \mathbf{e}_j \rangle$, $\langle \mathbf{e}_i, \mathbf{e}_j \rangle$ and looking at the multivariate normal of dimension $s + 2$.

3.2   A Control Variate Basis

In the bivariate normal case, the estimate for the inner product can be viewed as the following quadratic form $v_{1.} v_{2.} = \mathbf{w}^T A \mathbf{w}$ where $A = \begin{pmatrix} 0 & \frac{1}{2} \\ \frac{1}{2} & 0 \end{pmatrix}$. Any control variate with matrix C can be represented as $C := \begin{pmatrix} \alpha_{11} & 0 \\ 0 & \alpha_{22} \end{pmatrix} = \alpha_{11} \begin{pmatrix} 1 & 0 \\ 0 & 0 \end{pmatrix} +$ $\alpha_{22} \begin{pmatrix} 0 & 0 \\ 0 & 1 \end{pmatrix}$ where the basis vectors comprise the information which we know. In the bivariate normal case, we know that the norms $\|\mathbf{u}_1\|_2^2, \|\mathbf{u}_2\|_2^2$ are 1, hence we have $\alpha_{11} = \alpha_{22} = 1$.

For the multivariate normal in $s + 2$ dimensions, we can set $A_{s+2,s+2}$ to be the zero matrix, and have entries $a_{12} = a_{21} = \frac{1}{2}$. We write $C := \sum_i \alpha_i C_i$, where C_i are the basis vectors of C.

Theorem 1. *For the multivariate normal (described above) in $s+2$ dimensions, the number of basis vectors for the matrix C is $\frac{s^2+5s+4}{2}$.*

Proof. The number of basis vectors comprise the information we know. For the multivariate normal in $s + 2$ dimensions, we know $s + 2$ entries (diagonal of 1s), so we have $s + 2$ basis vectors. We also know all pairwise inner products except $\langle \mathbf{u}_1, \mathbf{u}_2 \rangle$, which gives us $\frac{(s+2)^2 - (s+2)}{2} - 1$ basis vectors. Summing them up gives $\frac{s^2+5s+4}{2}$.

3.3   The Proposed Method

We give the optimal coefficients α_i for the multivariate normal in 3 and 4 dimensions which give the most variance reduction. While Eq. 1 guarantees a variance reduction for *any* control variate, we want to pick the best control variate with most variance reduction. Therefore, we want to pick basis vectors C_i where the coefficients α_i are represented by the terms of Σ *except* for a (as this is what we want to estimate). From [13] we can get the following lemma.

Lemma 1. *Let $\mathbf{v} \sim N(\mathbf{0}, \Sigma)$, and A, B be symmetric matrices. We have*

$$Var[\mathbf{w}^T A \mathbf{w}] = 2\,Tr[A\Sigma A\Sigma] \qquad\qquad Cov[\mathbf{w}^T A \mathbf{w}, \mathbf{w}^T B \mathbf{w}] = 2\,Tr[A\Sigma B\Sigma] \quad (3)$$

Proof. The proof can be found in Muirhead [12].

We need to find control variates $\mathbf{w}^T C \mathbf{w} = \mathbf{w}^T \left(\sum_i \alpha_i C_i \right) \mathbf{w}$ which maximizes the expression $\frac{\mathrm{Cov}[\mathbf{w}^T B \mathbf{w}, \mathbf{w}^T C \mathbf{w}]^2}{\mathrm{Var}[\mathbf{w}^T C \mathbf{w}]}$ in Eq. 1. This is equivalent to finding α_i which optimizes the term $\frac{2 \mathrm{Tr}[A \Sigma C \Sigma]^2}{\mathrm{Tr}[C \Sigma C \Sigma]}$. We must also ensure that the optimal α_is are not in terms of a since we do not know this value. We take note of the cross terms in matrix multiplication, and set these α_is to zero before optimizing. The α_is set to zero would correspond to the basis matrices which have 1s in either of the first two rows or first two columns.

Theorem 2. *In three dimensions, the only non-zero coefficient for the basis matrix C is α_{33}. The value of α_{33} does not affect the optimality of the control variate so we can set $\hat{c} = -1$ corresponding to $\alpha_{33} = a_{13} a_{23}$.*

Proof. We apply Lemma 1 to $\frac{\mathrm{Cov}[\mathbf{w}^T A \mathbf{w}, \mathbf{w}^T C \mathbf{w}]}{\mathrm{Var}[\mathbf{w}^T C \mathbf{w}]}$ and get the optimal value of $\hat{c}$ to be $\frac{a_{13} a_{23}}{\alpha_{33}}$. We set $\alpha_{33} = a_{13} a_{23}$.

Theorem 3. *In four dimensions, the only non-zero coefficients for the basis matrix C is given by*

$$\alpha_{33} = -2(a_{13} - a_{14} a_{34})(-a_{23} + a_{24} a_{34}) \tag{4}$$

$$\alpha_{44} = -2(a_{14} - a_{13} a_{34})(-a_{24} + a_{23} a_{34}) \tag{5}$$

$$\alpha_{34} = a_{14} a_{23} + a_{13} a_{24} - 2 a_{13} a_{23} a_{34} - 2 a_{14} a_{24} a_{34} + a_{14} a_{23} a_{34}^2 + a_{13} a_{24} a_{34}^2 \tag{6}$$

with $\hat{c} = \frac{1}{2(a_{34}^2 - 1)^2}$. If $a_{34} = \pm 1$, this implies that $\mathbf{e}_1 = \pm \mathbf{e}_2$, and thus we fall back to the multivariate normal in 3 dimensions.

Proof. We apply Lemma 1 to $\frac{\mathrm{Cov}[\mathbf{w}^T B \mathbf{w}, \mathbf{w}^T C \mathbf{w}]}{\mathrm{Var}[\mathbf{w}^T C \mathbf{w}]}$ to get us a term in $\alpha_1, \alpha_2, \alpha_3$. We optimize over $\alpha_1, \alpha_2, \alpha_3$ (proof omitted), and substitute the optimal values to find $\hat{c}$, $\frac{\mathrm{Cov}[\mathbf{w}^T B \mathbf{w}, \mathbf{w}^T C \mathbf{w}]^2}{\mathrm{Var}[\mathbf{w}^T C \mathbf{w}]}$.

In fact, we can repeat the process to find optimal coefficients α_i for the basis vectors of the control variate in $s + 2$ dimensions.

3.4   Analysis of Algorithm and Computational Cost

Figure 2 shows how our method performs on vector pairs with varying a_{ij}. The variance reduction we achieve on uncorrelated pairs of vectors is substantially greater than Li's MLE as well as our prior method. As vectors get more and more correlated, our original control variate with the bivariate normal outperforms our algorithm. This is not so much of a problem in practice since the probability that two vectors are uncorrelated increases as the dimension p increases.

One concern of this algorithm from the plots is that we may not always get a variance reduction in the estimate of the true value of the inner product $\langle \mathbf{u}_i, \mathbf{u}_j \rangle$ when the known vectors $\mathbf{e}_j$ picked are orthogonal to one of the $\mathbf{u}_i$s.

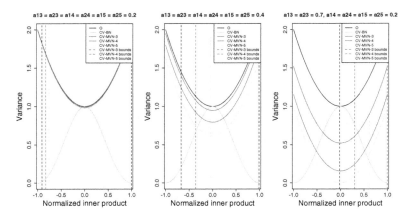

Fig. 2. Comparison of multivariate normal control variate. O denotes the original variance, BN stands for control variate in bivariate normal, and MVN-3, MVN-4, MVN-5 stands for the multivariate normal in the 3, 4, and 5 dimensions. The closer to zero, the better the variance reduction. The bounds (dotted lines) are the bounds for the inner product given a_{ij} values by triangle inequality. MVN-5 is included to show the trend even though its analysis is omitted.

However, we should make use of the data to generate our vectors $\mathbf{e}_j$, and we can pick $\mathbf{e}_j$s to be the first few principal components (PC) of our data. This actually allows us to easily use the multivariate normal in higher dimensions as we can set $a_{ij} = 0$ for $i, j > 2$ and get nice cancellation and easy representation for α_is. This motivates the naming of our algorithm COntrol Variates For Estimation via Few Eigenvectors (COVFEFE).

The analysis of the multivariate normal in higher dimensions is omitted in this paper to save space. Computing the top few eigenvectors is done via the Lanczos algorithm. If the data arrives in streams, methods such as [1,2,6] can be used to estimate the eigenvectors. Knowing the actual value of the inner products a_{ij} also helps to bound the true value of the inner product, given by the dotted lines in Fig. 2. We can use the triangle inequality to place a bound on the true inner product. If an estimate for the inner product is outside the bound, we can set the estimate to be the boundary.

We next look at the computational cost of our algorithm. We can choose to generate random $\mathbf{e}_1, \ldots, \mathbf{e}_s$, choose to generate orthogonal $\mathbf{e}_1, \ldots, \mathbf{e}_s$, or choose to generate the first s principal components. Each take an increasing amount of time. We may be able to ignore this cost if we plan to store the matrix V in memory and do computations on V for future use. Computing the extra inner products take $O(np)$ of time as $s < k$. $O(np)$ may sound large, but other pre-processing tasks like centering, scaling, or normalizing data will also take $O(np)$ time. Storing these extra inner products takes $O(n)$ extra space, which is negligible. The control variate only needs to be computed once in $O(k)$ time since it is the same for each pair. The theoretical value of $\hat{c}$ takes only $O(1)$ to compute per pair. Thus our algorithm takes the same order of time $O(npk+n^2k)$ as ordinary random projections after the pre-processing period.

4 Numerical Experiments

We verify that our theory works on the *Arcene* dataset [3] ($n = 900$, $p = 10000$, pairs = 404,550) and the *MNIST* test images [8] ($n = 10000$, $p = 784$, pairs = 49,995,000) dataset. We compute the average root mean square error (RMSE) of all pairwise inner product estimates in the above two datasets. All vectors were normalized to have a norm of 1. We look at the inner product estimates given by the random projection matrix R, $r_{ij} \sim N(0,1)$, with columns $k = \{10, 20, \dots, 100\}$.

We compute the average RMSE with eight different methods. For baseline comparisons, we look at the ordinary random projection method (Ord), Li's MLE [9] (Li), and our prior bivariate normal control variate [7] (CV-BN). We then compare this with the control variate using n extra eigenvectors, for $n = 4, 8, 12, 16, 20$. We denote the estimates by CV-MVN:4, CV-MVN:8, CV-MVN:12, CV-MVN:16, CV-MVN:20. We run our experiments for 1000 simulations on the *Arcene* dataset, and 100 simulations on the *MNIST* test images dataset (Fig. 3).

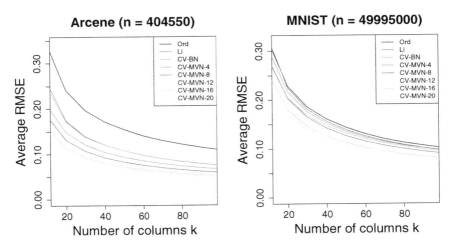

Fig. 3. Comparison of average RMSE on the *Arcene* dataset and the *MNIST* test images dataset.

We see that as we add more and more extra eigenvectors $\mathbf{e}_s$, the average RMSE of all pairwise estimates for the inner products decreases. In conclusion, our empirical results tend to verify our theory that using control variates with the multivariate normal improves our overall inner product estimates as more information is added.

5   Conclusion and Future Work

In this paper, we picked eigenvectors to represent known vectors $\mathbf{e}_i$. However, other optimal $\mathbf{e}_i$ may exist which give us greater variance reduction. More analysis and experiments can be done on how to choose optimal $\mathbf{e}_i$ to get optimal $\hat{c}$. It would also be worthwhile to look at how many $\mathbf{e}_i$ should be chosen before we get diminishing returns on the improvement of our inner product estimate.

There may also be a link between Li's MLE and our prior bivariate control variate estimator. The control variate estimate for the inner product $\langle \mathbf{x}_i, \mathbf{x}_j \rangle$ is written as

$$a \approx \mathbb{E}\left[\left(\mathbf{v}_i^T \mathbf{v}_j + \frac{a}{1+a^2} \left(\|\mathbf{v}_1\|_2^2 + \|\mathbf{v}_1\|_2^2 - 2 \right) \right) \right] \tag{7}$$

and the value of $c = \frac{a}{1+a^2}$ is computed empirically from the data since a is not known. But if the expectation is dropped and (7) treated as an equality, then we get an equation similar (but not identical) to Li's cubic equation. It would be exciting if we can get some relation between these two equations. If this is the case, we could consider estimators where α_i is in terms of a. Some estimators of this form outperform the control variate estimator with the bivariate normal for all values of inner product a.

Overall, much work can be done with control variates to further improve estimates of the inner product.

Acknowledgements. We thank the reviewers who provided us with much helpful comments. This research was supported by the SUTD Faculty Fellow Award.

References

1. Anaraki, F.P., Hughes, S.: Memory and computation efficient PCA via very sparse random projections. In: Proceedings of the 31st International Conference on Machine Learning (2014)
2. Fowler, J.: Compressive-projection principal component analysis. IEEE Trans. Image Process. **18**(10), 2230–2242 (2009)
3. Guyon, I., Gunn, S., Ben-Hur, A., Dror, G.: Result analysis of the nips 2003 feature selection challenge. In: Saul, L.K., Weiss, Y., Bottou, L. (eds.) Advances in Neural Information Processing Systems, vol. 17, pp. 545–552. MIT Press (2005). http://papers.nips.cc/paper/2728-result-analysis-of-the-nips-2003-feature-selection-challenge.pdf
4. Honda, K., Nonoguchi, R., Notsu, A., Ichihashi, H.: PCA-guided k-means clustering with incomplete data. In: 2011 IEEE International Conference on Fuzzy Systems (FUZZ), pp. 1710–1714. IEEE (2011)
5. Johnson, W.B., Lindenstrauss, J.: Extensions of Lipschitz mappings into a Hilbert space. Contemp. Math. **26**(189–206), 1 (1984)
6. Kang, K., Hooker, G.: Improving the recovery of principal components with semi-deterministic random projections. In: 2016 Annual Conference on Information Science and Systems, CISS 2016, Princeton, NJ, USA, 16–18 March 2016, pp. 596–601 (2016). https://doi.org/10.1109/CISS.2016.7460570

7. Kang, K., Hooker, G.: Random projections with control variates. In: Proceedings of the 6th International Conference on Pattern Recognition Applications and Methods, vol. 1, ICPRAM, pp. 138–147. INSTICC, ScitePress (2017)
8. Lecun, Y., Bottou, L., Bengio, Y., Haffner, P.: Gradient-based learning applied to document recognition. In: Proceedings of the IEEE, pp. 2278–2324 (1998)
9. Li, P., Hastie, T.J., Church, K.W.: Improving random projections using marginal information. In: Lugosi, G., Simon, H.U. (eds.) COLT 2006. LNCS (LNAI), vol. 4005, pp. 635–649. Springer, Heidelberg (2006). doi:10.1007/11776420_46
10. Li, P., Mahoney, M.W., She, Y.: Approximating higher-order distances using random projections. CoRR abs/1203.3492 (2012). http://arxiv.org/abs/1203.3492
11. Loia, V., Tomasiello, S., Vaccaro, A.: Using fuzzy transform in multi-agent based monitoring of smart grids. Inf. Sci. **388**, 209–224 (2017)
12. Muirhead, R.J.: Aspects of Multivariate Statistical Theory. Wiley-Interscience, Hoboken (2005)
13. Petersen, K.B., Pedersen, M.S.: The matrix cookbook. http://www2.imm.dtu.dk/pubdb/p.php?3274, version 20121115
14. Ross, S.M.: Simulation, 4th edn. Academic Press Inc., Orlando (2006)
15. Xu, Q., Ding, C., Liu, J., Luo, B.: PCA-guided search for k-means. Pattern Recogn. Lett. **54**, 50–55 (2015)

Automatic Motion Segmentation via a Cumulative Kernel Representation and Spectral Clustering

O.R. Oña-Rocha[1,2]([⊠]), O.T. Sánchez-Manosalvas[2], A.C. Umaquinga-Criollo[1], P.D. Rosero-Montalvo[1,4], L.E. Suárez-Zambrano[1], J.L. Rodríguez-Sotelo[5], and D.H. Peluffo-Ordóñez[1,3,6]

[1] Universidad Técnica del Norte, Ibarra, Ecuador
`oronia@utn.edu.ec`
[2] Universidad de las Fuerzas Armadas - ESPE, Sangolquí, Ecuador
[3] Corporación Universitaria Autónoma de Nariño, Pasto, Colombia
[4] Instituto Tecnológico Superior 17 de Julio, Yachay, Ecuador
[5] Universidad Autónoma de Manizales, Manizales, Colombia
[6] Universidad de Nariño, Pasto, Colombia

Abstract. Dynamic or time-varying data analysis is of great interest in emerging and challenging research on automation and machine learning topics. In particular, motion segmentation is a key stage in the design of dynamic data analysis systems. Despite several studies have addressed this issue, there still does not exist a final solution highly compatible with subsequent clustering/classification tasks. In this work, we propose a motion segmentation compatible with kernel spectral clustering (KSC), here termed KSC-MS, which is based on multiple kernel learning and variable ranking approaches. Proposed KSC-MS is able to automatically segment movements within a dynamic framework while providing robustness to noisy environments.

Keywords: Kernel spectral clustering · Motion segmentation · Time-varying data · Variable ranking

1 Introduction

The analysis of dynamic or time-varying data has increasingly taken an important place in scientific research, mainly in automation and pattern recognition, being video analysis [1] as well as motion identification for surveillance [2] and body movement classification [3] some of its remarkable applications. In this connection, kernel-based and spectral matrix analysis approaches have arisen as suitable alternatives. For instance, [4] proposes a continuos scheme of weighted kernel principal component analysis (WKPCA) able to capture the dynamic behavior of data. Another study [5] takes advantages of spectral clustering properties to explore the time-varying nature by adding the memory effect into a

H. Yin et al. (Eds.): IDEAL 2017, LNCS 10585, pp. 406–414, 2017.
https://doi.org/10.1007/978-3-319-68935-7_44

kernel spectral clustering (KSC) framework [6]. Also, there exists another alternative, known as multiple kernel learning (MKL), which has emerged to deal with different issues in machine learning, mainly, regarding support-vector-machines (SVM) [7] and KSC [8] formulations. The premise that underlies the use of MKL is that learning can be enhanced when using different kernels rather than a single one.

Followed from this premise, in this work, we present a KSC-based motion segmentation (KSC-MS) method, that is based on formulation for both MKL and variable ranking. Broadly speaking, KSC-MS works as follows: Firstly, MKL is used in such a manner that kernel matrices are computed from an input data sequence, in which each data matrix represents a frame at a different time instance. Secondly, weighting factors are obtained by ranking each sample contained in the frame in order to form a cumulative kernel matrix as a linear combination of the previously obtained kernels. Herein, to perform such a ranking procedure, we propose an improved alternative to previously introduced approaches used in weighted approaches for dimensionality reduction [8] and motion tracking [9]. Our approach consists of a variable ranking elegantly obtained from a spectral formulation, wherein the eigenvector are those calculated into the KSC process. The positivity of ranking values is clearly guaranteed given the square nature of the proposed formulation. The big advantage of our method is that there is no need for an extra calculation of eigenvectors as it can be formulated within a SVM approach. To show the ability of our approach to automatically segment movements within a dynamic framework, experiments on a moving-curve toy example are performed. As a result, it is demonstrated that our approach can determine the starting and ending of movements as well as its robustness to noisy environments.

The rest of this paper is as follows: Sect. 2 briefly outlines the so-called kernel spectral clustering. Next, in Sect. 3, we describe the proposed approach KSC-MS. Section 4 presents some experimental results. Finally, in Sect. 5, final concluding remarks are drawn.

2   Kernel Spectral Clustering

Let us consider $\boldsymbol{X} = [\boldsymbol{x}_1^\top, \ldots, \boldsymbol{x}_N^\top]^\top$, with $\boldsymbol{X} \in \mathbb{R}^{N \times d}$, as the input data matrix, which is to be divided into K disjoint subsets, being $\boldsymbol{x}_i \in \mathbb{R}^d$ the i-th d dimensional data point, N the number of data points, and K the number of desired groups. In this work, we employ the *Kernel Spectral Clustering* (KSC) [6] that is based on least squares- Support Vector Machines (LS-SVM) and can be seen as a weighted kernel principal component analysis, with primal-dual formulation wherein the following clustering model is assumed. Let us define $\boldsymbol{e}^{(l)} \in \mathbb{R}^N$ as the l-th projection vector as the latent variable model in the form:

$$\boldsymbol{e}^{(l)} = \boldsymbol{\Phi}\boldsymbol{w}^{(l)} + b_l \mathbf{1}_N, l \in \{1, \ldots, n_e\}, \tag{1}$$

where $\boldsymbol{w}^{(l)} \in \mathbb{R}^{d_h}$ is the l-th weighting vector, b_l is a bias term, n_e is the number of considered latent variables, notation $\mathbf{1}_N$ stands for a N dimensional all-ones vector, and the matrix $\boldsymbol{\Phi} = \left[\boldsymbol{\phi}(\boldsymbol{x}_1)^\top, \ldots, \boldsymbol{\phi}(\boldsymbol{x}_N)^\top\right]^\top$, $\boldsymbol{\Phi} \in \mathbb{R}^{N \times d_h}$, is a high dimensional representation of data. Then, vector $\boldsymbol{e}^{(l)}$ represents the latent variables from a set of n_e binary cluster indicators obtained by $\mathrm{sign}(\boldsymbol{e}^{(l)})$, which are further encoded to obtain the K resultant groups. Therefore, by aiming to cluster input data and considering a least-squares SVM formulation [10] for Eq. (1), we can pose, in matrix terms, the following optimization problem:

$$\max_{\boldsymbol{E}, \boldsymbol{W}, \boldsymbol{b}} \frac{1}{2N} \operatorname{tr}(\boldsymbol{E}^\top \boldsymbol{V} \boldsymbol{E} \boldsymbol{\Gamma}) - \frac{1}{2} \operatorname{tr}(\boldsymbol{W}^\top \boldsymbol{W}) \quad \text{s. t.} \quad \boldsymbol{E} = \boldsymbol{\Phi} \boldsymbol{W} + \boldsymbol{1}_N \otimes \boldsymbol{b}^\top \quad (2)$$

where $\boldsymbol{b} = [b_1, \ldots, b_{n_e}]$, $\boldsymbol{b} \in \mathbb{R}^{n_e}$, $\boldsymbol{\Gamma} = \mathrm{Diag}([\gamma_1, \ldots, \gamma_{n_e}])$, $\boldsymbol{W} = [\boldsymbol{w}^{(1)}, \cdots, \boldsymbol{w}^{(n_e)}]$, $\boldsymbol{W} \in \mathbb{R}^{d_h \times n_e}$, $\gamma_l \in \mathbb{R}^+$ is the l-th regularization parameter, $\boldsymbol{V} \in \mathbb{R}^{N \times N}$ is a weighting matrix for projections, and $\boldsymbol{E} = [\boldsymbol{e}^{(1)}, \cdots, \boldsymbol{e}^{(n_e)}]$, $\boldsymbol{E} \in \mathbb{R}^{N \times n_e}$. Notations $\operatorname{tr}(\cdot)$ and $\otimes$ stand for the trace and the Kronecker product, respectively. The Lagrangian of problem Eq. (2) can be expressed as follows:

$$\mathcal{L}(\boldsymbol{E}, \boldsymbol{W}, \boldsymbol{\Gamma}, \boldsymbol{A}) = \frac{1}{2N} \operatorname{tr}(\boldsymbol{E}^\top \boldsymbol{V} \boldsymbol{E}) - \frac{1}{2} \operatorname{tr}(\boldsymbol{W}^\top \boldsymbol{W}) - \operatorname{tr}(\boldsymbol{A}^\top (\boldsymbol{E} - \boldsymbol{\Phi} \boldsymbol{W} - \boldsymbol{1}_N \otimes \boldsymbol{b}^\top))$$

where matrix $\boldsymbol{A} \in \mathbb{R}^{N \times n_e}$ is formed by the Lagrange multiplier vectors, i.e., $\boldsymbol{A} = [\boldsymbol{\alpha}^{(1)}, \cdots, \boldsymbol{\alpha}^{(n_e)}]$, and $\boldsymbol{\alpha}^{(l)} \in \mathbb{R}^N$ represents the l-th vector of Lagrange multipliers. By solving the partial derivatives of Eq. (3), we yield the following Karush-Kuhn-Tucker (KKT) conditions:

$$\frac{\partial \mathcal{L}}{\partial \boldsymbol{E}} = 0 \quad \Rightarrow \boldsymbol{E} = N \boldsymbol{V}^{-1} \boldsymbol{A} \boldsymbol{\Gamma}^{-1}, \qquad \frac{\partial \mathcal{L}}{\partial \boldsymbol{W}} = 0 \quad \Rightarrow \boldsymbol{W} = \boldsymbol{\Phi}^\top \boldsymbol{A},$$

$$\frac{\partial \mathcal{L}}{\partial \boldsymbol{A}} = 0 \quad \Rightarrow \boldsymbol{E} = \boldsymbol{\Phi} \boldsymbol{W}, \text{and} \qquad \frac{\partial \mathcal{L}}{\partial \boldsymbol{b}} = 0 \quad \Rightarrow \boldsymbol{b}^\top \boldsymbol{1}_N = 0.$$

Now, by eliminating the primal variables from initial problem of Eq. (2), the following eigenvectors-based dual solution is obtained:

$$\boldsymbol{A} \boldsymbol{\Lambda} = \boldsymbol{A} \boldsymbol{V} (\boldsymbol{I}_N + (\boldsymbol{1}_N \otimes \boldsymbol{b}^\top)(\boldsymbol{\Omega} \boldsymbol{\Lambda})^{-1}) \boldsymbol{\Omega}, \quad (3)$$

where $\boldsymbol{\Lambda} = \mathrm{Diag}(\boldsymbol{\lambda})$, $\boldsymbol{\Lambda} \in \mathbb{R}^{N \times N}$, $\boldsymbol{\lambda} \in \mathbb{R}^N$ is the vector of eigenvalues with $\lambda_l = N/\gamma_l$, $\lambda_l \in \mathbb{R}^+$ and $\boldsymbol{\Omega} \in \mathbb{R}^{N \times N}$ is a given kernel matrix, such that $\boldsymbol{\Phi} \boldsymbol{\Phi}^\top = \boldsymbol{\Omega}$. In particular, we employ $\boldsymbol{V} = \boldsymbol{D}^{-1}$ and since $K - 1$ eigenvectors $\boldsymbol{A}$ are indicators of cluster assignment, value n_e is set to be $k - 1$ [11]. Since the condition $\boldsymbol{b}^\top \boldsymbol{1}_N = 0$ can be satisfied by centering vector $\boldsymbol{b}$, the bias term is chosen in the form $b_l = -1/(\boldsymbol{1}_N^\top \boldsymbol{V} \boldsymbol{1}_N) \boldsymbol{1}_N^\top \boldsymbol{V} \boldsymbol{\Omega} \boldsymbol{\alpha}^{(l)}$. Thus, the solution of problem of Eq. (2) is reformulated to the following eigenvector problem: $\boldsymbol{A} \boldsymbol{\Lambda} = \boldsymbol{V} \boldsymbol{H} \boldsymbol{\Omega} \boldsymbol{A}$, where matrix $\boldsymbol{H} \in \mathbb{R}^{N \times N}$ is the centering matrix that is defined as $\boldsymbol{H} = \boldsymbol{I}_N - 1/(\boldsymbol{1}_N^\top \boldsymbol{V} \boldsymbol{1}_N) \boldsymbol{1}_N \boldsymbol{1}_N^\top \boldsymbol{V}$, ($\boldsymbol{I}_N$ denotes a N-dimensional identity matrix) and $\boldsymbol{\Omega} = [\Omega_{ij}]$, $\boldsymbol{\Omega} \in \mathbb{R}^{N \times N}$, being $\Omega_{ij} = \mathcal{K}(\boldsymbol{x}_i, \boldsymbol{x}_j)$, $i, j \in 1, \ldots, N$. Notation $\mathcal{K}(\cdot, \cdot) : \mathbb{R}^d \times \mathbb{R}^d \to \mathbb{R}$ stands for the introduced kernel function. Consequently, we can calculate the set of projections as follows:

$$\boldsymbol{E} = \boldsymbol{\Omega} \boldsymbol{A} + \boldsymbol{1}_N \otimes \boldsymbol{b}^\top. \quad (4)$$

Finally, projections are encoded in order to determine the the cluster assignments. Since each cluster is represented by a single point in the $K-1$-dimensional eigenspace, such that those single points are always in different orthants, eigenvectors encoding can be done considering that two points are in the same cluster when they are in the same orthant in the corresponding eigenspace as explained in [11].

3   KSC-based Motion Segmentation

Proposed KSC-based motion segmentation (KSC-MS) approach works as follows: Similarly as the relevance analysis explained in [12] in which a functional regarding a non-negative matrix is introduced, we pose an optimization problem with the difference that our focus is obtaining the ranking values for samples instead of features, as follows: Consider a data matrix sequence $\{\boldsymbol{X}^{(1)}, \ldots, \boldsymbol{X}^{(N_f)}\}$, where N_f is the number of frames and $\boldsymbol{X}^{(t)} = [\boldsymbol{x}_1^{(t)\top}, \ldots, \boldsymbol{x}_N^{(t)\top}]^\top$ is the data matrix associated to time instance t in size $N \times d$. Also, consider the frame matrix $\boldsymbol{\mathcal{X}} \in \mathbb{R}^{N_f \times Nd}$ which is formed in such a way that each row is a frame by letting $\tilde{\boldsymbol{x}}_t \in \mathbb{R}^{Nd}$ be the vectorization of coordinates representing the t-th frame. In other words, $\boldsymbol{\mathcal{X}} = [\tilde{\boldsymbol{x}}_1^\top, \ldots, \tilde{\boldsymbol{x}}_{N_f}^\top]^\top$ and $\tilde{\boldsymbol{x}}_t = \text{vec}(\boldsymbol{X}^{(t)})$. The corresponding kernel matrix can be expressed as $\widetilde{\boldsymbol{\Omega}} \in \mathbb{R}^{N_f \times N_f}$ such that $\widetilde{\Omega}_{ij} = \mathcal{K}(\tilde{\boldsymbol{x}}_i, \tilde{\boldsymbol{x}}_j)$. Then, the high dimensional representation matrix $\widetilde{\boldsymbol{\Phi}} \in \mathbb{R}^{N_f \times d_h}$ is

$$\widetilde{\boldsymbol{\Phi}} = \left[\phi(\tilde{\boldsymbol{x}}_1)^\top, \ldots, \phi(\tilde{\boldsymbol{x}}_{N_f})^\top\right]^\top,$$

where $\phi(\cdot) : \mathbb{R}^{Nd} \rightarrow \mathbb{R}^{d_h}$. Assume a linear projection in the form $\boldsymbol{Z} = \widetilde{\boldsymbol{\Phi}}^\top \boldsymbol{U}$, where $\boldsymbol{U}$ is an orthonormal matrix in size $N_f \times N_f$. Also, a lower rank representation of $\boldsymbol{Z}$ is assumed in the form $\widehat{\boldsymbol{Z}} = \widetilde{\boldsymbol{\Phi}}^\top \widehat{\boldsymbol{U}}$, where $\widehat{\boldsymbol{U}}$ is in size $N_f \times c$ ($c < N_f$). Then, an energy maximization problem can be written as:

$$\max_{\widehat{\boldsymbol{U}}} \text{tr}(\widehat{\boldsymbol{U}}^\top \widetilde{\boldsymbol{\Omega}} \widehat{\boldsymbol{U}}) \quad \text{s. t.} \quad \widehat{\boldsymbol{U}}^\top \widehat{\boldsymbol{U}} = \boldsymbol{I}_c. \tag{5}$$

Indeed, by using the kernel trick, we have $\text{tr}(\widehat{\boldsymbol{Z}}^\top \widehat{\boldsymbol{Z}}) = \text{tr}(\widehat{\boldsymbol{U}}^\top \widetilde{\boldsymbol{\Phi}} \widetilde{\boldsymbol{\Phi}}^\top \widehat{\boldsymbol{U}}) = \text{tr}(\widehat{\boldsymbol{U}}^\top \widetilde{\boldsymbol{\Omega}} \widehat{\boldsymbol{U}})$.

Then, recalling the KSC dual problem explained in previous section, we can write the centering matrix for frame matrix $\boldsymbol{\mathcal{X}}$, so:

$$\widetilde{\boldsymbol{H}} = \boldsymbol{I}_{N_f} - \frac{1}{\boldsymbol{1}_{N_f}^\top \widetilde{\boldsymbol{V}} \boldsymbol{1}_{N_f}} \boldsymbol{1}_{N_f} \boldsymbol{1}_{N_f}^\top \widetilde{\boldsymbol{V}},$$

where $\widetilde{\boldsymbol{V}}$ is chosen as the degree matrix given by $\widetilde{\boldsymbol{D}} = \text{Diag}(\widetilde{\boldsymbol{\Omega}} \boldsymbol{1}_{N_f})$. Next, by normalizing regarding degree and centering both $\boldsymbol{Z}$ and $\widehat{\boldsymbol{Z}}$, which means to premultiply $\widetilde{\boldsymbol{\Phi}}$ by $\widetilde{\boldsymbol{L}} \widetilde{\boldsymbol{V}}^{-1/2}$, we can infer that $\text{tr}(\widehat{\boldsymbol{U}}^\top \widetilde{\boldsymbol{\Omega}} \widehat{\boldsymbol{U}}) = \sum_{t=1}^{c} \widetilde{\lambda}_t$, where $\widetilde{\boldsymbol{L}}$ comes from the Cholesky decomposition of $\widetilde{\boldsymbol{H}}$ such that $\widetilde{\boldsymbol{L}}^\top \widetilde{\boldsymbol{L}} = \widetilde{\boldsymbol{H}}$ and $\widetilde{\lambda}_l$ is the l-th eigenvalue obtained by KSC when applied over $\mathcal{X}$ with a determined number of clusters $\widetilde{K}$. Therefore, a feasible solution of the problem is $U = \widetilde{A}$, being $\widetilde{A} = [\tilde{\boldsymbol{\alpha}}^{(1)}, \ldots, \tilde{\boldsymbol{\alpha}}^{(c)}]$ the corresponding eigenvector matrix. Thus, c is the same number of considered support vectors $\tilde{n}_e$. Similarly as the MKL approach explained in [13], we introduce a tracking vector $\boldsymbol{\eta} \in \mathbb{R}^{N_f}$ as the solution of minimizing the dissimilarity term given by $\|\widetilde{\boldsymbol{\Phi}} - \widehat{\boldsymbol{\Phi}}\|_F^2$, subject to some orthogonality conditions, being $\widehat{\boldsymbol{\Phi}}$ a lower-rank representation of $\widetilde{\boldsymbol{\Phi}}$. Then, the ranking vector can be calculated by:

$$\boldsymbol{\eta} = \sum_{\ell=1}^{\tilde{n}_e} \tilde{\lambda}_\ell \tilde{\boldsymbol{\alpha}}^{(\ell)} \circ \tilde{\boldsymbol{\alpha}}^{(\ell)}, \tag{6}$$

where $\circ$ denotes Hadamard (element-wise) product. Accordingly, the ranking factor η_i is a single value representing an unique frame in a sequence. Notation $\tilde{a}$ means that variable a is related to $\tilde{\Omega}$. Since $\boldsymbol{\eta}$ comes from a linear combination of the squared eigenvectors being the coefficients the corresponding eigenvalues which are positive, the positivity of tracking vector is guaranteed. In addition, we can normalize the vector by multiplying it by $1/\max|\boldsymbol{\eta}|$ in order to keep the entries of $\boldsymbol{\eta}$ ranged into the interval $[0,1]$.

4   Experimental Results and Discussion

To illustrate the performance of the tracking vector, let us consider the following toy example of a moving-curve. At time instance t, the effect of a 2-D curve moving in an arc from down up is emulated by the XY coordinates:

$$\boldsymbol{x}^{(t)} = \begin{pmatrix} |\cos(2\pi\boldsymbol{\tau})|^\top \\ -|\cos(2\pi\boldsymbol{\tau})|^\top \end{pmatrix} \quad \text{and} \quad \boldsymbol{y}^{(t)} = \begin{pmatrix} |t\sin(2\pi\boldsymbol{\tau})|^\top \\ |t\sin(2\pi\boldsymbol{\tau})|^\top \end{pmatrix},$$

where each entry of vector $\boldsymbol{\tau}$ is $\tau_n = n/N$ with $n \in \{1, \ldots, N/2\}$, being N the number of samples per frame. Then, we can form the corresponding data matrix $\boldsymbol{X}^{(t)} \in \mathbb{R}^{N \times 2}$ as $\boldsymbol{X}^{(t)} = [\boldsymbol{x}^{(t)}, \boldsymbol{y}^{(t)}]$ as well as the frame sequence $\{\boldsymbol{X}^{(1)}, \ldots, \boldsymbol{X}^{(N_f)}\}$. Then the video effect until a certain frame T is done by keeping the previous frames to show the trace of path followed by the curve. Figure 1 depicts the arc moving effect, when considering $N = 100$, and $N_f = 10$.

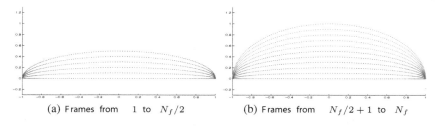

(a) Frames from   1 to $N_f/2$ (b) Frames from $N_f/2+1$ to N_f

Fig. 1. Views of toy 2-D moving-curve plotting divided into two level clusters.

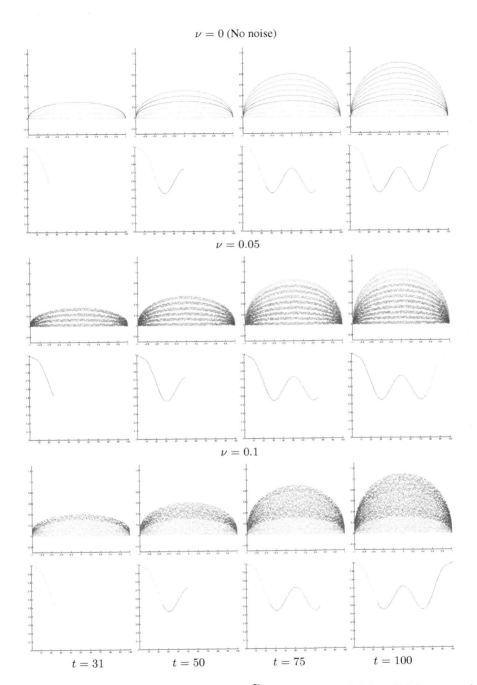

Fig. 2. Clustering of 2-D moving-curve into $\widetilde{K} = 4$ clusters with $N_f = 100$ frames and $N = 100$ samples per frame

The clustering is done by using KSC aimed to identify two natural movements or clusters ($\widetilde{K} = 2$). To carried out the clustering, KSC is used with a scaled Gaussian kernel $\mathcal{K}(\cdot, \cdot)$ by selecting the 7th ($m = 7$) nearest neighbor as scaling parameter. The whole frame matrix is used for training, then $\tilde{q}_{\text{train}}$ are the cluster assignment to color the frames according to the found groups, being $\tilde{q}_{\text{train}} = \text{KSC}(\boldsymbol{\mathcal{X}}, \mathcal{K}(\cdot, \cdot), \widetilde{K})$. To add noise to the moving-curve model, we consider an additive noise to be applied over the Y coordinate in the form $\nu\mathbf{n}$, where ν is the noise level and $\mathbf{n} \in \mathbb{R}^N$ is the introduced noise following a Gaussian distribution $\mathbf{n} \sim \mathcal{N}(0, 1)$, so: $\boldsymbol{y}_{\text{n}}^{(t)} = \boldsymbol{y} + \nu\mathbf{n}$. An example of the tracking performance is shown in Fig. 2 when setting the number of frames to be $N_f = 100$ and the number of samples per frame to be $N = 100$, representing an experiment changing the number of groups and noise conditions. It shows the behavior of tracking vector for clustering the frame matrix into 4 clusters ($\widetilde{K} = 4$) while Fig. 2. Tracking vector is calculated in both noisy and non-noisy environments by varying the noise level ν, by assuming that the frame matrix $\boldsymbol{\mathcal{X}}$ is to be split into $\widetilde{K}$ clusters. In both figures, for each considered level noise, clustered sequence at time instance t is shown on the top row as well as the behavior of the corresponding vector plotting on the bottom row. To assess the robustness of this approach, we assess the tracking vector over the noise levels $\nu \in \{0.05, 0.1\}$. No noise case is also considered ($\nu = 0$). For visualization, purposes some meaningful frames from the two experiments are selected. For the first experiments, selected frames are $t \in \{31, 71, 100\}$. Likewise, frames $t \in \{31, 50, 75, 100\}$ are selected for the second experiment. Also, in order to observe the behavior of tracking vector between clusters, frame matrices are clustered with KSC as the KSC output by considering the same conditions mentioned above. As it can be appreciated, the shape of the tracking vector has inflections on the transition from a cluster to the next one. Then, the vector plotting is directly related to the changes along the frame sequence. In addition, since there are no changes in the tracking vector when adding noise, it is possible to say that is less sensitive to noisy input data.

5   Final Remarks

When analyzing a sequence of frames represented by a single data matrix, aiming the identification of underlying dynamic events, kernel-based approaches represent a suitable alternative. Certainly, kernel functions come from an estimation of inner product of high-dimensional representation spaces where clusters are assumed to be separable, and are often defined as a similarity measure between data points from the original space. Such similarity is designed for a local data analysis. In other words, kernels allow a piecewise data exploring by means of an estimation of a generalized variance. Therefore, we can infer that the evolutionary behavior of the sequence can be tracked by some ranking values derived from a kernel-based formulation. Indeed, in this work, we demonstrate that a feasible tracking approach can be accomplished by maximizing an energy term regarding an approximated version of the high-dimensional representation space. We use a linear orthonormal model, being the aim of maximization problem the calculation of an optimal low rank projection or rotation matrix. Finally, taking advantage of the singular value decomposition of kernel matrix, we deduce a tracking vector as a linear combination of the squared eigenvectors. Tracking is done aimed to find an unique value representing adequately each single frame. The here proposed approach (KSC-MS) approach determines a vector that has a direct relationship with the underlying dynamic behavior of the analyzed sequence, allowing even to estimate the number of groups as well as the ground truth.

In further studies, the true meaning of the amplitude values of the tracking vectors and the dynamic behavior of data is to be formally revealed. As well, other kernel alternatives and properties are to be explored in order to reach low-computational-cost and efficient approaches for motion segmentation and tracking.

References

1. Lu, L., Zhang, X., Xu, X., Shang, D.: Video analysis using spatiotemporal descriptor and kernel extreme learning machine for lip reading. J. Electron. Imaging **24**(5), 053023–053023 (2015)
2. Yadav, S., Dubey, R., Ahmed, M.: An advanced motion detection algorithm with video quality analysis for video surveillance systems. Int. J. Adv. Res. Comput. Sci. **5**(8) (2014)
3. Saripalle, S.K., Paiva, G.C., Cliett, T.C., Derakhshani, R.R., King, G.W., Lovelace, C.T.: Classification of body movements based on posturographic data. Hum. Mov. Sci. **33**, 238–250 (2014)
4. Shanmao, G., Yunlong, L., Lijun, L., Ni, Z.: Weighted principal component analysis applied to continuous stirred tank reactor system with time-varying. In: 2015 34th Chinese Control Conference (CCC), pp. 6377–6381. IEEE (2015)
5. Langone, R., Alzate, C., Suykens, J.A.: Kernel spectral clustering with memory effect. Physica A: Statistical Mechanics and Its Applications (2013)
6. Alzate, C., Suykens, J.A.: Multiway spectral clustering with out-of-sample extensions through weighted kernel PCA. IEEE Trans. Pattern Anal. Mach. Intell. **32**(2), 335–347 (2010)
7. Bucak, S.S., Jin, R., Jain, A.K.: Multiple kernel learning for visual object recognition: a review. IEEE Trans. Pattern Anal. Mach. Intell. **36**(7), 1354–1369 (2014)
8. Peluffo, D., Garcia, S., Langone, R., Suykens, J., Castellanos, G.: Kernel spectral clustering for dynamic data using mulitple kernel learning. In: Proceedings of the International Joint Conference on Neural Networks, pp. 1085–1090 (2013)
9. Peluffo-Ordóñez, D., García-Vega, S., Castellanos-Domínguez, C.G.: Kernel spectral clustering for motion tracking: a first approach. In: Ferrández Vicente, J.M., Álvarez Sánchez, J.R., de la Paz López, F., Toledo Moreo, F.J. (eds.) IWINAC 2013. LNCS, vol. 7930, pp. 264–273. Springer, Heidelberg (2013). doi:10.1007/978-3-642-38637-4_27
10. Suykens, J.A.K., Van Gestel, T., De Brabanter, J., De Moor, B., Vandewalle, J.: Least Squares Support Vector Machines. World Scientific, Singapore (2002)
11. Alzate, C., Suykens, J.A.K.: Multiway spectral clustering with out-of-sample extensions through weighted kernel PCA. IEEE Trans. Pattern Anal. Mach. Intell. **32**(2), 335–347 (2010)

12. Peluffo, D., Lee, J., Verleysen, M., Rodríguez-Sotelo, J., Castellanos-Domínguez, G.: Unsupervised relevance analysis for feature extraction and selection: A distance-based approach for feature relevance. In: International Conference on Pattern Recognition, Applications and Methods-ICPRAM (2014)
13. Molina-Giraldo, S., Álvarez-Meza, A.M., Peluffo-Ordoñez, D.H., Castellanos-Domínguez, G.: Image segmentation based on multi-kernel learning and feature relevance analysis. In: Pavón, J., Duque-Méndez, N.D., Fuentes-Fernández, R. (eds.) IBERAMIA 2012. LNCS, vol. 7637, pp. 501–510. Springer, Heidelberg (2012). doi:10.1007/978-3-642-34654-5_51

Generation of Reducts and Threshold Functions and Its Networks for Classification

Naohiro Ishii[1]([⊠]), Ippei Torii[1], Kazunori Iwata[2], Kazuya Odagiri[3], and Toyoshiro Nakashima[3]

[1] Aichi Institute of Technology, Toyota, Japan
{ishii,mac}@aitech.ac.jp
[2] Aichi University, Nagoya, Japan
kazunori@vega.aichi-u.ac.jp
[3] Sugiyama Jyogakuen University, Nagoya, Japan
{odagiri,nakasima}@sugiyama-u.ac.jp

Abstract. Dimension reduction of data is an important issue in the data processing and it is needed for the analysis of higher dimensional data in the application domain. Rough set is fundamental and useful to reduce higher dimensional data to lower one for the classification. We develop generation of reducts by using partial data for the classification in which their operations derive reducts without using all the data. The nearest neighbor relation plays a fundamental role for generation of reducts and threshold functions using the Boolean reasoning on the discernibility and in discernibility matrices, in which the indiscernibility matrix is proposed here to test the sufficient condition for reduct and threshold function. Finally, reduct-threshold network is proposed for the higher classification accuracy.

Keywords: Reduct · Nearest neighbor relation · Discernibility matrix · Indiscernibility matrix · Generation of reduct and threshold function · Reduct-threshold networks

1 Introduction

Rough sets theory firstly introduced by Pawlak [1, 2] provides us a new approach to perform data analysis, practically. Up to now, rough set has been applied successfully and widely in machine learning and data mining. An important task in rough set based data analysis is computation of the attributes or feature reducts for the classification [1, 2]. By Pawlak's [1, 2] rough set theory, a reduct is a minimal subset of features, which has the discernibility power as using the entire features. Skowlon [3, 4] developed the reduct derivation by using the Boolean expression for the discernibility of data. But, generating reducts is a computationally complex task in which the computational complexity may grow non-polynomially with the number of attributes in data set [3]. So, a new concept for the efficient generation of reducts is expected. We develop generation of reducts from useful data for the classification in which their partial data computes reducts without using all the data in the conventional methods [1, 4]. Nearest neighbor relation with minimal distance between different classes proposed here has a basic

© Springer International Publishing AG 2017
H. Yin et al. (Eds.): IDEAL 2017, LNCS 10585, pp. 415–424, 2017.
https://doi.org/10.1007/978-3-319-68935-7_45

information for classification. For the classification of data, a nearest neighbor method [5–8] is simple and effective one. As a classification method, threshold function is well known [10]. Recent studies of threshold functions are of fundamental interest in circuit complexity, game theory and learning theory. We have developed further analysis for the generation of reducts and threshold functions by using the nearest neighbor relations and the Boolean reasoning on the discernibility and indiscernibility matrices. We propose here new generation method for reducts and threshold functions based on the nearest neighbor relation with minimal distance using discernibility and the indiscernibility matrices, in which the indiscernibility matrix tests sufficient conditions for them. Finally, reduct-threshold network is developed for the higher classification accuracy, in which a nonlinear mapping plays a fundamental role.

2   Boolean Reasoning of Reducts

Skowron proposed to represent a decision table in the form of the discernibility matrix [3, 4]. This representation has many advantages, in particular it enables simple computation of the core, reducts and other concepts [1–3].

Definition 1. The discernibility matrix $M(T)$ is defined as follows. Let $T = \{U, A, C, D\}$ be a decision table, with $U = \{x_1, x_2, \ldots x_n\}$, set of instances. A is a subset of C called condition, and D is a set of decision classes. By a discernibility matrix of T, denoted by $M(T)$, which is $n \times n$ matrix defined as

$$c_{ij} = \{a \in C : a(x_i) \neq a(x_j) \\ \wedge (d \in D, d(x_i) \neq d(x_j))\} \; i, j = 1, 2, \ldots n, \tag{1}$$

where U is the universe of discourse and C is a set of features or attributes.

The left side data in the column in Table 1 as shown in, $\{x_1, x_2, x_3, \ldots, x_7\}$ is a set of instances, while the data $\{a, b, c, d\}$ on the upper row, shows the set of attributes of the instance. In Table 2, the discernibility matrix of the decision table in Table 1 is shown.

Table 1. Decision table of data example (instances)

Attribute	a	b	c	d	class
x_1	1	0	2	1	+1
x_2	1	0	2	0	+1
x_3	2	2	0	0	−1
x_4	1	2	2	1	−1
x_5	2	1	0	1	−1
x_6	2	1	1	0	+1
x_7	2	1	2	1	−1

Table 2. Discernibility matrix of the decision table in Table 1

	x_1	x_2	x_3	x_4	x_5	x_6
x_2	-					
x_3	a,b,c,d	a,b,c				
x_4	b	b,d	-			
x_5	a,b,c	a,b,c,d	-	-		
x_6	-	-	b,c	a,b,c,d	c,d	
x_7	a,b	a,b,d	-	-	-	c,d

3 Generation of Reducts Based on Nearest Neighbor Relation and External Set

We can define a new concept, a nearest neighbor relation with minimal distance, δ. Instances with different classes are assumed to be measured in the metric distances for the nearest neighbor classification.

Definition 2. A nearest neighbor relation with minimal distance is a set of pair of instances and decision function d in Definition 1, which are described in

$$\{(x_i, x_j) : d(x_i) \neq d(x_j) \wedge |x_i - x_j| \leq \delta\}, \tag{2}$$

where $|x_i - x_j|$ shows the distance between x_i and x_j, and δ is the minimal distance. Then, x_i and x_j in the Eq. (2) are called to be in the nearest neighbor relation with minimal distance δ.

In Table 1, $(x_6, x_7), (x_5, x_6), (x_1, x_7)$ and (x_3, x_6) are elements of the relation with a distance $\sqrt{2}$. Thus, a nearest neighbor relation with minimal distance $\sqrt{2}$ becomes

$$\{(x_6, x_7), (x_5, x_6), (x_1, x_7), (x_3, x_6)\} \tag{3}$$

Here, we want to introduce the nearest neighbor relation on the discernibility matrix. Assume that the set of elements of the nearest neighbor relation are $\{nn_{ij}\}$.

Lemma 1. Respective Boolean term consisting of the set $\{nn_{ij}\}$ becomes a necessary condition to be reducts in the Boolean expression.

Lemma 2. Boolean product of respective terms corresponding to the set $\{nn_{ij}\}$ becomes a necessary condition to be reducts in the Boolean expression.

Lemma 3. Reducts in the Boolean expression are included in the Boolean term of Lemma 1 and the Boolean product in Lemma 2.

Theorem 4. If the distance δ is greater than the δ', i.e., $\delta > \delta'$ in the Eq. (3), the Boolean expression of the case of δ' includes that of δ.

Two sets of attributes(variables), set [A] and set [B] are defined to extract reducts from the nearest neighbor relation $\{nn_{ij}\}$.

Set [A]: Set of elements in the discernibility matrix includes those of any respective element in $\{nn_{ij}\}$ and those of elements absorbed by $\{nn_{ij}\}$.

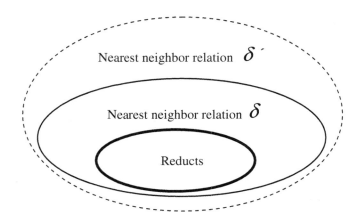

Fig. 1. Boolean condition of nearest neighbor relations and reducts

Set [B]: Set of elements in the discernibility matrix, which are not absorbed from those of any respective element in $\{nn_{ij}\}$. The set [B] is called here external set (Fig. 1).

Lemma 5. Within set [B], the element with fewer attributes(variables) plays a role of the absorption for the element with larger attributes(variables).

Theorem 6. Reducts are derived from the nearest neighbor relation by the absorption of variables in the external set [B] terms in the Boolean expression.

Variables of the set [A] in these relations are shown in shading in Table 2 of the discernible matrix. The Boolean product of these four terms becomes

$$(a+b) \cdot (b+c) \cdot (c+d) = b \cdot c + b \cdot d + a \cdot c, \tag{4}$$

which becomes a candidate of reducts. The third term in the Eq. (4) is absorbed by the product of variable $\{b\}$ of the external set [B] and the Eq. (4). The final reducts equation becomes

$$b \cdot c + b \cdot d \tag{5}$$

4   Generation of Reducts Based on Nearest Neighbor Relation and Indiscernibility Matrix

In this section, we propose another generation method of reducts, which is based on nearest neighbor relation and indiscernibility matrix proposed here. The external set [B] in the previous Sect. 3 is replaced to indiscernibility matrix.

Definition 3. Indiscernibility matrix is defined to be $IM(T)$, which is $n \times n$ matrix defined using notations in Definition 1 as

$$c_{ij} = \{a \in C : a(x_i) = a(x_j)$$
$$\wedge (d \in D, d(x_i) \neq d(x_j))\} \; i,j = 1, 2, \ldots n,$$
(6)

where U is the universe of discourse, C is a set of attributes.

The attributes pair $\{a, c\}$ is found between x_1 and x_4, also between x_2 and x_4. The minterm $a \cdot c$ cannot discriminate instances between x_1 and x_4, also between x_2 and x_4. Then, the minterm $a \cdot c$ is removed from the Eq. (4). Thus, the reducts

$$b \cdot c + b \cdot d$$
(7)

is obtained, which is the same equation derived in the Eq. (5) (Table 3).

Table 3. Indiscernibility matrix in Table 1

	x_1	x_2	x_3	x_4	x_5	x_6
x_2	—					
x_3	—	d				
x_4	a,c,d	a,c	—			
x_5	d	—	—	—		
x_6	—	—	a,d	—	a,b	
x_7	c,d	c	—	—	—	a,b

4.1   Relation Between External Set [B] and Indiscernibility Matrix

The variable b represents the external set [B]. The variable b multiplied by the Boolean form of the set [A] derived from the nearest neighbor relation becomes

$$b \cdot (a \cdot c + b \cdot c + b \cdot d) = b \cdot c + b \cdot d$$

The element variable (a, c, d) in indiscernibility table, which is placed in the same element in discernibility matrix, removes the Boolean implicant $a \cdot c$. Thus, the final Boolean reduct form becomes

$$b \cdot c + b \cdot d$$

Theorem 7. The Boolean absorption of the variables in the set [B] multiplied by the Boolean forms derived from the nearest neighbor relations generate reducts, which are also generated by directly removing Boolean variables in the indiscerniblity matrix as the sufficient condition from the Boolean forms of the nearest neighbor relations.

5 Generation of Threshold Functions Using Discernibility and Indiscernibility Matrices

The threshold function f is characterized by the hyperplane $WX - \theta$ with the weight vector $W(= (w_1, w_2, \ldots, w_n))$, threshold θ and input vector $X(= x_1, x_2, \ldots, x_n)$, where $x_i = 0$ or 1. Then, $f(X) = 1$ if $WX - \theta \geq 0$ and $f(X) = 0$ if $WX - \theta < 0$.

Definition 4. The nearest neighbor relation (X_i, X_j) on the threshold function is defined from the Eq. (2),

$$\{(X_i, X_j) : f(X_i) \neq f(X_j) \wedge |X_i - X_j| \leq \delta(= 1)\}, \tag{8}$$

where $\delta = 1$ shows one bit difference between X_i and X_j in the Hamming distance (also in the Euclidean distance).

Definition 5. The boundary vector X is defined to be the vector which satisfies

$$|WX - \theta| \leq |WY - \theta| \text{ for the } X(\neq Y) \tag{9}$$

Theorem 8. The boundary vector X becomes an element of nearest neighbor relation in the threshold function.

The 3-dimensional cube is shown in Fig. 2, in which the black circle belongs to +1 class, while the white circle belongs to −1 class. In Fig. 2, a true valued data (101) has nearest neighbor relations as $\{(101), (001)\}$ and $\{(101), (100)\}$ as shown in shaded cells in Table 4. In the discernibility matrix for making threshold function, it is necessary to perform AND operation of the terms in the column, while to perform OR operation among different columns as shown in Fig. 3. Thus, the Boolean reasoning in these relations becomes a Boolean product $x_1 \cdot x_3$ of the respective variable x_1 and x_3 in Table 4. Similarly, a Boolean product $x_1 \cdot x_2$ is generated in the 2nd column and x_1 is generated in the 3rd column in Table 4. These products becomes minterms.

But, the minterm x_1 is removed from the Boolean sum equation from the indiscernibility matrix of data in Fig. 3. Thus, the Boolean function obtained is

$$f = x_1 \cdot x_2 + x_1 \cdot x_3 \tag{10}$$

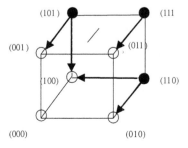

Fig. 2. Arrows show nearest neighbor relations in 3-dimensional cube

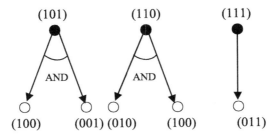

Fig. 3. Boolen operation for nearest neighbor relation for threshold function

Table 4. Discernibility matrix of nearest neighbor data in Fig. 3

		● (101)	●(110)	●(111)
○	(001)	x_1	...	...
○	(100)	x_3	x_2	...
○	(010)	...	x_1	...
○	(011)	...	...	x_1

The Boolean function f becomes a threshold function, since a hyperplane exists to satisfy the Eq. (10).

6   Reduct-Threshold Networks for Classification

Reduct-threshold networks are developed for classification. The graphical mapping of the nearest neighbor data is shown in Fig. 4 from the data in Table 1. Further, a nonlinear mapping is shown in Table 6 from the reduct data.

Table 5. Data of reduct {b, c}

	b	c	class
x_1	0	2	+1
x_2	0	2	+1
x_3	2	0	−1
x_4	2	2	−1
x_5	1	0	−1
x_6	1	1	+1
x_7	1	2	−1

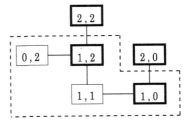

Fig. 4. Graphical mapping for the reduct {b,c}

6.1 Generation of Independent Vectors Based on Nearest Neighbor Relation

Table 6 shows a nonlinear mapping by extending variables of a reduct in Table 5. The nonlinear mapping of the reduct is realized as a reduct-threshold network in Fig. 5.

Table 6. Nonlinear mapping of reduct {b, c}

Var./NNR	b	c	bc	c^2	class
(0,2)	0	2	0	4	+1
(1,2)	1	2	2	4	−1
(1,1)	1	1	1	1	+1
(1,0)	1	0	0	0	−1

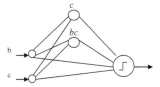

Fig. 5. Nonlinear mapping of reduct {b, c}

Theorem 9. A nonlinear mapping of the reduct is realized by the products of variables of the reduct and a threshold element.

6.2 Weighted Transformation of Nearest Neighbor Relation for Classification

Since the nearest neighbor relations must satisfy the following inequalities.

$$bw_1 + cw_2 + bcw_3 + c^2w_4 - \theta \geq 0 (\text{or} \leq 0), \tag{11}$$

where the value of the right side in each inequality, +1 or −1 indicate + class and −1 class, respectively. In case of the equality of (11), the weights $\{w_1, w_2, w_3, w_4\}$ of the hyperplane are derived as $\theta = 0$ and

$$\{w_1 = -1, \ w_2 = 9/2, \ w_3 = -1/2, \ w_4 = -2\} \tag{12}$$

Thus, the hyperplane is generated by the weight values of (12) and threshold θ. The variable $\{x_i\}$ of the hyperplane is replaced by $\{b, c, bc, c^2, \ldots\}$. The classification accuracy of the reducts, modified reducts and nonlinear mapping developed here is shown in Fig. 6. The nonlinear mapping shows higher classification accuracy [9].

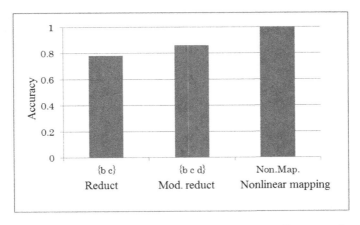

Fig. 6. Comparison of accuracy of classification by nonlinear mapping

6.3   Order Preserving of Data Relations by Nonlinear Mapping

A checking method is proposed here for the order preserving of data relation. An algorithm of order preserving of data relation for reduct {b, c} in Fig. 4 is shown in the following.

1. Nearest neighbor data X(x, y) is picked up.
2. Difference value between incremented data value of X(x + Δx, y + Δy) and data value X(x,y) is computed by the Eq. (12) with the values of Eq. (13).
3. From the difference value of the step 2, the order preserving of data relation is computed.

When the nearest neighbor data X = (1,2) is picked up, the incremental neighbor data X' is described as the $X' = (1 + \Delta b, 2 + \Delta c)$. Then, the difference between X and X' is computed by the Eq. (12) with the values in Eq. (13) in the following.

$$\text{Difference} = -2\Delta b(2 + 1/2(\Delta c)) - 2\Delta c(2 + \Delta c) \qquad (13)$$

Based on the nearest neighbor data X = (1,2), all the data $X' = (1 + \Delta b, 2 + \Delta c)$ is located in the minus value data than X = (1, 2). The difference Eq. (13) based on the nearest neighbor data is useful to check the data ordering relation by the nonlinear mapping.

7   Conclusion

Nearest neighbor relation developed here is the set of pair elements with minimal distance, which classify between different classes. Reduct is introduced as the minimal set for the data classification in the rough set theory. Threshold functions are of fundamental interest in the classification device. This paper develops the role of the nearest neighbor relations for the classification of data, which is proposed here. Then, generation of the reducts and threshold functions based on the nearest neighbor relations using the discernibility and the indiscernibility matrices is developed. The reduct-threshold networks are also developed here for the higher classification accuracy.

References

1. Sets, R.: Pawlak, Z. Int. J. Comput. Inf. Sci. **11**, 341–356 (1982)
2. Pawlak, Z.R., Slowinski, R.: Rough set approach to multi-attribute decision analysis. Eur. J. Oper. Res. **72**, 443–459 (1994)
3. Skowron, A., Rauszer, C.: The discernibility matrices and functions in information systems. In: Intelligent Decision Support - Handbook of Application and Advances of Rough Sets Theory, pp. 331–362. Kluwer Academic Publishers, Dordrecht (1992)
4. Skowron, A., Polkowski, L.: Decision algorithms, a survey of rough set theoretic methods. Fundamenta Informatica **30**(3–4), 345–358 (1997)
5. Cover, T.M., Hart, P.E.: Nearest neighbor pattern classification. IEEE Trans. Inf. Theory **13** (1), 21–27 (1967)
6. Ishii, N., Morioka, Y., Bao, Y., Tanaka, H.: Control of variables in reducts-kNN classification with confidence. In: König, A., Dengel, A., Hinkelmann, K., Kise, K., Howlett, R.J., Jain, L.C. (eds.) KES 2011. LNCS, vol. 6884, pp. 98–107. Springer, Heidelberg (2011). doi:10.1007/978-3-642-23866-6_11
7. Ishii, N., Torii, I., Bao, Y., Tanaka, H.: Modified reduct nearest neighbor classification. In: Proceedings of ACIS-ICIS, pp. 310–315. IEEE Computer Society (2012)
8. Ishii, N., Torii, I., Mukai, N., Iwata, K., Nakashima, T.: Classification on nonlinear mapping of reducts based on nearest neighbor relation. In: Proceedings of ACIS-ICIS. IEEE Computer Society, pp. 491–496 (2015)
9. Ishii, N., Torii, I., Iwata, K., Nakashima, T.: Generation and nonlinear mapping of reducts-nearest neighbor classification. In: Advances in Combining Intelligent Methods, Chap. 5, pp. 93–108. Springer (2017)
10. Hu, S.T.: Threshold Logic. University of California Press, Berkeley (1965)

Exploring Elitism in Genetic Algorithms for License Plate Recognition with Michigan-Style Classifiers

Dante Giovanni Sterpin Buitrago[1(✉)] and Fernando Martínez Santa[2]

[1] Corporación Unificada Nacional de Educación Superior, Bogotá, Colombia
dante_sterpin@cun.edu.co
[2] Universidad Distrital Francisco José de Caldas, Bogotá, Colombia
fmartinezs@udistrital.edu.co

Abstract. This document describes the application of Genetic Algorithms (GAs) in the recognition of the printed characters in Colombian vehicular license plates. First of all, the accuracy achieved by a genetic algorithm with simple elitism is contrasted with the accuracy of a population elitism-based genetic algorithm. Due to the notorious difficulty of using the standard technique of dedicating from, 70 to 80% of the available data to train the classifier, and the rest of data for its validation, here, two methods to generate the training data are described, as well as some other techniques to improve the classifier performance.

Keywords: License plate recognition · Optical character recognition · Genetic algorithm · Elitism · Michigan-style classifier · Collaboration

1 Introduction

In this paper, the importance of automatic License Plate Recognition (LPR) consists in the need of supporting access control systems, to get automatically registers of license plates, in relation with the video of the vehicular entrance to parking lots, because in case of some vehicular incident which involves a complaint, the burden of proof concerns to the surveillance company, according to the Colombian legislation.

Exploring the literature about LPR, it details that the standard methodology in this application consists of 3 fundamental processes: Detecting the area of interest where the plate is located, in photographs or video frames; the segmentation and normalization of the characters involved; and subsequently, the recognition of them [1].

In the implementation of those processes a lot of techniques are used. For the segmentation of the plate area in the images, for example, in [1] a genetic algorithm is used, in [2] fuzzy C-means is used, in [3] AdaBoost and Haar-like features are used, and in [4] rough set theory and neural networks are used. Regarding the Optical Character Recognition (OCR), it is quite common to find implementations based on neural networks [1,5,6] as well as other techniques such

© Springer International Publishing AG 2017
H. Yin et al. (Eds.): IDEAL 2017, LNCS 10585, pp. 425–435, 2017.
https://doi.org/10.1007/978-3-319-68935-7_46

as support vector machines [3], fuzzy logic-based classifiers [7], fuzzy C-means [8], Holland-style learning classifier systems [9], among other evolutionary algorithms [10–13], and some genetic algorithms specifically used to extract features [6,14,15]. Because of that, we have started a comparative study among several techniques, for the recognition of Colombian license plates, but this paper reports just an experimental exploration based on genetic algorithms, and Holland-style classifiers [16], known as Michigan-style classifiers more extensively, although it does not imply the use of learning classifier systems (LCS) [16], because this work just exploits the capability of GA to find good classifiers, thinking of a future work where it could be used for initializing an LCS more properly. Additionally, this paper demonstrates the effects of applying two kinds of elitism in that GA, because its potential to improve an evolutionary OCR system could slightly notice in [13], and here the effectiveness of one of them is explicitly emphasised.

Confronting the problem of recognizing the printed characters in the Colombian license plates, a data base with 1697 characters samples was used. Those characters were previously segmented and normalized in binary images of 32×16 pixels. In the Sect. 2 of this paper, the genetic classifiers design is described as well as its first results, obtained with an agent that used those classifiers to recognize the characters. Since the samples diversity, inside each class, complicates the optimization of such classifiers, in Sect. 3 the strategies used to reduce that diversity are detailed, along with other techniques, with which, according to the analyzed results in the Sect. 4, two meaningful goals were achieved: Increasing the recognition accuracy, and, to decrease both of involved times, the training one and the recognition one.

2   Genetic Learning of the Characters

This section describes the genetic algorithms used for learning the 36 printed characters in Colombian vehicular license plates. For that, two evolutionary approaches were used: a simple genetic algorithm with simple elitism [17], and a genetic algorithm with population elitism [18]. The first one was named: simply elitist Genetic Algorithm (seGA), and the second one: population elitist Genetic Algorithm (peGA). The Sects. 2.1 and 2.2 show, respectively, the genotype and the *fitness* landscape for both GA. The Sect. 2.3 details the elitism of the peGA, the Sect. 2.4 details the genetic operators, and the Sect. 2.5 analyses the initial results.

2.1   Genotype

Since the purpose of each individual in the population is to symbolize genetically a possible solution, and as the planted problem is to classify characters disposed on images of 32×16 pixels, the individuals were structured with 32 chromosomes of 16 genetic *loci* each one, with which, they are able to represent and generalize the own features of each class, using a ternary code: $\{0; 1; \#\}$ where "#" is a *don't-care* symbol that works as a *wildcard*, due to it allows to ignore its respective position (p), when the Hamming distance is calculated, using the Eq. 1, between the individual (i) and the training sample (e), where 512 is the amount of genetic *loci*. Thus, this kind of genotype corresponds to the conditional part of a Michigan-style classifier [16], because the reactive part, that is a classification label in this case, was defined as phenotype.

$$D(i;e) = \sum_{p=1}^{512} \left\{ \begin{array}{ll} 0 & ; i_p = \text{``\#''} \\ |i_p - e_p| & ; i_p \neq \text{``\#''} \end{array} \right\} \tag{1}$$

2.2 Fitness Landscape

The *fitness* of each individual is evaluated between [0,0; 1,0] with four indexes: Uniqueness, Certainty, Generality and Completeness. The uniqueness evaluates the fact of classifying certain character in a unique class. The certainty evaluates the similarity between certain character and any classifier. The generality evaluates the ability to group diverse characters of the same class; and the completeness evaluates the ability of a classifier population of collaboratively recognizing, all the samples of the same class, included in the training set $\{S_c\}$.

Classification as Phenotype. Each individual expresses its genotype in this way:

- The complete set of training $\{S_c\}$ is exposed to individual (i), and $(\forall e \in S_c)$ the distance $D(i;e)$ is calculated, using the Eq. 1.
- The lowest calculated distance is registered as: $[D(i;e)_{min}]$.
- The classification of each $(e \in S_c)$ is determined, verifying if its distance $D(i;e)$ is within $\{[D(i;e)_{min}]; [D(i;e)_{min}] + (\delta * 512)\}$. Where (δ) is a kind of tolerance with which (i) can classify characters slightly different.
- The number of classes with one or more recognized elements (RCN) is counted.

The Fig. 1 details two possible classification phenotypes for an individual in the 2nd and 11th generation, when the population is just learning the character "B", and it has 4 training samples by each class. It is possible to notice that in the 2nd generation the individual does not clearly differentiate which is the class that represents genetically, while in the 11th generation the individual achieves to classify the 4 "B" samples. It is important to notice, that each cell could have from 0 to 4 "classified" samples, and it is possible that an individual classifies a few of them, because the complete classification will depend on the collaboration of other classifiers.

Individual Indexes. The Uniqueness Index $U(i)$ and the Certainty Index $C(i)$ are calculated with the Eqs. 2 and 3, respectively, using the RCN value; while the Generality Index $G(i)$ is calculated using the Eq. 4, where WCN is the current number of *wildcards* in the genotype, and (θ) defines certain portion of the genetic *loci*. These three indexes take into account only the individual phenotype.

$$U(i) = \left\{ 1 - \left[\frac{RCN - 1}{36} \right] \right\}^8 \tag{2}$$

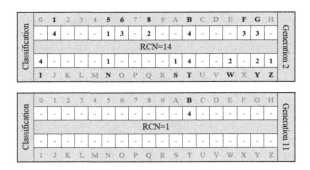

Fig. 1. Evolution of an individual learning to classify 4 samples of "B".

$$C(i) = \frac{512 - [D(i,e)_{min}]}{512 * RCN} \tag{3}$$

$$G(i) = \left\{ \begin{array}{ll} \left[\frac{WCN}{\theta*512}\right]^{0,2} & ; WCN < (\theta * 512) \\ \left[\frac{512-WCN}{512-(\theta*512)}\right]^{0,2} & ; WCN \geq (\theta * 512) \end{array} \right\} \tag{4}$$

Population Collaboration. The Completeness Index $K(P_i)$ is calculated with the Eq. 5, taking into account the classification phenotype of all individuals in a population (P_i). In the Eq. 5 SNC is the number of training samples by class, and RSN is the number of recognized samples among all the individuals. This index pretends that, for certain amount of training samples by class in $\{S_c\}$, each individual has the responsibility to recognize just some of them, thus among all individuals in a population (P_i) the responsibility of recognizing all the samples is distributed.

$$K(P_i) = \left[\frac{RSN}{SNC}\right]^2 \tag{5}$$

Pretending that among all the individuals of a population, all the samples of the same class are recognized, the learning of the 36 characters was performed using 36 separated populations, because it was considered that the classifiers of certain class must be genetically incompatible with the ones of another class, however, each population was trained with the whole data within $\{S_c\}$, in order to each population (P_i) achieves to clearly differentiate its own class from the rest. Finally, with the Eq. 6, the individual *fitness* is calculated, in each of the 36 populations, in both: seGA and peGA.

$$F(i) = U(i) * C(i) * G(i) * K(P_i) \tag{6}$$

2.3   Simple and Populational Elitism

The most known form of elitism in genetic algorithms is to allow the invariable pass of the best individual, from a generation to the next one [17], and this is the kind of elitism used in the seGA. On the other hand, the peGA used the kind of elitist selection of the CHC algorithm, proposed by Eshelman [18], considering that each individual in the current population is a parent, but unlike the CHC algorithm, here the selection by roulette was kept in order to define a mate for each parent, and also the same genetic operators used in the seGA were preserved; this with the purpose of exclusively studying the effects caused by the kind of used elitism.

In each one of the 36 training populations, a total of 6 populations were used: The current one or elite; another one for the mates; other two for the resultant individuals from the crossing of each couple; and other two for inserting several mutations in those successors. The *fitness* function was applied to the 4 populations of successors, and the new elite population was made-up of the fittest individuals, which were chosen to compare each current parent with its respective 4 successors.

2.4   Genetic Operators

The crossing is defined as double point in each of the 32 chromosomes independently. For the results here reported, a probability of 0,6 was used for defining the first point, and another one of 0,3 for defining the second point. Thus, around 432 crossing points in each couple were expected, that is the 90% of possible points.

The mutation is decided probabilistically in each chromosome of each individual. For the results here reported, a probability of 0,5 was used for deciding if the chromosome mutates, or not. In that case, it is chosen between mutation by negation, inversion or rotation, with a distributed probability like this: {0,5; 0,25; 0,25}, respectively, and later a gene with length of $[1; m]$ *loci* is defined for the mutation by negation, or a gene with length of $[2; m]$ *loci*, for the inversion or the rotation. For the results here reported $(m = 4)$ was defined. In the mutation by negation: "0"s are switched by {1; #}, "1"s by {0; #}, or "#"s by {0; 1}, with 50% for the alternatives in each case. For mutating the gene by inversion, the order of its genetic *loci* is switched, and for mutating the gene by rotation, its content is circularly shifted, one *locus* deciding with 50% if shifting by left or right. Since in the peGA the genetic diversity in the elite population is strongly reduced, by means of a mechanism of micro-mutation the individuals phenotypically equal to any other one in that population, are modified. Those individuals classify the training samples in the same way as another elite individual. In the genotype of such individuals, the algorithm decides by each *locus* if it mutates or not, with a probability of 0,1%, after which {0; 1} is switched by "#", or "#" by {0; 1}.

2.5   Initial Results

The values $\{\delta; \theta\}$, defined in the Sect. 2.2, are considered as experimental parameters in this work, because they are key variables, for the mechanism with which the genetic individuals classify the $(e \in S_c)$, during the training. In this first results, it is established: $\{(\delta = 0,02); (\theta = 0,1)\}$, and in the Sect. 4 it will be demonstrated that these values are the best possible ones, in the context of this work.

It was established that, each one of the 36 populations had 10 individuals, and once evolved the learning of the 36 characters, there are 10 classifiers by each class, and a unique agent use them to perform the recognition of the test samples set. To separate the data base in a training samples set and a test set, it was considered that the data are unbalanced, in this case. The higher number of samples by class is 143, but the least one is just 17. The 70% of the lower quantity is 12 and this was the amount of samples by class, dedicated for training the algorithms seGA and peGA. In this way, the 25,5% of the data base was used for training, and using crossing validation during several days, the result was that only some of the 36 populations achieved 0,9 of maximum *fitness* in the seGA, or 0,9 of average *fitness* in the peGA.

Since better results were appreciated when using lower amount of training samples, it was reduced to 4 samples by class, that is the 8,5% of the data base. In this way, the seGA was trained by an average of 12,2 h, restricting its 36 populations to 10000 generations, while the peGA was trained by an average of 12,57 h, restricting it to 1000 generations, because the last one required more time by generation. With the classifiers of the seGA the agent has an average accuracy of 82,2%, while with the classifiers of the peGA just a 67,95% of average accuracy was evidenced.

3   Improving the "Worst" Classifier

At first sight it seems obvious concluding that seGA is better than peGA. However, this section presents the techniques used to improve the apparently worst of such genetic classifiers. The respective improvements are shown in Sect. 4.

3.1   Reduction of Data Diversity

Calculating the Hamming distance among each sample and all the rest, in the same class, the diversity among them is quantified. For the 4*36 training samples discussed in Sect. 2.5 that diversity is within [6,9; 20,4]. In the testings with which the outcomes discussed in the same section were obtained, the populations for learning characters with high diversity were harder to train, while the populations for learning characters with low diversity were the simplest ones. Since a high data diversity makes difficult the training, it generated samples with less diversity than the original ones.

For that, a unique sample of each character was chosen from the data base, and noise was added to them, using two methods: a random one and the other one is a probabilistic type. For the random noise, it is decided that, each pixel of the character edge changes, or not, depending on the probability (P_s), and in that case, it is decided with 50% if "bulging" or "bitting" the character image. Using the probabilistic kernel shown in the Fig. 2b, it is decided which surrounding pixels are set in "1", for "bulging", or which ones are set in "0" for "bitting".

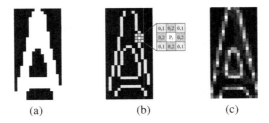

(a)   (b)   (c)

Fig. 2. (a) Unique sample of "A", chosen from the original data base, (b) Random generator of "A" samples, and (c) Probabilistic generator of "A" samples

The probabilistic noise is added according to the inconsistency of each pixel, in all of the samples of the same class, in the original data base. The inconsistency is the variation level concerning each pixel (x) according to its value in those samples. It is calculated with the Eq. 7, where (N_b) is the times that such pixel is "0", (N_w) is the times that it is "1", and SNK is the number of samples for the respective class in the data base. This is the negation of consistency in [7], and it models the variability of the samples in each class. The obtained model for the "A" is shown in the Fig. 2c, where the lighter pixels are the most variable.

$$V(x) = 1 - \left[\frac{|N_b - N_w|}{SNK} \right] \qquad (7)$$

Multiplying each value in that inconsistency matrix, by a general (P_s), a probability by pixel is obtained, with which the probabilistic noise is added to the unique sample, of each character, for which it is decided probabilistically, in each pixel of the original image, if it will switch its value, from "0" to "1", or vice versa, in the process of generating a new training sample. These new training samples, generated by means of these methods, are totally different to the samples of the original data base, due to that, the 100% of the original samples were dedicated to validate the classifiers.

3.2   Assigning Utility to the Classifiers

In order to improve the agent performance, an additional calculus was performed based on the Holland's *bucket brigade* [16], with which an Utility value is set to each classifier, and it is modified by means of the Eqs. 8–10, during only one test of the recognition agent, using the 360 classifiers given by seGA, or peGA, and confronting the 100% of the samples in the original data base $\{Z_c\}$.

$$U(t) = U(t-1) - P(t) + Q(t) \qquad (8)$$

$$P(t) = k * U(t); \quad 0 < k < 1 \qquad (9)$$

$$Q(t) = 2k * U(t) * R(t); \quad 0 < k < 1 \qquad (10)$$

In the Eqs. 8–10, each iteration (t) is equivalent to expose each validation sample $(e \in Z_c)$, to the agent. $P(t)$ is a "loss" calculated by means of Eq. 9, and $Q(t)$ is a value calculated by means of Eq. 10, which is a "profit" if the recognition is correct: $[R(t) = +1]$, since otherwise it is a second "loss": $[R(t) = -1]$. The initial value of $U(t)$ is 1,0 and its high and low value boundaries are: $[0,0; 1,0]$.

For each validation sample $(e \in Z_c)$, its Hamming distance: $D(i; e)$ is calculated, in relation to the 360 classifiers (i), and it modifies the $U(t)$ in the classifiers with the following criteria: The "loss" $P(t)$ is applied to every classifiers with the minimum distance $D(i; e)$ obtained, and additionally to every classifiers of respective (e) class, that has never been used for emitting the recognition response. Such response must be emitted for each (e), by the classifier that has the minimum $D(i; e)$ and the maximum $U(t - 1)$. The value $Q(t)$ only applies to this response-classifier. At the end, the classifiers into disused are eliminated. Like this, there are three ways to validate the classifiers with the samples $(e \in Z_c)$: (1) Without Utility (NU), (2) Assigning the Utility (UA), and (3) Re-assigning the Utility after eliminating the useless classifiers (CE).

3.3   Couple of Collaborative Agents

Until now, a unique agent has been used for recognizing as alphabet letters as numbers, but with the Colombian vehicular plates, it is possible to obtain, first the images of alphabet letters and then the images of numbers, from the pre-processing, with which it is possible to use two agents, one for alphabet recognition and the other one for numbers recognition. Like this, there is a second recognition system.

4   Results Analysis

From here, the set of 4*36 samples discussed in the Sect. 2.5 will be denoted as OS; the set of 4*36 randomly generated samples as RG; the set of 4*36 probabilistically generated samples as PG; the single agent as SA; and the collaborative agents as CA. In order to test the improvements described in the Sect. 3, some sets of 4*36 training samples, as randomly as probabilistically, were generated with $(P_s = 0,1)$. Establishing $\{(\delta = 0,02); (\theta = 0,1)\}$, the peGA was training with those sets, in order to find a RG and a PG, with which the best accuracy levels be achieved, applying NU, UA and CE, with $(k = 0,1)$. This was the way to do a cross-validation in this method. The diversity of the best RG (bRG) and the best PG (bPG) was within $[1,76; 5,01]$, and with these sets the seGA was also trained. Since OS-SA is the reference of the results discussed in the Sect. 2.5, the Table 1 details the increments achieved in the recognition accuracy (R_A), and the Table 2 details the decrements achieved in the training time (T_T).

The use of training sets $\{RG; PG\}$ does not demand to restrict the amount of generations, and thus, the peGA was trained 48,8% faster than seGA, with $\{(\delta = 0,02); (\theta = 0,1)\}$. More experiments were performed for exploring around of those values, combining $\delta : \{0,01; 0,02; 0,03\}$ with $\theta : \{0,05; 0,1; 0,15\}$ and

Table 1. Achieved increments in the R_A with respect to initial OS-SA results

Training and testing seGA with					Training and testing peGA with				
OS	bRG		bPG		OS	bRG		bPG	
CA	SA	CA	SA	CA	CA	SA	CA	SA	CA
9,2%	8,8%	14,2%	9,6%	14,3%	10,2%	23,4%	28,6%	25,5%	29,3%

Table 2. Achieved decrements in the T_T with respect to initial OS results

Training seGA with		Training peGA with	
bRG	bPG	bRG	bPG
−75,2%	−74,2%	−87,1%	−87,8%

training with $\{bRG; bPG\}$. As a result, it is observed that, using SA, the best performance was achieved training the peGA with bPG, and $\{(\delta = 0,02); (\theta = 0,1)\}$, because such classifier evidenced average R_A of 93,4%, and increased it to 97,2% using CA. In contrast, the best seGA was training with bPG too, but using $\{(\delta = 0,03); (\theta = 0,05)\}$, and such classifier just evidenced average R_A of 92,3% with SA, and increased it to 96,7% using CA. Training with bRG and $\{(\delta = 0,01); (\theta = 0,1)\}$, the peGA evidenced until 92,4% of average R_A with SA, and 96,7% with CA; while the seGA evidenced until 91,8% of average R_A with SA, and 96,6% with CA, using $\{(\delta = 0,01); (\theta = 0,15)\}$. Thus, it could affirm that generating samples with PG allows to achieve better results than RG.

The Fig. 3 shows the approximate normal distributions of the obtained results for $[T_T \times R_A]$, separating the combinations of $(\delta; \theta)$ in two sets: (a) $\{\delta : [0,01; 0,02]; \theta : [0,05; 0,15]\}$, and (b) $\{\delta : [0,02; 0,03]; \theta : [0,05; 0,15]\}$. The Fig. 3a shows that the results of peGA are more dispersed than the ones of seGA, and the Fig. 3b shows the contrary effect. With this, it is possible to predict that when

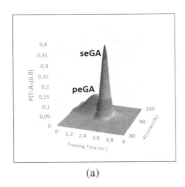

(a)

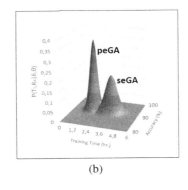

(b)

Fig. 3. Approximate normal distributions for the plane $[T_T \times R_A]$, training as seGA as peGA, with bRG and bPG, and only using SA. [See the text for details]

training seGA and peGA with the parameters (a) it is more probable that seGA spends more T_T and evidences less R_A; but with the parameters (b) it is more probable that peGA spends less T_T and evidences more R_A; thus, peGA is better than seGA, more frequently in (b).

Finally, the recognition time (R_T) for each test character was measured, using an Intel® CORE™ i7 CPU to 2,5 GHz. Applying UN and UA, the best peGA evidenced average RT of 60,7 ms with SA, and 33 ms with CA. Applying CE the best peGA reduced its R_T to 22,4 ms with SA, and to 12 ms with CA. In general, it was noticed that using CA the R_T is reduced to the half approximately.

5   Conclusions

In this paper, two forms of elitism were explored in genetic algorithms for the recognition of printed characters in Colombian vehicle plates. The simple elitism (seGA) is widely known in GA literature and, in this article it is contrasted with a mechanism of population elitism (peGA) that, along with two processes to generate training samples with added noise, achieved both: Reducing the training time and improving the recognition accuracy, in spite of consuming more time by generation, using a very low amount of training samples and using a more demanding *fitness*-based stop criterion. Among the previous works in OCR, only [12] is slightly similar to the presented in this paper, but it does not use Michigan-style classifiers, neither emphasize in elitism nor focus on LPR. In [6] also the accuracy of a LPR system is improved, adding noise in the training samples, and although the same database was not used here neither neural networks, 97,2% achieved by combining the peGA with the couple of collaborative agents is really notably, compared to 93,96% there reported for the recognition of the alphabet, along with 92,37% of recognition of the numbers. Additionally, 12 ms could be an appropriate time for on-line character recognition, taking into account the conditions of the vehicles entrance, in a typical Colombian real parking lot.

As future work, other improvements have been projected using probabilistic techniques. Any additional studies have been projected too, considering the complete CHC algorithm [18], the character encoding, the recognition method or the local search used in [13], other kinds of elitism [19], and maybe another mating method could be introduced. Additionally, it has been projected to use the peGA for properly initializing the classifiers of a Michigan-style LCS [16], in order to evaluate eventual improvements on both, the recognition accuracy and the involved times.

References

1. Thakur, M., Raj, I., Ganesan, P.: The cooperative approach of genetic algorithm and neural network for the identification of vehicle License Plate number. In: 2015 International Conference on Innovations in Information, Embedded and Communication Systems (ICIIECS), pp. 1–6. IEEE (2015)

2. Nijhuis, J.A.G., Ter Brugge, M.H., Helmholt, K.A., Pluim, J.P.W., Spaanenburg, L., Venema, R.S., Westenberg, M.A.: Car license plate recognition with neural networks and fuzzy logic. In: 1995 Proceedings of the IEEE International Conference on Neural Networks, vol. 5, pp. 2232–2236. IEEE (1995)

3. Arth, C., Limberger, F., Bischof, H.: Real-time license plate recognition on an embedded DSP-platform. In: 2007 IEEE Conference on Computer Vision and Pattern Recognition, CVPR 2007, pp. 1–8. IEEE (2007)

4. Yingyong, Z., Jian, Z., Yongde, Z., Xinyan, C., Guangbin, Y., Juhui, C.: Research on algorithm for automatic license plate recognition system. Int. J. Multimed. Ubiquit. Eng. **10**(1), 101–108 (2015)

5. Al-Zoubaidy, L.M.: Efficient genetic algorithms for Arabic handwritten characters recognition. In: Tiwari, A., Roy, R., Knowles, J., Avineri, E., Dahal, K. (eds.) Applications of Soft Computing. Advances in Intelligent and Soft Computing, vol. 36, pp. 3–14. Springer, Heidelberg (2006). doi:10.1007/978-3-540-36266-1_1

6. Qu, Z., Chang, Q., Chen, C., Lin, L.: An improved character recognition algorithm for license plate based on BP neural network. Open Electr. Electron. Eng. J. **8**, 202–207 (2014)

7. Ravi, B., Ang Jr., M.H.: Fuzzy logic based character recognizer. In: Proceedings of the Philippine Computing Science Congress (PCSC) (2000)

8. Vipindas, K., Ravindran, K.P., Ramachandran, N.: Character recognition using fuzzy c-means classification. http://www.ugcfrp.ac.in/images/userfiles/33971-Paper.pdf

9. Frey, P.W., Slate, D.J.: Letter recognition using Holland-style adaptive classifiers. Mach. Learn. **6**(2), 161–182 (1991). Springer

10. Pornpanomchai, C., Daveloh, M.: Printed Thai character recognition by genetic algorithm. In: 2007 International Conference on Machine Learning and Cybernetics, vol. 6, pp. 3354–3359. IEEE (2007)

11. Spivak, P.K.: Discovery of optical character recognition algorithms using genetic programming. In: Genetic Algorithms and Genetic Programming at Stanford, pp. 223–232 (2002)

12. Noaman, K.M.G., Saif, J.A.M., Alqubati, I.A.A.: Optical character recognition based on genetic algorithms. J. Emerg. Trends Comput. Inf. Sci. **6**(4), 203–208 (2012)

13. Welekar, R., Thakur, N.V.: Memetic algorithm used in character recognition. In: Panigrahi, B.K., Suganthan, P.N., Das, S. (eds.) SEMCCO 2014. LNCS, vol. 8947, pp. 636–646. Springer, Cham (2015). doi:10.1007/978-3-319-20294-5_55

14. Kim, G., Kim, S., Tek, T., Kyungki, S.: Feature selection using genetic algorithms for handwritten character recognition. Citeseer (2000)

15. Kimura, Y., Suzuki, A., Odaka, K.: Feature selection for character recognition using genetic algorithm. In: 2009 Fourth International Conference on Innovative Computing, Information and Control (ICICIC), pp. 401–404. IEEE (2009)

16. Golberg, D.E.: Genetic Algorithms in Search, Optimization, and Machine Learning. Addion Wesley, Reading (1989)

17. Reed, P.M., Minsker, B.S., Goldberg, D.E.: The practitioner's role in competent search and optimization using genetic algorithms. In: Bridging the Gap: Meeting the World's Water and Environmental Resources Challenges, pp. 1–9 (2001)

18. Eshelman, L.J.: The CHC adaptive search algorithm: how to have safe search when engaging. In: Foundations of Genetic Algorithms 1991 (FOGA 1), vol. 1, p. 265. Morgan Kaufmann (2014)

19. Liang, Y., Leung, K.: Genetic algorithm with adaptive elitist-population strategies for multimodal function optimization. Appl. Soft Comput. **11**(2), 2017–2034 (2011)

Comparison Among Physiological Signals for Biometric Identification

M. Moreno-Revelo[1], M. Ortega-Adarme[1], D.H. Peluffo-Ordoñez[2],
K.C. Alvarez-Uribe[3], and M.A. Becerra[3(✉)]

[1] Universidad de Nariño, Pasto, Colombia
[2] Universidad Técnica del Norte, Ibarra, Ecuador
[3] Instituto Tecnológico Metropolitano, Medellín, Colombia
migb2b@gmail.com

Abstract. The biometric is an open research field that requires analysis of new techniques to increase its accuracy. Although there are active biometric systems for subject identification, some of them are considered vulnerable to be fake such as a fingerprint, face or palm-print. Different biometric studies based on physiological signals have been carried out. However, these can be regarded as limited. So, it is important to consider that there is a need to perform an analysis among them and determine the effectivity of each one and proposed new multimodal biometric systems. In this work is presented a comparative study of 40 physiological signals from a multimodal analysis. First, a preprocessing and feature extraction was carried out using Hermite coefficients, discrete wavelet transform, and statistical measures of them. Then, feature selection was applied using two selectors based on Rough Set algorithms, and finally, classifiers and a mixture of five classifiers were used for classification. The more relevant results shown an accuracy of 97.7% from 3 distinct EEG signals, and an accuracy of 100% using 40 different physiological signals (32 EEG, and eight peripheral signals).

Keywords: Biometric · Classifiers mixture · Multimodal system · Physiological signals · Signal processing

1 Introduction

Biometrics are gaining increasing interest in many areas, including securing financial systems, telecommunications and healthcare applications [1]. Physiological characteristics such as fingerprint, iris and facial structure and behavioral characteristics such as voice, gait and signature have been used for identity recognition. However, the choice of right biometric depends on characteristics of application environment and usually is a tradeoff between factors such as performance, ease of collection, user acceptability and deployment cost [2]. Importantly, recent approaches to automated biometric analysis have sought to improve identification performance through the application of fusion methods. Fusion has been documented through three distinct approaches all of which can be applied when evaluating whether a particular sample donated by an individual is a match (or not) to a reference sample within a database [3].

© Springer International Publishing AG 2017
H. Yin et al. (Eds.): IDEAL 2017, LNCS 10585, pp. 436–443, 2017.
https://doi.org/10.1007/978-3-319-68935-7_47

At present, biometric is regarded an open research field, which studies methods for individual identification and authentication based on physiological signals, physical traits among others physical measures. Some acquisition of physiological signals can be uncomfortable [4] and require equipment of high cost [5], and they are considered hard to duplicate. However, with the advance of the technology, these become more vulnerable every day. Therefore, some biometric studies are focus on the combination of multiple signals but these can be regarded insufficient yet.

The biological signals, most of which have been studied in biometric on independently way [6–10], and some cases combining two signals [11–13]. The classical biometric signal processing is based on five principal stages such as acquisition, preprocessing, feature extraction, classification, and verification to finally decide if the answer is accept or reject [14].

Those stages are applied for both mono-modal and multimodal processing. This last one uses two type architectures (serial or parallel) which is described in [15]. The studies can be considered limited because they do not compare a lot physiological signals together. In addition, they do not consider the quality of the signal and contextual variables for increase and establish risk and vulnerabilities. These reasons make it difficult to identify the advantages, disadvantages and opportunities of each one and their combinations to propose new effective multimodal biometrics mechanisms based on physiological signals.

In this work, different physiological signals (32 EEG signals and 8 peripheral signals) of DEAP database [16] are analyzed. Some of them have been widely discussed in the literature and others not. An effective mechanism for biometric identification was achieved. Besides, the discriminant power and weaknesses of these physiological signals for this task were presented, identifying the more relevant signals applying Rough Set feature selection algorithms. Five classifiers (Support vector machine, Quadratic Bayes Classifier, PARZEN, k- Nearest neighbors, and Linear Bayes Classifier), and a mixture of them using 5 strategies (Product combiner, Mean combiner, Median combiner, Maximum combiner, Minimum combiner, and Voting combiner) were applied for individual identification from features obtained of Discrete Wavelet transform, statistical measures, and Hermite Coefficients.

The rest of this paper is structured as follows: Sect. 2 describes the Materials and methods over proposed local database used in this work. Section 2.1 presents the proposed experimental setup. Results and discussion are gathered in Sect. 3. Finally, some concluding remarks and future works are drawn in Sect. 4.

2 Proposed Procedure

The methodology proposed to comprise physiological signals for biometric identification take into account 5 stages (see Fig. 1). Information of 32 subjects is used, followed by stages as selection of signals, preprocessing, feature extraction, feature selection and finally classification.

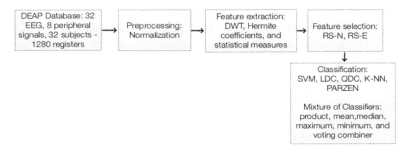

Fig. 1. Proposed procedure

2.1 Database

For evaluating the proposed procedure, the DEAP dataset is used. The dataset consists of 40 different measurements grouped in 32 EEG signals acquired with a 10–20 system positioning, and 8 peripheral signals. These last are GSR (resistance of the skin by positioning two electrodes on the distal phalanges of the middle and index fingers), respiration belt- Rb, skin temperature, plethysmograph, electromyograms of Zygomaticus and Trapezius muscles, and horizontal electrooculogram - HEOG and vertical electrooculogram VEOG. The acquisition of the signals is widely discussed in [16]. The position of the electrodes of each signals based on 10–20 are: EEG1 – Fp1, EEG2 – AF3, EEG3 – F3, EEG4 – F7, EEG5 –FC5, EEG6- FC1, EEG7-C3, EEG8-T7, EEG9-CP5, EEG10-CP1, EEG11-P3, EEG12-P7, EEG13-PO3, EEG14-O1, EEG15-Oz, EEG16-Pz, EEG17-Fp2, EEG18-AF4, EEG19-Fz, EEG20-F4, EEG21-F8, EEG22-FC6, EEG23-FC2, EEG24-Cz, EEG25-C4, EEG26-T8, EEG27-CP6, EEG28-CP2, EEG29-P4, EEG30-P8, EEG31-PO4, and EEG32-O2.

2.2 Preprocessing

Before submitting to the physiological signals to the characterization, the DC level is eliminated, and the signal is normalized.

With the above makes sure that the signal does not surpass a certain range and with it making easy posterior group analysis when obtaining more information of the same as exposes [17].

2.3 Feature Extraction

Characterization of the EEG signals is performed with a total of 25 features (see Table 1). Six features were obtained by making use of Shannon entropy and morphological features such as area and energy and statistics such as mean, median and standard deviation. 9 features are based on Hermite coefficient, and 10 based on coefficients of Wavelet transform.

Table 1. Features set.

Heading level	$x^{(1)}$	$x^{(2)}$	$x^{(3)}$	$x^{(4)}$	$x^{(5)}$	$x^{(6)}$	$x^{(7)}...x^{(15)}$	$x^{(16)}...x^{(25)}$
Feature	Area	Energy	Median	Mean	Standard deviation	Shannon entropy	Hermite coefficients	Variance and maximum of wavelet coefficients

- **Wavelet transform**

The wavelet transform of the function $f(t)$ is the decomposition of $f(t)$ in functions set $\varphi_{s,\tau}(t)$ that form a basis and they are name wavelets. The wavelets are generated from the translation and scale change of a mother function called "mother wavelet" [18]. One of the most known wavelet families are the Daubechies that characterize a signal with the use of two filters: one high passes filter and another low passjas filter to obtain the approximation and detail coefficients respectively [19]. Taking into account the above, the family of Daubuchies 10 with four levels of approximation are used. With the use of statistical measures of the coefficients wavelet as the variance and rate between minimum and maximum, a total of 10 features are obtained.

- **Hermite Coefficients**

To reconstruct the Hermite bases, a scale parameter of 0.25 ms and 9 polynomials are used as recommended [20]. The polynomials Hermite are defined as $H_n(z) = (-1)^n e^{z^2} \frac{d^n}{dz^n} e^{-z^2}$. The Hermite polynomials form an orthonormal set with respect to the function e^{-z^2}, that is to say: $\delta_{m,n} = \frac{1}{\sqrt{2^n n!\sqrt{\pi}}}\left(e^{-z^2}H_n(z), H_m(z)\right)$, doing the replacement $z = \frac{t}{\sigma}$, The bases can be reconstructed in the way: $\emptyset_n^\sigma = \frac{e^{-t^2/2\sigma^2}}{\sqrt{2^n \sigma n!\sqrt{\pi}}}H_n\left(\frac{t}{\sigma}\right)$, where σ is the scale parameter. The Hermite coefficients associated with the signal $x(t)$ are obtained with: $C_n^\sigma = \frac{1}{fs}\int_{t=-\infty}^{\infty}x(t)\emptyset_n^\sigma(t)dt$, and the reconstruction of the signal is performed as indicated by the following expression $x(t) = \sum_{n=0}^{\infty}C_n^\sigma\emptyset_n^\sigma(t)$.

2.4   Feature Selection

After preprocessing and feature extraction stages a variable selection procedure is employed to reduce the number of parameters per instance of a dataset and establish the more relevant features and signals for biometric identification into this study. In this work, two feature selection algorithms based on rough set theory were applied the first one Rough Set Neighbor RS-N, and the second one Rough Set Entropy RS-E, both algorithms are discussed widely in [21]. Their parameters were adjusted following the methodology shown in [21]. The inclusion rate was adjusted to 0.5, and neighbor distance tolerance was adjusted to 0.05.

2.5   Classification

This step was performed using 5 classifiers such as: *(i)* SVM is a model-based classifier that represents the sample points in space, separating the classes by a separation hyperplane; *(ii)* QDC Bayesian classification aims at assigning to a feature vector the class it belongs to with the highest probability; *(iii)* PARZEN Obtains estimates of probability densities. This method requires a smoothing parameter for the Gaussian distribution computation, which is optimized; *(iv)* k-NN is a method of non-parametric classification, which is based on the calculation of distances; *(v)* LDC classifier constructs models that predict the probability of possible results depends on the value of a linear combination of its features.

These classifiers were testing individuality and a mixture among them was carried out using the following techniques: *(i)* Product: a set of classifiers by selecting the class which yields the highest value of the product of the classifier outputs; *(ii)* Mean combiner: It selects the class with the mean of the outputs of the input classifiers; *(iii)* Median combiner: It chooses the class with the median of the outputs of the input classifiers; *(iv)* Maximum combiner: It selects the class that gives the maximal output of the input classifiers; *(v)* Minimum combiner: It chooses the class with the minimum of the outputs of the input classifiers; *(vi)* Voting combiner: It selects the class with the highest vote of the base classifiers.

3   Results

Table 2 shows the more relevant signals for biometric identification obtained from feature selection algorithms.

Table 2. Relevant signals

Signal	EEG1	EEG4	EEG6	EEG8	EEG9	EEG11
	EEG22	EEG23	EEG30	HEOG	GSR	Rb

Tables 3, 5, 7, and 9, show the accuracy of individual classifiers in terms of the error, and Tables 4, 6, 8 and 10 show the accuracy applying mixture of classifiers using the features selected by RS-N and RS-E algorithms. The best result 0% of error was obtained using the QDC classifier and the mixture of classifiers combining the 40 physiological signals. The best classifier was QDC for all tests. The more relevant peripheral signals achieved are HEOG, GSR, and Rb. However, their contributions can be considered not relevant a despite of it increase the results around 2%.

Table 3. Accuracy of classifier – feature selected by RS-N and RS-E from all signals

	SVC	QDC	Parzen	K-NN	LDC
RS-N	0.5898	0.0039	0.0547	0.0469	0.0352
RS-E	0.6328	0	0.0586	0.0508	0.0586

Table 4. Mixture of classifiers – feature selected by RS-N and RS-E from all signals

	prodc	meanc	medianc	maxc	minc	votec
RS-N	0.0039	0.0117	0.0156	0.0117	0.0312	0.0352
RS-E	0	0.0078	0.0117	0.0078	0.0195	0.0234

Table 5. Accuracy of classifier – feature selected by RS-N and RS-E from selected peripheral signals (HEOG, GSR, and Rb)

	SVC	QDC	Parzen	K-NN	LDC
RS-N	0.8906	0.3164	0.7188	0.7031	0.6055
RS-E	0.9297	0.5703	0.7266	0.7266	0.6953

Table 6. Mixture of classifiers – feature selected by RS-N and RS-E from selected peripheral signals (HEOG, GSR, and Rb).

	prodc	meanc	medianc	maxc	minc	votec
RS-N	0.3242	0.3203	0.4844	0.3242	0.7148	0.5508
RS-E	0.5625	0.5664	0.6250	0.5859	0.8555	0.6406

Table 7. Accuracy of classifiers – feature selected by RS-N and RS-E from EEG signals (EEG6, EEG8, EEG30)

	SVC	QDC	Parzen	K-NN	LDC
RS-N	0.8320	0.0469	0.4141	0.4727	0.2539
RS-E	0.8555	0.0352	0.4609	0.5039	0.2617

Table 8. Mixture of classifiers – feature selected by RS-N and RS-E EEG signals (EEG6, EEG8, and EEG30)

	prodc	meanc	medianc	maxc	minc	votec
RS-N	0.0586	0.0625	0.0977	0.0703	0.2344	0.2539
RS-E	0.0273	0.0820	0.0664	0.0859	0.1719	0.2422

Table 9. Accuracy of classifier – feature selected by RS-N and RS-E from EEG signals (EEG6, EEG8, EEG30) and selected peripheral signals (HEOG, GSR, and Rb).

SVC	QDC	Parzen	K-NN	LDC
0.7539	0.0430	0.3047	0.3477	0.0781

Table 10. Mixture of classifiers – feature selected by RS-N and RS-E EEG signals (EEG6, EEG8, EEG30) and selected peripheral signals (HEOG, GSR, and Rb)

prodc	meanc	medianc	maxc	minc	votec
0.0234	0.0977	0.1602	0.0977	0.0586	0.1211

4   Conclusions

In this work, a comparison among 40 physiological signals for biometric identification was carried out. A total of 32 subjects were taken into account for this analysis, and according to the results obtained, it is possible to conclude that with the combination of the 40 physiological signals a smaller error is obtained than when using only some signals like the peripherals, however, this involves a greater computational cost. Besides, the results demonstrated after applied relevant analysis using RS-N and RS-E, the capability of three EEG signals (EEG6, EEG8, EEG30), the Hermite and Wavelet coefficients, and the mixture of classifiers for individual identification, which allows obtaining an objective and accurate mechanism of biometric identification in terms of error. The success rate of classifiers mixture depends on the signals, but it should be noted that the product combiners allow achieving the lowest errors in the most of the tests. As future work, we proposed to extend this study with other features and classifiers techniques to accomplish a reliable biometric identification system based on the relevant EEG and peripheral (HEOG, GSR, and Rb) signals obtained in this work to reduce the complexity.

References

1. Merone, M., Soda, P., Sansone, M., Sansone, C.: ECG databases for biometric systems. A systematic review. Expert Syst. Appl. **67**, 189–202 (2017)
2. Komeili, M., Louis, W., Armanfard, N., Hatzinakos, D.: Feature selection for nonstationary data: application to human recognition using medical biometrics. IEEE Trans. Cybern. (99), 1–14 (2017)
3. Stevenage, S., Guest, R.: Combining forces: Data fusion across man and machine for biometric analysis. Image Vis. Comput. **55**, 18–21 (2016)
4. Lourenço, A., Hugo, S., Ana, F.: Unveiling the biometric potential of finger-based ECG signals. Comput. Intell. Neurosci. **2011**, 5 (2011)
5. Liwen, F., Cai, X., Ma, J.: A dual-biometric-modality identification system based on fingerprint and EEG. In: Fourth IEEE International Conference on Biometrics: Theory Applications and Systems (BTAS), pp. 1–6. IEEE (2010)
6. Campisi, P., La Rocca, D.: Brain waves for automatic biometric-based user recognition. IEEE Trans. Inf. Forensics Secur. **9**(5), 782–800 (2014)
7. Tseng, K., Luo, J., Hegarty, R., Wang, W., Haiting, D.: Sparse matrix for ecg identification with two-lead features. Sci. World J. **2015** 1–9 (2015)
8. Beritelli, F., Serrano, S.: Biometric identification based on frequency analysis of cardiac sounds. IEEE Trans. Inf. Forensics Secur. **2**(3), 596–604 (2007)
9. Wang, J., Wang, C., Chin, Y., Liu, Y., Chen, E., Chang, P.: Spectral-temporal receptive fields and MFCC balanced feature extraction for robust speaker recognition. Multimed. Tools Appl. **76**(3), 1–14 (2016)
10. Lee, A., Kim, Y.: Photoplethysmography as a form of biometric authentication. IEEE Sensors, pp. 1–2. IEEE (2015)
11. Abo-Zahhad, M., Ahmed, S., Abbas, S.: A new multi-level approach to EEG based human authentication using eye blinking. Pattern Recogn. Lett. **82**, 216–225 (2016)

12. Belgacem, N., Fournier, R., Nait-Ali, A., Bereksi-Reguig, F.: A novel biometric authentication approach using ECG and EMG signals. J. Med. Eng. Technol. **39**(4), 226–238 (2015)
13. Bugdol, M., Mitas, A.: Multimodal biometric system combining ECG and sound signals. Pattern Recogn. Lett. **38**, 107–112 (2014)
14. Dinca, L., Hancke, G.: The fall of one, the rise of many: a survey on multi-biometric fusion methods. IEEE Access **5**, 6247–6289 (2017)
15. Liu, Y., Hatzinakos, D.: Earprint: transient evoked otoacoustic emission for biometrics. IEEE Trans. Inf. Forensics Secur. **9**(12), 2291–2301 (2014)
16. Koelstra, S., Muhl, C., Soleymani, M., Lee, S., Yazdani, A., Ebrahimi, T., Pun, T., Nijholt, A., Patras, I.: Deap: a database for emotion analysis; using physiological signals. IEEE Trans. Affect. Comput. **3**(1), 18–31 (2012)
17. Byrne, C.: Signal Processing: A Mathematical Approach. CRC Press, Boca Raton (2014)
18. Xie, X., Wang, S., Juang, S., Lee, S., Lin, K., Wang, X., Deng, N.: An ECG feature extraction with wavelet algorithm for personal healthcare. In: Bioelectronics and Bioinformatics, pp. 128–131 (2015)
19. Banerjee, S., Mitra, M.: A cross wavelet transform based approach for ECG feature extraction and classification without denoising. In: Control, Instrumentation, Energy and Communication, pp. 162–165 (2014)
20. Peluffo, D., Rodríguez, J., Castellanos, C.: Metodología para la reconstrucción y extracción de características del complejo QRS basada en el modelo parametrico de Hermite. Semana Técnica de ingenierias eléctrica y electrónica, pp. 1–5 (2008)
21. Orrego, D., Becerra, M., Delgado, E.: Dimensionality reduction based on fuzzy rough sets oriented to ischemia detection. In: Engineering in Medicine and Biology Society (EMBC), pp. 5282–5285 (2012)

A Pay as You Use Resource Security Provision Approach Based on Data Graph, Information Graph and Knowledge Graph

Lixu Shao[1], Yucong Duan[1(✉)], Lizhen Cui[2], Quan Zou[3], and Xiaobing Sun[4]

[1] College of Information Science and Technology, State Key Laboratory
of Marine Resource Utilization in the South China Sea,
Hainan University, Haikou, China
751486692@qq.com, duanyucong@hotmail.com
[2] College of Computer Science and Technology, Shandong University, Jinan, China
clz@sdu.edu.cn
[3] College of Computer Science, Tianjin University, Tianjin, China
zouquan@tju.edu.com
[4] School of Information Engineering, Yangzhou University, Yangzhou, China
sundomore@163.com

Abstract. With the development of data mining technology, lack of private resource protection has become a serious challenge. We propose to clarify the expression of Knowledge Graph in three layers including Data Graph, Information Graph and Knowledge Graph and illustrate the representation of Data Graph, Information Graph and Knowledge Graph respectively. We elaborate a pay as you use resource security provision approach based on Data Graph, Information Graph and Knowledge Graph in order to ensure that resources will not be used, tampered with, lost and destroyed in unauthorized situations.

Keywords: Resource modelling · Knowledge graph · Security provision

1 Introduction

Disclosure risk of data, information, knowledge and other resources has always been the focus of statistics and computer science fields. Some statistical disclosure control techniques have been proposed to limit the risk of resource disclosure [7]. Knowledge graph has become a powerful tool to represent knowledge in the form of a labelled directed graph and to give semantics to textual information. In our previous work [3], we specified Knowledge Graph in a progressive manner as four basic forms including Data Graph, Information Graph, Knowledge Graph and Wisdom Graph. Development of software service system can be divided into stages of data sharing, information transfer, and knowledge creation in terms of data, information and knowledge [6]. From the perspective of extending the existing concept of Knowledge Graph, we present an architecture that can automatically abstract and adjust resources based on DataGraph$_{DIK}$,

© Springer International Publishing AG 2017
H. Yin et al. (Eds.): IDEAL 2017, LNCS 10585, pp. 444–451, 2017.
https://doi.org/10.1007/978-3-319-68935-7_48

InformationGraph$_{DIK}$ and KnowledgeGraph$_{DIK}$ where DIK is an abbreviation for typed resources that are Data$_{DIK}$, Information$_{DIK}$ and Knowledge$_{DIK}$. Through reasonable organization and storage of massive resources, we can achieve the goal of satisfying protection requirements for private resources of stakeholders according to their investment.

In the rest of the paper, we firstly illustrate definitions of resource types and DataGraph$_{DIK}$, InformationGraph$_{DIK}$ and KnowledgeGraph$_{DIK}$ in Sect. 2. Then we elaborate a pay as you use resource security provision approach in Sect. 3. We give examples of crossing type storage programs of resources in Sect. 4. The related works are presented in Sect. 5 and we conclude our work in Sect. 6.

2   Definitions of Resources and Graphs

We propose to clarify the expression of Knowledge Graph as a whole and extend the existing concept of Knowledge Graph to three levels that are DataGraph$_{DIK}$, InformationGraph$_{DIK}$, KnowledgeGraph$_{DIK}$. Data$_{DIK}$ is not specified for any stakeholder or machine. Data$_{DIK}$ consists of a collection of discrete elements. Data$_{DIK}$ is used for statistic and transmission. Information$_{DIK}$ is specified for stakeholder and machine. Information$_{DIK}$ is conveyed through conceptual mapping and related relationship combination of Data$_{DIK}$. Knowledge$_{DIK}$ is the probabilistic or categorization of elements and their relationships and is utilized for reasoning and predicting unknown resources.

Definition 1. *(Resource elements). Resource elements including data, information and knowledge can be expressed as:*
   Elements$_{DIK}$: = < Data$_{DIK}$, Information$_{DIK}$, Knowledge$_{DIK}$ >;
   And we use D to represent Data$_{DIK}$, I to represent Information$_{DIK}$ and K to represent Knowledge$_{DIK}$ for convenient description.

Definition 2. *(Graphs). We propose to specify the existing concept of Knowledge Graph in three layers. Graphs can be expressed as:*
   Graph$_{DIK}$: = <(DataGraph$_{DIK}$), (InformationGraph$_{DIK}$), (Knowledge Graph$_{DIK}$)>.

Definition 3. *(DataGraph$_{DIK}$). We define DataGraph$_{DIK}$ as:*
   DataGraph$_{DIK}$: = collection array, list, stack, queue, tree, graph.
   DataGraph$_{DIK}$ is a collection of discrete elements expressed in the form of various data structures including arrays, lists, stacks, trees, graphs, and so on. DataGraph$_{DIK}$ is used to model the temporal and spatial features of Data$_{DIK}$.

Definition 4. *(InformationGraph$_{DIK}$). We define InformationGraph$_{DIK}$ as:*
   InformationGraph$_{DIK}$: = combination related Data$_{DIK}$. Information Graph$_{DIK}$ is a combination of related Data$_{DIK}$. InformationGraph$_{DIK}$ expresses the interaction and transformation of InformationDIK between entities in the form of a directed graph.

Definition 5. *(KnowledgeGraph$_{DIK}$). We define KnowledgeGraph$_{DIK}$ as:*
 KnowledgeGraph$_{DIK}$: = collection statistical rules.
 KnowledgeGraph$_{DIK}$ is of free schema and expresses rich semantic relation-ships which is conductive to have a completing mapping towards users' require-ments described through natural language. KnowledgeGraph$_{DIK}$ is a collection of statistical rules summarized from known resources.

3   Overview of Our Approach

3.1   Input of the Problem

Firstly, we give the input of the problem discussed in our work.

Definition 6. *(Private resources). Private resources of resource owners that need protection service are defined as a tuple $PR =< PR_{type}, PR_{sca} >$, where PR_{type} is the type set of the private resources represented by a triad $< PTY_1, PTY_2, PTY_3 >$ and PR_{sca} is the scale of different kinds of resources represented by a triad $< PScale_1, PScale_2, PScale_3 >$ where each $PScale_i$ denotes the scale of resources in the form of PTY_i.*

Definition 7. *(Resource protection space). The resource protection space is defined as a tuple $RPS =< RPS_{type}, RPS_{sca} >$, where RPS_{type} is the type set of the private resources represented by a triad $< RTY_1, RTY_2, RTY_3 >$ and RPS_{sca} is the scale of different kinds of resources represented by a triad $< RScale_1, RScale_2, RScale_3 >$ where each $RScale_i$ denotes the scale of resources in the form of RTY_i.*

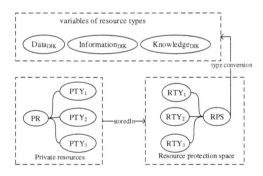

Fig. 1. Assigning values to types of resources in PR and RPS.

From the perspective of resource owners, security concerns and privacy pro-tection issues, remain the primary inhibitor for adoption of cloud computing services. Figure 1 shows the process of assigning values to types of resources in PR and RPS. The security level of privacy protection service is determined by a resource owner's investment.

3.2 Calculation of Type Conversion Cost

Characteristic of a pay as you use resource security provision approach based on DataGraph$_{DIK}$, InformationGraph$_{DIK}$ and KnowledgeGraph$_{DIK}$ lies in providing resources owners with security protection services to ensure that their resources will not be used, tampered with, lost or destroyed in unauthorized situations.

We assign values from Elements$_{DIK}$ to each element of $RPStype$ forming the combination case. The atomic type conversion cost of resource in RPS, denoted as $RCost$, is shown in Table 1. Type conversion cost of all resources their original types to RPS_{type}, denoted as $CostTF$, can be calculated according to Eq. 1:

$$CostTF = \sum_{i=1}^{3} (RCost * RScale_i) \tag{1}$$

Table 1. Atomic type conversion cost of resources in RPS

	Data$_{DIK}$	Information$_{DIK}$	Knowledge$_{DIK}$
Data$_{DIK}$	$RCost_{D-D}$	$RCost_{D-I}$	$RCost_{D-K}$
Information$_{DIK}$	$RCost_{I-D}$	$RCost_{I-I}$	$RCost_{I-K}$
Knowledge$_{DIK}$	$RCost_{K-D}$	$RCost_{K-I}$	$RCost_{K-K}$

3.3 Calculation of Storage Cost of Resources

Types of resources in PR can be changed and resources after being transformed are stored in RPS of which type is inconsistent with the original resources. Storage cost of resources with different storage scenarios, denoted as $STCost$ can be calculated according to Eq. 2:

$$STCost = \sum_{i=1}^{3} (CCost * PScale_i + PScale_i') \tag{2}$$

where $CCost$ represents the type conversion cost of resources from PR to RPS. We show the atomic conversion cost of resource element from PR to RPS in Table 2. $STCost$ is related to $CCost$ and scale of resources after being transformed that is denoted as $PScale_i'$.

Table 2. Atomic type conversion cost of resources from PR to RPS

	Data$_{DIK}$	Information$_{DIK}$	Knowledge$_{DIK}$
Data$_{DIK}$	$CCost_{D-D}$	$CCost_{D-I}$	$CCost_{D-K}$
Information$_{DIK}$	$CCost_{I-D}$	$CCost_{I-I}$	$CCost_{I-K}$
Knowledge$_{DIK}$	$CCost_{K-D}$	$CCost_{K-I}$	$CCost_{K-K}$

3.4   Calculation of Ratio of Investment and Benefit

Costs of providing protection services for private resources consist of two parts that are type converting cost and spatial cost. After computing $CostTF$ and $STCost$, we can calculate the total cost, denoted as $Total\_Cost$, and the corresponding inquired investment that is denoted as $Inves$ of each resource protection program according to Eqs. 3 and 4:

$$Total\_Cost = CostTF + STCost \tag{3}$$

$$Inves = \mu * Total\_Cost \tag{4}$$

where μ indicates the atomic investment that can be obtained through data training and set up in advance.

Private resource protection models usually depend on one or several parameters that determine how much disclosure risk of resources is acceptable. We measure the security level of different resource storage schemes according to the searching cost. We denote searching cost as $SECost$ which can be calculated according to Eq. 5:

$$SECost = \sum_{i=1}^{3}(RScale_i + CCost * RScale_i') \tag{5}$$

where $CCost'$ indicates the atomic type conversion cost of resource from RPS to PR which is shown in Table 3. $SECost$ is related to $CCost'$ and scale of initial resources of PR. And after computing the investment and $SECost$ of each resource protection program we can get the ratio of investment and corresponding benefit according to Eq. 6:

$$Ratio_{IO} = \frac{Inves}{SECost} \tag{6}$$

In the same amount of investment, resource protection program with greater searching cost is better. Algorithm 1 describes the process of assigning values to types of resources in RPS and calculations of varieties of costs of private resources protection programs. A flowchart of a pay as you use resource security provision approach based on DataGraph$_{DIK}$, InformationGraph$_{DIK}$ and KnowledgeGraph$_{DIK}$ is shown in Fig. 2.

Table 3. Atomic type conversion cost of resources from PR to RPS

	Data$_{DIK}$	Information$_{DIK}$	Knowledge$_{DIK}$
Data$_{DIK}$	$CCost'_{D-D}$	$CCost'_{D-I}$	$CCost'_{D-K}$
Information$_{DIK}$	$CCost'_{I-D}$	$CCost'_{I-I}$	$CCost'_{I-K}$
Knowledge$_{DIK}$	$CCost'_{K-D}$	$CCost'_{K-I}$	$CCost'_{K-K}$

Algorithm 1. Calculating $Ratio_{IO}$ and $Inves$ of each resource protection program

Input: PR, RPS, $Ratio_{IO0}$, $Inve_0$

Output: The minimum $Ratio_{IO}$

    For: each RTY Do

        Assign value from Elements$_{DIK}$;

        Compute $CostTF$;Compute $STCost$;

        Compute $Total\_Cost$;Compute $Inves$;

        Compute $SECost$;Compute $Ratio_{IO}$;

        If $(Ratio_{IO} < Ratio_{IO0} \& Inves < Inves_0)$

           $Ratio_{IO0} = Ratio_{IO}$;

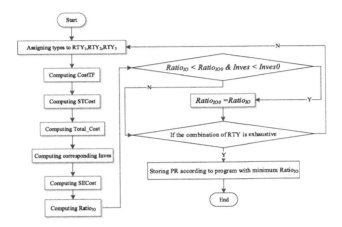

Fig. 2. Flowchart of private resources security provision approach.

4 Case Study

We give an example to illustrate the rationality and verify the feasibility of our proposed approach. For convenience, we assign values to parameters in the equations. Table 4 illustrate the assignment of scale changes of resources after being transformed. But the actual value of each parameter should be obtained through data learning so as to compare the actual differences between results. Through comparing calculated results of $Ratio_{IO}$, we choose a program with the smallest $Ratio_{IO}$ and the required investment should less than $Inve_0$.

Example 1: $PR_{type} = \{\text{Data}_{DIK}, \text{Information}_{DIK}, \text{Information}_{DIK}\}$, $PR_{sca} = \{20, 50, 80\}$;

    $RPS_{type} = \{\text{Data}_{DIK}, \text{Information}_{DIK}, \text{Knowledge}_{DIK}\}$, $RPS_{sca} = \{100, 200, 300\}$.

Figure 3 illustrates times of each indicator in $CostTF$, $CCost$ and $CCost'$ used in each resource security provision program. Edges with the same color in (a)

Table 4. Scale changes of PR after type conversion

	Data$_{DIK}$	Information$_{DIK}$	Knowledge$_{DIK}$
Data$_{DIK}$	1	1/2	1/3
Information$_{DIK}$	2	1	2/3
Knowledge$_{DIK}$	3	3/2	1

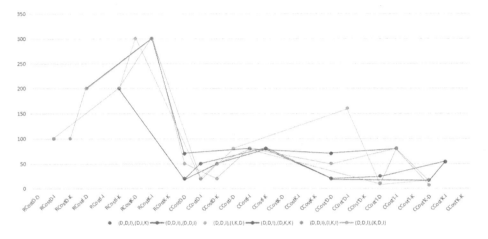

Fig. 3. Usage of indicators of $CostTF$, $CCost$ and $CCost'$ of different programs.

and (b) respectively indicate the usage of indicators of $CostTF$, $CCost$ and $CCost'$ under the same resource security provision program. In fact there are 27 security provision programs because each element of RRS can be converted into one of the three types including Data$_{DIK}$, Information$_{DIK}$ and Knowledge$_{DIK}$.

5   Related Work

Scientists have been proposing numerous models for defining anything "as a service (aaS)", including discussions of products, processes, data and information management, and security as a service [2]. In [8] the author analyzed the security protection of personal health record systems. In [5] the authors specified Knowledge Graph in a progressive manner as four basic forms including Data Graph, Information Graph, Knowledge Graph and Wisdom Graph. In [1] the authors proposed and evaluated an optimization algorithm for the powerful de-identification procedure known as k-anonymization. In [4] the author present a scheme based on probabilistic distortion of user data that could simultaneously provide a high degree of privacy to the user.

6 Conclusion

Today's technical and legal landscape presents formidable challenges to personal resource privacy. Increasing reliance on Web services causes personal data, information and knowledge to be cached, copied, and archived by third parties without people's control. Disclosure of private resource has become commonplace due to carelessness, theft, or legal actions. The contributions of our work are proposing a pay as you use resource security provision approach based on DataGraph$_{DIK}$, InformationGraph$_{DIK}$ and KnowledgeGraph$_{DIK}$. Through reducing the probability of resources being obtained by attackers and maximizing searching cost under the premise of resource owners' investment, we propose to combine storage cost and searching cost to comprehensively consider resource storage optimization scheme. In the next step we will be committed to the development of corresponding prototype system and expand the scale of datasets to verify our proposed pay as you use resource security provision approach.

Acknowledgment. This paper is supported by NSFC under Grant (No.61363007, No.61662021, No. 6161019), NSF of Hainan No.ZDYF2017128.

References

1. Bayardo, R.J., Agrawal, R.: Data privacy through optimal k-anonymization. In: ICDE 2005 Proceedings, pp. 217–228 (2005)
2. Duan, Y., Fu, G., Zhou, N., Sun, X., Narendra, N.C., Hu, B.: Everything as a service (xaas) on the cloud: Origins, current and future trends. In: IEEE International Conference on Cloud Computing, pp. 621–628 (2015)
3. Duan, Y., Shao, L., Hu, G., Zhou, Z., Zou, Q., Lin, Z.: Specifying architecture of knowledge graph with data graph, information graph, knowledge graph and wisdom graph. In: IEEE International Conference on SERA, pp. 327–332 (2017)
4. Rizvi, S.J., Haritsa, J.R.: Maintaining data privacy in association rule mining. In: Proceedings of the 28th VLDB Conference, pp. 682–693 (2002)
5. Shao, L., Duan, Y., Sun, X., Gao, H.: Answering who/when, what, how, why through constructing data graph, information graph, knowledge graph and wisdom graph. In: Proceedings of the International Conference on SEKE, pp. 1–7 (2017)
6. Shao, L., Duan, Y., Sun, X., Zou, Q., Jing, R., Lin, J.: Bidirectional value driven design between economical planning and technical implementation based on data graph, information graph and knowledge graph. In: IEEE International Conference on SERA, pp. 339–344 (2017)
7. Soria-Comas, J., Domingo-Ferrer, J.: Big data privacy: challenges to privacy principles and models. Data Sci. Eng. **1**(1), 21–28 (2016)
8. Win, K.T., Susilo, W., Yi, M.: Personal health record systems and their security protection. J. Med. Syst. **30**(4), 309–315 (2006)

An Investment Defined Transaction Processing Towards Temporal and Spatial Optimization with Collaborative Storage and Computation Adaptation

Yucong Duan[1(✉)], Lixu Shao[1], Xiaobing Sun[2], Donghai Zhu[1], Xiaoxian Yang[3], and Abdelrahman Osman Elfaki[4]

[1] College of Information Science and Technology, State Key Laboratory
of Marine Resource Utilization in the South China Sea,
Hainan University, Haikou, China
duanyucong@hotmail.com, 751486692@qq.com, 466971642@qq.com
[2] School of Information Engineering, Yangzhou University, Yangzhou, China
sundomore@163.com
[3] School of Computer and Information, Shanghai Polytechnic University,
Shanghai, China
yangxiaoxian@shu.edu.cn
[4] Faculty of Information Sciences and Engineering, MSU, Shah Alam, Malaysia
abdelrahmanelfaki@gmail.com

Abstract. Transaction processing technology is the key technology of reporting information consistency and reliability, and determines whether web services can be applied to e-commerce. We propose an investment defined transaction processing approach towards temporal and spatial optimization with collaborative storage and computation adaptation approach aiming at satisfying requirements of different users according to their investment. We studies resource modelling, resource processing, processing optimization and resource management and puts forward a three-tier architecture consisting of Data Graph, Information Graph and Knowledge Graph.

Keywords: Resource modelling · Knowledge graph · Transaction processing · Resource optimization

1 Introduction

Due to the unreliability of web service platform, capability of providing transaction processing is one of the key technologies determining whether a web service can be applied to commercial applications. Scientists have been proposing numerous models for defining anything "as a service (aaS)" [1]. A Knowledge Graph is constructed through representing each item, entity and user as nodes, and linking those nodes that interact with each other via edges. In our previous work [2], we specified Knowledge Graph in a progressive manner as four basic

© Springer International Publishing AG 2017
H. Yin et al. (Eds.): IDEAL 2017, LNCS 10585, pp. 452–460, 2017.
https://doi.org/10.1007/978-3-319-68935-7_49

forms including Data Graph, Information Graph, Knowledge Graph and Wisdom Graph and answering 5 W questions through constructing the architecture composing of Data Graph, Information Graph, Knowledge Graph and Wisdom Graph [6]. Development of software service system can be divided into stages of data sharing, information transfer, and knowledge creation in terms of data, information and knowledge [7]. Through reasonable organization and storage of massive resources, we can achieve the goal of satisfying resource requirements of users with optimized spatial and temporal efficiency.

The rest of the paper is structured as follows. Section 2 illustrates definitions of typed resources and graphs. Section 3 elaborates the proposed investment defined transaction processing approach. Section 4 gives an example of resource type conversion. The related works are presented in Sect. 5 and we summarize the conclusion of our work in Sect. 6.

2   Definitions of Resources and Graphs

Resources that exist on Internet are mixed and are difficult to be dealt with. We identify and divide types of resources into three kinds that are $Data_{DIK}$, $Information_{DIK}$ and $Knowledge_{DIK}$ where DIK is an abbreviation for $Data_{DIK}$, $Information_{DIK}$ and $Knowledge_{DIK}$. We elaborate conceptual approaches for defining $Data_{DIK}$, $Information_{DIK}$, $Knowledge_{DIK}$ in a progressive manner and specify architecture of Knowledge Graph in three levels including $DataGraph_{DIK}$, $InformationGraph_{DIK}$ and $KnowledgeGraph_{DIK}$. We show the progressive forms of resource types in Table 1. Definitions of resource element and the three graphs are as follows.

Table 1. Progressive forms of typed resources

(Rightside: general forms)	$Data_{DIK}$	$Information_{DIK}$	$Knowledge_{DIK}$
Semantic load	Not specified for stakeholders/ machine	Settled for stakeholders/machine	Abstracting on known resources
Format	Discrete elements	Related elements	Probabilistic or Categorization
Knowledge answer	Who/When/Where	What	How
Usage	Transmission	Communication, Interaction	Reasoning, Predicting
Subgraph	$DataGraph_{DIK}$	$InformationGraph_{DIK}$	$KnowledgeGraph_{DIK}$

Definition 1 *(Resource element). We define resource element as a tuple RE = $< RE_{type}, RE_{amount} >$ where RE_{type} is the type of resource element including $Data_{DIK}$, $Information_{DIK}$ and $Knowledge_{DIK}$. Let $Type_{DIK}$ be the type set of resources and $Type_{DIK}$ is a collection of $Data_{DIK}$, $Information_{DIK}$ and $Knowledge_{DIK}$. RE_{amount} is the scale of resource element.*

Definition 2 *(Graphs). We propose to specify the existing concept of Knowledge Graph in three layers. Graphs can be expressed as:*
$Graph_{DIK}$: = $<(DataGraph_{DIK})$, $(InformationGraph_{DIK})$, $(Knowledge Graph_{DIK})>$.

Definition 3 *($DataGraph_{DIK}$). We define $DataGraph_{DIK}$ as:*
$DataGraph_{DIK}$: = collectionOf {array, list, stack, queue, tree, graph}.
$DataGraph_{DIK}$ is a collection of discrete elements expressed in the form of various data structures including arrays, lists, stacks, trees, graphs, and so on. $DataGraph_{DIK}$ is used to model the temporal and spatial features of $Data_{DIK}$.

Definition 4 *($InformationGraph_{DIK}$). We define $InformationGraph_{DIK}$ as:*
$InformationGraph_{DIK}$: = combinationOf {related $Data_{DIK}$}.
$InformationGraph_{DIK}$ is a combination of related $Data_{DIK}$. Information $Graph_{DIK}$ expresses the interaction and transformation of $Information_{DIK}$ between entities in the form of a directed graph.

Definition 5 *($KnowledgeGraph_{DIK}$). We define $KnowledgeGraph_{DIK}$ as:*
$KnowledgeGraph_{DIK}$: = collectionOf {statistical rules}.
$KnowledgeGraph_{DIK}$ is of free schema and expresses rich semantic relationships which is conductive to have a completing mapping towards users' requirements described through natural language.

3   Optimization of Temporal and Spatial Efficiency

3.1   Input of the Problem

Firstly, we give the input of the problem discussed in our work.

Definition 6 *(Transaction processing resources). We define transaction processing resources as a triple $TPR = < TPR_1, TPR_2, TPR_3 >$. Each TPR is a collection of RE with the same type. The type set of TPR is $TTS = \{TTS_1, TTS_2, TTS_3\}$ that can be $Data_{DIK}$, $Information_{DIK}$ and $Knoweldge_{DIK}$ and the scale set of TPR is $TSS = \{TSS_1, TSS_2, TSS_3\}$.*

Definition 7 *(Resources on $Graph_{DIK}$). We define resources on $Graph_{DIK}$ as:*
RoG:= RoG_1, RoG_2, RoG_3. The type set of RoG is $RTS = \{RTS_1, RTS_2, RTS_3\}$ where RTS represents a collection of RE with the same type and the scale set of RoG is $RSS = \{RSS_1, RSS_2, RSS_3\}$.

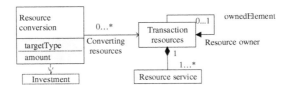

Fig. 1. Meta model of transaction processing resources.

3.2   Illustration of Resource Type Conversion

Figure 1 shows a meta-model of transaction processing resources. Resource sharing, personalized recommendation can be provided through constructing resource processing architecture. We define resource conversion as a tuple $< targetType, amount >$ where $targetType$ is the type of resources after being transformed and $amount$ indicates the scale of resources being transformed that is determined by users' investment.

• Conversion from $Data_{DIK}$ to $Information_{DIK}$

$Data_{DIK}$ is obtained through direct observation of numbers or other basic individual items. For example, $Data_{DIK}$ about authors, papers and conferences can be transformed into $Information_{DIK}$ through clustering algorithm. If the phenomenon where papers of the same author are published in different conferences occurs frequently, we recognize the conferences and cluster them into a class with specific theme.

• Conversion from $Information_{DIK}$ to $Knowledge_{DIK}$

$Knowledge_{DIK}$ is useful resources that are obtained through filtering, refining and processing relevant $Information_{DIK}$. In a special semantic context, $Knowledge_{DIK}$ is conductive to establish a meaningful link between application of interactions between $Information_{DIK}$, which embodies the nature, principles, and experience of $Information_{DIK}$. Through analyzing and summarizing various $Information_{DIK}$ about people's travel records, $Knowledge_{DIK}$ such as people's consumption habits or travelling preference can be obtained.

• Conversion from $Data_{DIK}$ to $Knowledge_{DIK}$

When the scale of $Data_{DIK}$ that a user wants to query is very large, we consider searching resources on $Knowledge_{DIK}$ and obtain $Data_{DIK}$ through reasoning $Knowledge_{DIK}$. For example, transaction processing resources are the natural numbers from 1 to 10000. Answers can be inferred through reasoning a static rule that is an arithmetic progression stored in the form of $Knowledge_{DIK}$.

The characteristics of our proposed approach lie in that storage schema is determined by calculation. Table 2 shows the searching method of resources of TPR in RoG with different type combinations.

3.3   Calculation of Resource Type Conversion Cost

Assumption 1. We make the hypothesis that resources of TPR have been organized on $Graph_{DIK}$.

Table 2. Conversion method of typed transaction resources

	Data$_{DIK}$	Information$_{DIK}$	Knowledge$_{DIK}$
Data$_{DIK}$	Traversing	Analyzing meaning of Data$_{DIK}$	Conclusion obtained from Data$_{DIK}$
Information$_{DIK}$	Analysis	Traversing	Application of Information$_{DIK}$
Knowledge$_{DIK}$	Reasoning	Reasoning	Traversing

Transaction resources have the original type and scale. We assign values from $Type_{DIK}$ to each element in the type set TTS of transaction processing resources. The atomic type conversion cost of RE in TPR, denoted as $TCost$, is shown in Table 3(a). Type conversion cost of all resources from their original types to corresponding types after conversion, denoted as $CostMT_1$, can be calculated according to Eq. 1:

$$CostMT_1 = \sum_{i=1}^{3}(TCost * TSS_i + TSS_i')$$ (1)

where TSS_i' is the scale of original transaction resources after being transformed. We assign values from $Type_{DIK}$ to each element of the type set RTS of resources on Graph$_{DIK}$ to form the combination case. The atomic type conversion cost of RE in RSS, denoted as $RCost$, is shown in Table 3(b). Type conversion cost of all resources from their original types to corresponding types after conversion, denoted as $CostMT_2$, can be calculated according to Eq. 2:

$$CostMT_2 = \sum_{i=1}^{3}(RCost * RSS_i + RSS_i')$$ (2)

where RSS_i' is the scale of resources on Graph$_{DIK}$ after being transformed.

3.4   Computation Cost of Processing TPR in RoG

An investment defined transaction processing approach aims at satisfying requirements of different users according to their investment through calculating the computation cost of searching needed resources and storage cost of transaction resources. Table 4 shows the atomic type conversion cost unit resource element that is denoted as SC. Computation cost of searching TPR in RoG consists of traversing cost and type conversion cost from found resources to the initial transaction resources. We calculate the computation cost of processing TPR in RoG, denoted as $Cost_c$, according to Eq. 3:

$$Cost_c = \sum_{i=1}^{3}(RSS_i + SC * RSS_i')$$ (3)

where RSS' is scale of found resources that may be of different type with the type of initial needed resources.

Table 3. Atomic conversion cost

(a) Atomic type conversion cost of RE in TPR

	Data$_{DIK}$	Information$_{DIK}$	Knowledge$_{DIK}$
Data$_{DIK}$	$TCost_{D-D}$	$TCost_{D-I}$	$TCost_{D-K}$
Information$_{DIK}$	$TCost_{I-D}$	$TCost_{I-I}$	$TCost_{I-K}$
Knowledge$_{DIK}$	$TCost_{K-D}$	$TCost_{K-I}$	$TCost_{K-K}$

(b) Atomic type conversion cost of RE in RoG

	Data$_{DIK}$	Information$_{DIK}$	Knowledge$_{DIK}$
Data$_{DIK}$	$RCost_{D-D}$	$RCost_{D-I}$	$RCost_{D-K}$
Information$_{DIK}$	$RCost_{I-D}$	$RCost_{I-I}$	$RCost_{I-K}$
Knowledge$_{DIK}$	$RCost_{K-D}$	$RCost_{K-I}$	$RCost_{K-K}$

Table 4. Atomic type conversion cost of RE

	Data$_{DIK}$	Information$_{DIK}$	Knowledge$_{DIK}$
Data$_{DIK}$	SC_{D-D}	SC_{D-I}	SC_{D-K}
Information$_{DIK}$	SC_{I-D}	SC_{I-I}	SC_{I-K}
Knowledge$_{DIK}$	SC_{K-D}	SC_{K-I}	SC_{K-K}

3.5   Calculation of Total Cost and Corresponding Investment

The expected investment of users that is denoted as $Inve_0$ is pre-set. After computing $CostMT_1$, $CostMT_2$ and $Cost_c$, we calculate the total cost denoted as $Total\_Cost$ of each type conversion program according to Eq. 4:

$$Total\_Cost = CostMT_1 + Cost_c + CostMT_2 \tag{4}$$

And the corresponding required investment that is denoted as $Inve$ is computed according to Eq. 5:

$$Inve = \gamma * Total\_Cost \tag{5}$$

where γ represents the required average investment that can be obtained through data training. We use the ratio of investment and $Cost_c$ to measure the suitability of different type conversion program that can be calculated according to Eq. 6:

$$Inve\_Cos = \frac{Inve}{Cost_c} \tag{6}$$

Then we compare $Inve\_Cos$ of different program to find the optimal program with maximum $Inve\_Cos$. Algorithm 1 elaborates the whole process of our proposed investment defined transaction processing approach.

Algorithm 1. Calculating $Inve\_Cos$ and $Inve$ of each resource type conversion program

Input: TPR, RoG, $Inve\_Cos_0$, $Inve_0$
Output: The maximum $Inve\_Cos$
    For: each TTS Do
        Assign value from $Type_{DIK}$;Compute $CostMT_1$;
        For: each RTS Do
            Assign value from $Type_{DIK}$;
            Compute $CostMT_2$;Compute $Cost_c$;
            Compute $Total\_Cost$;Compute $Inve\_Cos$;
            If $(Inve\_Cos > Inve\_Cos_0 \& Inve < Inve_0)$
                $Inve\_Cos_0 = Inve\_Cos$;

4 Examples of Collaborative Storage and Computation Adaptation

We give an example to illustrate the rationality of our proposed approach. Here we have listed seven cases of TPS and RoG assignment and the corresponding calculations of $CostMT_1$, $CostMT_2$ and $Cost_c$. For convenience, we assign values to parameters in the equations. Table 5 illustrates the assignment of scale changes

Table 5. Scale changes of RE after type conversion

	Data$_{DIK}$	Information$_{DIK}$	Knowledge$_{DIK}$
Data$_{DIK}$	1	1/2	1/3
Information$_{DIK}$	2	1	2/3
Knowledge$_{DIK}$	3	3/2	1

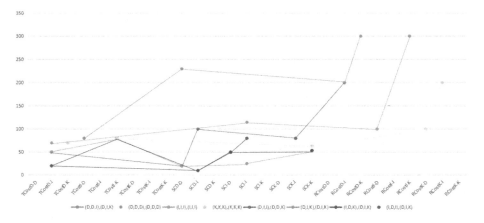

Fig. 2. Usage of indicators of $CostMT_1$, $Cost_c$ and $CostMT_2$.

of resource element. But the actual value of each parameter should be obtained through data learning so as to compare the actual differences between results.

Example: $TPR = \{\text{Data}_{DIK}, \text{Information}_{DIK}, \text{Information}_{DIK}\}$, $TSS = \{20, 50, 80\}$;

$RoG = \{\text{Data}_{DIK}, \text{Information}_{DIK}, \text{Knowledge}_{DIK}\}$, $RSS = \{100, 200, 300\}$.

Figure 2 illustrates usage of each indicator in $CostMT_1$, $Cost_c$ and $CostMT_2$ in each collaborative storage and adaptation program. Edges with different colors represent different programs. In fact there are 27 * 27 programs because each element of TPR and RoG can be transformed into one of the three types including Data_{DIK}, Information_{DIK} and Knowledge_{DIK}.

5 Related Work

A hierarchical model composing data, information, knowledge and wisdom is usually represented in the form of a pyramid [5]. The dynamic reconstruction of computation resources and storage resources not only improves the utilization of resources but also simplifies management. [4] introduced a representation format for relational data and components schema based on the Web Ontology Language. In [3] the authors described a knowledge representation schema form design that supported the initiation and continuation of the act of designing. We divide transaction processing resources into Data_{DIK}, Information_{DIK} and Knowledge_{DIK} since mix resources are hard to be dealt with. Capabilities to model software real-time embedded resources will allow describing software multitasking platform characteristics explicitly [8].

6 Conclusion

In the environment of web service, transaction processing must provide the ability to coordinate short-term operations and long-term business activities. The contributions of our work are to improve the spatial and temporal efficiency through optimizing processing mechanism and storage scheme of typed transaction resources including Data_{DIK}, Information_{DIK} and Knowledge_{DIK} and rationally deal with massive resources on Internet through the analysis and abstraction of transaction processing. Different resource requirements are searched on the most matching resource level which can improve the search efficiency effectively. In the next stage we will expand the scale of data to verify our proposed architecture.

Acknowledgment. This paper is supported by NSFC under Grant (No. 61363007, No. 61662021, No. 61661019), NSF of Hainan No. ZDYF2017128.

References

1. Duan, Y., Fu, G., Zhou, N., Sun, X., Narendra, N.C., Hu, B.: Everything as a service (xaas) on the cloud: origins, current and future trends. In: IEEE International Conference on Cloud Computing, pp. 621–628 (2015)

2. Duan, Y., Shao, L., Hu, G., Zhou, Z., Zou, Q., Lin, Z.: Specifying architecture of knowledge graph with data graph, information graph, knowledge graph and wisdom graph. In: IEEE International Conference on SERA, pp. 327–332 (2017)

3. Gero, J.S.: Design prototypes: a knowledge representation schema for design. AI Magazine **11**(4), 26–36 (1990)

4. Laborda, C.P.D., Conrad, S.: Relational.owl - a data and schema representation format based on owl. In: Conceptual Modelling 2005, pp. 89–96 (2005)

5. Rowley, J.: The wisdom hierarchy: representations of the DIKW hierarchy. J. Inf. Sci. **33**(2), 163–180 (2010)

6. Shao, L., Duan, Y., Sun, X., Gao, H.: Answering who/when, what, how, why through constructing data graph, information graph, knowledge graph and wisdom graph. In: Proceedings of the International Conference on SEKE, pp. 1–7 (2017)

7. Shao, L., Duan, Y., Sun, X., Zou, Q., Jing, R., Lin, J.: Bidirectional value driven design between economical planning and technical implementation based on data graph, information graph and knowledge graph. In: IEEE International Conference on SERA, pp. 339–344 (2017)

8. Thomas, F., Gerard, S., Delatour, J., Terrier, F.: Software real-time resource modeling. In: FDL 2007, pp. 231–236 (2008)

Interactive Data Visualization Using Dimensionality Reduction and Dissimilarity-Based Representations

D.F. Peña-Unigarro[1]([✉]), P. Rosero-Montalvo[2,3], E.J. Revelo-Fuelagán[1],
J.A. Castro-Silva[4], J.C. Alvarado-Pérez[5,6], R. Therón[6],
C.M. Ortega-Bustamante[2], and D.H. Peluffo-Ordóñez[2,5]

[1] Universidad de Nariño, Pasto, Colombia
diferpun@gmail.com
[2] Universidad Técnica del Norte, Ibarra, Ecuador
[3] Instituto Tecnológico Superior 17 de Julio, Ibarra, Ecuador
[4] Universidad Surcolombiana, Neiva, Colombia
[5] Coorporación Universitaria Autónoma de Nariño, Pasto, Colombia
[6] Universidad de Salamanca, Salamanca, Spain

Abstract. This work describes a new model for interactive data visualization followed from a dimensionality-reduction (DR)-based approach. Particularly, the mixture of the resulting spaces of DR methods is considered, which is carried out by a weighted sum. For the sake of user interaction, corresponding weighting factors are given via an intuitive color-based interface. Also, to depict the DR outcomes while showing information about the input high-dimensional data space, the low-dimensional representations reached by the mixture is conveyed using scatter plots enhanced with an interactive data-driven visualization. In this connection, a constrained dissimilarity approach define the graph to be drawn on the scatter plot.

Keywords: Data visualization · Dimensionality reduction · Pairwise dissimilarity

1 Introduction

In math terms, the goal of dimensionality reduction is to embed a high-dimensional data matrix $\mathbf{Y} = [\mathbf{y}_i]_{1 \leq i \leq N}$, such that $\mathbf{y}_i \in \mathbb{R}^D$ into a low-dimensional, latent data matrix $\mathbf{X} = [\mathbf{x}_i]_{1 \leq i \leq N}$, being $\mathbf{x}_i \in \mathbb{R}^d$, where $d < D$ [1,2], this issue allows both improving the performance of a pattern recognition system and reaching intelligible data representation trough the generation of a low-dimensional space from a high-dimensional space which preserves the structure of raw data as much as possible [2,3].

Due to mostly DR methods are executed as black box processes, they are usually not provided with user properties like synchrony, interactivity and controllability for the execution stages [4]. Such properties are characteristic of the

© Springer International Publishing AG 2017
H. Yin et al. (Eds.): IDEAL 2017, LNCS 10585, pp. 461–469, 2017.
https://doi.org/10.1007/978-3-319-68935-7_50

field of information visualization [5]. A number of works have been proposed attempting to import some properties of InfoVis into DR scenarios, which are mainly based on interaction models using geometric [6,7], and color-based [8] approaches. Nonetheless, data representation is still an open problem since there are many methods and models to explore and this work outlines a novel visualization approach based on an interactive dimensionality reduction model which is carried out by means of a mixture of data representations resulting from a three different DR methods. Such a mixture is performed via a weighted sum whose weighting factors are interactively given by users within a color-based framework. As a remarkable contribution of this work, it is important to highlight the incorporation of an interactive graph-based visualization able to illustrate the relationship between the DR outcomes and the structure of high-dimensional data space. Specifically, a constrained dissimilarity approach is proposed to define the graph to be drawn on the scatter plot. For experiments, the visualization approach is tested on two artificial data sets and the quality of the low-dimensional spaces is quantified by a scaled version of the average agreement rate between K-ary neighborhoods [9].

The rest of this paper is structured as follows: In Sect. 2, Data visualization via dimensionality reduction is outlined. Section 3 introduces the proposed interactive data visualization scheme. The experimental setup and results are presented in Sects. 4 and 5, respectively. Finally, Sect. 6 gathers some final remarks as conclusions and future work.

2   Data Visualization via Dimensionality Reduction

In data analysis, visualization may be considered as the first stage whose goal is to make sense or provide initial hints of the data before proceeding with others steps like modeling, classification and analysis [10]. Given a large set of measured variables, an intuitive idea to visualize is to reduce the attributes or features in the measurements by representing them with a smaller set of more condensed variables [11]. In this sense dimensionality reduction is the process of discovering the explanatory variables that account for the greatest changes in the response variable, due to working with thousands or even millions of explanatory variables is computationally expensive and impossible to visualize [12]. Therefore, dimensionality reduction can also be used to visualize important patterns, which might be inside of the raw data.

3   Proposed Model for Interactive Dimensionality Reduction

Human beings have a visual nature due to the fact that the bandwidth of vision is greater than all others senses combined [13]. By taking into account this fact, a chromatic model is proposed to take advantage of color's visual nature. In this sense, the mixture of DR methods is not directly given by numerical values but through colors, so the interaction becomes more direct and easy for the user.

3.1 Chromatic Model Implementation

The combination of three DR methods is made with a technique known as **barycentric interpolation** because weighted factors can be represented by colors [14] due to intensity values of the three channels are made clicking on or picking up a point inside the triangle Fig. 1 so a primary color represents a specific DR method.

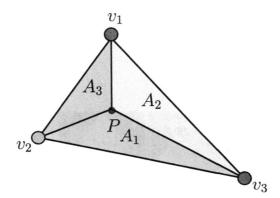

Fig. 1. Diagram that illustrates the proposed color-based model for DR. The term v_1 is the vertex (pixel) that contains the greatest value of the color red, in this point the intensity values of the two remaining colors have a intensity value equal to 0. Similarly, this logic is applied when a point is chosen in the two remaining vertices v_2 (green) y v_3 (blue). (Color figure online)

From the Fig. 1 the amount of one specific color can be defined as $A_\ell/A = \alpha_m$, where, A_ℓ denotes the area of the $\ell-$th triangle, A is the total area and α_m is the $m-$th weighted factor with the property shown in 1.

$$\sum_{i=1}^{3} \alpha_i = 1. \tag{1}$$

The area of the triangle A_ℓ can be calculated through the coordinates of its vertices with the next expression:

$$A_1 = \frac{1}{2} \left| \det \begin{pmatrix} 1 & p_x & p_y \\ 1 & v2x & v2y \\ 1 & v3x & v3y \end{pmatrix} \right|, \quad A_2 = \frac{1}{2} \left| \det \begin{pmatrix} 1 & p_x & p_y \\ 1 & v1x & v1y \\ 1 & v3x & v3y \end{pmatrix} \right|, \quad \text{and} \tag{2}$$

$$A_3 = \frac{1}{2} \left| \det \begin{pmatrix} 1 & p_x & p_y \\ 1 & v1x & v1y \\ 1 & v2x & v2y \end{pmatrix} \right|,$$

where the operator $\det(\cdot)$ is related to the determinant of a matrix and $|\cdot|$ means absolute value and sub-indexes x and y denote coordinates in the Cartesian plane.

For the chromatic model creation the equation (2) is applied over two grids with 256 rows by 256 columns (256×256 pixels), that represents the possible ordered pairs x, y in the Cartesian coordinate system, as a result of this procedure three images are created which represent channels red, green, and blue. Furthermore, the Hadamard product (Element-wise) is applied between a triangular binary mask with vertices in $v_1 = \{0, 255\}, v_2 = \{255, 0\}, v_3 = \{127, 255\}$ and the three channels which are bounded in a triangular plane. Finally the three triangular channels are superposed in order to obtain the proposed chromatic model. In addition, it is important take into account that intensity values or weighted factors are normalized (Fig. 2).

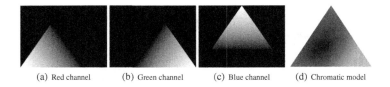

(a) Red channel   (b) Green channel   (c) Blue channel   (d) Chromatic model

Fig. 2. Channels of the primary colors (a) to (c) and the chromatic model used for the interactive dimensionality reduction mixture (c).

3.2 Mixture

Let us suppose that the input matrix $\boldsymbol{Y}$ is reduced by using M different DR methods, yielding then a set of normalized lower-dimensional representations: $\{\boldsymbol{X}^{(1)}, \cdots, \boldsymbol{X}^{(M)}\}$. Now, a lineal combination of this embedded spaces can be obtained as follows:

$$\bar{\boldsymbol{X}} = \sum_{m=1}^{M} \alpha_m \boldsymbol{X}^{(m)}, \tag{3}$$

where $\{\alpha_1, \cdots, \alpha_M\}$ are the weighting factors.

3.3 Dissimilarity-Based Visualization

In this work, a dissimilarity-based visualization approach is introduced with the aim to provide a visual hint about the structure of the high-dimensional input data matrix $\boldsymbol{Y}$ into the scatter plot of its representation in a lower-dimensional space $\boldsymbol{X}$ in order to find clues that allow finding the best low-dimensional representation. To do so, we use a pairwise dissimilarity matrix $\mathcal{D} \in \mathbb{R}^{N \times N}$, such that $\mathcal{D} = [d_{ij}]$ where, the entries of d_{ij} defines the dissimilarity between the i-th and j-th data point from $\boldsymbol{Y}$.

3.4   General Scheme

See Fig. 3.

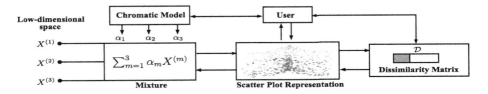

Fig. 3. The general interface's scheme which involves the mixture of DR outcomes, where the color selected defines the embedded data and the scatter plot visualization with the high-dimensional information provided by the pairwise-dissimilarity-based scheme. (Color figure online)

4   Experimental Setup

Database: In order to visually evaluate the performance of the Data-visualization approach, two data-sets are used: The first one is known as artificial spherical shell ($N = 700$ data points and $D = 3$) (Fig. 4(a)) and the second one is a toy set called here Swiss roll ($N = 700$ data points and $D = 3$) (Fig. 4(b)).

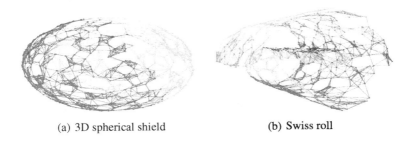

(a) 3D spherical shield   (b) Swiss roll

Fig. 4. 3D spherical shield and Swiss roll data-sets.

Parameter settings and methods: In order to capture the local structure for visualization, the euclidean distance given by $\|\boldsymbol{y}_{(i)} - \boldsymbol{y}_{(j)}\|^2$ is used also a parameter which defines the number of K-nearest neighbors is taken into account, this number is provided by a slider-bar in the interface with maximum K-nearest neighbors of 20. To perform the dimensionality reduction three spectral DR methods $M = 3$ are considered: CMDS [4], LE [15], LLE [16] which can be combine through the chromatic model in a shorter time response than other kernel approaches such as [6,7], due to there is not need to find kernel matrices, eigenvalues and eigenvectors. All of DR methods are intended to obtain spaces in two dimensions $(d = 2)$.

Performance measure: To quantify the performance of the studied methods, the scaled version of the average agreement rate $R_{NX}(K)$ introduced in [9] is used, which is ranged within the interval $[0, 1]$, since $R_{NX}(K)$ is calculated at each perplexity value from 2 to $N - 1$ and a performance indicator can be obtained by calculating its area under the curve (AUC).

Experiment description: To assess the performance of the interactive visualization interface and illustrate that some combinations of DR methods might have a better performance than the application of individuals DR methods, a testings were done by clicking inside the colored surface. Here, particularly a test is made in the vertexes and a random point inside the surface.

5   Results and Discussion

A first visual result is obtained from the dissimilarity graph, due to the chosen color (weighting factors) is affected the data topology and the k-nearest neighbors preservation, an example of this is shown in the Fig. 5 where in the first embedded space (Fig. 5(a)), the black data point bounded by the red circle do not maintain the properties of the original space due to the fact that the edge is not connected with the nearest point. Nevertheless, in the second embedded space (Fig. 5(b)) with other mixture the black data point in the red circle keeps the topology of the original space because in this case the edge is connected with the nearest neighbor.

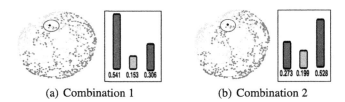

(a) Combination 1   (b) Combination 2

Fig. 5. Visual example of data topology conservation according to the embedded spaces mixture. Inside the red circle in both figures it can be appreciated a black point with one edge that connect his nearest neighbor in the high-dimensional space in order to give a graphical notion of neighbor preservation in the low-dimensional space. (Color figure online)

Nonetheless, the lineal combination of embedded spaces may improve the quality of resultant low-dimensional data. In fact, the process of combining the results of three different DR methods naturally obtained from exploring the triangular plane with an approximate resolution of $\alpha_{res} \approx \frac{1}{255}$ may yield better embedding representations when the effects of DR methods are adequately blended. For example, Figs. 6(e) and 7(e) shows an instance where the weighted sum of the embedded spaces with $\alpha_1 = 0.039$, $\alpha_2 = 0.004$, $\alpha_3 = 0.957$ and $\alpha_1 = 0.320$, $\alpha_2 = 0.316$, $\alpha_3 = 0.363$ respectively, reaches better performance in terms of the considered quality measure [9]. Such as performance is associated

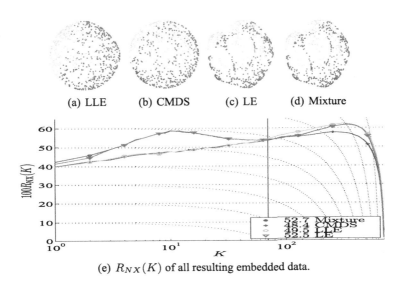

(e) $R_{NX}(K)$ of all resulting embedded data.

Fig. 6. Embedded spaces and quality assessment curve for the spherical shell data-set.

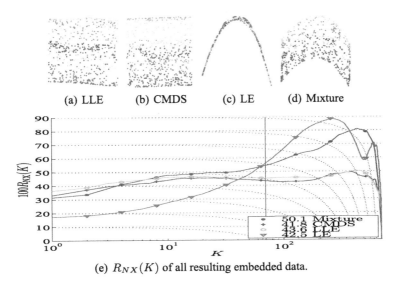

(e) $R_{NX}(K)$ of all resulting embedded data.

Fig. 7. Embedded spaces from the experiment and quality assessment curve for the Swiss roll data-set.

with the ability to unfold in 2D the spherical shell (Fig. 6(a)–(d)) and the Swiss roll (Fig. 7(a)–(d)).

6   Conclusions and Future Work

The proposed visualization methodology represents an alternative way of data visualization, in which dimensionality reduction outcomes -meaning embedded spaces- obtained by different methods can be mixed through a direct manipulation of a chromatic model in a interactive and friendly-user fashion. Thereby, an user can take advantage of DR methods and choose the data representation that best meets the requirements for a specific data analysis task. Additionally, our methodology incorporates a dissimilarity-based visualization provides a graphic and intelligible notion of the data topology preservation. Finally, as a future work, other dimensionality reduction methods are to be integrated into the data-driven dimensionality reduction frameworks with the aim of reaching a good trade-off between preservation of the data structure and performance of visualization results. Also, more mathematical properties will be explored to design data-driven schemes that best approximate the data topology as well as represent in alternative manners dimensionality reduction methods so that a generalized framework can be formulated.

Acknowledgments. This work is supported by the "Grupo de Investigación en Ingeniería Eléctrica y Electrónica - GIIEE" from Universidad de Nariño. As well, the authors acknowledge to the research project "Desarrollo de una metodología de visualización interactiva y eficaz de información en Big Data" supported by Agreement No. 180 November 1st, 2016 by VIPRI from Universidad de Nariño, as well as Universidad Técnica del Norte – Ecuador.

References

1. Peluffo-Ordóñez, D.H., Lee, J.A., Verleysen, M.: Recent methods for dimensionality reduction: a brief comparative analysis. In: European Symposium on Artificial Neural Networks (ESANN). Citeseer (2014)
2. Peluffo-Ordóñez, D.H., Lee, J.A., Verleysen, M.: Short review of dimensionality reduction methods based on stochastic neighbour embedding. In: Villmann, T., Schleif, F.-M., Kaden, M., Lange, M. (eds.) Advances in Self-Organizing Maps and Learning Vector Quantization. AISC, vol. 295, pp. 65–74. Springer, Cham (2014). doi:10.1007/978-3-319-07695-9_6
3. Lee, J.A., Verleysen, M.: Nonlinear Dimensionality Reduction. Springer Science & Business Media, New York (2007)
4. Borg, I.: Modern Multidimensional Scaling: Theory and Applications. Springer, New York (2005)
5. Ward, M., Grinstein, G., Keim, D.: Interactive Data Visualization: Foundations, Techniques, and Applications. AK Peters Ltd., Natick (2010)
6. Salazar-Castro, J., Rosas-Narváez, Y., Pantoja, A., Alvarado-Pérez, J.C., Peluffo-Ordóñez, D.H.: Interactive interface for efficient data visualization via a geometric approach. In: 2015 20th Symposium on Signal Processing, Images and Computer Vision (STSIVA), pp. 1–6. IEEE (2015)

7. Peluffo-Ordónez, D.H., Alvarado-Pérez, J.C., Lee, J.A., Verleysen, M.: Geometrical homotopy for data visualization. In: European Symposium on Artificial Neural Networks (ESANN 2015). Computational Intelligence and Machine Learning (2015)

8. Peña-ünigarro, D.F., Salazar-Castro, J.A., Peluffo-Ordóñez, D.H., Rosero-Montalvo, P.D., Oña-Rocha, O.R., Isaza, A.A., Alvarado-Perez, J.C., Theron, R.: Interactive visualization methodology of high-dimensional data with a color-based model for dimensionality reduction. In: 2016 XXI Symposium on Signal Processing, Images and Artificial Vision (STSIVA), pp. 1–7. IEEE (2016)

9. Lee, J.A., Renard, E., Bernard, G., Dupont, P., Verleysen, M.: Type 1 and 2 mixtures of kullback-leibler divergences as cost functions in dimensionality reduction based on similarity preservation. Neurocomputing **112**, 92–108 (2013)

10. Sedlmair, M., Munzner, T., Tory, M.: Empirical guidance on scatterplot and dimension reduction technique choices. IEEE Trans. Visual. Comput. Graphics **19**, 2634–2643 (2013)

11. Sedlmair, M., Aupetit, M.: Data-driven evaluation of visual quality measures. In: Computer Graphics Forum, vol. 34, pp. 201–210. Wiley Online Library (2015)

12. Hackeling, G.: Mastering Machine Learning with scikit-learn. Packt Publishing Ltd., Birmingham (2014)

13. Ware, C.: Information Visualization: Perception for Design. Elsevier, Amsterdam (2012)

14. Meyer, M., Barr, A., Lee, H., Desbrun, M.: Generalized barycentric coordinates on irregular polygons. J. Graph. Tools **7**, 13–22 (2002)

15. Roweis, S.T., Saul, L.K.: Nonlinear dimensionality reduction by locally linear embedding. Science **290**, 2323–2326 (2000)

16. Belkin, M., Niyogi, P.: Laplacian eigenmaps for dimensionality reduction and data representation. Neural Comput. **15**, 1373–1396 (2003)

Applying Random Forest to Drive Recommendation

Le Zhan[1], Jingwei Zhang[2,3], Qing Yang[1(✉)], and Yuming Lin[2]

[1] Guangxi Key Laboratory of Automatic Measurement Technology and Instrument,
Guilin University of Electronic Technology, Guilin 541004, China
1217093432@qq.com, gtyqing@hotmail.com
[2] Guangxi Key Laboratory of Trusted Software,
Guilin University of Electronic Technology, Guilin 541004, China
gtzjw@hotmail.com, ymlinbh@163.com
[3] Guangxi Cooperative Innovation Center of Cloud Computing and Big Data,
Guilin University of Electronic Technology, Guilin 541004, China

Abstract. Accurate information to users, which is required by online shopping, self-help travel etc. is very important to improve user experience. Recommendation is an important mechanism to match useful information to users with needs. Existing recommendation methods generally rely on massive similarity computation between users and recommended objects, which do not consider some fine-grained information and are not suitable for online recommendation. This paper introduces a novel model for recommendation, which merges classification strategy into recommendation and transforms classification rules into recommendation rules. Random forest is integrated with the proposed model for classification and then a ranking processing is carried out to find top-k users for recommendation. The proposed method makes full use of classification output and the relationships between users and recommended things, it is more suitable for online recommendation. Extensive experiments on different kinds of datasets indicate that the proposed method is effective.

Keywords: Recommendation · Random forest · Attribute information

1 Introduction

Abundant information and diverse applications arise out the problems of matching information to users. For example, online shopping needs to present intentional goods for users in the limited Web page, self-help travellers hope to receive tour information nearby them timely. Accurate information to users is an important requirement for improving user experience, and is also a big challenge under information overloading and users' diversified needs.

Recommendation is a popular mechanism to present expected information for people with different requirements. Existing methods for recommendation mostly depend on the similarity computation between users and recommended

© Springer International Publishing AG 2017
H. Yin et al. (Eds.): IDEAL 2017, LNCS 10585, pp. 470–480, 2017.
https://doi.org/10.1007/978-3-319-68935-7_51

things, which often adopt a strategy of collaborative recommendation and pay more attention on behavior-level information, such as collaborative filtering. Those methods are not suitable for online applications because they only produce the relationships among users by a large amount of computation, especially attribute-level information are not covered well for recommendation, which are often very valuable and can bring recommendation gains on effectiveness.

The contribution of this paper is to present a different recommendation strategy from those existing recommendation methods. First, we establish a recommendation model, which is responsible for holding behavior-level information and attribute-level information. Second, we merge classification strategy into recommendation, a classification method, random forest, is borrowed to produce the initial recommendation rules. At last, a recommendation decision based on scoring and ranking is defined. This proposed method is suitable for online recommendation for its simplicity on recommendation decisions.

This paper is organized as following. Section 2 summarizes the recommendation technologies. Section 3 defines the recommendation problem and introduces our recommendation model, Sect. 4 details the concrete recommendation method. Experimental results and analysis are provided in Sects. 5 and 6 concludes the whole work.

2   Related Work

For data are diversified and can provide richer information, recommendation is playing a very important role in many fields, such as smart tourism, online advertisements, and so on. Based on the different implementation technologies, recommendation can be grouped into the following types, recommendation on collaborative technologies, recommendation on classification strategies, and recommendation on neural networks.

Recommendation on classification strategies usually pushes the recommended objects to users according to the classification results, in which ID3 [1], C4.5 [2]and CART [3] are often applied for classification. Quinlan put forwards ID3 for classification in 1986, which uses information gain to choose test attributes for classification. ID3 only supports single variable, does not consider the relationships between attributes, and does not support continuous values. C4.5 was advised to improve ID3 in 1993, which exploits information gain ratio as attribute evaluation to overcome the shortcoming of ID3. Considering too many branches in decision trees produced by ID3 and C4.5, Breiman et al. used Gini index to evaluate attributes and put forward CART, which can reduce the scale of decision trees effectively and improve the problem on overfitting. In this paper, we will introduce random forest into our recommendation model and produce the corresponding recommendation rules for online recommendation.

Recommendation on collaborative technologies includes user-based collaborative recommendation [4], item-based collaborative recommendation [5,6] and model-based collaborative recommendation [7]. Those methods model the relationships between users and recommended objects into a matrix and construct user vectors to compute user similarity. Gong et al. [8] proposed a recommendation method based on user clustering and item clustering, which computes the similarity between users and centroid of user clustering for accurate recommendation. Zhang et al. [9] put forwards a collaborative recommendation method based on smooth slope, which predicts missing scores for some items and introduces Pearson Correlation Coefficient to find the neighbors of the target user. Jiang et al. [10] realized an item-based collaborative recommendation on MapReduce, which presents a good efficiency under a well-designed partition. Xing et al. [11] used time-based data weighting and similarity-based data weighting to capture the change of users' interests, and improved the recommendation performance.

For recommendation on neural networks, Mai et al. [12] took the advantage of both neural networks and collaborative filtering to establish a classifier based on BP neural network, which can deal with sparse data well and has a good performance on electronic commerce. Li et al. [13] exploited users' short-term actions and long-term behaviors by deep convolutional neural networks for predictions, which makes performance improvement on mobile recommendation. Lee et al. [14] proposed a recommendation system on deep learning, which focuses on the context information to construct sequence model for recommendation. Wu et al. [15] used a deep forward neural network to model users' shopping behaviors, and contributed a recommendation model named DeepSession.

Both recommendation on neural networks and recommendation on collaborative technologies need massive and complex computation, whose strategies are not suitable for online applications. Recommendation on classification strategy can work well for online recommendation since it can contribute concise rules.

3   Problem Definition and Recommendation Model

In this work, recommendation focuses on the matching problems between a user set and an item set, which is based on the users' behaviors and items' attributes. The recommendation problem can be formalized as Definition 1,

Definition 1. *Given an item set I, a user-item pair set U, and a scoring function S. Each element of I is described by m attributes, each element of U is a pair $< uid, item >$, uid represents a specific user and $item \in I$. For a new item x, the recommendation requires an output O consisting of $k(k > 0)$ users, which should satisfy $\forall o \in O, \neg \exists uid \in (U - O), S(uid) > S(o)$.*

The complete recommendation model is presented in Fig. 1, in which Fig. 1(a) and (c) correspond to I, U respectively. The whole model is composed of three stages. Stage 1 is responsible for classifying those items into different groups according to their attributes. Stage 2 merges U and the output of stage 1 to build the initial recommendation rules for different users. When a new item comes, stage 3 will compute the matching scores between users and the new item by those initial recommendation rules, and then applies these scorings and ranking function to find the top-k users.

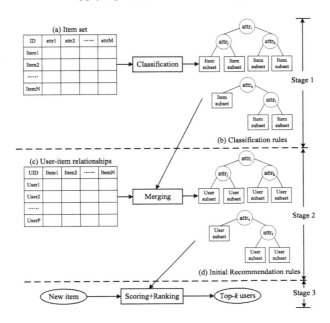

Fig. 1. Recommendation model

4 Proposed Method

In this section, we present a novel method for goods recommendation, which incorporates classification strategy for inducing recommendation rules and is suitable for online recommendation for its simplicity. The whole recommendation process is supported by three functions. The first is responsible for introducing random forest to classify the given dataset, the second is to transform the classification rules into recommendation rules with the help of the relationships between users and items, and the third is to rank users when a new item arrived. The framework for recommendation is presented in Fig. 1. In the following, we will detail the concrete recommendation process.

4.1 Selecting Attributes and Constructing Decision Trees

In this stage, we focus on inducing the classification rules on the given dataset, which will lay a foundation of recommendation. We apply random forest for classification, which needs to construct multiple decision trees. Gini index is adapted to choose attributes for decision trees. Supposing the whole item set I (see Fig. 1(a)) are divided into multiple subsets I_i on a set of attributes A, the Gini index of A can be computed as Formulas 1 and 2. In Formula 2, C_j represents the set of items that belong to the j_{th} class under a right classification.

$$Gini_A(I) = \sum_i \frac{I_i}{I} Gini(I_i) \tag{1}$$

$$Gini(I_i) = 1 - \sum_j \left(\frac{|I_i \cap C_j|^2}{|I_i|} \right) \tag{2}$$

In order to construct different decision trees, we sample a half from the whole data set in a random mode for M times. For each subset output by one independent sampling, we use Gini index to choose optimal attributes as split attributes and to construct a decision tree. The output M decision trees will be transformed into the initial recommendation rules.

4.2   Constructing Initial Recommendation Rules

The above section only divides the data set into different groups, this section will form recommendation rules with the help of classification output, which can be applied for online recommendation directly.

Each item in I belongs to one single group after the first stage, each item has some relationships with some users, which are recorded in U(Fig. 1(c)). For example, if a user(uid) bought an item($item$), then $< uid, item >\in U$. Based on the above facts, if $< uid, item >\in U$, we can substitute $item$ in each decision tree with its corresponding users uid. Obviously, a user may occur in different groups when this user is associated with many items in I. In order to distinguish the different possibility for one user to be in a group, we let a weight value accompany with a user, namely $< uid, weight >$ replaces its corresponding item. Then those decision trees can be used as initial recommendation rules for a new item. The value of $weight$ depends on the number of items related with the specific user.

The computation for $weight$ is illustrated in Table 1 and Fig. 2. Table 1 is a snippet of dataset from U. Figure 2(a) illustrates a decision tree built on Table 1, Fig. 2(b) is its corresponding recommendation rules. Taking $user_1$ as an example, those items related with $user_1$ are assigned to two categories, but there are 2 items($item_1$,$item_3$) in the first category, and only one in the second category. So the weight contributed by $user_1$ to the first category is 2, and the weight to the second category is 1.

Table 1. Dataset snippet

UserID	$item_1$	$item_2$	$item_3$	$item_4$	$item_5$
$user_1$	1	0	1	1	0
$user_2$	1	1	1	0	1

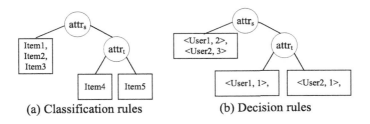

Fig. 2. Illustration of wight computation

4.3 User Scoring and Ranking for Recommendation

Now, we can build different decision trees based on random forest, and those trees have been merged with the user set. The following needs to determine the specific set of users that should be recommended when a new item comes. We use the strategy of majority voting to confirm the target users. For example, when a new item i needs to be recommended, i will be tested by those decision trees in Sect. 4.2 to get some corresponding user sets, namely $ru_1, ru_2, \cdots$. Especially, $ru_i = \{< uid_{i1}, weight_{i1} >, < uid_{i2}, weight_{i2} >, \cdots, < uid_{ij}, weight_{ij} >, \cdots \}$. For a single user set ru_i, weights related with all users will be normalized, which can guarantee that users will be recommended existing in a small-scale group. The scoring of a user is the sum of all related normalized weights in all output user set. We can rank users according to their scorings, and then top-k users are output. The complete recommendation algorithm is presented in Algorithm 1.

5 Experiments

In this section, we designed and carried out experiments on real datasets to verify the proposed method. All experiments are run on a computer with dual-core CPU @1.90GHz and 4GB memory.

5.1 Datasets and Evaluation Metrics

In order to test the effectiveness of the proposed method, we crawled data from *www.jd.com* to construct the datasets, which are about different kinds of goods contributed by users. The whole data set is from three categories of goods, namely movies, books and clothes, each category consists of 6,000 records. For each category, we chose six attributes for classification. Precision, recall and F1-score on recommendation are taken as the evaluation metrics, which are defined in Formulas 3, 4 and 5.

$$Precision = \frac{RightNum}{RecNum} \tag{3}$$

Algorithm 1. Recommendation driven by classification

Require:
  a set of items , $I = \{item_1, item_2, \cdots\}, item_i = \{attr_{i1}, attr_{i2}, \cdots, attr_{im}\}$
  a set of relationships between users and items, $U = \{< user_1, item_1 >, < user_2, item_2 >, \cdots\}$
  the number of recommended users, k
  a new item, $newitem$
Ensure: top-k users, E_n;
 1: sampling I randomly and getting M subsets $\{I_1, I_2, \cdots, I_M$, each subset has equal number of elements;
 2: **for** each subset I_i **do**
 3:    **for** each attribute $attr_{ij}$ **do**
 4:       computing Gini index of $attr_{ij}$
 5:    **end for**
 6:    ranking those attributes on their Gini indexes;
 7:    choose the top-x attributes for building decision trees dt_i;
 8: **end for**
 9: scanning U, substituting the content in leaf nodes of dt_i by $< user, weight >$;
10: **for** each dt_i **do**
11:    applying $newitem$ on dt_i and outputting $O = \{\cdots, < user_k, weight_k >, \cdots\}$;
12:    $UW = UW \cup O$;
13: **end for**
14: scanning UW and computing the scores of each user, inserting them into $UR = \{\cdots, < user, score >, \cdots\}$;
15: assigning the top-k users on $score$ in UR to E_n;
16: **return** E_n;

$$Recall = \frac{RightNum}{ExpectedNum} \tag{4}$$

$$F1 - score = \frac{2 * RightNum}{RecNum + ExpectedNum} \tag{5}$$

$RecNum$ represents the number of the proposed items by our method, $ExpectedNum$ is the number of items that should be recommended. $RightNum$ is the cardinality of the intersection of the proposed item set and the expected item set.

5.2   Experimental Results and Analysis

In order to get an accurate classification, attributes should be considered carefully. We checked different groups of attributes for the above datasets, which are listed in Table 2. These attributes are ranked according to their gini index, and then some of them are filtered out to construct decision trees. Especially, different datasets use different group of attributes as the input of random forest.

Parameter Testing We designed a group of experiments to choose the appropriate attributes for random forest. First, we focused on the dataset related with

Table 2. Datasets and their attributes

Dataset	Attributes
Movies	score, price, type(action, fiction, etc.), length, kind(2D,3D, etc.) and nation
Books	score, price, type, package, language and nation
Clothes	score, price, material, color, style and brand

books, which are used to create 4 subsets by a random sampling, each of which consists of 2,000 records. Second, we compute the Gini index of all attributes in each subset and all attributes are ranked on the Gini index. Finally, those attributes are added gradually to construct decision trees for testing the accuracy of classification. The experimental results are presented in Fig. 3, which show that the top-4 attributes on Gini index make the greatest contributions to the precision for classification. Figure 4 also presents the same conclusions, which shows the comparison of the classification precision on the three whole datasets, the precision contributed by top-4 attributes in each dataset covers the precision contributed by all attributes. The following experiments will use the top-4 attributes on Gini index for classification by random forest.

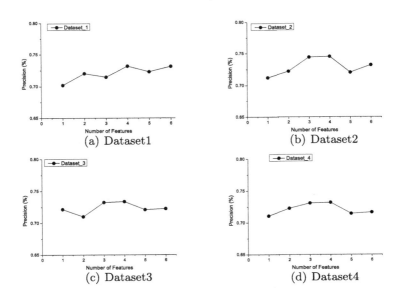

(a) Dataset1

(b) Dataset2

(c) Dataset3

(d) Dataset4

Fig. 3. Parameter test

Cross Verification and Experimental Comparison In order to verify the recommendation accuracy of our proposed method, we fixed the top-4 attributes on Gini index for random forest to construct decision trees and to induce recommendation rules. First, the cross verification is applied, each dataset in Table 2

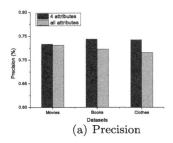
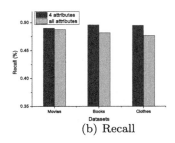

(a) Precision   (b) Recall

Fig. 4. Parameter verification

is divided into five groups randomly that are numbered from 1 to 5, one single group are used for testing and the other four groups are for training. The experimental results are presented in Table 3. For each test, the proposed method has a steady recommendation precision, which almost keeps over 70%. The proposed method has a steady performance.

Table 3. Precision on different testing sets

Training set	1,2,3,4	1,2,3,5	1,2,4,5	1,3,4,5	2,3,4,5	Mean
Testing set	5	4	3	2	1	-
Precision on Movies	0.707	0.713	0.722	0.715	0.733	0.718
Precision on Books	0.734	0.732	0.724	0.717	0.735	0.728
Precision on Clothes	0.691	0.723	0.714	0.732	0.725	0.717

We also conducted an experimental comparison on recommendation performance between our proposed method on random forest (abbr. RF) and the recommendation on CART (abbr. CART). The experimental results are presented in Fig. 5, which show that our proposed method performs better on precision, recall and F1-score. The achieved performance gains benefit from the appropriate

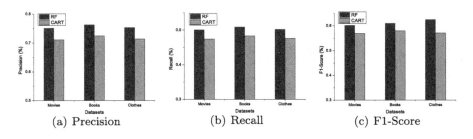

(a) Precision   (b) Recall   (c) F1-Score

Fig. 5. Comparison experiments on recommendation performance

attribute selection for random forest, the subtle production of recommendation rules and the improved scoring for recommended objects.

6   Conclusions

This paper contributes a novel method for recommendation, which provides a cooperative mechanism between classification and recommendation. An efficient strategy is applied for transforming classification rules into recommendation rules. Experiments on real datasets for different scenarios show the effectiveness of the proposed method. Especially, the proposed method can perform well for online recommendation since it uses rules to obtain the recommended objectives.

Acknowledgment. This work is supported by the National Natural Science Foundation of China (No.61363005, 61462017, U1501252), Guangxi Natural Science Foundation of China(No.2014GXNSFAA118353, 2014GXNSFAA118390), Guangxi Key Laboratory of Automatic Detection Technology and Instrument Foundation(YQ15110), Guangxi Cooperative Innovation Center of Cloud Computing and Big Data.

References

1. Quinlan, J.R.: Introduction of decision trees. Mach. Learn. **1**(1), 81–86 (1986)
2. Quinlan, J.R.: C4.5: Programs for Machine Learning. Morgan Kaufmann, San Metro (1993).
3. Breiman, L., Friedman, J., Stone, C.J., Olshenv, R.A.: Classification and Regression Trees. Chapman&HallCRC, Boca Raton (1984)
4. Goldberg, D., Nichols, D., Brain, M.O., et al.: Using collaborative filtering to weave an information tapestry. Commun. ACM **35**(12), 61–67 (1992)
5. Sarwar, B., Karypis, G., Konstan, J., et al.: Item-based collaborative filtering recommendation algorithms. In: Proceedings of the 10th International World Wide Web Conference, pp. 285–295 (2001)
6. Zhang, J., Lin, Z., Xiao, B., et al.: An optimized item-based collaborative filtering recommendation algorithm. In: IEEE International Conference on Network Infrastructure and Digital Content, pp. 414–418. IEEE (2009)
7. Jia, D., Zhang, F., Liu, S.: A robust collaborative filtering recommendation algorithm based on multidimensional trust model. J. Softw. **8**(1), 806–809 (2013)
8. Gong, S.: A Collaborative filtering recommendation algorithm based on user clustering and item clustering. J. Softw. **5**(7), 745–752 (2010). 2013
9. Zhang, D.J.: An Item-based collaborative filtering recommendation algorithm using slope one scheme smoothing. In: Second International Symposium on Electronic Commerce and Security, IEEE Computer Society, pp. 215–217 (2009)
10. Jiang, J., Lu, J., Zhang, G., et al.: Scaling-up item-based collaborative filtering recommendation algorithm based on hadoop. In: Services, pp. 490–497. IEEE (2011)
11. Xing, C., Gao, F., Zhan, S., et al.: A collaborative filtering recommendation algorithm incorporated with user interest change. J. Comput. Res. Dev. **44**(2), 296–301 (2007)
12. Mai, J., Fan, Y., Shen, Y.: A neural networks-based clustering collaborative filtering algorithm in e-commerce recommendation system. In: International Conference on Web Information Systems and Mining, pp. 616–619. IEEE (2009)

13. Li, X., Qian, S., Peng, F., et al.: Deep convolutional neural network and multi-view stacking ensemble in ali mobile recommendation algorithm competition: the solution to the winning of ali mobile recommendation algorithm. In: IEEE International Conference on Data Mining Workshop, pp. 1055–1062. IEEE (2015)

14. Lee, H., Ahn, Y., et al.: Quote recommendation in dialogue using deep neural network. In: International ACM SIGIR Conference on Research and Development in Information Retrieval, pp. 957–960. ACM (2016)

15. Wu, S., Ren, W., Yu, C., et al.: Personal recommendation using deep recurrent neural networks in NetEase. In: IEEE International Conference on Data Engineering, pp. 1218–1229. IEEE (2016)

Linguistic Truth-Valued Multi-Attribute Decision Making Approach Based on TOPSIS

Yuanyuan Shi[2]([✉]), Li Zou[1], Yingying Xu[1], Siyuan Luo[2], and Jia Meng[1]

[1] School of Computer and Information Technology,
Liaoning Normal University, Dalian, China
[2] School of Mathematics, Liaoning Normal University, Dalian, China
shiyuanyuanwww@163.com

Abstract. In order to solve multi-attribute decision making (MADM) problem with fuzzy linguistic-valued information, a linguistic truth-valued MADM approach based on TOPSIS is proposed in combination with traditional TOPSIS approach. Based on linguistic truth-valued lattice implication algebra, the approach uses linguistic truth values to express fuzzy linguistic-valued information which is both comparable and incomparable. Then we define the normalized distance algorithm of linguistic truth values, and its related properties are discussed. The definitions of linguistic truth-valued positive and negative ideal points are proposed, the distances between every alternative and positive ideal alternative or the distances between every alternative and negative ideal alternative are calculated, obtaining the relative degree of closeness between every alternative and ideal alternatives. Come out the best alternative according to the ranking result of relative degree of closeness. The validity and the feasibility of the proposed decision making approach is illustrated by an example.

Keywords: TOPSIS · Linguistic values · Normalized distance · Ideal point · Multi-attribute decision making

1 Introduction

Since Zadeh proposed fuzzy sets, the fuzzy sets theory has been highly concerned [1]. Experts applied fuzzy sets theory to the fields of decision making, reasoning, fuzzy clustering, pattern recognition and expert systems. In the real life, people prefer to use fuzzy language to evaluate things [2]. Based on the structural features of tone algebra, Xu Y proposed linguistic truth-valued lattice implication algebra to deal with comparable and incomparable linguistic-valued information [3].

In the field of decision making, there are many ranking methods, and one of the commonly used ranking methods is the Technique for Order Preference by Similarity to an Ideal Solution (TOPSIS). This method is an ideal solution should be close to the positive ideal solution and at the same time be away from the negative ideal solution [4]. Since the method was proposed, it has been aroused the concern of researchers in various fields. Joshi D proposed intuitionistic fuzzy entropy and distance measure based on TOPSIS method in MADM [5]; Chen studied a new intuitionistic fuzzy

© Springer International Publishing AG 2017
H. Yin et al. (Eds.): IDEAL 2017, LNCS 10585, pp. 481–488, 2017.
https://doi.org/10.1007/978-3-319-68935-7_52

MADM method [6]. Biswas P presented TOPSIS method to single-valued neutrosophic environment for multi-attribute group decision making [7]. Chen C T used the form of linguistic values to express attribute values and weights, then transformed linguistic values to triangular fuzzy numbers in group decision making model based on TOPSIS [8]. Mahdavi I converted the language values into a trapezoidal fuzzy numbers and gave the positive and negative ideal points in a new multi-attribute group decision making model based on TOPSIS [9].

In the representation and processing of uncertain information, we often get cross some information represented with linguistic values in natural language. Linguistic truth-valued lattice implication algebra can deal with both comparable and incomparable linguistic-valued information. The expression can close to the human thinking process and realize the intelligent decision making process. Based on the above analysis, we propose the following approach.

The structure of this paper is organized as follows: Sect. 2 briefly reviews some basic concepts on lattice implication algebra, linguistic truth-valued lattice implication algebras and the relative operations. In Sect. 3, we propose normalized distance between linguistic truth values and discuss the relative properties, then we define linguistic truth-valued positive and negative ideal points. Section 4 establishes linguistic truth-valued MADM approach based on TOPSIS. In Sect. 5, an illustrative example is provided to demonstrate the reasonability and the validity of the new decision making approach. At the same time, some characters about this method are given. The conclusions are drawn in Sect. 6.

2 Preliminaries

In this section, we briefly review some relevant knowledge including lattice implication algebra, linguistic truth-valued lattice implication algebra and the relative operations, which will provide a basis of this paper.

Definition 1 [3]. Let $L_n = \{d_1, d_2, \ldots, d_n\}$, $d_1 < d_2 < \ldots < d_n$, $L_2 = \{b_1, b_2\}$, $b_1 < b_2$, $\left(L_n, \vee_{(L_n)}, \wedge_{(L_n)}, '^{(L_n)}, \rightarrow_{(L_n)}, d_1, d_n\right)$ and $\left(L_2, \vee_{(L_2)}, \wedge_{(L_2)}, '^{(L_2)}, \rightarrow_{(L_2)}, b_1, b_2\right)$ be two Lukasiewicz implication algebras. For any $(d_i, b_j), (d_k, b_m) \in L_n \times L_2$, define

$$
\begin{aligned}
&(1)\, (d_i, b_j) \vee (d_k, b_m) = \left(d_i \vee_{(L_n)} d_k, b_j \vee_{(L_2)} b_m\right); \\
&(2)\, (d_i, b_j) \wedge (d_k, b_m) = \left(d_i \wedge_{(L_n)} d_k, b_j \wedge_{(L_2)} b_m\right); \\
&(3)\, (d_i, b_j)' = \left(d_i'^{(L_n)}, b_j'^{(L_2)}\right); \\
&(4)\, (d_i, b_j) \rightarrow (d_k, b_m) = \left(d_i \rightarrow_{(L_n)} d_k, b_j \rightarrow_{(L_2)} b_m\right)
\end{aligned}
\tag{1}
$$

Then $\left(L_n \times L_2, \vee, \wedge, ', \rightarrow, (d_1, b_1), (d_n, b_2)\right)$ is a lattice implication algebra, denoted by $L_n \times L_2$.

Definition 2 [3]. Let $AD_n = \{h_1, h_2, \ldots, h_n\}$ be a set with n modifiers and $h_1 < h_2 < \ldots < h_n$, $MT = \{f, t\}$ be a set of meta truth values and $f < t$. Denote as $L_{V(n \times 2)} = AD_n \times MT$.

Define a mapping g as $g: L_{V(n\times2)} \to L_n \times L_2$,

$$g((h_i, mt)) = \begin{cases} (d_i', b_1), mt = f \\ (d_i, b_2), mt = t \end{cases} \tag{2}$$

then g is a bijection, denote its inverse mapping as g^{-1}. For any $x, y \in L_{V(n\times2)}$, define:

$$\begin{aligned} &(1) x \vee y = g^{-1}(g(x) \vee g(y)); \\ &(2) x \wedge y = g^{-1}(g(x) \wedge g(y)); \\ &(3) x' = g^{-1}((g(x))'); \\ &(4) x \to y = g^{-1}(g(x) \to g(y)) \end{aligned} \tag{3}$$

then $L_{V(n\times2)} = (L_{V(n\times2)}, \vee, \wedge, ', \to, (h_n, f), (h_n, t))$ is called a linguistic truth- valued lattice implication algebra generated by AD_n and MT.

Let $L_5 = \{h_i \mid i = 1,2,3,4,5, \ h_1 = \text{"slightly"}, \ h_2 = \text{"rather"}, \ h_3 = \text{"extremely"}, h_4 = \text{"very"}, \ h_5 = \text{"absolutely"}, \ h_1 < h_2 < h_3 < h_4 < h_5\}$ be a set of five hedge operators, element linguistic truth-valued set $L_2 = \{t, f \mid t = \text{"true"}, f = \text{"false"}\}$, we get 10-element linguistic truth-valued lattice implication algebra $L_{V(5\times2)}$. For example, the linguistic truth- valued element (h_1, t) means "slightly true" and (h_1, f) means "slightly false". The Hasse diagram of 10-element linguistic truth-valued lattice implication algebra $L_{V(5\times2)}$ is shown in Fig. 1.

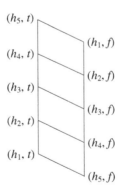

Fig. 1. Hasse diagram of 10-element linguistic truth-valued lattice implication algebra $L_{V(5 \times 2)}$

Definition 3 [10]. Let $L_{V(n\times2)} = AD_n \times MT$, $AD_n = \{h_1, h_2,\ldots, h_n\}$ be a set with n modifiers and $h_1 < h_2 < \ldots < h_n$, $MT = \{f, t\}$ be a set of meta truth values. For any $(h_i, mt_1), (h_j, mt_2) \in L_{V(n\times2)}$, define the operation "$\vee$" and "$\wedge$" as:

$$(h_i, mt_1) \vee (h_j, mt_2) = \begin{cases} \left(h_{\min\{i,j\}}, f\right) & , mt_1 = mt_2 = f \\ \left(h_{\max\{i,n-j+1\}}, t\right), mt_1 = t, mt_2 = f \\ \left(h_{\max\{j,n-i+1\}}, t\right), mt_1 = t, mt_2 = f \\ \left(h_{\max\{i,j\}}, t\right) & , mt_1 = mt_2 = t \end{cases} \tag{4}$$

$$(h_i, mt_1) \wedge (h_j, mt_2) = \begin{cases} \left(h_{\max\{i,j\}}, f\right) & , mt_1 = mt_2 = f \\ \left(h_{\max\{j,n-i+1\}}, f\right), mt_1 = t, mt_2 = f \\ \left(h_{\max\{i,n-j+1\}}, f\right), mt_1 = t, mt_2 = f \\ \left(h_{\min\{i,j\}}, t\right) & , mt_1 = mt_2 = t \end{cases} \tag{5}$$

3   Normalized Distance Algorithm and Positive and Negative Ideal Point of Linguistic Truth Values

In order to rank alternatives in the process of decision making, learning from the idea of traditional TOPSIS method, we need to calculate distances between every alternative and positive ideal point or calculate distances between every alternative and negative ideal point. Thus, the normalized distance among linguistic truth values and linguistic truth-valued positive and negative ideal points as follows:

Definition 4. For any $(h_i, mt_1), (h_j, mt_2) \in L_{V(n \times 2)}$, the normalized distance $d_{LV}((h_i, mt_1), (h_j, mt_2))$ between linguistic truth value (h_i, mt_1) and (h_j, mt_2) is defined as:

$$d_{LV}((h_i, mt_1), (h_j, mt_2)) = \begin{cases} \frac{|i-j|}{n} & , mt_1 = mt_2 \\ \frac{|i+j-n-1|+1}{n} & , mt_1 \neq mt_2 \end{cases} \tag{6}$$

Theorem 1. For any $(h_i, mt_1), (h_j, mt_2), (h_k, mt_3) \in L_{V(n \times 2)}$, Definition 4 has the following properties:

(1) $0 \leq d_{LV}((h_i, mt_1), (h_j, mt_2)) \leq 1$;
(2) $d_{LV}((h_i, mt_1), (h_j, mt_2)) = 0$, if and only if $(h_i, mt_1) = (h_j, mt_2)$;
(3) $d_{LV}((h_i, mt_1), (h_j, mt_2)) = d_{LV}((h_j, mt_2), (h_i, mt_1))$;
(4) If $(h_i, mt_1) \geq (h_j, mt_2) \geq (h_k, mt_3)$, then $d_{LV}((h_i, mt_1), (h_k, mt_3)) \geq d_{LV}((h_i, mt_1), (h_j, mt_2))$ and $d_{LV}((h_i, mt_1), (h_k, mt_3)) \geq d_{LV}((h_j, mt_2), (h_k, mt_3))$.

Definition 5. Let decision matrix be $R = (r_{ij})_{m \times n}, r_{ij} \in L_{V(n \times 2)}$, the linguistic truth-valued positive ideal point r_j^+ and negative ideal point r_j^- are defined as:

$$r_j^+ = \begin{cases} \bigvee_i r_{ij}, j \in \Omega_1 \\ \bigwedge_i r_{ij}, j \in \Omega_2 \end{cases} \tag{7}$$

$$r_j^- = \begin{cases} \bigwedge_i r_{ij}, j \in \Omega_1 \\ \bigvee_i r_{ij}, j \in \Omega_2 \end{cases} \tag{8}$$

where $i = 1, 2, \ldots, m$, $j = 1, 2, \ldots, n$, Ω_1 and Ω_2 respectively represent benefit attribute and cost attribute index set.

4 The Steps of Linguistic Truth-Valued MADM Approach Based on TOPSIS

In this subsection, based on the $2n$-element linguistic truth-valued lattice implication algebra $L_{V(n \times 2)}$, an approach of linguistic truth-valued MADM based on TOPSIS is proposed, where Ω_1 and Ω_2 respectively represent benefit attribute and cost attribute index set. The specific steps are as follows:

Step 1. For a MADM, let $A = \{A_1, A_2, \ldots, A_m\}$ be a set of m possible alternatives, $G = \{G_1, G_2, \ldots, G_n\}$ be a set of n attributes. Let $w = \{w_1, w_2, \ldots, w_n\}$ be a attribute weighted vector, where $w_j \in [0, 1]$, $\sum_{j=1}^{n} w_j = 1$. Let linguistic assessment values of alternative $A_i \in A$ on an attribute $G_j \in G$ be r_{ij}, and we get a decision matrix $R = (r_{ij})_{m \times n}$, where $r_{ij} \in L_{V(n \times 2)}$, $i = 1, 2, \ldots, m$, $j = 1, 2, \ldots, n$.

Step 2. Utilizing formula (7) and (8) to determine linguistic truth-valued positive and negative ideal alternatives:

(1) Linguistic truth-valued positive ideal alternative is

$$A^+ = \left(r_1^+, r_2^+, \ldots, r_n^+ \right) \tag{9}$$

(2) Linguistic truth-valued negative ideal alternative is

$$A^- = \left(r_1^-, r_2^-, \ldots, r_n^- \right) \tag{10}$$

Step 3. Giving weights of attributes, respectively calculating distances between every alternative and linguistic truth-valued positive ideal points or calculating distances between every alternative and linguistic truth-valued negative ideal points:

(1) The distance between alternative A_i and linguistic truth-valued positive ideal alternative is

$$d(A_i, A^+) = \sum_{j=1}^{n} w_j d_{LV} \left(r_{ij}, r_j^+ \right) \tag{11}$$

(2) The distance between alternative A_i and linguistic truth-valued negative ideal alternative is

$$d(A_i, A^-) = \sum_{j=1}^{n} w_j d_{LV}\left(r_{ij}, r_j^-\right) \qquad (12)$$

Step 4. Respectively calculating relative degree of closeness between alternative A_i and ideal alternatives:

$$C(A_i) = \frac{d(A_i, A^-)}{d(A_i, A^+) + d(A_i, A^-)} \qquad (13)$$

Step 5. Utilizing relative degree of closeness $C(A_i)$ to rank alternative A_i, and the larger the $C(A_i)$, the better the alternative A_i.

5   Illustrative Example

With the rapid development of mobile networks, mobile learning software based on the mobile phone (mobile learning app) is gradually used by people. For a certain area, there are four mobile learning software can be chosen: A_1, A_2, A_3, A_4. From the function of mobile learning software, we develop six evaluation indicators:G_1— Resource rich indicator; G_2—Resource quality indicator; G_3—Download support indicator; G_4—Interaction degree indicator; G_5—Learning feedback indicator; G_6—Language diversification indicator; Learner gives evaluation by using linguistic truth values on the 10-element linguistic truth-valued lattice implication algebra $L_{V(5 \times 2)}$, as shown in Table 1 Determine the best mobile learning software.

Utilizing linguistic truth-valued ideal points approach based on TOPSIS to determine the best software, the specific steps are as follows:

Step 1. Transform Table 1 into matrix R.

$$R = \begin{bmatrix} (h_5, t) & (h_2, f) & (h_3, t) & (h_4, t) & (h_3, f) & (h_2, t) \\ (h_4, t) & (h_3, t) & (h_3, f) & (h_5, t) & (h_2, f) & (h_1, f) \\ (h_1, f) & (h_4, t) & (h_4, t) & (h_2, f) & (h_5, t) & (h_3, t) \\ (h_2, f) & (h_1, f) & (h_5, t) & (h_4, t) & (h_5, t) & (h_3, t) \end{bmatrix}$$

Step 2. Determine linguistic truth-valued positive and negative ideal alternative: Because the six attributes are benefit attributes, so from Definition 4 we can get:

(1) Linguistic truth-valued positive ideal alternative is:

$$r_1^+ = (h_5, t),\ r_2^+ = (h_5, t),\ r_3^+ = (h_5, t),\ r_4^+ = (h_5, t),\ r_5^+ = (h_5, t),\ r_6^+ = (h_5, t)$$

Table 1. Assessment of mobile learning software indexes

Software	Assessment indexes					
	G_1	G_2	G_3	G_4	G_5	G_6
A_1	Absolutely true	Rather false	Extremely true	Very true	Extremely false	Rather true
A_2	Very true	Extremely true	Extremely false	Absolutely true	Rather false	Slightly false
A_3	Slightly false	Very true	Very true	Rather false	Absolutely true	Extremely true
A_4	Rather false	Slightly false	Absolutely true	Very true	Absolutely true	Extremely true

(2) Linguistic truth-valued negative ideal alternative is:

$$r_1^- = (h_2,f), \ r_2^- = (h_3,f), \ r_3^- = (h_3,f), \ r_4^- = (h_2,f), \ r_5^- = (h_3,f), \ r_6^- = (h_4,f)$$

Step 3. Let weights be $w = \{0.2, 0.15, 0.1, 0.2, 0.25, 0.1\}$, then

(1) Calculate the distance between every alternative and linguistic truth-valued positive ideal alternative:

$$d(A_1, A^+) = 0.35, d(A_2, A^+) = 0.28, d(A_3, A^+) = 0.21, d(A_4, A^+) = 0.19$$

(2) Calculate the distance between every alternative and linguistic truth-valued negative ideal alternative:

$$d(A_1, A^-) = 0.19, d(A_2, A^-) = 0.26, d(A_3, A^-) = 0.33, d(A_4, A^-) = 0.35$$

Step 4. Calculate relative degree of closeness between every alternative and ideal alternatives:

$$C(A_1) = 0.35185, C(A_2) = 0.48148, C(A_3) = 0.61111, C(A_4) = 0.64815$$

Step 5. Utilizing relative degree of closeness $C(A_i)$ to rank alternatives:$A_4 > A_3 > A_2 > A_1$

Thus, the learner chooses mobile learning software A_4 to learn.

According to this example, we can get some characters about this approach:

(1) The approach uses the form of linguistic truth-valued lattice implication algebra to express linguistic values, which can evaluate from positive and negative sides to deal with comparable and incomparable fuzzy linguistic-valued information. It can make decision result more reasonable.

(2) The paper gives decision making method of linguistic truth-valued positive ideal point r_j^+ and negative ideal point r_j^-, and the method can determine both ideal points of comparable and incomparable linguistic-valued information.

(3) The approach proposes normalized distance algorithm of linguistic truth values, which can be used to get relative degree of closeness. We can rank the relative degree of closeness to obtain the best alternative. The approach has higher practical value.

6   Conclusion

In order to deal with MADM problem with fuzzy linguistic values on the linguistic truth-valued lattice implication algebra, this paper proposes an approach of linguistic truth-valued MADM based on TOPSIS. We define normalized distance among linguistic truth values and propose linguistic truth-valued positive and negative ideal points. We use an example of the learner to choose mobile learning software to illustrate the validity and feasibility of the proposed approach. The decision making approach provided by this paper has practical value. Comparing with traditional linguistic-valued processing approaches, this approach can avoid the loss of information. It is also provides a theoretical basis for expert system. For the future research, we will deeply research approach of linguistic-valued weights.

Acknowledgments. This work is partially supported by the National Natural Science Foundation of P. R. China (Nos.61772250,61673320,61672127) , the Fundamental Research Funds for the Central Universities (No. 2682017ZT12) , and the National Natural Science Foundation of Liaoning Province (No. 2015020059).

References

1. Zadeh, L.: Fuzzy sets. In: CONFERENCE 1996. LNCS, vol. 8, pp. 394–432. World Scientific Publishing, Singapore (1996)
2. Rodriguez, R.: A group decision making model dealing with comparative linguistic expressions based on hesitant fuzzy linguistic term sets. Inf. Sci. **241**, 28–42 (2013)
3. Xu, Y.: Determination of alpha-resolution for lattice-valued first-order logic based on lattice implication algebra. Int. J. Comput. Intell. Syst. **17**(2), 178–181 (2007)
4. Xu, Y., Chen, S.: Linguistic truth-valued lattice implication algebra and its properties. In: CONFERENCE 2006. Computational Engineering in Systems Applications, vol. 2, pp. 1413–1418. IEEE (2006)
5. Joshi, D.: Intuitionistic fuzzy entropy and distance measure based TOPSIS method for multi-criteria decision making. Egypt. Inform. J. **15**(2), 97–104 (2014)
6. Chen, S.: Multicriteria decision making based on the TOPSIS method and similarity measures between intuitionistic fuzzy values. Inf. Sci. (2016)
7. Biswas, P.: TOPSIS method for multi-attribute group decision-making under single-valued neutrosophic environment. Neural Comput. Appl. **27**(3), 727–737 (2016)
8. Chen, C.: Extensions of the TOPSIS for group decision-making under fuzzy environment. Fuzzy Sets Syst. **114**(1), 1–9 (2000)
9. Mahdavi, I.: A general fuzzy TOPSIS model in multiple criteria decision making. Int. J. Adv. Manuf. Technol. **45**(3–4), 406–420 (2009)
10. Zou, L.: Implication operator on the set of v-irreducible element in linguistic truth-valued intuitionistic fuzzy lattice. Int. J. Mach. Learn. Cybernet. **4**, 365–372 (2013)

A Comparative Study on Lagrange Ying-Yang Alternation Method in Gaussian Mixture-Based Clustering

Weijian Long[1], Shikui Tu[1(✉)], and Lei Xu[1,2(✉)]

[1] Department of Computer Science and Engineering, and Center for Cognitive Machines and Computational Health, Shanghai Jiao Tong University, Shanghai, China
`weijianlong7@126.com`, {`tushikui,leixu`}`@sjtu.edu.cn`
[2] Department of Computer Science and Engineering, The Chinese University of Hong Kong, Hong Kong, China

Abstract. Gaussian Mixture Model (GMM) has been applied to clustering with wide applications in image segmentation, object detection and so on. Many algorithms were proposed to learn GMM with appropriate number of Gaussian components automatically determined. Lagrange Ying-Yang alternation method (LYYA) is one of them and it has advantages of no priors as well as the posterior probability bounded by traditional probability space. This paper aims to investigate the performance of LYYA, in comparisons with other methods including Bayesian Ying-Yang (BYY) learning, Rival penalized competitive learning (RPCL), hard-cut Expectation Maximization (EM) method, and classic EM with Bayesian Information Criterion (BIC). Systematic simulations show that LYYA is generally more robust than others on the data generated by varying sample size, data dimensionality and real components number. Unsupervised image segmentation results on Berkeley datasets also confirm LYYA advantages when comparing to the Mean shift and Multiscale graph decomposition algorithms.

Keywords: Gaussian Mixture Model · Lagrange Ying-Yang alternation method · Unsupervised image segmentation · Lagrange coefficient

1 Introduction

Gaussian Mixture Model (GMM) is a classic probabilistic model and has been widely used in clustering analysis, image segmentation, speaker identification [1]. Parameters learning and model selection are two essential parts of conventional

This work was supported by the Zhi-Yuan chair professorship start-up grant (WF220103010) from Shanghai Jiao Tong University.
Shikui Tu was supported by the Tenure-track associate professorship start-up grant from Shanghai Jiao Tong University.

© Springer International Publishing AG 2017
H. Yin et al. (Eds.): IDEAL 2017, LNCS 10585, pp. 489–499, 2017.
https://doi.org/10.1007/978-3-319-68935-7_53

learning in GMM. Parameter learning is often implemented by Expectation-Maximization (EM) algorithm to maximize the likelihood, while determining the number, denoted as k, of Gaussian components is a model selection problem which is traditionally selected by a criterion, e.g., Bayesian Information Criterion (BIC), via a two-stage implementation which runs EM for all possible candidate component numbers.

However, this traditional model selection method is time-consuming. Efforts have been made on automatic model selection. Rival penalized competitive learning (RPCL) [2] is an early attempt on automatic model selection. Proposed in [3], BYY combines parameter learning with model selection and provides the general learning framework as well as specific algorithms. Moreover, automatic model selection can also be implemented via Bayesian approach with proper priors. Minimum message length (MML) [4] and variational Bayes(VB) [5] in GMM learning are two instances of this roadway. Readers can refer to [6] for a detailed analysis and comparison among the three Bayesian approaches.

Further improvements on Ying-Yang two-step alternation algorithm indicate that BYY learning methods may be improved without any help of priors if they can restrict covariance matrices as positive definite matrices [7] which is not considered in traditional BYY method. One recent method given in [8] is called Lagrange Ying-Yang alternation method (LYYA), which ignores influences from priors. It considers the Kullback-Leibler divergence between Ying structure and Yang structure as a Lagrange constraint and uses the coefficient η to control this restriction. There is still lack of a detailed investigation of LYYA in comparisons with other methods.

In this paper, we provide such a detailed investigation. A wide scope of configurations of experiments are considered to generate simulated data sets, with varying sample size, data dimensionality, number of clusters, overlap degree of clusters, and so on. LYYA shows best performance in simulations, comparing with the classic EM with BIC, RPCL, hard-cut EM. Even if repeating 5 times, BIC is still worse than LYYA when processing data with high overlapping degree or high dimensionality. We also study the impact of the coefficient η on model selection and clustering, and suggest an optimal scope of η. Real world applications are also considered. Unsupervised image segmentation results on Berkeley show that LYYA is better than other methods including hard-cut EM, RPCL, BYY and Mean shift algorithm [9].

2   Gaussian Mixture Model and EM Algorithm

For an item $x \in R^n$, Gaussian Mixture Model (GMM) supposes that it comes from a linear combination of k Gaussian distributions:

$$q(x|\theta) = \sum_{j=1}^{k} \alpha_j G(x|\mu_j, T_j), \alpha_j \geq 0, \sum_{j=1}^{k} \alpha_j = 1, \theta = \{\alpha_j, \mu_j, T_j\}_{j=1}^{k}, \quad (1)$$

where $G(x|\mu_j, T_j)$ represents a Gaussian density with mean μ_i and covariance matrix T_i, α_j is the mixing weight of the j-th Gaussian component. GMM can be treated as a latent variable model by introducing a latent binary vector $y = \{y_1, \ldots, y_k\}$ to mark the Gaussian component the data x belongs to, where $\forall j, y_j \in \{0, 1\}$, $\sum_{j=1}^{k} y_{ij} = 1$. Then, we have

$$q(x|\theta) = \sum_y q(x, y|\theta), q(x, y|\theta) = \prod_{j=1}^{k} [\alpha_j G(x|\mu_j, T_j)]^{y_j} \qquad (2)$$

Parameters can be estimated from a set of observations $X_N = \{x\}_{i=1}^{N}$ which is assumed to be independently identically distributed (i.i.d.) following GMM, by maximizing the likelihood function, i.e., $\max_\theta q(X_N|\theta) = \prod_{t=1}^{N} q(x_t|\theta)$, with the help of Expectation-Maximization (EM) algorithm, which iterates between the expectation step (E-step) and the maximization step (M-step):
E-step:

$$p_{ij} = p(j|x_i, \theta) = \frac{\alpha_j^{old} G(x_i|\mu_j^{old}, T_j^{old})}{\sum_{j=1}^{k} \alpha_j^{old} G(x_i|\mu_j^{old}, T_j^{old})} \qquad (3)$$

M-step:

$$\alpha_j^{new} = \frac{\sum_i p_{ij}}{N}, \mu_j^{new} = \frac{\sum_i x_i p_{ij}}{\sum_i p_{ij}}, T_j^{new} = \frac{\sum_i p_{ij}(x_i - \mu_j)(x_i - \mu_j)^T}{\sum_i p_{ij}}. \qquad (4)$$

3   Model Selection Methods

3.1   Traditional Two-Stage Model Selection Method

Maximum likelihood (ML) is not a good principle to determine the number k of Gaussian components in GMM because its value increases as k grows, leading to the overfitting problem. A conventional way is to select the component number according to a model selection criterion such as Bayesian Information Criterion (BIC):

$$k^* = \arg \max_k J_{BIC}(k), J_{BIC}(k) = \ln q(X_N|\hat{\theta}_{ML}) - \frac{1}{2} d_k \ln N, \qquad (5)$$

where $\hat{\theta}_{ML}$ is the ML estimate of parameters, d_k is the number of free parameters in GMM, and N is the sample size.

3.2   RPCL and Hard-Cut EM

Model selection by Eq. (5) is time-consuming because it requires running EM for a set of candidate component numbers. Efforts have been made on selecting k automatically during parameter learning. An early attempt is RPCL [2], in which not only the winner (i.e., the one with maximum posterior) is learned but also its rival (i.e., the second winner) is repelled a little bit from the sample to reduce a duplicated information allocation. Thus, a batch version of RPCL learning is to replace Eq. (3) by:

$$
p_{ij}^{new} = \begin{cases} 1 & j = j^*, j^* = \max_j p(j|x_i, \theta) \\ -\gamma & j = r, r = \max_{j \neq j^*} p(j|x_i, \theta) \\ 0 & otherwise \end{cases} \tag{6}
$$

where $p(j|x_i, \theta)$ is the posterior probability computed by Eq. (3), and j^*,r represents the winner and the rival respectively, and γ controls the de-learning strength. When $\gamma = 0$, it degenerates to the so called hard-cut EM algorithm, see Eqs. (19) and (20) in [3].

3.3   Ying-Yang Alternation Method in BYY System

Firstly proposed in [3] and systematically developed in the past two decades, Bayesian Ying-Yang (BYY) harmony learning on typical structures leads to a class of algorithms that approach automatic model selection during parameter learning. Readers can refer to [8] for recent systematic introduction about BYY harmony learning. Briefly, BYY considers best harmony between two types of decomposition, namely Yang machine $p(R|X)p(X)$ and Ying machine $q(X|R)q(R)$, where data X is regarded to be generated from its inner representation $R = \{Y, \theta\}$ with latent variables Y and parameters θ. Mathematically, the BYY harmony learning is to maximize the following function, which is called harmony measure [3]:

$$
H(p\|q) = \int p(R|X)p(X) \ln[q(X|R)q(R)] dX dR \tag{7}
$$

For GMM given in Eq. (1), if ignoring prior distributions over parameters, we have

$$
H(p\|q) = \sum_{j=1}^{k} \sum_{i=1}^{n} p(j|x_i, \theta) \ln[\alpha_j G(x_i|\mu_j, T_j)] \tag{8}
$$

Maximizing the Eq. (8), subject to the structure $p(j|x_i, \theta)$ as the posterior distribution, leads to a BYY algorithm, iterating between Ying-Step which is the same as M-step in EM algorithm by Eq. (4) and Yang-Step given by

$$
p_{ij}^{new} = p(j|x_i, \theta)(1 + \delta_{ij}(\theta))
$$

$$
\delta_{ij}(\theta) = \ln[\alpha_j G(x_i|\mu_j, T_j)] - \sum_{j=1}^{k} p(j|x_i, \theta) \ln[\alpha_j G(x_i|\mu_j, T_j)] \tag{9}
$$

where $p(j|x_i, \theta)$ is calculated by Eq. (3) and $\delta_{ij}(\theta)$ is the adjustment on posterior probability $p(j|x_i, \theta)$. When $\delta_{ij}(\theta) > 0$, it will award the effect of the j_{th} component on sample x_i by enhancing the value of p_{ij}. When $\delta_{ij}(\theta) < 0$, it will give a punishment on p_{ij} and reduce the degree that the j_{th} component evolves toward sample x_i.

3.4   Lagrange Ying-Yang Alternation Method

The existing algorithms for maximizing Eq. (7) directly impose the equal-covariance constraint between Ying machine and Yang machine. Posterior probability p_{ij} calculated in Eq. (9) may be negative, which makes learning suffer from local optimum problem and learning instability [8]. To tackle this problem, the equal-covariance constraint can be indirectly considered as a Lagrange constraint [8], i.e.,

$$H_L(\theta) = H(\theta) - \eta KL(p(Y|X)p(X)||q(X|Y,\theta)q(Y|\theta))$$
$$H(\theta) = \int p(Y|X)p(X)\ln[q(X|Y,\theta)q(Y|\theta)]dYdX \tag{10}$$

where η is a Lagrange coefficient bounded by $\eta \geq 0$.

Maximizing Eq. (10) for GMM in Eq. (1) gives an algorithm called Lagrange Ying-Yang alternation (LYYA), in which Ying step is the same as the M-step in EM algorithm by Eq. (4), while Yang-step is given by:

$$p_{ij} = \frac{[\alpha_j^{old} G(x_i|\mu_j, T_j)]^{\frac{1+\eta}{\eta}}}{\sum_{j=1}^{k} [\alpha_j^{old} G(x_i|\mu_j, T_j)]^{\frac{1+\eta}{\eta}}} \tag{11}$$

when $\eta \to \infty$, $\frac{(1+\eta)}{\eta} \to 1$, Lagrange Ying-Yang alternation method will be equivalent to EM algorithm. When $\eta \to 0$, $\frac{(1+\eta)}{\eta} \to \infty$, Lagrange Ying-Yang alternation method will be extremely closed to hard-cut EM algorithm.

3.5   Rules to Trim a Gaussian Component During Automatic Model Selection

Automatic model selection is achieved during learning with the component being discarded when their mixing weights and determinants of covariance matrices are small enough. In this paper, we adopt the same trimming rule for all algorithms, i.e., for each iteration, among all the components with small enough mixing weights and covariance matrix determinants, the one with least determinant is discarded.

4   Simulation Experiment

4.1   Illustration on 2-D Datasets

To demonstrate how automatic model selection algorithms work, we give two synthetic 2-D datasets as shown in Fig. 1. We run EM+BIC, hard-cut EM, RPCL($\gamma = 0.0001$), BYY and LYYA($\eta = 2$) algorithms on both datasets, for 500 independent trials, respectively. All algorithms are initialized after the first round of K-means algorithm.

We use Rand Index (RI), Normalize Mutual Information (NMI) as well as Correct selection rate (CSR) to evaluate performances of algorithms. CSR is the

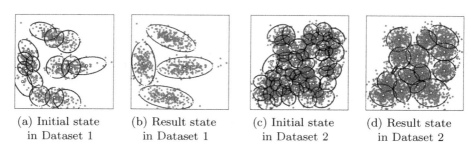

| (a) Initial state in Dataset 1 | (b) Result state in Dataset 1 | (c) Initial state in Dataset 2 | (d) Result state in Dataset 2 |

Fig. 1. (a) Dataset 1 is generated by a 4-component GMM with equal weights $\alpha_j^* = 0.25$, with 800 points and the component number is initialized to be $k = 15$. (c) Dataset 2 (taken from http://cs.joensuu.fi/sipu/datasets/) consists of 5000 points in 15 categories and is initialized with $k = 50$. (b) and (d) Successful cases with correct number of components determined. (Red indicates data points, blue represents means and black boundary indicates a contour of density of a Gaussian component.) (Color figure online)

frequency of correct number of clusters obtained by algorithms. All three criteria are at range $[0, 1]$ and the larger value, the better performance. Results in Table 1 show that LYYA gets best result except in CSR column of Dataset 2 where BYY obtains best result. In this two datasets, the performances between LYYA algorithm and BYY algorithm are closed and both are better than the other three algorithms.

Table 1. Performances of BIC, hard-cut EM, RPCL, BYY and LYYA on two datasets, where numbers in bold type indicate the best within columns

Algorithms	Dataset 1			Dataset 2		
	RI	NMI	CSR	RI	NMI	CSR
BIC	0.9895	0.9688	0.6980	0.9469	0.7466	0.0360
hard-cut EM	0.9791	0.9519	0.5300	0.9634	0.7905	0.2240
RPCL	0.9797	0.9530	0.5440	0.9609	0.7882	0.2520
BYY	0.9957	0.9822	**0.9940**	0.9638	0.7949	**0.6400**
LYYA	**0.9961**	**0.9826**	**0.9940**	**0.9640**	**0.7951**	0.2340

4.2 Systematic Comparisons

We compare all algorithms with extensive experiments as in [6] which cover a wide scope of conditions by varying sample size n, real components number k^*, data dimensionality d and overlap degree β. The synthetic datasets are generated by GMMs, and their mean vectors μ_j and covariance matrices T_j are randomly generated according to the joint Normal-Wishart distribution $G(\mu_j|m_j, T_j/)W(T_j|\phi, \gamma)$ with $\phi = I$, $\gamma = 50$, $m_j = 0$. The weights of components in all synthetic datasets are equal.

For each configuration $\{n, d, k^*, \beta\}$ in Table 2, all algorithms are run on 500 randomly generated datasets, starting from $k = 20$. The BIC value may not be reliable because EM suffers from local optimum problem. Therefore, we also repeatedly implement EM for 5 times and select the one with the largest likelihood for BIC calculation, denoted as BIC(5). It can be noted from the results in Fig. 2 that BIC(5) is much better than BIC(1) at the cost of 4 times more computation. Two main observations in Fig. 2 can be summarized as follows:

(1) Compared with RPCL, hard-cut EM and BYY algorithms, LYYA algorithm has better performance in all series experiments except for $\beta > 0.4$ in series 4.
(2) Even if ignoring the huge computational cost by BIC(5), LYYA is still more robust than BIC(5) for the cases with the data dimensionality exceeding 25 and the overlapping degree growing high.

Table 2. Four series comparison experiments

Starting cases	$\{n, d, k^*, \beta\} = \{500, 5, 5, 0.02\}$
Series 1	n varies in $[50, 100, 150 \cdots 500]$ with fixed d, k^* and β
Series 2	d varies in $[5, 6, 7 \cdots 40]$ with fixed n, k^* and β
Series 3	k^* varies in $[5, 6, 7 \cdots 16]$ with fixed n, d and β
Series 4	β varies in $[0.1, 0.2, 0.3 \cdots 2]$ with fixed n, k^* and d

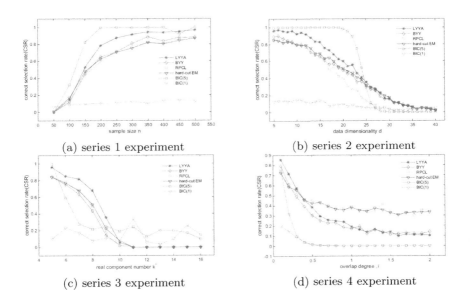

(a) series 1 experiment   (b) series 2 experiment

(c) series 3 experiment   (d) series 4 experiment

Fig. 2. Results from BIC, hard-cut EM, RPCL, BYY, LYYA in four series experiments

4.3   Investigation on η in LYYA

We choose series 1 experiment in Table 2 to study the change of η in LYYA, which details are illustrated in Table 3. The differences of datasets among various series experiments mainly come from their weights(α^*) of Gaussian components. From series 1.1 to series 1.4, the data volume difference of Gaussian components increases gradually. Each configuration $\{n, d, k^*, \beta\}$ consists of 500 cases and Initializes with 20 components. All cases are processed with 11 different η.

The result is shown in Fig. 3. When datasets share equal weights like series 1.1 experiment in Fig. 3(a), the performance of LYYA algorithm gets better as η grows and remains stable after $\eta = 10$. However, when data volume difference of Gaussian components increases in Fig. 3(b)–(d), $\eta = 1$ and $\eta = 10$ two red lines

Table 3. Four series experiments of $\eta \in \{10^{-5}, \cdots, 1, \cdots, 10^5\}$ in LYYA

Starting cases	$\{n, d, k^*, \beta\} = \{n, 5, 5, 0.02\}$
Series 1.1	$\alpha^* = \{\frac{1}{5}, \frac{1}{5}, \frac{1}{5}, \frac{1}{5}, \frac{1}{5}\}$, n varies in $[50, 100, \cdots 500]$ with fixed d,k^*,β
Series 1.2	$\alpha^* = \{\frac{1}{7}, \frac{2}{7}, \frac{1}{7}, \frac{2}{7}, \frac{1}{7}\}$, n varies in $[70, 140, \cdots 700]$ with fixed d,k^*,β
Series 1.3	$\alpha^* = \{\frac{1}{9}, \frac{1}{3}, \frac{1}{9}, \frac{1}{3}, \frac{1}{9}\}$, n varies in $[90, 180, \cdots 900]$ with fixed d,k^*,β
Series 1.4	$\alpha^* = \{\frac{1}{11}, \frac{4}{11}, \frac{1}{11}, \frac{4}{11}, \frac{1}{11}\}$, n varies in $[110, 220, \cdots 1100]$ with fixed d,k^*,β

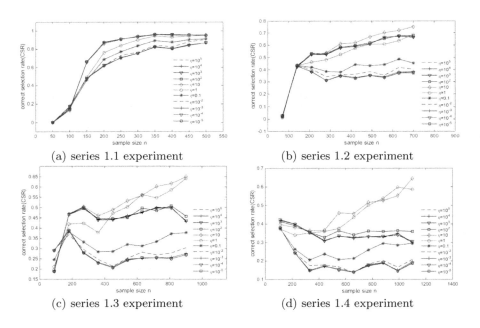

(a) series 1.1 experiment

(b) series 1.2 experiment

(c) series 1.3 experiment

(d) series 1.4 experiment

Fig. 3. Performances among 11 various η in LYYA from four series experiments

Table 4. Average score of criteria of 100 test images on BSDS300 dataset(k = 30), where numbers in bold type indicate the best within rows

Algorithms	Mean shift	MN-Cut	Hard-cut EM	RPCL	BYY	LYYA
PRI	0.4037	0.4214	0.4806	0.4960	0.4866	**0.5023**
RI	0.7242	0.7327	0.7289	0.7304	0.7282	**0.7359**
NMI	0.5352	**0.5765**	0.4799	0.4801	0.4827	0.4836

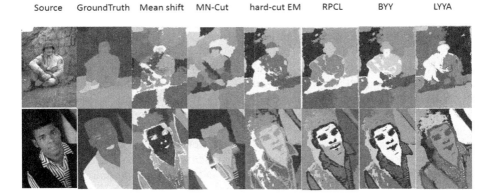

Fig. 4. Segmentation results of two images in BSDS300 dataset among six algorithms

obtain outstanding performances. As a result, we recommend that the value of η should be controlled within $[1, 10]$ when using LYYA method.

5   Application on Image Segmentation

We apply algorithms to unsupervised image segmentation on 100 test images in Berkerly dataset (http://www.eecs.berkeley.edu/Research/Projects/CS/vision/bsds/). We use Blobworld feature [10,11] plus position information to represent information of pixels. Blobworld feature of a pixel is a 6-D vector which consists of its color information from Lab space and 3-D texture information. Position information of a pixel is a 2-D vector of image coordinate and total feature is an 8-D vector per pixel.

We compare LYYA with hard-cut EM, RPCL, BYY, Mean shift [9] and Multiscale graph decomposition(MN-Cut) [12] algorithms on the dataset. The former four algorithms as well as MN-Cut algorithm initialize with 30 components and the bandwidth of Mean shift method is 0.15. Since the real component numbers (k^*) of images are uncertain, we evaluate segmentation results with RI, NMI as well as Probabilistic Rand Index(PRI) [13]. Each value of criteria is the average score of ground truth segmentations per image and the result is shown in Table 4. In this experiment, no data post-processing is applied, and thus the original clustering is kept.

Table 4 shows that LYYA method has best result in PRI as well as RI and MN-Cut algorithm owns the best score of NMI. On the vision of the segmentation examples in Fig. 4, the results of hard-cut EM, RPCL, BYY and LYYA are closed and they have better description than Mean shift as well as MN-Cut algorithms in image details though their dividing lines of different regions tend to be more rough.

6   Conclusion

In this paper, based on GMM, we provide a comparative study on several automatic model selection algorithms including LYYA, BYY, hard-cut EM and RPCL, together with BIC model selection criterion, through systematic experiments. Results indicate that LYYA algorithm is generally more robust than others on the data generated by varying sample size, data dimensionality and real components number. Described in [7], BYY algorithm may be unstable for little constrain on the range of posterior probability. The calculation of p_{ij} can be negative and may be not in the traditional probability space. Different from it, LYYA algorithm is easy computation and doesn't need extra adjustment to solve above problem. We provide an investigation on the Lagrange coefficient η in LYYA algorithm and the result indicates that as the amount of data in Gaussian components is increasingly unbalanced, the choice of various η is increasingly important and the ideal scope of η should be controlled in $[1, 10]$. Moreover, on image segmentation, LYYA outperforms hard-cut EM, RPCL as well as the previous BYY algorithm and also is generally better than Mean shift and MN-Cut algorithms.

References

1. Constantinopoulos, C., Titsias, M.K., Likas, A.: Bayesian feature and model selection for Gaussian mixture models. IEEE Trans. Pattern Anal. Mach. Intell. **28**(6), 1013–1018 (2006)
2. Xu, L., Krzyzak, A., Oja, E.: Rival penalized competitive learning for clustering analysis, RBF net and curve detection. IEEE Trans. Neural Netw. **4**(4), 636–649 (1993)
3. Xu, L.: Bayesian-Kullback coupled Ying-Yang machines: unified learnings and new results on vector quantization. In: Proceedings of International Conference on Neural Information Processing, pp. 977–988 (1995)
4. Figueiredo, M.A.T., Jain, A.K.: Unsupervised learning of finite mixture models. IEEE Trans. Pattern Anal. Mach. Intell. **24**(3), 381–396 (2002)
5. Jaakkola, T.S., Jordan, M.I.: Bayesian parameter estimation via variational methods. Stat. Comput. **10**(1), 25–37 (2000)
6. Shi, L., Tu, S., Xu, L.: Learning Gaussian mixture with automatic model selection: A comparative study on three Bayesian related approaches. A special issue on Machine learning and intelligence science: IScIDE2010 (B). J. Front. Electr. Electron. Eng. China **6**(2), 215–244 (2011)
7. Chen, G., Heng, P.A., Xu, L.: Projection-embedded BYY learning algorithm for Gaussian mixture-based clustering. SpringerOpen J. Appl. Inform. **1**(2) (2014)

8. Xu, L.: Further advances on Bayesian Ying-Yang harmony learning. SpringerOpen J. Appl. Inform. 2(5), (2015)
9. Comaniciu, D., Meer, P.: Mean shift: a robust approach toward feature space analysis. IEEE Trans. Pattern Anal. Mach. Intell. **24**(5), 603–619 (2002)
10. Carson, C., Belongie, S., Greenspan, H., Malik, J.: Blobworld: image segmentation using expectation-maximization and its application to image querying. IEEE Trans. Pattern Anal. Mach. Intell. **24**(8), 1026–1038 (2002)
11. Nikou, C., Likas, A.C., Galatsanos, N.P.: A Bayesian framework for image segmentation with spatially varying mixtures. IEEE Trans. Image Process. Publ. IEEE Sig. Process. Soc. **19**(9), 2278–2289 (2010)
12. Cour, T., Bènèzit, F., Shi, J.: Spectral segmentation with multiscale graph decomposition. IEEE Comput. Soc. Conf. Comput. Vis. Pattern Recogn. **2**(2), 1124–1131 (2005)
13. Carpineto, C., Romano, G.: Consensus clustering based on a new probabilistic rand index with application to subtopic retrieval. IEEE Trans. Pattern Anal. Mach. Intell. **34**(12), 2315–2326 (2012)

Convolutional Neural Networks for Unsupervised Anomaly Detection in Text Data

Oleg Gorokhov, Mikhail Petrovskiy$^{(\boxtimes)}$, and Igor Mashechkin

Computer Science Department of Lomonosov Moscow State University, MSU,
Vorobjovy Gory, Moscow 119899, Russia
owlman995@gmail.com, {michael,mash}@cs.msu.su

Abstract. In this paper, we discuss the problem of anomaly detection in text data using convolutional neural network (CNN). Recently CNNs have become one of the most popular and powerful tools for various machine learning tasks. CNN's main advantage is an ability to extract complicated hidden features from high dimensional data with complex structure. Usually CNNs are applied in supervised learning mode. On the other hand, unsupervised anomaly detection is an important problem in many applications, including computer security, behavioral analytics, etc. Since there is no specified target in unsupervised mode, traditional CNN's objective functions cannot be used. In this paper, we develop a specific CNN architecture. It consists of one convolutional layer and one subsampling layer, we use RBF activation function and logarithmic loss function on the final layer. Minimization of the corresponding objective function helps us to calculate the location parameter of the features' weights discovered on the last network layer. We use l_2-regularization to avoid degenerate solution. Proposed CNN has been tested on anomalies discovering in a stream of text documents modeled with well-known Enron dataset, where proposed method demonstrates better results in comparison with the traditional outlier detection methods based on one-class SVM and NMF.

Keywords: Anomaly detection · Text mining · Convolutional neural network · One-class classification · Regularization · SVM · NMF

1   Introduction

Nowadays the interest to anomaly detection is increasing. The essence of anomaly detection problem is in searching the data whose behavior or characteristics are different from the normal ones [2]. Thus, solving the anomaly detection problem, we define criteria, describing standard behavior. If the data correspond to the criteria we call them normal. Otherwise, the data are considered as abnormal (anomalous). Anomaly detection is widely used in different areas. Recently anomaly detection in text data began to be of particular interest. This is because

© Springer International Publishing AG 2017
H. Yin et al. (Eds.): IDEAL 2017, LNCS 10585, pp. 500–507, 2017.
https://doi.org/10.1007/978-3-319-68935-7_54

the most of information, used by humans, are texts, and finding anomalies in such information helps to solve computer security problems, to prevent information leaks, to understand clients' interests and expectations, etc.

Each word in a user's document is interpreted only in its context, therefore some traditional document-to-document similarity measures demonstrate poor quality in anomaly detection tasks [6]. Thus, there is an open problem of anomaly detection in text data the user interacts with. This paper is devoted to this topic. Classical approaches to anomaly detection in text data use the following technique. A classifier is trained on normal data only, and then it is used for separating normal and abnormal data [2]. It is worth noting, that methods based on support vector machine (SVM) and latent semantic analysis (LSA) are widely used nowadays [9,13]. The usage of convolutional neural networks (CNN) becomes popular for the text analysis. However, CNNs are mostly used for supervised classification tasks [1,7]. In this paper, we propose adaptation of the existing CNN approach for the unsupervised anomaly detection.

2   Anomaly Detection in Text Data

According to various researches [1,2,6,7,9,13] the anomaly detection scheme for text data consists of the stages shown in the Fig. 1.

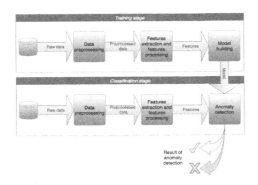

Fig. 1. General scheme for anomaly detection

At the first stage the data preprocessing [13] is used. After that a document is represented in the form of a sequence of words or basic word forms. Next stage is features extraction. Traditional methods represent a document as a vector of frequency characteristics of document's terms. The size of the vector is equal to the size of the dictionary [9,13]. But these methods do not take into account the initial order of words, though the order of words is important in case of using CNNs. In our approach, the alternative method based on vector representation of terms is applied [1,7,8].

At the final stage the one-class classification model is to be built. Usually it leads to the reduction of the anomaly detection problem to the optimization problem [2]. Traditional approaches to the model construction are presented by one-class SVM and LSA-based methods, discussed in [3,9,13]. Recently there is a growing interest in application of neural networks to the problem of unsupervised anomaly detection. For now, autoencoders neural networks such as replicator neural networks (RNN) presented in [4] are used for these purposes. In this research, we adapt convolution neural networks traditionally used for supervised learning to the problem of unsupervised anomaly detection [1,7].

3   Proposed Approach

Initially convolutional neural networks were widely used in computer vision tasks. However, it has been noticed that they also give fairly good results in the natural language processing [7]. Standard CNN is the multilayer neural network, where one or a few hidden layers are convolutional, i.e. they allow getting features from the input data by applying a filter (convolution operation). The proposed version of the architecture of a convolutional neural network is based on the research presented in [1,7].

The classical convolutional neural network has two basic operations: convolution and subsampling. The convolution operation involves the transformation of a consistent group of features according to a particular rule.

Consider a text corpus consisting of m documents. As it was mentioned earlier, instead of bag-of-words features vector of the document, we use a sequence of vectors corresponding to each term in the document. In this case, each document is represented as the following vector:

$$x_{1:n} = x_1 \oplus x_1 \oplus ... \oplus x_n, \tag{1}$$

where $x_j \in \mathbb{R}^k$, $j = 0, 1, ..., n$, is the vector corresponding to the jth term, $k \in \mathbb{N}$ is fixed feature size, and $\oplus$ is concatenation operator.

Let h be a window size, that is the length of the subsequence of features, which each filter is applied to. This length is set at the stage of the network architecture design. Now consider a filter that is a vector $w \in \mathbb{R}^{hk}$. This filter is applied to a window consisting of h terms. Application of the filter to the window is described in the following way:

$$c_i = f(w \cdot x_{i:i+h-1} + b), \tag{2}$$

where $b \in \mathbb{R}$ is a bias term and f is a non-linear function. Thus, by applying this filter to one window, we get an abstract feature c_i. The specified filter is applied to all possible windows of successive terms in the document. Thus, we obtain the following feature map:

$$c = (c_1, c_2, ..., c_n), \tag{3}$$

where $c_i \in \mathbb{R}$ is a feature extracted from $x_{i:i+h-1}$ using filer w.

The pooling (subsampling) operation is performed after convolution. It implements a non-linear compression of the feature map. As a rule, the maximum function is used for pooling operation. In this case, the subsampling extracts the most significant feature from the resulting map. So we get only one, the most important feature for each filter. Thus, there is the only one most significant element from the feature map corresponding to each individual filter. To determine several features from the document, we use a variety of different filters of different length. In general case, a consecutive alternation of convolution-subsampling pairs is used in the convolutional network. In our case, only one convolution and one subsampling layers are used.

The final layer follows convolutional and subsampling layers. At the last layer, it is necessary to produce a certain mapping of the feature space into a space of smaller dimension.

The considered architecture of the neural network consists of the layer that maps words into low-dimensional vectors, the convolution layer, the subsampling layer, and the output layer that uses the dropout to avoid overfitting. Convolutional, pooling, and final layers are shown in Fig. 2.

In the described approach, the index of the term in the internal vocabulary (i.e., $k = 1$) stands for the value of the corresponding term. The first layer of the neural network represents the algorithm that compresses the terms by combining

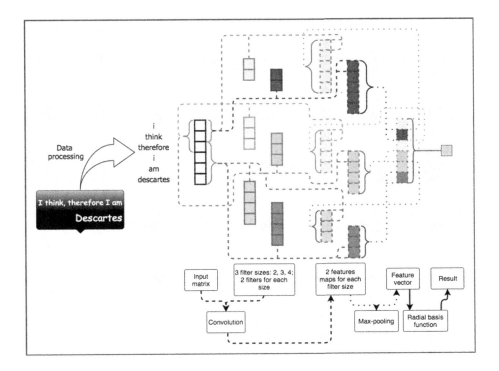

Fig. 2. CNN for text data analysis

them into groups according to the remainders after division of their indices by the number equal to the size of the resulting space of the processing data. For example, if the vocabulary contains 10 terms, and it is needed to be compressed to 5 elements, the indices are combined into the following five groups: [[0, 5], [1, 6], [2, 7], [3, 8], [4, 9]].

The convolutional layer is a collection of convolutional "sub-layers", each of which uses filters of a certain size. A rectifier is used as an activation function on each "sublayer". All weights on this layer are initialized using the truncated normal distribution. Biases are initialized as constants. These initial assumptions are based on the experiments carried out in [7]. Each convolutional "sublayer" is followed by a layer that compresses the feature map with subsampling operation. Features with maximum value are selected in the chosen neural network architecture. The convolution layer and the subsampling layer use several filters for windows of different sizes, which allow obtaining a vector of features. The outputs of all "sublayers" which perform subsampling are combined into one vector. Thus, at this stage we have the features vector, used as the basis of the anomaly detection in the user's work. The final layer detects abnormal data basing on the constructed features.

Dropout [1,7] is used for regularization at the last layer. Its sense is the following. At the beginning, a set of synaptic connections (SC) for dropout is chosen. Then the probability p of s_i is removed on the training stage (the synaptic weight of the corresponding connection is equal to zero) is assigned to each $s_i \in SC$. This method avoids the co-adaptation of hidden layer neurons in the learning stage, since some connections may be accidentally broken during the training process, and all the scales cannot be adjusted to each other. This regularization method is described in details in [12].

We also use l_2-regularization for the coordinates of the scatter of feature values center, that is weights of synaptic connections on the last layer. As it is said before, the last layer unites all the features into one vector, which is used for the anomalous data identification.

The neural network is designed to solve the problem of a one-class classification, therefore the last layer has only one output in the proposed solution. We use the Hypothesis of compactness, assuming that all feature vectors values for the normal training samples must be close to each other and form compact cloud in the discovered feature space. According to this assumption, we construct the output layer in such a way that the obtained result depends directly on the distance of the feature vector to a center of the distribution of features found in the learning process. The distance is constructed on the convolutional layer. We propose to use the radial-basis function as the activation function on the last layer and logarithmic loss function $- \ln(\cdot)$. As a result, we obtain a simple form of the quadratic objective function with l_2-regularization:

$$g(t) = (t - c)^2 + \alpha ||t||_2, \tag{4}$$

where c is the feature vector (3), t are coordinates of the center of distribution of the convolutional features c, and α is a l_2-regularization parameter.

Thus, in the learning process, minimizing the objective function (4), we find a location parameter, that is the center of distribution of convolutional features c with l_2-regularization α. Regularization is applied to the coordinates of the center of distribution, i.e. to the vector t.

4   Experiments

Experimental evaluation of our method is presented in the form of comparison with the most popular classical methods, such as the one-class SVM [3,9], RNN [4], and a method based on the orthonormal non-negative matrix factorization (ONMF) [11]. ROC AUC is used for the performance evaluation.

4.1   Experimental Setup

Comparison of the anomaly detection algorithms is carried out on well-known ENRON dataset (see [5] for the detailed description). It includes emails obtained from 150 ENRON employees in 2000 and 2001 years. In this paper, we use the ENRON dataset with all attachments. Documents in text formats (DOC, RTF, PDF) attached to letters are considered as text data for the experiment. We select only those users, who have the cumulative size of all text files not less than 1 GB, to provide representative samples. Thus, 15 users are considered, they have 11941 text documents from 2000 to 2001. Data for each user are divided into 6-week experimental periods with 2-week step. Two weeks are used for the user model construction, two weeks for metaparameters selection and another two weeks for the testing set. We also impose an additional restriction that the number of documents in the training sample should be greater than 20. Thus, 118 experimental periods are obtained.

4.2   Results

To compare the performance of the proposed method with one-class SVM, we choose radial basis function as kernel function and tuned SVM meta-parameters. The best results are obtained for binary term weights (i.e., 0 if the term is not found in the document and 1 otherwise), as well as for tf-idf weights [9]. We have discovered that the most accurate results are achieved when only terms with the largest average weights in the text corpus are taken into account. We select 30–50 terms with the largest average weights.

An experimental estimation of the method based on the nonnegative ortho-normal factorization is given in [10]. RNN with architecture based on [4] is also used for comparison. RNN consists of input and output layers of the same size equal to the size of the input space. Besides, there are several hidden layers organized by "hourglass" scheme. In our experiments, we choose the number of the hidden layers and their activation functions. The best results are obtained for one hidden layer with tanh activation function [4] and for three hidden layers with linear activation functions on the outer layers and tanh activation function on the hidden (middle) layer.

In the proposed solution, it is necessary to vary the following CNN parameters: the number of filters used for feature vectors construction, number of learning epochs and the value of the l_2-regularization coefficient α. In a series of experiments, it is found that optimal number of learning epochs is 400. For the chosen number of learning epochs, a series of experiments is carried out for selection of the l_2-regularization coefficient, as well as the number of filters of a certain size [1,7]. The results of all the experiments carried out are shown in Table 1.

Thus, it should be noted that the best result for the proposed method we have with l_2-regularization factor of 10,000, and with 70 filters corresponding to the window sizes of 3, 4 and 5.

Table 1. Results of experiments. *bin* means binary weights for terms, *tf-idf* means tf-idf weights, l_2 is l_2-regularization coefficient, *filters* stands for number of filters.

Classification method	Median ROC AUC
SVM (bin, 30 words)	0.8862
SVM (bin, 50 words)	*0.8921*
SVM (tf-idf, 30 words)	0.89
SVM (tf-idf, 50 words)	0.8783
ONMF (Euclidean norm)	0.8996
ONMF (max norm)	*0.9065*
RNN (one hidden layer)	*0.7576*
RNN (three hidden layers)	0.7298
Proposed method ($\alpha = 0.5$, filters $= 70$)	0.8088
Proposed method ($\alpha = 100$, filters $= 70$)	0.8801
Proposed method ($\alpha = 10000$, filters $= 50$)	0.9
Proposed method ($\alpha = 10000$, filters $= 70$)	**0.9207**

5   Conclusion

In the paper, the new algorithm for anomaly detection is developed. The algorithm solves one-class classification problem with the use of convolutional neural networks. Experimental evaluation of the proposed method shows better result in comparison with several methods traditionally used for anomaly detection problem.

Acknowledgment. This research is supported by the RFBR Grant No. 16-29-09555.

References

1. Britz, D.: Implementing a CNN for text classification in tensorflow (2015). http://www.wildml.com/2015/12/implementing-a-cnn-for-text-classification-in-tensorflow/
2. Chandola, V., Banerjee, A., Kumar, V.: Anomaly detection: a survey. ACM Comput. Surv. **41**(3), 15:1–15:58 (2009)
3. Clifton, L., Clifton, D.A., Zhang, Y., Watkinson, P., Tarassenko, L., Yin, H.: Probabilistic novelty detection with support vector machines. IEEE Trans. Reliab. **63**(2), 455–467 (2014)
4. Hawkins, S., He, H., Williams, G., Baxter, R.: Outlier detection using replicator neural networks. In: Kambayashi, Y., Winiwarter, W., Arikawa, M. (eds.) DaWaK 2002. LNCS, vol. 2454, pp. 170–180. Springer, Heidelberg (2002). doi:10.1007/3-540-46145-0_17
5. Enron email dataset. www.cs.cmu.edu/./enron/
6. Kannan, R., Woo, H., Aggarwal, C.C., Park, H.: Outlier detection for text data: An extended version. CoRR abs/1701.01325 (2017)
7. Kim, Y.: Convolutional neural networks for sentence classification. CoRR abs/1408.5882 (2014)
8. Lee, J.Y., Dernoncourt, F.: Sequential short-text classification with recurrent and convolutional neural networks. CoRR abs/1603.03827 (2016). http://arxiv.org/abs/1603.03827
9. Manevitz, L.M., Yousef, M.: One-class SVMS for document classification. J. Mach. Learn. Res. **2**, 139–154 (2001)
10. Mashechkin, I.V., Petrovskii, M.I., Tsarev, D.V.: Machine learning methods for analyzing user behavior when accessing text data in information security problems. Mosc. Univ. Comput. Math. Cybern. **40**(4), 179–184 (2016)
11. Mirzal, A.: Converged algorithms for orthogonal nonnegative matrix factorizations. CoRR abs/1010.5290 (2010)
12. Srivastava, N., Hinton, G., Krizhevsky, A., Sutskever, I., Salakhutdinov, R.: Dropout: a simple way to prevent neural networks from overfitting. J. Mach. Learn. Res. **15**, 1929–1958 (2014)
13. Tsarev, D.V., Petrovskiy, M.I., Mashechkin, I.V., Korchagin, A.Y., Korolev, V.Y.: Applying time series to the task of background user identification based on their text data analysis. Proc. Inst. Syst. Program. **27**(1), 151–172 (2015)

Solving the Bi-criteria Max-Cut Problem with Different Neighborhood Combination Strategies

Li-Yuan Xue[1], Rong-Qiang Zeng[2,3(✉)], Zheng-Yin Hu[3], and Yi Wen[3]

[1] EHF Key Laboratory of Science, School of Electronic Engineering,
University of Electronic Science and Technology of China,
Chengdu 611731, Sichuan, People's Republic of China
`xuely2013@gmail.com`
[2] School of Mathematics, Southwest Jiaotong University,
Chengdu 610031, Sichuan, People's Republic of China
`zrq@swjtu.edu.cn`
[3] Chengdu Documentation and Information Center, Chinese Academy
of Sciences, Chengdu 610041, Sichuan, People's Republic of China
{`huzy,wenyi`}`@clas.ac.cn`

Abstract. Local search is known to be a highly effective metaheuristic framework for solving a number of classical combinatorial optimization problems, which strongly depends on the characteristics of neighborhood structure. In this paper, we integrate different neighborhood combination strategies into the hypervolume-based multi-objective local search algorithm, in order to deal with the bi-criteria max-cut problem. The experimental results indicate that certain combinations are superior to others and the performance analysis sheds lights on the ways to further improvements.

Keywords: Multi-objective optimization · Hypervolume contribution · Neighborhood combination · Local search · Max-cut problem

1 Introduction

Local search is a simple and effective metaheuristic framework for solving a number of classical combinatorial optimization problems, which proceeds from an initial solution with a sequence of local changes by defining the proper neighborhood structure for the considered problem. In order to study different neighborhood combination strategies during the local search process, we present the experimental analysis of different neighborhoods to solve the bi-criteria max-cut problem.

Given an undirected graph $G = (V, E)$ with the vertex set $V = \{1, \ldots, n\}$ and the edge set $E \subset V \times V$. Each edge $(i, j) \in E$ is associated with a weight w_{ij}. The max-cut problem is to seek a partition of the vertex set V into two disjoint subsets V_1 and V_2, which is mathematically formulated as follows [1]:

$$f_k(V_1, V_2) = max \sum_{i \in V_1, j \in V_2} w_{ij}^k \tag{1}$$

© Springer International Publishing AG 2017
H. Yin et al. (Eds.): IDEAL 2017, LNCS 10585, pp. 508–515, 2017.
https://doi.org/10.1007/978-3-319-68935-7_55

where w_{ij}^k is the weight of the kth $(k \in \{1, 2\})$ graph. As one of Karps 21 NP-complete problems with numerous practical applications [4], a large number of metaheuristics have been proposed to tackle this problem, including scatter search [6], global equilibrium search [7], tabu search [8], etc.

In this paper, we integrate different neighborhood combination strategies into the hypervolume-based multi-objective local search algorithm, in order to study the search capability of different neighborhood combinations on the bi-criteria max-cut problem. The experimental results indicate that certain combinations are superior to others. The performance analysis explains the behavior of the algorithms and sheds lights on the ways to further enhance the search.

The remaining part of this paper is organized as follows. In the next section, we briefly introduce the basic notations and definitions of bi-objective optimization. In Sect. 3, we present the hypervolume-based multi-objective local search algorithm with different neighborhood combination strategies for solving bi-criteria max-cut problem. Section 4 indicates that the experimental results on the benchmark instances of max-cut problem. The conclusions are provided in the last section.

2 Bi-objective Optimization

In this section, we briefly introduce the basic notations and definitions of bi-objective optimization. Without loss of generality, we assume that X denotes the search space of the optimization problem under consideration and $Z = \Re^2$ denotes the corresponding objective space with a maximizing vector function $Z = f(X)$, which defines the evaluation of a solution $x \in X$ [5]. Specifically, the dominance relations between two solutions x_1 and x_2 are presented below [9]:

Definition 1 *(Pareto Dominance). A decision vector x_1 is said to dominate another decision vector x_2 (written as $x_1 \succ x_2$), if $f_i(x_1) \geq f_i(x_2)$ for all $i \in \{1, 2\}$ and $f_j(x_1) > f_j(x_2)$ for at least one $j \in \{1, 2\}$.*

Definition 2 *(Pareto Optimal Solution). $x \in X$ is said to be Pareto optimal if and only if there does not exist another solution $x' \in X$ such that $x' \succ x$.*

Definition 3 *(Non-Dominated Solution). $x \in S$ $(S \subset X)$ is said to be non-dominated if and only if there does not exist another solution $x' \in S$ such that $x' \succ x$.*

Definition 4 *(Pareto Optimal Set). S is said to be a Pareto optimal set if and only if S is composed of all the Pareto optimal solutions.*

Definition 5 *(Non-dominated Set). S is said to be a non-dominated set if and only if any two solutions $x_1 \in S$ and $x_2 \in S$ such that $x_1 \nsucc x_2$ and $x_2 \nsucc x_1$.*

Actually, we are interested in finding the Pareto optimal set, which keeps the best compromise among all the objectives. However, it is very difficult or even impossible to generate the Pareto optimal set in a reasonable time for the NP-hard problems. Therefore, we aim to obtain a non-dominated set which is as close to the Pareto optimal set as possible. That's to say, the whole goal is to identify a Pareto approximation set with high quality.

3   Neighborhood Combination Strategies

In this work, we integrate different neighborhood combination strategies into the hypervolume-based multi-objective local search algorithm, in order to solve bi-criteria max-cut problem. The general scheme of Hypervolume-Based Multi-Objective Local Search (HBMOLS) algorithm [3] is presented in Algorithm 1, and the main components of this algorithm are described in detail in the following subsections.

Algorithm 1. Hypervolume-Based Multi-Objective Local Search Algorithm

---

Input: N (Population size)
Output: A: (Pareto approximation set)
Step 1 - Initialization: $P \leftarrow N$ randomly generated solutions
Step 2: $A \leftarrow \Phi$
Step 3 - Fitness Assignment: Assign a fitness value to each solution $x \in P$
Step 4:
while Running time is not reached **do**
    repeat
        Hypervolume-Based Local Search: $x \in P$
    until all neighbors of $x \in P$ are explored
        $A \leftarrow$ Non-dominated solutions of $A \bigcup P$
end while
Step 5: Return A

In HBMOLS, each individual in an initial population is generated by randomly assigning the vertices of the graph to the two vertex subsets V_1 and V_2. Then, we employ a Hypervolume Contribution (HC) indicator proposed in [3] to achieve the fitness assignment for each individual. Based on the dominance relation and two objective function values, the HC indicator calculate the hypervolume contribution of each individual in the objective space.

For the hypervolume-based local search procedure, we implement the f-flip ($f \in \{1,2\}$) move based neighborhood strategy with the combinations. Afterwards, each solution is optimized by the hypervolume-based local search procedure, which will repeat until the termination criterion is satisfied, so as to obtain a high-quality Pareto approximation set.

3.1   One-Flip Move

In order to deal with the max-cut problem, one-flip move is realized by moving a randomly selected vertex to the opposite set, which is calculated as follows:

$$\Delta_i = \sum_{x \in V_1, x \neq v_1} w_{v_i x} - \sum_{y \in V_2} w_{v_i y}, \ v_i \in V_1 \tag{2}$$

$$\Delta_i = \sum_{x \in V_2, y \neq v_1} w_{v_i y} - \sum_{y \in V_1} w_{v_i x}, \ v_i \in V_2 \tag{3}$$

Let Δ_i be the move gain of representing the change in the fitness function, and Δ_i can be calculated in linear time by the formula above, more details about this formula can be found in [8]. Then, we can calculate the objective function values high efficiently with the streamlined incremental technique.

3.2 Two-Flip Move

In the case of two-flip move, we can obtain a new neighbor solution by randomly moving two different vertices v_i and v_j from the set V_1 to another set V_2. In fact, two-flip move can be seen as a combination of two single one-flip moves. We denote the move value by δ_{ij}, which is derived from two one-flip moves Δ_i and Δ_j $(i \neq j)$ as follows:

$$\delta_{ij} = \Delta_i + \Delta_j \qquad (4)$$

Especially, the search space generated by two-flip move is much bigger than the one generated by one-flip move. In the following, we denote the neighborhoods with one-flip move and two-flip move as N_1 and N_2 respectively.

4   Experiments

In this section, we present the experimental results of 3 different neighborhood combination strategies on 9 groups of benchmark instances of max-cut problem. All the algorithms are programmed in C++ and compiled using Dev-C++ 5.0 compiler on a PC running Windows 7 with Core 2.50 GHz CPU and 4 GB RAM.

4.1   Parameters Settings

In order to conduct the experiments on the bi-objective max-cut problem, we use two single-objective benchmark instances of max-cut problem with the same dimension provided in [4][1] to generate one bi-objective max-cut problem instance. All the instances used for experiments are presented in Table 1 below.

In addition, the algorithms need to set a few parameters, we only discuss two important ones: the running time and the population size, more details about the parameter settings for multi-objective optimization algorithms can be found in [2,8]. The exact information about the parameter settings in our work is presented in the following Table 2.

4.2   Performance Assessment Protocol

In this paper, we evaluate the efficacy of 3 different neighborhood combination strategies with the performance assessment package provided by Zitzler et al.[2].

---

[1] More information about the benchmark instances of max-cut problem can be found on this website: http://www.stanford.edu/~yyye/yyye/Gset/.

[2] More information about the performance assessment package can be found on this website: http://www.tik.ee.ethz.ch/pisa/assessment.html.

Table 1. Single-objective benchmark instances of max-cut problem used for generating bi-objective max-cut problem instances.

	Dimension	Instance 1	Instance 2
bo_mcp_800_01	800	g1.rud	g2.rud
bo_mcp_800_02	800	g11.rud	g12.rud
bo_mcp_800_03	800	g15.rud	g19.rud
bo_mcp_800_04	800	g17.rud	g21.rud
bo_mcp_2000_01	2000	g22.rud	g23.rud
bo_mcp_2000_02	2000	g32.rud	g33.rud
bo_mcp_2000_03	2000	g35.rud	g39.rud
bo_mcp_1000_01	1000	g43.rud	g44.rud
bo_mcp_3000_01	3000	g49.rud	g50.rud

Table 2. Parameter settings used for bi-objective max-cut problem instances: instance dimension (D), vertices (V), edges(E), population size (P) and running time (T).

	Dimension (D)	Vertices (V)	Edges (E)	Population (P)	Time (T)
bo_mcp_800_01	800	800	19176	20	40″
bo_mcp_800_02	800	800	1600	20	40″
bo_mcp_800_03	800	800	4661	20	40″
bo_mcp_800_04	800	800	4667	20	40″
bo_mcp_2000_01	2000	2000	19990	50	100″
bo_mcp_2000_02	2000	2000	4000	50	100″
bo_mcp_2000_03	2000	2000	11778	50	100″
bo_mcp_1000_01	1000	1000	9990	25	50″
bo_mcp_3000_01	3000	3000	6000	75	150″

The quality assessment protocol works as follows: First, we create a set of 20 runs with different initial populations for each strategy and each benchmark instance of max-cut problem. Then, we generate the reference set RS^* based on the 60 different sets $A_0, \ldots, A_{59}$ of non-dominated solutions.

According to two objective function values, we define a reference point $z = [r_1, r_2]$, where r_1 and r_2 represent the worst values for each objective function in the reference set RS^*. Afterwards, we assign a fitness value to each non-dominated set A_i by calculating the hypervolume difference between A_i and RS^*. Actually, this hypervolume difference between these two sets should be as close to zero as possible [10].

4.3   Computational Results

In this subsection, we present the computational results on 9 groups of bi-objective max-cut problem instances, which are obtained by three different neighborhood combination strategies. The information about these algorithms are described in the following table:

Table 3. The algorithms with different neighborhood combination strategies.

	Algorithm description
HBMOLS_N_1	One-flip move based local search
HBMOLS_N_2	Two-flip move based local search
HBMOLS_$(N_1 \bigcup N_2)$	f-flip move based local search ($f \in \{1, 2\}$)

In Table 3, the algorithms HBMOLS_$(N_1 \bigcup N_2)$ selects one of the two neighborhoods to be implemented at each iteration during the local search process, choosing the neighborhood N_1 with a predefined probability p and choosing N_2 with the probability $1 - p$. In our experiments, we set the probability $p = 0.5$.

The computational results are summarized in Table 4. In this table, there is a value both **in bold** and **in grey box** at each line, which is the best result obtained on the considered instance. The values both **in italic** and **bold** at each line refer to the corresponding algorithms which are **not** statistically outperformed by the algorithm obtaining the best result (with a confidence level greater than 95%).

Table 4. The computational results on bi-criteria max-cut problem obtained by the algorithms with 4 different neighborhood combination strategies.

Instance	Algorithms		
	N_2	$N_1 \bigcup N_2$	N_1
bo_mcp_800_01	0.176056	0.131597	**0.115592**
bo_mcp_800_02	0.165386	0.138711	**0.104922**
bo_mcp_800_03	0.151565	0.147590	**0.117074**
bo_mcp_800_04	0.198746	*0.142533*	**0.134102**
bo_mcp_2000_01	0.243824	0.231131	**0.176594**
bo_mcp_2000_02	0.152572	*0.098441*	**0.090894**
bo_mcp_2000_03	0.377184	0.352036	**0.304157**
bo_mcp_1000_01	0.279807	0.260862	**0.229748**
bo_mcp_3000_01	0.099988	0.098586	**0.092150**

From Table 4, we can observe that all the best results are obtained by N_1, which statistically outperforms the other two algorithms on all the instances. Moreover, the results obtain by $N_1 \bigcup N_2$ is close to the results obtained by N_1. Especially, the most significant result is achieved on the instance bo_mcp_2000_01, where the average hypervolume difference value obtained by N_1 is much smaller than the values obtained by the other two algorithms.

Nevertheless, N_2 does not perform as well as N_1, although the search space of them is much bigger than N_1. We suppose that there exists some key vertices in the representation of the individuals, which means these vertices should be assigned in some set in order to search the local optima effectively. Two-flip moves change the positions of the key vertices much more frequently than the one-flip move, then the efficiency of local search is obviously affected by this neighborhood strategy. Actually, $N_1 \bigcup N_2$ provides a possibility to keep the positions of the key vertices unchanged and broaden the search space. Thus, the combination of one-flip move and two-flip move is very potential to obtain better results.

5   Conclusion

In this paper, we have presented different neighborhood combination strategies to deal with the bi-criteria max-cut problem, which are based on one-flip, two-flip and the combination. For this purpose, we have carried out the experiments on 9 groups of benchmark instances of max-cut problem. The experimental results indicate that the better outcomes are achieved with the simple one-flip move based neighborhood and the neighborhood combination with two-flip is very potential to escape the local optima for further improvements.

Acknowledgments. The work in this paper was supported by the Fundamental Research Funds for the Central Universities (Grant No. A0920502051722-53) and supported by the West Light Foundation of Chinese Academy of Science (Grant No: Y4C0011001).

References

1. Angel, E., Gourves, E.: Approximation algorithms for the bi-criteria weighted max-cut problem. Discrete Appl. Math. **154**, 1685–1692 (2006)
2. Basseur, M., Liefooghe, A., Le, K., Burke, E.: The efficiency of indicator-based local search for multi-objective combinatorial optimisation problems. J. Heuristics **18**(2), 263–296 (2012)
3. Basseur, M., Zeng, R.-Q., Hao, J.-K.: Hypervolume-based multi-objective local search. Neural Comput. Appl. **21**(8), 1917–1929 (2012)
4. Benlic, U., Hao, J.-K.: Breakout local search for the max-cut problem. Eng. Appl. Artif. Intell. **26**, 1162–1173 (2013)
5. Coello, C.A., Lamont, G.B., Van Veldhuizen, D.A.: Evolutionary Algorithms for Solving Multi-Objective Problems (Genetic and Evolutionary Computation). Springer, Secaucus (2007)

6. Marti, R., Duarte, A., Laguna, M.: Advanced scatter search for the max-cut problem. INFORMS J. Comput. **21**(1), 26–38 (2009)
7. Shylo, V.P., Shylo, O.V.: Solving the maxcut problem by the global equilibrium search. Cybern. Syst. Anal. **46**(5), 744–754 (2010)
8. Wu, Q., Wang, Y., Lü, Z.: A tabu search based hybrid evolutionary algorithm for the max-cut problem. Appl. Soft Comput. **34**, 827–837 (2015)
9. Zitzler, E., Künzli, S.: Indicator-based selection in multiobjective search. In: Yao, X., Burke, E.K., Lozano, J.A., Smith, J., Merelo-Guervós, J.J., Bullinaria, J.A., Rowe, J.E., Tiňo, P., Kabán, A., Schwefel, H.-P. (eds.) PPSN 2004. LNCS, vol. 3242, pp. 832–842. Springer, Heidelberg (2004). doi:10.1007/978-3-540-30217-9_84
10. Zitzler, E., Thiele, L.: Multiobjective evolutionary algorithms: a comparative case study and the strength pareto approach. Evol. Comput. **3**, 257–271 (1999)

Semi-supervised Regularized Discriminant Analysis for EEG-Based BCI System

Yuhang Xin[1], Qiang Wu[1(✉)], Qibin Zhao[2], and Qi Wu[3]

[1] School of Information Science and Engineering, Shandong University,
Jinan 250010, Shandong, China
wuqiang@sdu.edu.cn
[2] Tensor Learning Unit, Center for Advanced Intelligence Project, RIKEN,
Chuo-ku, Tokyo 103-0027, Japan
[3] Qilu Hospital, Shandong University, Jinan 250012, Shandong, China

Abstract. Brain-Computer interface (BCI) is a new technique which allows direction connection between human and computer or other external device. It employs the classification of event-related potential to control the equipment rather than using language or limb movement. Currently, one of the most important issues for ERP-based BCIs is that the ERP classification performance degrades when the training number of samples is small. In order to solve this problem, semi-supervised regularized discriminant analysis (SRDA) was proposed to extract features and classify ERP patterns by integrating semi-supervised learning and regularization approach. The labeled data was used to maximize the separability between different classes and calculate the within-class covariance matrix by regularization. The labeled and unlabeled data were employed to construct the penalty term by neighbor graph. Our proposed approach was evaluated on the BCI Competition Challenge Dataset and the simulation results indicated that it achieved a better accuracy than the traditional algorithms.

1 Introduction

Brain-Computer Interface provides a direct connection between the brain and the computer by EEG signal. With in-depth study of the human brain, BCI as a human and computer communication system has developed rapidly and been applied in many fields. BCI system extracts the useful features from EEG signal and translates it into commands or messages. This makes it possible for those completely disabled people to communicate with the external environment. And the healthy subjects can also use it to improve the quality of their life. According to the way of the EEG signal generated, the BCI system can be achieved by the following paradigms, the event-related potential (ERP) [1,14], the steady-state visual evoked potential (SSVEP) [12,13] and sensorimotor rhythm (SMR) [7,11]. P300 is one of important ERP-based paradigm for BCI system. The P300 signal is a significant peak signal that occurs about 300 ms after the surprising task-relevant stimuli. One of its major advantages is that it does not require additional training for subject to achieve good performance.

© Springer International Publishing AG 2017
H. Yin et al. (Eds.): IDEAL 2017, LNCS 10585, pp. 516–523, 2017.
https://doi.org/10.1007/978-3-319-68935-7_56

The key problem for ERP-based BCI system is to improve the accuracy of classification. Linear discriminant analysis (LDA) is one of widely used classification algorithm for ERP-based BCI. It assumes that the two classes are Gaussian with equal covariance and the data labels are known [4]. This assumption means that a large number of training samples need to satisfy assumptions, and more time is expected to initialize the BCI system. When the number of training samples is less than the number of features dimensionality, the covariance matrix within class is singular, i.e. the undersample problem. Stepwise linear discriminant analysis (SWLDA) [4] is another efficient algorithm to reduce the undersample problem. The feature vectors were added one by one and each time one feature was added, the previously feature vectors should be examined until the discriminant function is satisfied.

In this paper, we introduced a novel classification algorithm based on semi-supervised regularized discriminant analysis (SRDA) to improve the performance of classification for BCI system based on ERP. SRDA algorithm used the labeled data to calculate the between-class scatter matrix and the within-class scatter matrix. The unlabeled data and labeled data were used to construct the penalty term [2]. By investigating the geometry structure of the labeled data and unlabeled data, a new objective function based regularized/shrinkage method was constructed. Lagrange multiplier method was employed to optimize the objective function. The simulation results showed that our proposed method outperformed the traditional methods.

2   Methods

2.1   Linear Discriminant Analysis

Linear discriminant analysis [6] is a traditional approach for P300-based BCI system. The goal of the LDA algorithm is to find a projection vector which can make the projected samples have the largest distance between different classes and the smallest distance within classes. Supposed the EEG data matrix have N samples $x_1, x_2, \ldots, x_N \in \mathbb{R}^l$ belongs to one of c classes. The objective function of LDA is as follows:

$$a_{opt} = \arg\max_{a} \frac{a^T S_b a}{a^T S_w a}. \tag{1}$$

$$S_b = \sum_{k=1}^{c} N_k \left(\bar{x}^{(k)} - \bar{x} \right) \left(\bar{x}^{(k)} - \bar{x} \right)^T \tag{2}$$

$$S_w = \sum_{k=1}^{c} \left(\sum_{i=1}^{N_k} \left(x_i^{(k)} - \bar{x} \right) \left(x_i^{(k)} - \bar{x} \right)^T \right) \tag{3}$$

where $\bar{x}$ is the mean vector of all samples, N_k is the number of samples in class k, $\bar{x}^{(k)}$ is the mean vector of samples in class c, $x_i^{(k)}$ is the i-th sample of the k-th class, S_b is the between-class scatter matrix and S_w is the within-class scatter matrix. However, the computational complexity of S_w is high. One of solution is to replace S_w by the total scatter matrix $S_t = \sum_{i=1}^{N} (\boldsymbol{x}_i - \bar{\boldsymbol{x}})(\boldsymbol{x}_i - \bar{\boldsymbol{x}})^T$. Then the objective function can be written as follows:

$$\boldsymbol{a}_{opt} = \arg\max_{\boldsymbol{a}} \frac{\boldsymbol{a}^T S_b \boldsymbol{a}}{\boldsymbol{a}^T S_t \boldsymbol{a}}. \tag{4}$$

Thus Eq. (4) can be written as $S_b \boldsymbol{a} = \lambda S_t \boldsymbol{a}$. Obviously it will be transformed into solving the inverse eigenvalue problem.

2.2 Semi-supervised Regularized Discriminant Analysis

In order to reduce the undersample problem, the unlabeled data and a regularized/shrinkage approach proposed in [5,8] are used to improve the performance of discriminant analysis algorithm. When the number of samples is less than feature dimensionality, Eq. (4) will not be able to calculate the eigenvector accurately. One of efficient solution is to reconstruct a new objective function by adding a penalty term [9] as follows:

$$\boldsymbol{a}_{opt} = \arg\max_{\boldsymbol{a}} \frac{\boldsymbol{a}^T S_b \boldsymbol{a}}{\boldsymbol{a}^T S_t \boldsymbol{a} + \alpha \mathrm{J}(\boldsymbol{a})}. \tag{5}$$

where α is the weighted parameter to tune the penalty term and $\mathrm{J}(\boldsymbol{a})$ is the penalty term. As stated in [2], we use the relationship between the unlabeled data and the labeled data to define the penalty term $\mathrm{J}(\boldsymbol{a})$. The nearby points are assumed to have the same label. Supposed a set of data $\{\boldsymbol{x}_i\}_{i=1}^{m}$ is given, the k-nearest neighbor graph is used to define the relationship among the nearby data points. Specifically, if $\boldsymbol{x}_i$ and $\boldsymbol{x}_j$ are among k-nearest neighbors of each other, they are considered to be in the same class. The relational matrix S is defined as follows [2]

$$S_{ij} = \begin{cases} e^{-|\boldsymbol{x}_i - \boldsymbol{x}_j|^2} & \text{if } \boldsymbol{x}_i \in N_k(\boldsymbol{x}_j) \text{ or } \boldsymbol{x}_j \in N_k(\boldsymbol{x}_i) \\ 0 & \text{others} \end{cases} \tag{6}$$

where $N_k(\boldsymbol{x}_j)$ is the k nearest neighbor of $\boldsymbol{x}_j$. Then the objective function can be written as follow

$$\boldsymbol{a}_{opt} = \arg\max_{\boldsymbol{a}} \frac{\boldsymbol{a}^T S_b \boldsymbol{a}}{\boldsymbol{a}^T (S_t \boldsymbol{a} + \alpha \boldsymbol{X} \boldsymbol{H} \boldsymbol{X}^T) \boldsymbol{a}}. \tag{7}$$

where $D_{ii} = \sum_j S_{ij}$, $\boldsymbol{H} = \boldsymbol{D} - \boldsymbol{S}$. The derivation of Eq. (7) in detail can be found in [2]. Equation (7) can be transformed into a problem of solving the inverse eigenvalue by Lagrange multiplier method as follows:

$$S_b \boldsymbol{a} = \lambda (S_t \boldsymbol{a} + \alpha \boldsymbol{X} \boldsymbol{H} \boldsymbol{X}^T) \boldsymbol{a} \tag{8}$$

Since the covariance matrix cannot be calculated accurately for the undersample problem. Specifically, the smaller eigenvalues of the covariance matrix are estimated to be too small and the larger eigenvalues are estimated to be too large. So a shrinkage approach was employed to regularize the covariance matrix. The small and large eigenvalue were averaged by regularization. The shrinkage covariance matrix is defined as follows [1]:

$$\tilde{\Sigma} = (1 - \gamma)\,\Sigma + \gamma \nu \boldsymbol{I} \tag{9}$$

where $\nu = tr(\Sigma)/d$ is the average of original eigenvalue. d is the dimension of feature space, $\boldsymbol{I}$ is an identity matrix. γ is the parameter to control the eigenvalue. The optimal γ can be estimated as follows [1]:

$$\gamma = \frac{N_k}{(N_k - 1)^2} \frac{\sum\limits_{i,j=1}^{d} var_k\left(z_{ij}\left(k\right)\right)}{\sum\limits_{i \neq j} s_{ij}^2 + \sum\limits_{i=j} (s_{ii} - \nu)^2} \tag{10}$$

where N_k is the number of the k-th class, the s_{ij} denote the element in the i-th row and j-th column of Σ, and $z_{ij}(k) = ((\boldsymbol{x}^{(k)})_i - (\bar{\boldsymbol{x}})_i)((\boldsymbol{x}^{(k)})_j - (\bar{\boldsymbol{x}})_j)$, here $(\boldsymbol{x}^{(k)})_i$ denotes the i-th element of the vector $\boldsymbol{x}^{(k)}$ and $(\bar{\boldsymbol{x}})_i$ denote the i-th element of the vector $\bar{\boldsymbol{x}}$. Then Eq. (8) can be rewritten as follows:

$$S_b a = \lambda(\tilde{S}_t a + \alpha \boldsymbol{X H X}^T)a \tag{11}$$

The SRDA algorithm in detail is described in Table 1.

Table 1. Algorithm for semi-supervised regularized discriminant analysis

---

Algorithm for Semi-supervised Regularized Discriminant Analysis

---

Input :
Data matrix $\boldsymbol{X} \in \mathbb{R}^{l \times N}$, label matrix $\boldsymbol{Y} \in \mathbb{R}^{N}$, unlabeled data matrix $\boldsymbol{U} \in \mathbb{R}^{l \times M}$,
tuning parameter α
Output: projection vector $\boldsymbol{a}$
1: Construct the relationship matrix by Eq.(6), and calculate the
$D_{ii} = \sum_j S_{ij}, \boldsymbol{H} = \boldsymbol{D} - \boldsymbol{S}$.
2: Calculate the between-class scatter matrix S_b .
3: Calculate the within-class scatter matrix S_t and optimize it by $\tilde{\Sigma} = (1 - \gamma)\,\Sigma + \gamma \nu \boldsymbol{I}$,

The parameter γ can be solve by $\gamma = \frac{N_k}{(N_k-1)^2} \frac{\sum\limits_{i,j=1}^{d} var_k(z_{ij}(k))}{\sum\limits_{i \neq j} s_{ij}^2 + \sum\limits_{i=j}(s_{ii}-\nu)^2}$

4: Solve the equation $S_b a = \lambda(\tilde{S}_t a + \alpha \boldsymbol{X H X}^T)a$ to calculate $\boldsymbol{a}$.

---

3   Experimental Results

In this paper, we used the BCI competition III data set to evaluate the performance of SRDA algorithm. Compared with LDA, SWLDA, Semi-supervised Discriminant Analysis (SDA) [2,8], our algorithm provided a better accuracy with limited number of training samples. Figure 1 shows feature extraction and classification framework based on SRDA algorithm.

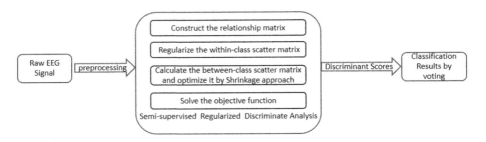

Fig. 1. The feature extraction and classification framework by Semi-supervised regularized discriminant analysis

3.1   BCI Competition III Data Set

This data set was obtained from the BCI Competition III Challenge 2004[1]. The P300 evoked potentials were recorded by BCI2000 using a paradigm described by Donchin [3]. The sampling rate of EEG signal was 240 Hz, and the signal was filtered from 0.1 to 60 Hz. In this experiment, the subjects were presented with a 6 by 6 matrix of characters as described in Fig. 2 (available online from http://www.bbci.de/competition/iii/). The subjects focused attention on a character that was prompt by investigator. The matrix was intensified for100 ms randomly by rows and columns. There was 75 ms between two intensifications. 6 rows and 6 columns were block randomized in blocks of 12. This process was repeated 15 times for each character epoch. There are two subjects A and B in this experiment. The dataset was separated into training data set (85 characters) and test data set (100 characters) for each subject. 16 channels were selected to extract the efficient features. We employed a 168-points window to separate the intensifications. The signal was filtered from 0.1 to 30 Hz by a sixth order forward-backward Butterworth bandpass filter [10]. Because the aim of this experiment is to evaluate the performance of classification with limited samples, we used the training data set of two subjects to evaluate the performance of our algorithm.

3.2   Results

In the experiment, we use 5, 10, 15, 20, 25, 30, 35, 40 characters as training and 45 characters for testing. In the training data set, 70% of the characters were

---

[1] Available online: http://www.bbci.de/competition/iii/.

Fig. 2. The paradigm displayed to the subjects for recording the BCI Competition III dataset

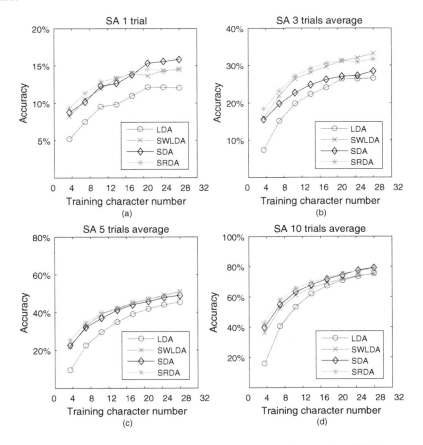

Fig. 3. Classification accuracy for SA on the BCI Competition III dataset

labeled and the last 30% were unlabeled. We compared the classification accuracy of SRDA, SDA, LDA, and SWLDA. We selected the training characters randomly, and repeated the procedure 200 times. The results of the subject A and subject B were presented in Figs. 3 and 4. It was obvious that the SRDA algorithm achieve better performance than the other algorithms when the number of training samples was small especially with few trails.

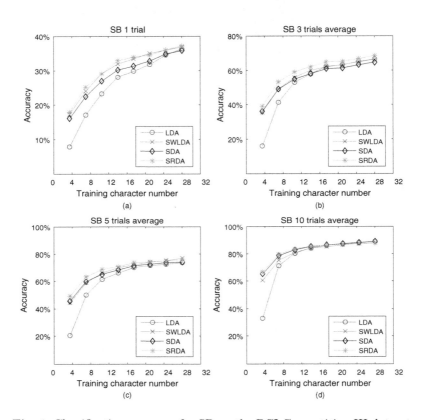

Fig. 4. Classification accuracy for SB on the BCI Competition III dataset

4   Conclusion

In this paper, we present a new algorithm called SRDA by integrating the semi-supervised learning and shrinkage approach. The undersample problem in P300 detection task was investigated to improve the classification performance. Experiment results showed that our proposed algorithm performs better than the LDA, SWLDA and SDA with limited training samples.

Acknowledgements. This work is supported by the Fundamental Research Funds of Shandong University (Grant no. 2017JC013, 2016JC021), Shandong Province Key

Innovation Project (Grant no. 2017CXGC1504), Shandong Provincial Science and Technology Major Project (Emerging Industry) (Grant no. 2015ZDXX0801A01), University Independent Innovation Foundation of Jinan Science & Technology Bureau (Grant no. 201401208).

References

1. Blankertz, B., Lemm, S., Treder, M., Haufe, S., Müller, K.R.: Single-trial analysis and classification of ERP components l a tutorial. Neuroimage **56**(2), 814–825 (2011)
2. Cai, D., He, X., Han, J.: Semi-supervised discriminant analysis. In: IEEE International Conference on Computer Vision, pp. 1–7 (2007)
3. Donchin, E., Spencer, K.M., Wijesinghe, R.: The mental prosthesis: assessing the speed of a p300-based brain-computer interface. IEEE Trans. Rehabil. Eng. **8**(2), 174–179 (2000)
4. Krusienski, D.J., Sellers, E.W., Cabestaing, F., Bayoudh, S., McFarland, D.J., Vaughan, T.M., Wolpaw, J.R.: A comparison of classification techniques for the p300 speller. J. Neural Eng. **3**(4), 299 (2006)
5. Ledoit, O., Wolf, M.: A Well-Conditioned Estimator for Large-Dimensional Covariance Matrices. Academic Press Inc., New York (2004)
6. Li, M., Yuan, B.: 2D-LDA: a statistical linear discriminant analysis for image matrix. Pattern Recogn. Lett. **26**(5), 527–532 (2005)
7. Pfurtscheller, G., Neuper, C.: Motor imagery and direct brain-computer communication. Proc. IEEE **89**(7), 1123–1134 (2001)
8. Schäfer, J., Strimmer, K.: A shrinkage approach to large-scale covariance matrix estimation and implications for functional genomics. Stat. Appl. Genet. Mol. Biol. **4**(1), 1175–1189 (2005)
9. Webb, A.R., Copsey, K.D.: Introduction to Statistical Pattern Recognition. John Wiley & Sons Ltd., Chichester (1990)
10. Wu, Q., Zhang, Y., Liu, J., Sun, J., Li, J.: Sparse optimal score based on generalized elastic net model for brain computer interface. In: 2016 Sixth International Conference on Information Science and Technology (ICIST), pp. 66–71. IEEE (2016)
11. Zhang, Y., Wang, Y., Jin, J., Wang, X.: Sparse Bayesian learning for obtaining sparsity of EEG frequency bands based feature vectors in motor imagery classification. Int. J. Neural Syst. **27**(02), 1650032 (2017)
12. Zhang, Y., Zhou, G., Jin, J., Wang, M., Wang, X., Cichocki, A.: L1-regularized multiway canonical correlation analysis for SSVEP-based BCI. IEEE Trans. Neural Syst. Rehabil. Eng. **21**(6), 887–896 (2013)
13. Zhang, Y., Zhou, G., Jin, J., Zhang, Y., Wang, X., Cichocki, A.: Sparse Bayesian multiway canonical correlation analysis for EEG pattern recognition. Neurocomputing **225**, 103–110 (2017)
14. Zhang, Y., Zhou, G., Jin, J., Zhao, Q., Wang, X., Cichocki, A.: Sparse Bayesian classification of EEG for brain-computer interface. IEEE Trans. Neural Netw. Learn. Syst. **27**(11), 2256–2267 (2016)

Predicting Learning Effect by Learner's Behavior in MOOCs

Ye Tian[1], Yimin Wen[1(✉)], Xinhe Yi[1], Xi Yang[2], and Yuqing Miao[1]

[1] School of Computer Science and Information Security,
Guilin University of Electronic Technology, 1 Jinji Road, Guilin 541004, China
ymwen2004@aliyun.com
[2] Guangxi District Education Department, 69 Zhuxi Avenue,
Nanning 5530021, China
yangxi@gxedu.gov.cn

Abstract. With the fast development of MOOCs in recent years, more and more people start to take MOOCs to perfect themselves. However, there exist high dropout rate and low passing rate of examination in many courses. So it is very important to predict learners' learning effect exactly. For learners who predicted good learning effect, teachers can impose intervention to help these learners to stick to the end of courses, while for predicted bad learning effect, teachers can take measures to help these learners to study harder to improve their learning. In this paper, we first analyze learners' learning behavior data to explore the differences among learners with different categories, then a cascade prediction model is proposed to predict whether a learner can earn certificate in a course. Experiments conducted on a real-world dataset illustrated the effectiveness of the proposed model.

Keywords: MOOCs · Learning effect · Learning behavior · Prediction model · Education data mining

1 Introduction

As an extension of existing online learning technologies, MOOCs have become increasingly popular and offered learners the opportunity to take online courses from prestigious universities. In recent years, MOOCs platform like edX, Courera and Udacity have provided hundreds of courses, and most courses may attract more than 100,000 registered learners [1]. However, due to the open and online nature of MOOCs, learners may enroll and drop a course freely, result in high dropouts rate and low passing rate of examinations [2]. It is reported that completion rate on Coursera platform is only 7%–9% [3]. And in anther report, the passing rate of examinations in first year of courses offered by MITx and HarvardX on edX platform is only 1%–12% [4].

This work was supported in part by the National Natural Science Foundation of China 61363029, Online Education Research Fund (QTone Education) of Ministry of Education of China 2016YB155, and the Natural Science Foundation of Guangxi District 2014GXNSFAA118395.

H. Yin et al. (Eds.): IDEAL 2017, LNCS 10585, pp. 524–533, 2017.
https://doi.org/10.1007/978-3-319-68935-7_57

Therefore, it is very important to analyze and predict learners' learning effect exactly. For the learners of predicted good learning effect, teachers can impose intervention to help these learners to stick to the end of courses, while for the predicted bad learning effect, teachers can take measures to push these learners to study harder to improve their learning.

In this paper, we focus on understanding learners' learning behaviors and predicting whether a learner will earn a certificate in a course based on their learning behaviors. The main contribution of this paper has two aspects. On the one hand, we make analysis for learners' learning behaviors. Firstly, we divide learners into three categories based on their grades and make statistics on several behavior features like events, days, videos, chapters, forum for the learners with different categories in all courses, then we apply K-means algorithm to learners' behavior features to find the separability between different categories. On the other hand, we construct a cascade prediction model to predict whether a learner can earn a certificate in a course. The proposed model can alleviate the problem of class imbalance by utilizing class imbalance classification techniques and separability between different categories. Experiments conducted on a real-world dataset illustrated the effectiveness of the proposed model.

2 Related Work

MOOCs have attracted millions of learners in recent years. Analyzing and mining the big data generated from online courses becomes an important and hot topic. Some researches focused on dropout prediction in MOOCs. Kloft et al. [5] used the click-stream data to extract behavior features and took SVM to predict whether a student will drop out or not. Fei et al. [6] used the dataset from the MOOCs offered by the Coursera and edX, and extracted behavior features from the click-stream data pre-week and took recurrent neural network (RNN) model and RNN with long short-term memory (LSTM) cells model to predict whether learner will dropout from course in next week. Tang et al. [2] developed a framework based on decision tree, to automatically identify the potential dropout students in MOOCs.

Some researches paid attention to learning effect in MOOCs. Jiang et al. [7] made a thorough study about the relationship between learning behaviors and learning effect, then used linear discriminant analysis, logisitic Regression and SVM to predict whether a learner will earn a certificate in a course base on behavior features. Qiu et al. [8] used the dataset from XuetangX, and conducted in-depth analysis for student demographics, learning behaviors in course forums, videos and assignments, then used latent dynamic factor graph (LadFG) to incorporate students' demographics and learning behaviors into a unified framework. Elbadrawy et al. [9] utilized personalized multiregression and matrix factorization approaches based on recommender systems to forecast learners' grades in future courses as well as on in-class assessments. Sinha et al. [10] utilized three interaction features from learners to train the probabilistic framework of Conditional Random Fields (CRF) to predict the sequence of grades achieved by a student in different MOOCs.

Except predicting learners' dropout and learning effect, Ramesh et al. [11] used learners' activities such as posting in discussion forums, timely submission of assignments, linguistic features from forum content and structural features from forum interaction to identify student's engagement is passive or active in MOOCs. Shankar et al. [12] took K-means clustering method to analyse the behavior performance of learners based on different attributes with respect to their country.

In this work, we predict whether a learner can pass examination and earn certificate in courses. In MOOCs, the proportion of learners who do not earn certificate is much larger than the learners who earn certificate. The problem of class imbalance will lead predict inaccurately. But this problem has not be concerned in existing researches. To solve the problem, we construct a model which can reduces the impact of class imbalance in dataset by utilizing class imbalance classification techniques and distribution characteristic of learners with different categories in behavior feature space.

3   Dataset Introduction

The dataset used in this work is from the first year (Fall 2012, Spring 2013, and Summer 2013 semesters) of MITx and HarvardX courses offered on the edX platform [13]. The dataset includes 641138 records, and each record represents several learner's activities in a course. In each record, we focus on the information showed in Table 1. The dataset consists of learners' activity data in 11 completed courses. There are 57400 records in each of which learner's grade is null, so these records are deleted.

Table 1. Variables in each record of dataset.

Variable	Definitions
Course_id	Course ID
Userid_DI	User ID
Certified	Anyone who earned a certificate. Certificates are based on course grades, and depending on the course, the cutoff grade for a certificate varies from 0.5–0.8
Grade	Final grade in the course, ranges from 0 to 1
Events	Number of interactions with the course, recorded in the tracking logs
Days	Number of days student interacted with course
Videos	Number of playing video events within the course
Chapters	Number of chapters (within the Courseware) with which the student interacted, represent learning progress
Forum	Number of posts to the Discussion Forum

Based on Chapters variable, all learners can be divided into three groups including only registered, general, and active. The group of only registered represents the learners who never access courseware, the group of general represents the learners who access courseware but access less than half of the available courseware chapters, and the group of active represents the learners who access more than half of the available courseware chapters. Nearly 37%, 57% and 6% of the learners belong to the groups of only registered, general and active respectively. So most of the learners are not enthusiastic for learning in MOOCs, and only few have learnt most course content.

4   Learning Behavior Analysis

Both grade and certificate variable are important indicator for evaluating learning effect for learners in MOOCs. So we take a statistic on grade to show the distribution of learners in 11 courses. From the statistic we can know that learners can be divide into three categories in each course, like most of the learners with grade of 0, about 10%–20% learners with grade over 0, but didn't earn certificate in course, and the ratio of learners who earned certificate is about 3%–10%. For convenience, we call the cases of grade = 0, grade > 0 but no certificated, and certificated as three categories like class 0, class 1 and class 2 respectively.

Based on the above analysis, we try to understand the difference between the three categories. We calculated the mean, minimum, quarter quantile, half quantile, three quarters quantile, maximum for learner's behavior features like events, days, videos, chapters, forum respectively. Based on these statistics, we can observe that the learners who get high grade are more active than the learners with low grade in MOOCs. For example, for the case of class 0, the mean of event, days, videos, and chapters are respectively round 100, 3, 17 and 1, but for the case of class 1 the mean numbers are respectively larger than 1000, 13, 100, and 5, and for the case of class 2 the mean numbers are respectively larger than 5000, 40, 400 and 10.

In order to understand the distribution of the learners with different categories in behavior feature space. We applied K-means algorithm to the learners' behavior features like Events, Days, Videos, Chapters, Forum in Table 1 for a course. All the learners are clustered into two clusters or three clusters, and then we calculated the mean, minimum, quarter quantile, half quantile, three quarters quantile, maximum for learners' grade in each cluster showed in Tables 2 and 3.

From the results in Table 2, we can confidently guess that the cluster A and the cluster B can represent the learners who have a poor performance or better performance

Table 2. The grade statistics of two clusters.

	Mean	Min	25%	50%	75%	Max
Cluster A	0.0158	0	0	0	0	1
Cluster B	0.5572	0	0.22	0.62	0.86	1

Cluster A: 4609, Cluster B: 382

Table 3. The grade statistics of three clusters.

	Mean	Min	25%	50%	75%	Max
Cluster A'	0.0149	0	0	0	0	1
Cluster B'	0.3643	0	0.02	0.1	0.77	1
Cluster C'	0.5401	0	0.2	0.62	0.81	1

Cluster A': 4560, Cluster B': 76, Cluster C': 355

in online learning respectively. And further the cluster A and the cluster B can be regarded as the learners with the grade of 0 and the learners with grade over 0 respectively. The average distance between the points of the cluster A and the cluster B is 6.23 in behavior feature space.

From the results in Table 3, we can observe that the cluster A' and the cluster C' can be regarded as class 0 or class 2 respectively. The cluster B' includes the records come from the cluster A and the cluster B in Table 2 and represents the learners whose learning performance generally and have medium grade in examination. So we can regard the cluster B' in Table 3 represents class 1. The average distance between the points of the cluster A' and the cluster B', the cluster B' and the cluster C', the cluster A' and the cluster C' is 5.91, 2.72 and 7.87 respectively in behavior feature space. It means that there are overlapping between class 1 and class 2, and better separability between class 0 and not class 0.

5 Learning Effect Prediction

5.1 Class Imbalance

Imbalanced data refers to some classes are much smaller than other classes. We can know from the statistics of each course that the total number of the records of class 0 and class 1 is about 9–28 times for the number of the records of class 2, the number of the records of class 0 is about 2.5–5.5 times for the total number of the records of class 1 and class 2, and the number of the records of class 1 is about 2–5.6 times for the number of the records of class 2. So, the dataset used is very imbalanced.

5.2 Cascade Prediction Model

In this paper, the problem of predicting learner's learning effect can be expressed as a classification problem of two classes which includes the no certificated and the certificated. Considering the better separability between class 0 and not class 0, we transform the problem of classifying the certificated and the no certificated into two cascade binary classification problem. The prediction process of the proposed model is showed in Fig. 1. In Fig. 1, classifier 1 is used to differentiate class 0 from class 1 and 2. If the classifier 1 give the prediction output of class 0, the input will be labeled as class 0 which means the learner will get a grade of 0, if classifier 1 give the prediction output of

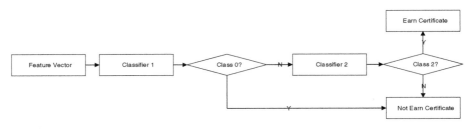

Fig. 1. The proposed cascade prediction model.

not class 0, the input will be classified by classifier 2, and further the input will be classified as class 1 or class 2 which means the learner will get a grade over 0 but no certificated or certificated. However, the class imbalance in classifier 1 and classifier 2 is only around 2.5–5.5 and 2–5.6 respectively.

For classifier 1, its training dataset includes the learners with class 0 or not class 0. Although the number of these two types of samples is imbalanced, there are better separability between these two types of learners. Class separability is a key factor for classification of imbalanced data [14]. Therefore it is not necessary to consider the class imbalance when training classifier 1. For classifier 2, the training dataset includes the learners with class 1 or class 2.

We choose SVM algorithm to train classifier 1 and classifier 2. For classifier 1, ordinary SVM is used, while for classifier 2, we take SVM-W to implements cost-sensitive learning by set different weights (w_n, w_p) for different classes. The basic idea behind SVM-W is to assign a larger penalty value to misclassification of minority class (learners who earned certificate) than misclassification of majority class (learners who didn't earn certificate) [15]. These two kinds of SVM have been implemented in LIBSVM [16].

6   Experiments and Discussions

6.1   Evaluation Metrics

In order to evaluate the effectiveness of the proposed cascade prediction model, accuracy, precision, recall and F-measure are used as evaluation criteria. we take the learners earned certificate as positive class and the learners didn't earn certificate as negative class.

6.2   Baselines

We first compare the proposed model with several alternative predictive models: linear discriminant analysis (LDA), Logisitic Regression (LR), linear support vector machine (l-SVM), rbf keanel support vector machine (rbf-SVM).

Then we compare the proposed model with three other methods using 'rebalance' techniques. SVM-SM adopts the SMOTE algorithm [17] to generate more pseudo positive samples and then builds an SVM on the oversampling dataset [15]. SVM-RD randomly selects a few of negative samples and then builds an SVM on the under sampling dataset [18]. SVM-W implements a cost-sensitive learning solution to overcome class imbalance problem in SVM.

6.3   Experiment Setup

We conducted experiments for the dataset from 11 courses. We do a 9:1 split of the collected data into training and test sets. The results of the experiment for each method are based on the mean of 5 times 10 cross-validation. We set the parameters in Table 4

Table 4. Experiment parameters of all methods.

Parameters	Definition	Value
w_n, w_p	Weight parameter in classifier 2 for the proposed model	w_n = 1, w_p = 1.5
r_o	Oversampling multiple in SVM-SM	10
r_u	Undersampling ratio in SVM-RD	60%
w1, w2	Weight parameter in SVM-W	w1 = 1, w2 = 2

for the proposed model and the baseline methods. These parameters are proved to make a great effectiveness by experiments.

6.4   Experiment Results and Discussions

For convenience, we call the proposed model as DW-SVM. The experiment results are showed in Tables 5, 6, 7 and 8. It can be observed that all the methods have great performance for accuracy in Table 5, and the accuracy of all methods in all courses is all over 96%. Then we can further find the nonlinear model (rbf-SVM) have greater performance than the linearity models (LDA, LR, I-SVM) for the prediction task in Tables 5, 6, 7 and 8. The models take an imbalanced classification techniques (SVM-W, SVM-RD, SVM-SM) have greater performance for the recall in Table 7. Experiment results also show that the proposed model has greater performance for F1-score and at the same time keeps greater accuracy for the most courses in Tables 5 and 8.

Table 5. Experiment results of accuracy (%).

	C1	C2	C3	C4	C5	C6	C7	C8	C9	C10	C11
LDA	96.55	97.64	98.16	97.91	98.78	97.77	98.25	98.66	97.79	98.73	98.23
LR	96.65	97.98	98.27	**97.99**	98.91	98.11	98.52	99.15	**98.1**	**99.08**	98.32
I-SVM	96.83	98.02	98.29	**97.99**	98.97	98.15	98.53	99.17	98.07	**99.08**	98.37
rbf-SVM	97.37	**98.23**	98.31	97.95	99.17	98.42	98.74	99.26	98.06	99.05	**98.38**
SVM-W	97.18	98.21	98.19	97.93	99.11	98.38	98.74	99.21	98.06	98.97	98.29
SVM-RD	**97.38**	98.13	98.31	97.96	99.08	98.42	98.73	99.2	**98.1**	98.94	98.21
SVM-SM	96.84	98.11	98.06	97.84	99.17	98.29	98.71	99.21	98.04	98.9	98.32
DW-SVM	97.32	98.2	**98.34**	97.93	**99.2**	**98.44**	**98.75**	**99.26**	**98.1**	99.05	**98.38**

In the end, for the learners are predicted wrongly, we conduct an error analysis on the results of the proposed model and three major types of errors are observed.

1. Some learners interact actively with a course, but they don't earn a certificate in the end of the course. Maybe it is because the aim of them is not to earn a certificate in a course, they may just want to learn this course. These learners would be misclassified into the learners who can earn certificate.
2. Some learners have very few learning activities in a course. However they got a good grade in the courses' examination and earned the course's certificate.

Table 6. Experiment results of precision (%).

	C1	C2	C3	C4	C5	C6	C7	C8	C9	C10	C11
LDA	65.89	62.1	87.76	76.98	81.82	75.29	73.78	75.49	72.3	77.49	70.34
LR	70.59	73.92	**91.21**	**81.27**	86.73	81.15	**82.35**	87.56	80.49	**86.39**	**78.32**
I-SVM	72.5	74.87	90.83	80.75	85.15	80.16	80.42	87.81	**79.12**	86.03	77.48
rbf-SVM	**74.04**	**75.9**	90.23	79.59	86.97	**82.69**	82.04	**88.5**	78.66	85.74	75.85
SVM-W	68.11	71.33	86.4	77.27	85.22	80.03	79.28	85.3	74.18	81.21	70.1
SVM-RD	72.8	68.88	90.08	78.14	84.76	82.29	79.61	85.42	76.98	80.93	69.49
SVM-SM	66.57	70.23	86.44	75.6	85.04	79.74	79.35	85.51	75.24	80.17	71.53
DW-SVM	69.25	70.82	90.34	74.65	**87.23**	81.66	79.39	86.89	75.3	83.59	72.03

Table 7. Experiment results of recall (%).

	C1	C2	C3	C4	C5	C6	C7	C8	C9	C10	C11
LDA	80.06	79.97	94.33	82.59	94.15	90.57	87.5	92.62	87.45	91.72	91.15
LR	69.25	63.18	91.12	77.32	90	87.08	80.4	89.16	79.9	88.8	76.19
I-SVM	70.17	63.46	91.74	78.13	93.89	89.82	83.95	89.29	81.24	89.17	79.59
rbf-SVM	81.96	70.66	92.7	78.94	96.23	91.48	88	91.28	81.72	88.56	83.33
SVM-W	93.64	79.68	**96.68**	82.59	97.4	**95.31**	93.07	**94.34**	**91.35**	**93.06**	**94.89**
SVM-RD	85.12	**82.79**	92.94	81.61	**97.35**	92.03	92.09	93.9	86.21	92.82	92.99
SVM-SM	87.84	77.6	94.99	83.31	93.29	93.92	91.79	93.94	88.13	92.57	91.63
DW-SVM	**93.97**	80.81	92.99	**88.25**	96.75	93.65	**93.07**	93.4	89.89	91.72	92.85

Table 8. Experiment results of F1-score (%).

	C1	C2	C3	C4	C5	C6	C7	C8	C9	C10	C11
LDA	72.29	69.91	90.93	79.68	87.56	82.23	80.06	83.18	79.16	84.01	79.4
LR	69.92	68.13	91.16	79.25	88.33	84.01	81.36	88.35	80.19	**87.58**	77.24
I-SVM	71.32	68.7	91.28	79.42	89.31	84.72	82.14	88.54	80.16	87.57	78.52
rbf-SVM	77.8	73.19	91.45	79.26	91.36	86.86	84.92	89.87	80.16	87.13	79.41
SVM-W	78.86	75.28	91.25	79.84	90.9	87.01	85.62	89.59	81.87	86.73	80.63
SVM-RD	78.48	75.19	91.49	79.84	90.61	86.89	85.4	89.46	81.33	86.46	79.54
SVM-SM	75.74	73.73	90.51	79.27	88.97	86.25	85.11	89.53	81.18	85.93	80.34
DW-SVM	**79.74**	**75.49**	**91.65**	**80.89**	**91.74**	**87.25**	**85.69**	**90.03**	**81.95**	87.47	**81.12**

Maybe it is because they have taken similar courses offline, and the aim of selecting a course is only to test their skills in an area.

3. Each course has a minimum passing score, above which the student would be certificated. There are always some learners whose scores are hovering around the minimum passing score. The proposed model may misclassify these learners.

7 Conclusion

In this paper, we first made an analysis about learning behaviors of learners in MOOCs and explored the differences and characteristic of learning behavior features between the learners with different grades. Then we proposed a cascade prediction model to predict whether a learner can pass examination and earn certificate in a course. Experiments conducted on a real-world dataset showed the effectiveness of the proposed method.

MOOCs learning is one kind of high-level behavior, learners maybe influenced by many incentive factors of ultimate goal. The passing rate maybe not enough incentive for some learners. Therefore, for the proposed prediction model, the selected five behavior features seem not to describe the learning behavior completely and accurately. In future research, we plan to collect more detail behaviors to find more influencing factors, and apply these influencing factors to construct more effective prediction model.

References

1. Seaton, D.T., Bergner, Y., Chuang, I., Mitros, P., Pritchard, D.E.: Who does what in a massive open online course? Commun. ACM **57**(4), 58–65 (2014)
2. Tang, Jeff K.T., Xie, H., Wong, T.-L.: A big data framework for early identification of dropout students in MOOC. In: Lam, J., Ng, K.K., Cheung, Simon K.S., Wong, T.L., Li, K. C., Wang, F.L. (eds.) ICTE 2015. CCIS, vol. 559, pp. 127–132. Springer, Heidelberg (2015). doi:10.1007/978-3-662-48978-9_12
3. Massive open online course. https://en.wikipedia.org/wiki/Massive_open_online_course
4. Ho, A.D., Reich, J., Nesterko, S.O., Seaton, D.T., Mullaney, T., Waldo, J., Chuang, I.: HarvardX and MITx: The First Year of Open Online Courses, Fall 2012-Summer 2013. MIT Office of Digital Learning, HarvardX Research Committee, pp. 1–33 (2014)
5. Kloft, M., Stiehler, F., Zheng, Z., Pinkwart, N.: Predicting MOOC dropout over weeks using machine learning methods. In: Proceedings of EMNLP 2014 Workshop on Analysis of Large Scale Social Interaction in MOOCs, pp. 60–65. ACL, Stroudsburg (2014)
6. Fei, M., Yeung, D.Y.: Temporal models for predicting student dropout in massive open online courses. In: Proceedings of 2015 IEEE International Conference on Data Mining Workshop (ICDMW), pp. 256–263. IEEE, Washington (2015)
7. Jiang, Z.X., Zhang, Y., Li, X.M.: Learning behavior analysis and prediction based on MOOC data. J. Comput. Res. Dev. **52**(3), 614–628 (2015)
8. Qiu, J., Tang, J., Liu, T.X., Gong, J., Zhang, C.H., Zhang, Q., Xue, Y.F.: Modeling and predicting learning behavior in MOOCs. In: Proceedings of International Conference on Web Search and Data Mining, pp. 93–102. ACM, New York (2016)
9. Elbadrawy, A., Polyzou, A., Ren, Z.: Predicting student performance using personalized analytics. Computer **49**(4), 61–69 (2016)
10. Sinha, T., Cassell, J.: Connecting the dots: predicting student grade sequences from Bursty MOOC interactions over time. In: Proceedings of the Second ACM Conference on Learning at Scale, pp. 249–252. ACM, New York (2015)
11. Ramesh, A., Goldwasser, D., Huang, B., Daum, H., Getoor, L.: Modeling learner engagement in MOOCs using probabilistic soft logic. In: Proceedings of NIPS Workshop on Data Driven Education, pp. 1–7. MIT Press, Massachusetts (2013)

12. Shankar, S., Sarkar, B.D., Sabitha, S.: Performance analysis of student learning metric using K-mean clustering approach K-mean cluster. In: Proceedings of the 6th International Conference on Cloud System and Big Data Engineering, pp. 341–345. IEEE, New York (2016)
13. Harvardx-mitx person-course academic year 2013 de-identified dataset, version 2.0. http://dx.doi.org/10.7910/DVN/26147
14. Prati, R.C., Batista, G.E.A.P.A., Monard, M.C.: Class imbalances *versus* class overlapping: an analysis of a learning system behavior. In: Monroy, R., Arroyo-Figueroa, G., Sucar, L.E., Sossa, H. (eds.) MICAI 2004. LNCS, vol. 2972, pp. 312–321. Springer, Heidelberg (2004). doi:10.1007/978-3-540-24694-7_32
15. Akbani, R., Kwek, S., Japkowicz, N.: Applying support vector machines to imbalanced datasets. In: Boulicaut, J.-F., Esposito, F., Giannotti, F., Pedreschi, D. (eds.) ECML 2004. LNCS, vol. 3201, pp. 39–50. Springer, Heidelberg (2004). doi:10.1007/978-3-540-30115-8_7
16. Chang, C.C., Lin, C.J.: LIBSVM: a library for support vector machines. ACM Trans. Intell. Syst. Technol. 2(3), 1–27 (2012)
17. Chawla, N.V., Bowyer, K.W., Hall, L.O., Kegelmeyer, W.P.: SMOTE: synthetic minority over-sampling technique. J. Artif. Intell. Res. 16(1), 321–357 (2011)
18. Kubat, M., Matwin, S.: Addressing the curse of imbalanced training sets: one-sided selection. In: Proceedings of the 14th International Conference on Machine Learning, pp. 179–186. ACM, New York (1997)

Standardised Reputation Measurement

Peter Mitic[1,2,3]($\boxtimes$)

[1] Santander UK, 2 Triton Square, Regents Place, London NW1 3AN, UK
peter.mitic@santandergcb.com
[2] Department of Computer Science, UCL, Gower Street, London WC1E 6BT, UK
[3] Laboratoire d'Excellence sur la Régulation Financière (LabEx ReFi), Paris, France

Abstract. Well-defined formal definitions for sentiment and opinion are extended to incorporate the necessary elements to provide a formal quantitative definition of reputation. This definition takes the form of a time-based index, in which each element is a function of a collection of opinions mined during a given time period. The resulting formal definition is validated against informal notions of reputation. Practical aspects of data procurement to support such a reputation index are discussed. The assumption that all mined opinions comprise a complete set is questioned. A case is made that unexpressed positive sentiment exists, and can be quantified.

Keywords: Reputation · Reputation index · Reputation definition · Opinion · Unexpressed positive sentiment

1 Introduction

Reputation measurement has gained increasing prominence in recent years as organisations become much more aware that their reputation matters. A formal mathematical definition for reputation has been elusive, and reputation has been expressed as a subjective concept. The aim of this paper is to provide a clear distinction between reputation and two closely related concepts (sentiment and opinion), and then to formalise the definition of reputation in quantitative terms.

1.1 Informal Definitions for Sentiment and Reputation

The Oxford English Dictionary defines sentiment as "A view or opinion that is held or expressed" (https://en.oxforddictionaries.com/definition/sentiment). Other dictionaries concur. Liu [1] uses the word *sentiment* in the sense of a positive or negative *feeling*, and also introduces the word *opinion* to indicate a broad context covering sentiment (by the Liu definition), evaluation, appraisal,

P. Mitic—The opinions, ideas and approaches expressed or presented are those of the author and do not necessarily reflect Santanders position. The values presented are just illustrations and do not represent Santander data.

attitude and associated information such as the opinion holder and the opinion target.

Reputation is more than a collection of opinions. Reputation expresses the relationship between opinion holders and the performance of the target, with respect to the expectation of the opinion holder. An informal definition is given in Mitic [2], but is amended below to be consistent with the Liu definition of opinion. Therefore we formulate these informal definitions.

- *Sentiment:* A view that is held or expressed
- *Opinion:* Sentiment expressed by a holder of a target at a particular time
- *Reputation:* Collective opinions, established over time, that can conflict with the expectation that the opinion holders have of the target

This informal definition of reputation encapsulates the idea that reputation is a difference between what you expect and what you get. It corresponds broadly to the definition of reputation from the Basel Committee [3].

1.2 Formal Definitions for Sentiment, Opinion and Reputation

The following formal definitions of sentiment and opinion are based on the definitions in Liu [1]. Liu defines three separate components of sentiment: polarity (positive, negative or zero); intensity (a measure of the extent of polarity numeric or otherwise); and type (rational or emotional). In practice, polarity and intensity emerge from data mining as a single entity, and it more convenient to subsume the type into a more general categorisation vector C (see below). Consequently, we will define a **standard measure of sentiment**, S, as a real number in the range $[-1, 1]$.

$$S \in \mathbb{R} : -1 \leq S \leq 1 \qquad (1)$$

An account of methods which may be used to quantify sentiment may be found in, for example [4] or [5]. They include methodologies such as Naive Bayes, Artificial Neural Nets and Support Vector Machines.

Opinion, O, is minimally a 5-dimensional vector which extends the concept of sentiment to include: a unique identifier i; a timestamp t; a target G; its holder (originator) H; and the sentiment $S_{i;H}$ (referred to its identifier and holder since it is often used in isolation). It is useful, although not necessary, to add a sixth component to O: a categorisation vector C which can be used to classify the opinion (for example, as social/business, or to indicate the influence of the holder). This component is useful for analysis of factors that affect reputation, and subsumes Liu's sentiment "type".

$$O = (i, t, G, H, S_{i;H}, C) \qquad (2)$$

Reputation at time t can then be defined in terms of a collection of opinions $\{O\}_{t \in T}^{i \in I}$, where T is an indexing set for time and I is an indexing set for unique identifiers. The reputation of an organisation G at time t, $R_G(t)$, can then be given as some generic function ρ of the collection of opinions:

$$R_G(t) = \rho(\{O\}_{t \in T}^{i \in I}). \qquad (3)$$

The function ρ could be, for example, a weighted average of the sentiments expressed in the collection of opinions. This is often the approach adopted in practice. With this approach, if w_i is the weight assigned to the opinion O with unique identifier i and sentiment $s_{i;H}$, and the indexing set I has n elements, then the reputation at time t takes the form in Eq. 4. This formulation ensures that $-1 \leq R_G(t) \leq 1$.

$$R_G(t) = \frac{\sum_{i=1}^{n} w_i s_{i;H}}{\sum_{i=1}^{n} w_i} \qquad (4)$$

The informal definition of reputation in the previous sub-section includes the reference *established over time*. To extend the definition $R_G(t)$ to cover the times in the indexing set T, we define the reputation $\hat{R}_G$ of the target G as the time series

$$\hat{R}_G = \{R_G(t)\}_{t \in T}. \qquad (5)$$

The definition in Eq. 5 depends on reputation measurements taken over an extended period. It is not sufficient to deal with cases where a potential opinion holder notes a small number (perhaps one only) of isolated comments, and formulates his/her own opinion based solely on that.

2   Data Mining for Reputation

In this section we give a brief overview of the practicalities of the data mining processes needed to implement viable reputational analysis. Full details may be found in [2]. The process is, for a given time period (typically 24 h):

1. Receive 'contents' (corresponding to Eq. 2) by electronic feeds from relevant public sources of opinion: news reports, radio and TV broadcasts, press releases, reports from trade events, comments on social media (Twitter, Facebook, blogs etc.).
2. Analyse each content for sentiment, define a weight (e.g. to reflect the influence of the opinion holder), resulting in a standardised sentiment (Eq. 1) for each.
3. Compose a reputation index component using all content received in the time period (Eq. 4, or more generally 3).

A reputation index, as defined in Eq. 5 can then be compiled by accumulating the results of the above process for a sequence of intervals t in a set T. Figure 1 summarises the reputation index procurement process.

The sequence shown in Fig. 1 is intended to remove subjectivity from the reputation index procurement process. This is possible provided that the sources for the data mining stage form a complete and comprehensive set and that the analysis of the contents is sufficient to determine sentiment accurately. The completeness assumption will be questioned in the next section. The definitions of reputation given can be said to induce bias by not quantifying contents that have not been received.

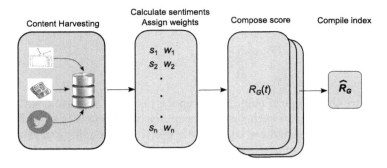

Fig. 1. Reputation procurement process summary

The possible existence of 'missing' positive sentiment is summed up in a quote by Donald Rumsfeld, who was US Secretary of Defence in the George W. Bush administration [6]: *The message is that there are no "knowns". There are things we know that we know. There are known unknowns. That is to say there are things that we now know we don't know. But there are also unknown unknowns. There are things we don't know we don't know.*". 'Missing' positive sentiment corresponds to "unknown unknowns".

3   Negative Opinion Bias: Methodology

The definitions in Eqs. 3 and 4 apply to all contents that are received. It clearly cannot be applied to comments that are not received. Such comments have been noted informally. The reason is bias towards negative sentiment, summed up in the phrase "No news is good news". Agents (individuals and groups/corporates may be forthright in expressing *strong* sentiment, either positive or negative. They might also express, relative to the ambient sentiment (i.e. the mean reputation score), very mild negative sentiment if they are slightly annoyed, but might not bother with corresponding positive sentiment if they are only mildly satisfied. Therefore some contents that express positive sentiments might be missing.

There are indications that such sentiment bias exists. Aktolga and Allan [7], and Cook and Ahmad [9] both detected positive and negative sentiment bias. Kelly and Ahmad [8], found statistically significant predictive power due to only negative sentiment when predicting stock market returns.

3.1   Bias Measurement

To try to estimate how many positive sentiments might be missing, we look specifically at reputation scores near the truncated mean m (i.e. the mean excluding the largest and the smallest to avoid extremes) of a set of reputation scores. This reflects the idea that sentiments that are mildly above average might be

'missing'. The argument proceeds first by considering the cumulative reputation
score S_T over a given period (i.e. the sum of elements in $\hat{R}_G$, Eq. 5).

$$S_T = \sum_{t \in T} R_G(t) \tag{6}$$

Figure 2 shows three typical cumulative reputation score profiles: positive, negative and zero trending. In each case, the cumulative reputation score from time
0 to t, $S_{[0,t]}$, is plotted against time. With positively trending cumulative reputation, missing positive sentiment is measured by seeking a negative trend in
a neighbourhood of m. For a negative trending cumulative sum, we reverse the
sign of the scores, and continue as though it were positive trending.

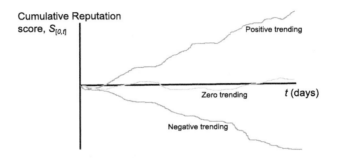

Fig. 2. Typical examples of cumulative reputation scores

First we compare the skewness of scores in a neighbourhood of m with the
skewness of all the scores. To do this we calculate the skewness of elements $R_G(t)$
in the range $[m - w, m + w]$ where w is the semi-width of a band, calculated as a
percentage of the range of scores (i.e. the difference between the maximum score
and the minimum score). A counter skew in a neighbourhood of m is a pointer
to 'missing' positive sentiment. Skewness is not a useful measure for quantifying
the extent of 'missing' positive sentiment as the numeric skewness values are not
directly related to numbers of contents. Counting reputation scores is.

To estimate the number of missing contents that express positive sentiment,
we calculate the *number* of elements in the range $[m-w, m+w]$. A counter
trend is measured by calculating the two ratios α and β in Eq. 7, in which N_w^+
and N_w^- are the number of scores in the intervals $(m, m+w]$ and $[m-w, m)$
respectively, and N^+ and N^- are the total numbers of positive and negative
scores respectively.

$$\alpha = \frac{N_w^-}{N_w^+}; \beta = \frac{N^-}{N^+} \tag{7}$$

Figure 3 shows the numbers N_w^+, N_w^-, N^+ and N^- for normally distributed
scores $R_G(t)$. Such Normal distributions are quite typical for reputation scores.

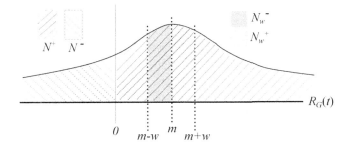

Fig. 3. Normally distributed reputation scores: positively trending cumulative sum

The case $\alpha > 1$ represents the counter trending case: there are more negative sentiments than positive in a neighbourhood of the mean score, despite an overall positive trend. We compare this ratio with β. We define a measure of the percentage of missing positive sentiments, M, by

$$M = 100(\frac{\alpha - \beta}{\beta}). \tag{8}$$

4   Negative Opinion Bias: Results

The data used in the calculations of this paper comprise values of the reputation index for ten UK retail banks, originating from the business intelligence consultancy Alva (www.alva-group.com). For confidentiality reasons, they are labelled Bank1, Bank2, ... Bank10. Alva's reputation index from January 2014 to December 2015 has been used, scaled linearly from its native range [1,10] to [−1,1]. In addition, the scores for the ten banks has been averaged on a per-day basis to produce a reputation score representing retail banking.

The values of w used reflect the target range of sentiment that correspond to the idea that contents that express "small positive" sentiment, relative to the ambient sentiment, are "missing". These cover one third of the range of score values, apart from the 5% nearest to zero, the results from which are unstable. Therefore we take w in a window $W = [2.5\%, 16.5\%]$.

Skewness Results
When there is a positive trend in cumulative reputation, the skewness corresponding to w in the window $W = [2.5\%, 16.5\%]$ is contrary to the overall skewness of the data. When there is a negative trend a contrary skewness is absent. The inference is that there are missing positive sentiments for positive trending cumulative reputation, but not if there is a negative trend. Table 1 shows the results.

Negative Sentiment Bias Measurement Results
A count of reputation scores is more useful for quantifying the extent of 'missing' positive sentiment than skewness because it is expressed in terms of actual

Table 1. Negative sentiment bias skewness results: *Skew-W* is the skewness in the window *W* and *Skew-All* is the skewness of all the data

Bank	Skew-W	Skew-All	Trend
1	−0.34	0.20	Positive
2	1.7	0.48	Negative
3	−1.44	0.55	Positive
4	0.00	0.13	Positive
5	1.87	0.61	Negative
6	−1.24	0.11	Positive
7	1.96	0.18	Negative
8	−0.48	−0.01	Positive
9	−0.07	0.22	Positive
10	2.16	0.28	Negative
Mean	−0.53	0.04	Positive

numbers. Table 2 shows the calculated values of M (Eq. 8) for each bank that shows a positive trending cumulative reputation score as w varies in the range indicated. Those are the ones that, according to the skewness analysis, have 'missing' positive sentiment.

Table 2. Negative sentiment bias measurement results

Bank	M
1	11.85
3	23.59
4	4.99
6	7.45
8	3.48
9	13.57
All	4.28

The results in Table 2 indicate a high dependence on the data used. Consequently, the 'All' result "4–5%" should be used to represent 'missing' positive sentiment. The highest value results are for the banks that exhibit the steepest positive cumulative score trends. The opposite effect has been noted in the context of the effect of all banks on the reputation of any particular bank: the ones with extreme reputations (either positive or negative) are influenced the least (see [10] for details).

5   Discussion

The formal definition of reputation presented in this paper reflects an informal view that reputation refers to collective opinion on the difference between what happens and what is expected. In order to test the view that agents do not always express positive sentiment, we have considered the skewness of the reputation score distribution in a neighbourhood of the mean score. There is some evidence for the existence of 'missing' contents that express positive sentiment, as shown by skewness in the neighbourhood of the mean score, counter-trending the overall trend. However, such evidence depends on the measure used, so one should remain sceptical. If the view is that such contents are, indeed, missing, there are two conclusions. First, 'missing' positive sentiment exists for positive trending cumulative sentiment, but not for negative. Second, the reputation score should be inflated by 4–5% to account for such 'missing' sentiment.

Quantifying 'missing' sentiment could be tackled by survey, although sampling bias would have to be considered carefully. It should also be recognised that the definitions in Eq. 5 depends on a long-term history. That conflicts with an alternative informal view of reputation: a received opinion prompted by noting a small number of comments (perhaps only one) from a *personally* trusted source. The proposed long-term view should prevail. Finally, the suggested range [−1,1] makes less sense in business terms. A mapping to [0%,100%] is sensible for risk managers because those limits translate as 'worst' and 'best', and 50% translates as 'neutral'.

Acknowledgments. I am grateful for the support of Alva-Group for their continued interest, support, and assistance in the preparation of this paper.

References

1. Liu, B.: Sentiment Analysis: Mining Opinions, Sentiments and Emotions, CUP (2015)
2. Mitic, P.: Reputation risk: measured. In: Proceedings of Complex Systems. WIT, May 2017
3. Bank for International Settlements, Enhancements to the Basel II framework (2009). www.bis.org/publ/bcbs157.pdf
4. Jurafsky, D., Martin, J.H.: Speech and Language Processing. Pearson, Upper Saddle River (2008)
5. Bishop, C.: Pattern Recognition and Machine Learning (Information Science and Statistics). Springer (2007)
6. NATO, Press Conference by Donald Rumsfeld, NATO HQ Brussels, 6–7 June 2002. http://www.nato.int/docu/speech/2002/s020606g.htm
7. Aktolga, E., Allan, J.: Sentiment diversification with different biases. In: Proceedings of SIGIR 2013 (36th International ACM SIGIR Conference on Research and Development in Information Retrieval, pp. 593–602. ACM, New York (2013)
8. Kelly, S., Ahmad, K.: The impact of news media and affect in financial markets. In: Jackowski, K., Burduk, R., Walkowiak, K., Woźniak, M., Yin, H. (eds.) IDEAL 2015. LNCS, vol. 9375, pp. 535–540. Springer, Cham (2015). doi:10.1007/978-3-319-24834-9_62

9. Cook, J.A., Ahmad, K.: Behaviour and markets: the interaction between sentiment analysis and ethical values? In: Jackowski, K., Burduk, R., Walkowiak, K., Woźniak, M., Yin, H. (eds.) IDEAL 2015. LNCS, vol. 9375, pp. 551–558. Springer, Cham (2015). doi:10.1007/978-3-319-24834-9_64

10. Mitic, P.: Reputation risk contagion. J. Netw. Theory Finan. **3**(1), 1–34 (2017)

Is a Reputation Time Series White Noise?

Peter Mitic[1,2,3](✉)

[1] Santander UK, 2 Triton Square, Regents Place, London NW1 3AN, UK
`peter.mitic@santandergcb.com`
[2] Department of Computer Science, UCL, Gower Street, London WC1E 6BT, UK
[3] Laboratoire d'Excellence sur la Régulation Financière (LabEx ReFi), Paris, France

Abstract. The plots of some reputation time series superficially resemble plots of white noise. This raises the question of whether or not the analysis of sentiment to produce a reputation index actually generates nothing more than noise. The question is answered by using the Box-Ljung statistical test to establish that the reputation time series considered in this analysis cannot be viewed as white noise. This result is supported by applying a new test based on cross-correlations of reputation time series with white noise time series.

Keywords: Reputation · Reputation index · White noise · Box-Hjung · Cross correlation · Auto correlation

1 Introduction

The mathematical basis of reputation measurement were described in these proceedings [1], and the process of procuring a reputation score was described in [2]. Procurement is a physical process which is well-defined and is objective, subject to possible error in the elucidation of sentiment. The result is simply a sequence of numbers. Unfortunately, such series can superficially resemble random number sequences. Figure 1 shows a genuine reputation time series (Mercedes-Benz) and a randomly generated time series from a normal distribution with mean and standard deviation approximately equal to the mean and standard deviation of the Mercedes-Benz time series. The question then arises "Is the method of procurement of a reputation time series merely a complicated physical method of generating random numbers?"

In this paper we test the conjecture that a reputation time series is randomly generated. We then consider a more subtle problem, which is to decide whether or not a reputation time series constitutes white noise. Implicit in this discussion is the concept that reputation time series has what could be described as a "memory" - the reputation score on one day depends on prior reputation scores. Standard statistical tests are used to test the question posed above, and an additional new statistical test is proposed that confirms the conclusion by using cross-correlations in an intuitive way.

The opinions, ideas and approaches expressed or presented are those of the author and do not necessarily reflect Santanders position. The values presented are just illustrations and do not represent Santander data.

© Springer International Publishing AG 2017
H. Yin et al. (Eds.): IDEAL 2017, LNCS 10585, pp. 543–550, 2017.
https://doi.org/10.1007/978-3-319-68935-7_59

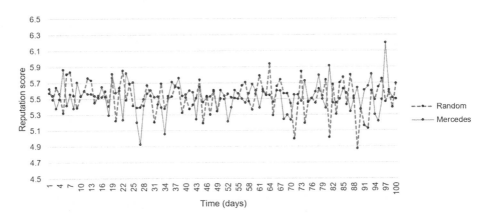

Fig. 1. Contrasting time series: Reputation and Random

1.1 Randomness and White Noise

A precise definition of randomness is surprisingly difficult, and opinions on a suitable definition differ. The essential elements are that the outcome of any given trial should be unpredictable, that it cannot subsequently be reliably reproduced, and that a sequence of random numbers does not exhibit autocorrelation. Knuth [3] argues that a sequence of random numbers is a sequence of independent numbers with a specified distribution and that there is a specified probability of selecting a number in any given range of values.

The concept of "white noise" extends the idea of randomness and presents a more precise definition in terms of a statistical process. Specifically, if x_t is a random variable with independent variable t, each term in the sequence $\{x_t\}$ should be a stochastic term ϵ_t with mean 0 and constant variance σ^2, and any two such terms ϵ_t and ϵ_s $(t \neq s)$ are independent. An instructive illustration of white noise may be found on the Wolfram Research website at https://reference.wolfram.com/language/tutorial/AudioProcessing.html. The two sound clips defined by the following Mathematica statements show the effect of reducing ("filtering") the white noise in the broadcast of Neil Armstrong's "One small step for man..." speech on the moon. The former has a marked background hiss.

- a = ExampleData["Audio", "Apollo11SmallStep", "Audio"]
- WienerFilter[a, 25]

Noise sounds like hiss, and conveys no information. We would like to think that a reputation time series is not just hiss!

2 Tests for Randomness and White Noise: Review

In this section we review some standard tests for white noise and randomness, prior to introducing a new test that makes auto-correlations central and intuitive to the decision process. Test usage is discussed in Sect. 3.

2.1 The Runs Test

The *Runs test* [4] examines randomness by counting runs of similar numbers. Having centred data about the data mean, a *run* is a set of sequential values that are either all above the mean (a "positive run") or below the mean (a "negative run"). The number of positive and negative runs, n_+ and n_- respectively, are counted. The total number of runs, $n = n_+ + n_-$, has the following Normal distribution for large n:

$$n \sim N(1 + \frac{2n_+ n_-}{n_+ + n_-}, \frac{2n_+ n_-(2n_+ n_- - n_+ - n_-)}{(n_+ + n_- - 1)(n_+ + n_-)^2}) \tag{1}$$

The test asks if the number of positive and negative runs are distributed equally in time. The null and alternative hypotheses are (respectively H_0 and H_a):

- H_0: The data are random
- H_a: The data are not random

2.2 The Box-Hjung Test

The Box-Ljung test [5] is an unusual statistical test because it is used to test the *lack* of fit of a time series model rather than how well a model fits the data. It is a specific test for white noise. The test is applied by first fitting an ARMA model to the data and then calculating residuals. Autocorrelations covering a range of lags, rather than any one lag, of the residuals are then examined. For this reason, it is often referred to as a "portmanteau" test. The Ljung-Box test is formulated with the following null and alternative hypotheses.

- H_0: The time series exhibits autocorrelation
- H_a: The time series does not exhibit autocorrelation

The Box-Ljung test statistic, Q_k, Eq. 2, is

$$Q_k = n(n+2) \sum_{i=1}^{k} \frac{r_i^2}{n-i}, \tag{2}$$

where n is the time series length, k is the maximum number of lags considered, and r_i is the autocorrelation at lag i. Q_k has a $\chi^2(k)$ distribution. At significance level α, H_0 is rejected if the calculated Q_k is less than the $(1-\alpha)\%$ value of $\chi^2(k)$. A loose interpretation is that small values of Q_k indicate "not white noise".

2.3 Cross-Correlation

Cross-correlation is an alternative way to assess serial autocorrelation, and is discussed in [6]. For large sample size n, the distribution of the cross-correlation coefficient for two time series $\{x_t\}$ and $\{y_t\}$ at lag k, $\rho_{xy}(k)$, is Normal, provided that at least one of the time series is white noise. Specifically,

$$\rho_{xy}(k) \sim N(0, \frac{1}{n}) \tag{3}$$

The adaptation described in Sect. 3 gives an amendment to Eq. 3 due to the Central Limit Theorem. The null and alternative hypotheses for both are:

- H_0: The mean cross-correlation coefficient $= 0$
- H_a: The mean cross-correlation coefficient $\neq 0$

2.4 Other Tests

This sub-section has a brief overview of some other tests for white noise. They are not implemented in this study because they are more complex, and the simpler tests used in this study suffice.

Bartlett's formula [7] tests for white noise using a matrix of autocorrelations, and assumes a stationary time series with IID noise. Autocorrelations are shown to be normally distributed for large sample size. The method is known to be inaccurate for non-linear time series because it assumes that all autocorrelations with lag greater than 1 are near zero.

Francq and Zacoïan [8] discuss an amendment based on the Bartlett formula, applicable for the asymptotic distribution of the sample autocorrelations of nonlinear processes. The first term is the same as the Bartlett linear formula. A second term introduces the kurtosis of the linear innovation process and the autocorrelation function of its square.

All the tests discussed so far are based in the time domain. An alternative is to use the frequency domain. A set of tests, starting with that of Hong [9] are based on a comparison between the spectral density of the time series and the spectral density of a white noise time series. There have been subsequent further developments of this approach.

3 Tests for Randomness and White Noise: Application

Details of the way in which the tests described in the previous section are applied are described in this section. Particular attention is paid to a comparison of reputation data with randomly generated data. There is no special adaptation of the Runs Test.

3.1 The Box-Hjung Test: Application

An implementation in the R statistical language, the function $Box.test$, was used to return the p-value of the embedded χ^2 test. Rather than just apply this test to the reputation data alone, it was also applied to randomly generated data (using an ARIMA(0,0,0) process) that resembles the reputation data. The Box-Hjung test was applied to 250 instances of this stochastic process and the mean p-value was calculated. The results for the reputation data and the randomly generated data were then compared.

3.2 The Cross-Correlation Test: Application

As it stands, Eq. 3 applies for each cross-correlation value in isolation. In order to derive a single-figure measure of the cross-correlation across all lags, the cross correlation test was applied 1000 times (i.e. enough to provide a consistent result), and the mean cross-correlation at each lag was calculated. Then, if the number of lags is L, the mean cross-correlation, R_{xy}, has the following distribution by the Central Limit Theorem.

$$R_{xy} \sim N(0, \frac{1}{nL}) \tag{4}$$

A measure of the significance of the empirical value of the statistic R_{xy} (the p-value, p) can then be obtained using (in R-pseudo code):

$$p = 2(1 - \Phi(|m|), 0, \frac{1}{nL}), \tag{5}$$

where Φ is the right-hand tail of the standard Normal distribution and m is the empirical mean cross-correlation value. In Eq. 5 the term $|m|$ accounts for both positive and negative values of m. The implication of the hypotheses for this test is that one time series either is, or is not, a predictor of the other. In particular, we will test whether or not a white noise time series can be a predictor of a reputation time series. If it can, we would consider a reputation time series to be white noise.

4 Results

Seventeen reputation time series were considered. Ten of them comprised reputation data for a 720-day period, and the remaining seven covered a 400-day period. Both are sufficiently long to provide reliable conclusions. The results of applying the tests described in the previous section are remarkably consistent. Therefore there is no need to present each indvidually. General comments for each test suffice.

Runs Test Results
With the null hypothesis that the input data are random, a "small" p-value indicates that the null hypothesis should be rejected (i.e. the data are not random). This is the case for all 17 reputational time series except three, for which the p-values were 0.314, 0.159 and 0.421. The p-values for the other 14 time series were all less than 0.006. The conclusion is that 14 of the time series are *most likely* not random, and 3 of them *could be*.

Box-Hjung Test Results
Using the Box-Hjung test on the 17 reputation time series resulted in p-values of less than 10^{-5} in all cases. This indicates that the reputation time series all exhibit serial correlation. In contrast, applying the Box-Hjung test to 250 instances of random data produced Normally distributed p-values with mean

0.498, and standard deviation 0.293. This indicates a lack of serial correlation. The conclusion is that none of the reputation time series corresponds to white noise. Figure 2 shows the lag plot for one of the reputation time series (Ford motors), and illustrates the correlation between the scores with no lag and the scores lagged by 1 day. The plot resembles a random scatter, indicating very low correlation. Plots for other lags and other organisations are similar.

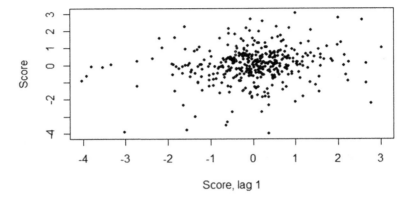

Fig. 2. Typical Lag Plot - Ford

Cross-Correlation Test Results

The output of the cross-correlation test is a p-value which measures the degree to which one input time series is a predictor of the other. Using any of the available reputation time series with a gaussian white noise time series, the p-values obtained were nearly always non-significant values in the range (0.942, 0.995). The only exception was one reputation time series for which the p-value was 0.792, also not significant This indicates that a white noise time series cannot be used as a predictor of a reputation time series (and vice versa). Using two gaussian white noise time series the results were much the same. The p-values obtained were always non-significant, in the range (0.940, 0.999). Clearly, one white noise time series should not be a predictor of another. Figure 3 shows a typical cross-correlogram (Ford against white noise) in which two (very marginal!) breaches of the 5% confidence limits (the dashed lines) are apparent. It should be noted that these confidence limits apply only if the underlying time series data are normally distributed (which applies in the case of Ford). If not, a Box-Cox transformation with parameter $\lambda \sim 6$ can be used to transform raw data to a near Normal distribution. When many such results are run and averaged, the result is as shown in Fig. 4: there are no significant breaches of the 5% confidence limits.

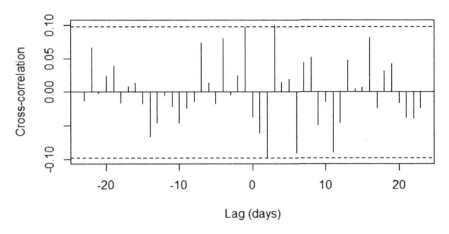

Fig. 3. Typical single cross-correlation: Ford

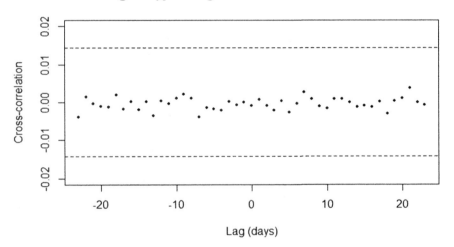

Fig. 4. Mean cross-correlations: 1000 trials: Ford

5 Discussion

This work was prompted by the observation that some reputation time series resemble white noise, and also because some of them exhibit very weak autocorrelations. Typically, the reputation time series that fall into this category tend to have empirical normal distributions with mean greater than the nominal neutral score. These are the suspect ones. Others exhibit distinct negative skews (i.e. there is a tail of low value scores). For these, correlations are more apparent, although they can be weak.

As a preliminary, the Runs Test indicates that only a few of the reputation time series can be considered as random. The resemblance shown in Fig. 1

is therefore superficial. The Box-Ljung test shows very clearly that there is a very distinct difference between a white noise signal and a reputation signal: the former has no auto-correlation and the latter does. This view is confirmed by an examination of cross correlations between white noise signals and reputation/noise pairings. A cross correlation plot such as Fig. 3 provides a good visual impression of a series of correlations at different lags, and the new test developed in Subsect. 3.2 quantifies that impression.

Overall, the answer to the question "is a reputation time series the same as white noise?" is a very clear "no". Establishing that implies that a reputation time series conveys useful information. Ways to interpret that are a matter for current study.

Acknowledgments. I am grateful for the support of Alva-Group for their continued interest, support, and assistance in the preparation of this paper.

References

1. Mitic, P.: Standardised reputation measurement. In: Yin, H., et al. (eds.) IDEAL 2017. LNCS, vol. 10585, pp. 1–9. Springer, Heidelberg (2017). doi:10.1007/978-3-319-68935-7_58
2. Mitic, P. Reputation Risk: Measured. In: Proceedings of Complex Systems. Wessex Institute of Technology Press, May 2017
3. Knuth, D.: The Art of Computer Programming, vol. 2, 2nd edn. Addison-Wesley (1981)
4. Bradley, J.V.: Distribution-Free Statistical Tests, chap. 12. Prentice-Hall (1968)
5. Ljung, G.M., Box, G.E.P.: On a measure of a lack of fit in time series models. Biometrika **65**(2), pp. 297–303 (1978). doi:10.1093/biomet/65.2.297
6. Shumway, R.H., Stoffer, D.S.: Time Series Analysis and its Applications, 3rd edn. chap. 1. Springer, Cham (2010)
7. Bartlett, M.S.: On the theoretical specification and sampling properties of autocorrelated time series. Suppl. J. Roy. Stat. Soc. **8**, 27–41 (1946)
8. Francq, C., Zacoïan, J.-M.: Bartlett's formula for a general class of nonlinear processes. J. Time Ser. Anal. **30**(4), 449–465 (2009)
9. Hong, Y., Consistent testing for serial correlation of unknown form. Econometrica **64**, 837–864 (1996)

Chaotic Brain Storm Optimization Algorithm

Eva Tuba[1], Edin Dolicanin[2], and Milan Tuba[3(✉)]

[1] Faculty of Mathematics, University of Belgrade, Belgrade, Serbia
[2] Department of Technical Sciences, State University of Novi Pazar,
Novi Pazar, Serbia
[3] Graduate School of Computer Science, John Naisbitt University,
Belgrade, Serbia
tuba@ieee.org

Abstract. Swarm intelligence algorithms are stochastic optimization algorithms that are very successfully used for hard optimization problems. Brain storm optimization is a recent swarm intelligence algorithm that has been proven successful in many applications but is still not researched enough. Many swarm intelligence algorithm have been recently improved by introduction of chaotic maps that better than random sequences contributed to search quality. In this paper we propose an improvement of the brain storm optimization algorithm by introducing chaotic maps. Two one-dimensional chaotic maps were incorporated into the original brain storm optimization algorithm. The proposed algorithms were tested on 15 standard benchmark functions from CEC 2013 and compared to the original brain storm optimization algorithm and particle swarm optimization. Our proposed chaos based methods obtained better results where for this set of benchmark functions circle maps were superior.

Keywords: Brain storm optimization algorithm · BSO · Metaheuristic algorithms · Chaos · Global optimization problems · Swarm intelligence

1 Introduction

Numerous real life problems can be represented as optimization problems where global optimum (minimum or maximum) of the objective function needs to be found. In most cases these optimization problems are hard optimization problems and often highly nonlinear. Standard deterministic algorithms are incapable to find the solution for such problems due to computational complexity and numerous local optima, hence different approaches are needed. In the past decades various stochastic optimization algorithms were proposed. One group of such algorithms are nature inspired algorithms where the idea is to mimic some processes from the nature.

Nature inspired algorithms can be divided into two main categories: evolutionary and swarm intelligence algorithms. Evolutionary algorithms are inspired by the biological evolution processes such as reproduction, recombination, mutation and selection. One of the oldest and the most famous member of this group

© Springer International Publishing AG 2017
H. Yin et al. (Eds.): IDEAL 2017, LNCS 10585, pp. 551–559, 2017.
https://doi.org/10.1007/978-3-319-68935-7_60

is genetic algorithm. Swarm intelligence algorithms are inspired by collective behavior of different agents in the nature. Each individual follows simple rules and interacts with other members of the swarm. These simple agents collectively exhibit remarkable intelligence that is used for solving optimization problems.

Swarm intelligence algorithms are inspired by processes from nature such as ant colonies, animal herding, food harvesting, nesting and others. By mimicking these processes two main parts of the algorithms have to be implemented: exploration and exploitation. Exploitation represents a local search around promising solutions that were found while exploration is global search where better solutions are looked for in different areas of the search space in order to prevent the algorithm to get stacked in some local optimum.

Among the oldest swarm intelligence algorithms are particle swarm optimization (PSO) and ant colony optimization. Consequently numerous different swarm intelligent algorithms have been proposed such as firefly algorithm [10,18,20], fireworks algorithm [13], krill herd algorithm [6,11], and others. Swarm intelligence algorithms have been applied for solving different problems such as traveling salesman problem [19], multilevel thresholding [14], support vector machine optimization [15,16], image registration [17], etc.

Optimization algorithms are constantly being improved by different modifications and hybridization. In recent years advances in theory and applications of chaos have been widely used in numerous fields and one of them is optimization algorithms. Chaotic maps such as circle, Gauss/mouse, logistic, piecewise, sine, sinusoidal and others were introduced in swarm intelligence algorithms instead of some random parameters [4]. For example, parameters of the bat algorithm were replaced by different chaotic maps and the results were compared to the original bat algorithm [5]. Results have shown that some chaotic bat algorithms can outperform the version with random numbers. Chebyshev map was introduced into fruit fly optimization algorithm and modified chaotic version of the algorithm has shown superior and more reliable behavior compared to the original one in [8]. In [7] ten different one-dimensional chaotic maps were used for improving fireworks algorithm. The best performance was when circle maps were used.

In this paper we introduce chaos into recent brain storm optimization algorithm [9] and propose chaos based BSO (CBSO). Since different chaotic maps can lead to different behavior of the optimization algorithm, we proposed two chaotic maps. The proposed algorithm was tested on standard benchmark functions proposed in CEC 2013 and it was favorably compared to the original BSO and the standard PSO.

The rest of the paper is organized as follows. Our proposed brain storm optimization algorithm with chaotic maps is presented in Sect. 2. Simulation results along with comparison with other algorithms are given in Sect. 3. At the end the conclusion and proposition for further work is presented in Sect. 4.

2   Chaos Based Brain Storm Optimization Algorithm

Brain storm optimization algorithm (BSO) was proposed by Yuhui Shi in 2011 [9]. This algorithm has been applied to numerous problems [2,12]. During the last few years different improved and modified BSO versions were proposed [1,3]. Inspiration for the algorithm was human idea generation process or brainstorming process. Brainstorming was summarized in several steps which were transformed into the brain storm optimization algorithm.

Brainstorming process contains step of generating initial ideas followed by choosing more promising ideas and generating new idea based on better solutions as well as new ones regardless of the previous ideas. It is expected that after a several iteration some good solution will be obtained. Brain storm optimization algorithm is presented in Algorithm 1.

In the Algorithm 1 several parameters of the BSO are mentioned. The first one is n, the number of solutions or individuals in each generation of the ideas and the second one is parameter m which represents the number of clusters. Beside these two parameters, four different parameters need to be set, parameters p_{5a}, p_{6b}, p_{6bi} and p_{6c} that determine how a new solution will be created according to the Algorithm 1.

Algorithm 1. Pseudo-code of the BSO algorithm

1: **Initialization**
2: Randomly generate n potential solutions.
3: **repeat**
4:     Cluster n solutions into m clusters.
5:     Rank solutions in each cluster and set the best one as cluster center.
6:     Randomly generate a value r between 0 and 1.
7:     **if** $r < p_{5a}$ **then**
8:         Randomly select a cluster center.
9:         Randomly generate an individual to replace the selected cluster.
10:    **end if**
11:    **repeat**
12:        Generate new solutions.
13:        Randomly generate a value r between 0 and 1.
14:        **if** $r < p_{6b}$ **then**
15:            Randomly select a cluster with probability p_{6bi}.
16:            Randomly generate a value r_1 between 0 and 1.
17:            **if** $r_1 < p_{6bii}$ **then**
18:                Select the cluster center and add random values to it to generate new individual.
19:            **else**
20:                Randomly select a solution from the chosen cluster and add random value to the solution to generate new one.
21:            **end if**
22:        **else**
23:            Randomly select two clusters.
24:            Generate random value r_2 between 0 and 1
25:            **if** $r_2 < p_{6c}$ **then**
26:                Two cluster centers are combined to generate new solution.
27:            **else**
28:                Two solutions from each selected cluster are randomly chosen to be combined to generate new individual.
29:            **end if**
30:        **end if**
31:        The newly generated solution is compared with the same solution index and the better one is kept.
32:    **until** n new solution is generated.
33: **until** Maximal iteration number is reached.

New solutions are generated by the following equation:

$$x_{new} = x_{selected} + \zeta * n(\mu, \sigma) \tag{1}$$

where x_{new} is a new solution in the d-dimensional space, $x_{selected}$ represents solution selected to be potentially changed, $n(\mu, \sigma)$ is a random number generated from Gaussian distribution with mean μ and variance σ, while ζ is the coefficient that controls the influence of the Gaussian random value. Parameter ζ is calculated in each generation by the following expression:

$$\zeta = \text{logsig}((0.5 * maxIteration - currentIteration)/k) * rand() \tag{2}$$

where $maxIteration$ and $currentIteration$ represent maximal number of iterations and the number of the current iteration, respectively. Parameter k changes the logsig() function's slope where logsig is a logarithmic sigmoid transfer function. Finally, $rand()$ represents random value from uniform distribution within [0,1]. In this paper we propose using chaotic maps instead of random values.

2.1 Chaotic Maps

Chaotic optimization algorithms are optimization algorithms that use chaotic variables rather then random values. The characteristics of chaotic maps such as non-repetition and ergodicity may improve search in the optimization algorithms [21]. In this paper two different one-dimensional maps were considered: circle map and sinusoidal map. Circle map is defined by the following equation:

$$x_{k+1} = \left[x_k + b - \frac{a}{2\pi} \sin(2\pi x_k) \right] \mod 1 \tag{3}$$

where for $a = 0.5$ and $b = 0.2$ the generated chaotic sequence is within $(0, 1)$.
    Sinusoidal map is defined as:

$$x_{k+1} = ax_k^2 \sin(\pi x_k) \tag{4}$$

where for $a = 2.3$ and $x_0 = 0.7$ the following simplified form can be used:

$$x_{k+1} = \sin(\pi x_k) \tag{5}$$

The proposed chaotic maps were used to generate chaos sequence of numbers that were used in Eq. 2.

3   Simulation Results

Our proposed method was implemented in Matlab R2016a and simulations were performed on the platform with Intel ® Core^{TM} i7-3770K CPU at 4 GHz, 8 GB RAM, Windows 10 Professional OS.
    The proposed algorithms with chaotic maps were tested on 15 well-known benchmark functions proposed in CEC 2013. We tested on 5 unimodal and 10

multimodal 10-dimensional functions and the details about these functions are presented in Table 1. Parameters for brain storm algorithm were set empirically. Number of individuals was set to 100 and number of the clusters was 5. Probabilities p_{5a}, p_{6b}, p_{6bi} and p_{6c} were set to 0.2, 0.8, 0.4 and 0.5, respectively. Maximal number of iterations was 5000. All tests were run 30 times.

Table 1. Benchmark function details

No	Function	Optimal
	Unimodal functions	
1	Sphere function	-1400
2	Rotated high conditioned elliptic function	-1300
3	Rotated bent cigar function	-1200
4	Rotated discus function	-1100
5	Different powers function	-1000
	Basic multimodal functions	
6	Rotated Rosenbrock's function	-900
7	Rotated Schaffers F7 function	-800
8	Rotated Ackley's function	-700
9	Rotated Weierstrass function	-600
10	Rotated Griewank's function	-500
11	Rastrigin's function	-400
12	Rotated Rastrigin's function	-300
13	Non-Continuous rotated Rastrigin's function	-200
14	Schwefel's Function	-100
15	Rotated Schwefel's Function	100

We compared the original brain storm optimization algorithm with two versions when chaotic maps were introduced. In the first version we used circle map and we named this chaotic brain storm optimization algorithm CBSO-C. The second version had implemented sinusoidal map and it was named CBSO-S. We also compared the results with results of standard PSO algorithm [22]. In [22] maximal number of objective function evaluation was set to 100,000 while in our proposed algorithm it was 50,000 since larger number of evaluations did not improve results. For each function algorithms were executed 30 times and median, standard deviation, the best and the worst solutions were calculated as in [22]. The obtained results are presented in Table 2.

For benchmark function f_1, the sphere, all algorithms successfully found global minimum in all cases since standard deviation was 0. Similarly, all algorithms found the global optimum for function f_5 in almost all cases. The smallest standard deviation was achieved by sinusoidal chaotic BSO. From the results presented in Table 2 it can be seen that PSO achieved the same best solutions in

Table 2. Comparison of PSO, BSO, CBSO-C and CBSO-S

Fun.		PSO	BSO	CBSO-C	CBSO-S
f_1	median	−1.400E+03	−1.400E+03	−1.400E+03	−1.400E+03
	std	0.000E+00	0.000E+00	0.000E+00	0.000E+00
	best	−1.400E+03	−1.400E+03	−1.400E+03	−1.400E+03
	worst	−1.400E+03	−1.400E+03	−1.400E+03	−1.400E+03
f_2	median	3.504E+04	**8.614E+03**	3.769E+05	1.470E+04
	std	7.356E+04	**1.240E+04**	3.054E+05	2.572E+04
	best	7.597E+02	**4.859E+02**	1.808E+05	2.963E+03
	worst	4.755E+05	**2.790E+04**	9.384E+05	6.751E+04
f_3	median	2.670E+05	**5.629E+04**	1.067E+05	7.762E+04
	std	1.656E+07	**3.062E+05**	1.514E+07	6.967E+05
	best	**−1.200E+03**	**−1.200E+03**	3.844E+04	5.471E+03
	worst	8.251E+07	**7.235E+05**	3.505E+07	1.611E+06
f_4	median	7.769E+03	−7.443E+02	−9.728E+02	**−9.953E+02**
	std	4.556E+03	2.078E+02	1.406E+04	**1.818E+02**
	best	2.454E+02	−1.097E+03	2.353E+03	**−1.099E+03**
	worst	1.856E+04	−6.070E+02	3.874E+04	**−6.570E+02**
f_5	median	−1.000E+03	−1.000E+03	−1.000E+03	−1.000E+03
	std	3.142E−05	1.574E−04	2.378E+01	**4.957E−05**
	best	−1.000E+03	−1.000E+03	−1.000E+03	−1.000E+03
	worst	−1.000E+03	−1.000E+03	−9.433E+02	−1.000E+03
f_6	median	−8.902E+02	**−8.997E+02**	−8.953E+02	−8.995E+02
	std	4.974E+00	4.164E+00	7.499E−01	**7.893E−02**
	best	−9.000E+02	**−9.000E+02**	−8.954E+02	−8.996E+02
	worst	−8.898E+02	−8.904E+02	−8.935E+02	**−8.994E+02**
f_7	median	−7.789E+02	−7.349E+02	−7.640E+02	**−7.885E+02**
	std	1.327E+01	2.639E+01	1.976E+01	**4.589E+00**
	best	−7.974E+02	−7.646E+02	−7.723E+02	**−7.987E+02**
	worst	−7.434E+02	−7.036E+02	−7.232E+02	**−7.831E+02**
f_8	median	−6.789E+02	−6.797E+02	**−6.800E+02**	−6.798E+02
	std	6.722E−02	9.599E−02	**1.873E−03**	5.318E−02
	best	−6.789E+02	−6.799E+02	**−6.800E+02**	−6.799E+02
	worst	−6.796E+02	−6.797E+02	**−6.800E+02**	−6.797E+02
f_9	median	−5.952E+02	−5.911E+02	**−5.959E+02**	−5.923E+02
	std	1.499E+00	1.114E+00	1.112E+00	**7.699E−01**
	best	−5.987E+02	−5.936E+02	**−5.989E+02**	−5.926E+02
	worst	−5.929E+02	−5.911E+02	**−5.931E+02**	−5.910E+02

(continued)

Table 2. (*continued*)

Fun.		PSO	BSO	CBSO-C	CBSO-S
f_{10}	median	−4.997E+02	**−4.999E+02**	**−4.999E+02**	**−4.999E+02**
	std	2.713E−01	8.620E−02	**1.631E−02**	1.795E−02
	best	−4.999E+02	**−5.000E+02**	**−5.000E+02**	−4.999E+02
	worst	−4.989E+02	−4.998E+02	**−4.999E+02**	−4.998E+02
f_{11}	median	−3.891E+02	−3.582E+02	**−3.940E+02**	−3.473E+02
	std	5.658E+00	8.425E+00	**8.899E−01**	1.777E+01
	best	**−3.970E+02**	−3.682E+02	**−3.970E+02**	−3.781E+02
	worst	−3.731E+02	−3.473E+02	**−3.940E+02**	−3.353E+02
f_{12}	median	−2.861E+02	−2.552E+02	**−2.920E+02**	−2.682E+02
	std	6.560E+00	2.464E+01	**2.725E+00**	1.054E+01
	best	**−2.970E+02**	−2.881E+02	**−2.970E+02**	−2.731E+02
	worst	−2.682E+02	−2.323E+02	**−2.891E+02**	−2.483E+02
f_{13}	median	−1.792E+02	−1.371E+02	**−1.834E+02**	−1.398E+02
	std	9.822E+00	2.110E+01	**5.168E+00**	9.588E+00
	best	−1.946E+02	−1.637E+02	**−1.952E+02**	−1.434E+02
	worst	−1.523E+022	−1.134E+02	**−1.802E+02**	−1.201E+02
f_{14}	median	7.338E+02	1.180E+03	**3.858E+02**	1.093E+03
	std	1.282E+02	1.687E+02	**9.365E+01**	1.444E+02
	best	2.228E+02	8.796E+02	**1.677E+02**	9.702E+02
	worst	1.109E+03	1.340E+03	**5.526E+02**	1.303E+03
f_{15}	median	8.743E+02	9.160E+02	**4.329E+02**	9.160E+02
	std	2.507E+02	**1.409E+02**	2.329E+02	2.323E+02
	best	4.372E+02	7.272E+02	**3.585E+02**	1.136E+03
	worst	1.705E+03	1.124E+03	**8.540E+02**	1.346E+03

the case of benchmark functions f_3, f_{11} and f_{12}. In all other cases standard PSO was outperformed by BSO or our proposed chaos based BSO.

For function f_2 original BSO achieved the best results, however with a large error. It can be said that all algorithms failed to find global optimum. Similarly to f_2, for f_3 again not nearly good solutions were found by any algorithm but BSO managed to find the closest solutions. For functions f_4 and f_7 the best results were achieved when BSO with sinusoidal map was used. For f_4 where optimal solution is −1100, the best found one was −995.3 which is significantly different, but the improvement over the original BSO is noticeable. Original BSO had −744.3 as median in 30 runs which is worse then the results obtained by both chaotic based BSO. Chaotic based BSO improved performance of the original BSO for function f_7, from −734.9 to 764.0 by CBSO-C and −788.5 by CBSO-S.

For functions from f_8 to f_{15} the best solutions were obtained by CBSO-C, chaotic BSO with circle map. Improvements are in some cases smaller, such as for functions f_8 and f_{10}. In the case of f_8 median with the original BSO was -679.7 while this solution is improved by CBSO-S to -679.8 and finally, CBSO-C obtained median -680.0. Function f_{10} was successfully solved by all three BSO versions, but the most robust one was CBCO-C since it had smallest standard deviation. In all other cases chaos based brain storm optimization with circle map improved results of original BSO significantly and also outperformed PSO. For functions f_9–f_{15}, except for the function f_{10}, BSO with circle map obtained results better than the original BSO and also better than BSO with sinusoidal map. This shows that different maps are more suitable for some functions.

4   Conclusion

In this paper chaos based brain storm optimization algorithm was proposed. Two different one-dimensional chaotic maps were implemented into the original BSO: circle and sinusoidal maps. Proposed algorithms were tested on 15 benchmark functions from CEC 2013 and the results were compared to the original BSO and standard PSO. The best results in the most cases were obtained by BSO algorithm with circle map. In other cases the best performance was achieved by the original BSO or BSO with sinusoidal map. BSO as well as two modifications outperformed standard PSO in all cases. In further work different chaotic maps can be used and compared to other chaotic optimization algorithms.

Acknowledgment. This research is supported by the Ministry of Education, Science and Technological Development of Republic of Serbia, Grant No. III-44006.

References

1. Cao, Z., Shi, Y., Rong, X., Liu, B., Du, Z., Yang, B.: Random grouping brain storm optimization algorithm with a new dynamically changing step size. In: Tan, Y., Shi, Y., Buarque, F., Gelbukh, A., Das, S., Engelbrecht, A. (eds.) ICSI 2015. LNCS, vol. 9140, pp. 357–364. Springer, Cham (2015). doi:10.1007/978-3-319-20466-6_38
2. Chen, J., Cheng, S., Chen, Y., Xie, Y., Shi, Y.: Enhanced brain storm optimization algorithm for wireless sensor networks deployment. In: Tan, Y., Shi, Y., Buarque, F., Gelbukh, A., Das, S., Engelbrecht, A. (eds.) ICSI 2015. LNCS, vol. 9140, pp. 373–381. Springer, Cham (2015). doi:10.1007/978-3-319-20466-6_40
3. Chen, J., Wang, J., Cheng, S., Shi, Y.: Brain storm optimization with agglomerative hierarchical clustering analysis. In: Tan, Y., Shi, Y., Li, L. (eds.) ICSI 2016. LNCS, vol. 9713, pp. 115–122. Springer, Cham (2016). doi:10.1007/978-3-319-41009-8_12
4. Gandomi, A., Yang, X.S., Talatahari, S., Alavi, A.: Firefly algorithm with chaos. Commun. Nonlinear Sci. Numer. Simul. **18**(1), 89–98 (2013)
5. Gandomi, A.H., Yang, X.S.: Chaotic bat algorithm. J. Comput. Sci. **5**(2), 224–232 (2014)
6. Gandomi, A.H., Alavi, A.H.: Krill herd: a new bio-inspired optimization algorithm. Commun. Nonlinear Sci. Numer. Simul. **17**(12), 4831–4845 (2012)

7. Gong, C.: Chaotic adaptive fireworks algorithm. In: Tan, Y., Shi, Y., Niu, B. (eds.) ICSI 2016. LNCS, vol. 9712, pp. 515–525. Springer, Cham (2016). doi:10.1007/978-3-319-41000-5_51

8. Mitic, M., Vukovic, N., Petrovic, M., Miljkovic, Z.: Chaotic fruit fly optimization algorithm. Knowl.-Based Syst. **89**, 446–458 (2015)

9. Shi, Y.: Brain storm optimization algorithm. In: Tan, Y., Shi, Y., Chai, Y., Wang, G. (eds.) ICSI 2011. LNCS, vol. 6728, pp. 303–309. Springer, Heidelberg (2011). doi:10.1007/978-3-642-21515-5_36

10. Strumberger, I., Bacanin, N., Tuba, M.: Enhanced firefly algorithm for constrained numerical optimization. In: IEEE Congress on Evolutionary Computation (CEC), pp. 2120–2127. IEEE (2017)

11. Strumberger, I., Bacanin, N., Tuba, M.: Hybridized krill herd algorithm for large-scale optimization problems. In: 15th International Symposium on Applied Machine Intelligence and Informatics (SAMI), pp. 473–478. IEEE (2017)

12. Sun, C., Duan, H., Shi, Y.: Optimal satellite formation reconfiguration based on closed-loop brain storm optimization. IEEE Comput. Intell. Mag. **8**(4), 39–51 (2013)

13. Tan, Y., Zhu, Y.: Fireworks algorithm for optimization. In: Tan, Y., Shi, Y., Tan, K.C. (eds.) ICSI 2010. LNCS, vol. 6145, pp. 355–364. Springer, Heidelberg (2010). doi:10.1007/978-3-642-13495-1_44

14. Tuba, E., Alihodzic, A., Tuba, M.: Multilevel image thresholding using elephant herding optimization algorithm. In: 14th International Conference on Engineering of Modern Electric Systems (EMES), pp. 240–243. IEEE (2017)

15. Tuba, E., Mrkela, L., Tuba, M.: Support vector machine parameter tuning using firefly algorithm. In: 26th International Conference Radioelektronika, pp. 413–418. IEEE (2016)

16. Tuba, E., Tuba, M., Beko, M.: Support vector machine parameters optimization by enhanced fireworks algorithm. In: Tan, Y., Shi, Y., Niu, B. (eds.) ICSI 2016. LNCS, vol. 9712, pp. 526–534. Springer, Cham (2016). doi:10.1007/978-3-319-41000-5_52

17. Tuba, E., Tuba, M., Dolicanin, E.: Adjusted fireworks algorithm applied to retinal image registration. Stud. Inform. Control **26**(1), 33–42 (2017)

18. Tuba, M., Bacanin, N.: Improved seeker optimization algorithm hybridized with firefly algorithm for constrained optimization problems. Neurocomputing **143**, 197–207 (2014)

19. Tuba, M., Jovanovic, R.: Improved ACO algorithm with pheromone correction strategy for the traveling salesman problem. Int. J. Comput. Commun. Control **8**(3), 477–485 (2013)

20. Yang, X.-S.: Firefly algorithms for multimodal optimization. In: Watanabe, O., Zeugmann, T. (eds.) SAGA 2009. LNCS, vol. 5792, pp. 169–178. Springer, Heidelberg (2009). doi:10.1007/978-3-642-04944-6_14

21. Yuan, X., Zhao, J., Yang, Y., Wang, Y.: Hybrid parallel chaos optimization algorithm with harmony search algorithm. Appl. Soft Comput. **17**, 12–22 (2014)

22. Zambrano-Bigiarini, M., Clerc, M., Rojas, R.: Standard particle swarm optimisation 2011 at CEC-2013: a baseline for future PSO improvements. In: IEEE Congress on Evolutionary Computation (CEC), pp. 2337–2344. IEEE (2013)

Universum Discriminant Canonical Correlation Analysis

Xiaohong Chen[1,2(✉)], Hujun Yin[2], Menglei Hu[3], and Liping Wang[1]

[1] College of Science, Nanjing University of Aeronautics and Astronautics,
Nanjing, China
lyandcxh@nuaa.edu.cn
[2] School of Electrical and Electronic Engineering,
The University of Manchester, Manchester, UK
hujun.yin@manchester.ac.uk
[3] College of Computer Science and Technology,
Nanjing University of Aeronautics and Astronautics, Nanjing, China

Abstract. Over the past decades, extensive studies on multi-view learning and Universum learning have been witnessed in pattern recognition and machine learning. Incorporating multi-view learning and Universum learning together, we propose a novel supervised dimensionality reduction method for multi-view data accompanied by Universum data, termed Universum discriminant canonical correlation analysis (UDCCA). UDCCA exploits inter-view information by means of the within-class correlation as well as the within-Universum correlation between different views, and at the same time, utilizes intra-view discriminant information captured from the target samples and Universum data of each view. In the low-dimensional discriminant space, the within-class correlation of the target samples is maximized, and the correlation of the Universum data is minimized, and simultaneously the scatter among target samples and Universum data is also maximized. Experimental results on real-world multi-view datasets show its effectiveness compared to other related state-of-art dimensionality reduction methods.

Keywords: Multi-view learning · Universum learning · Canonical correlation analysis · Dimensionality reduction

1 Introduction

In recent years, a number of methods for learning from multi-view data have been proposed. These views may include multiple sources or different feature subsets. For example, a gene can be represented by the genetic activity and text information [1], a webpage can be identified by the text in the page and the hyperlinks [2], multilingual documents have one view in each language [3], while an image can be shown by its color or texture features, regarded as different feature subsets. Data in different views may lie in completely different spaces and contain knowledge that other views may not have. How to make full use of the complementary knowledge hidden in different views is the topic of multi-view learning [4, 5]. Canonical correlation analysis (CCA) [6, 7] is the most representative technique in early stage of multi-view learning. It aims to find

© Springer International Publishing AG 2017
H. Yin et al. (Eds.): IDEAL 2017, LNCS 10585, pp. 560–570, 2017.
https://doi.org/10.1007/978-3-319-68935-7_61

two transformation matrices, one for each view, to respectively project the samples from the two views into a common subspace with maximal correlation. Discriminative CCA (DCCA) [8] was proposed as a supervised version of CCA by combining the class information into dimensionality reduction process. It seeks optimal linear transformations that map two view data into a common subspace, in which the within-class correlation is maximized and the between-class correlation is minimized simultaneously. Besides, there are several extensions of CCA, such as Kernel CCA, Sparse CCA and Deep CCA et al.

Supervised or not, CCA and its variants are performed under a common assumption that each training sample must belong to one of target classes (even the true class label is unknown in the unsupervised scenario). These kinds of training samples are named as target samples. But in practice, training data may include not only target samples, but also samples that do not belong to any target class but drawn from the same application domain of the learning task. For instance, for the task of classifying animal images "bat" and "beaver", we may have images of other animals. Although the latter are not the target samples, but they can reflect the image characters of animals and can potentially help improve the classification accuracy. These additional samples are termed as Universum data, first coined by Vapnik as an alternative capacity concept for large margin classifiers [9]. Discarding Universum data may result in waste of resources. Recently, there are a number of methods for utilizing these additional sources to improve the classification performance in the so-called Univesum learning [9–20]. Universum data may contain some complementary or contrary information. It is the idea of inference through contradictions that led to the Univesum support vector machine (USVM) [10], which outperforms the SVM without considering Universum data. Sinz et al. [11] put forward the least squares version of USVM with a closed-form solution. Dhar et al. [12] proposed multi-class Universum SVM (MU-SVM) for multi-class classification. Zhang et al. [13] designed a graph-based semi-supervised classifier with the help of labeled, unlabeled and Universum data. Pan et al. [14] proposed a new Universum graph classification framework which leverages additional non-example graphs to help improve classification accuracy. Recently, Wang et al. generalized Universum learning to multi-view data classification [15] and matrix-pattern-oriented classification [16], achieving higher accuracy.

Chen et al. incorporated Universum learning into dimensionality reduction, and obtained Universum linear discriminant analysis (ULDA) [17] and Universum principal component analysis (UPCA) [18]. Qiu et al. [19] generalized Universum learning to semi-supervised feature extraction. Accompanied by the success of Universum learning, some questions and challenges were raised. Does every Universum data provide useful information? If not, how to select informative Universum data so as to achieve improvement? Sinz et al. [11] stated that the USVM is equivalent to searching for a hyperplane that has its normals lying in the orthogonal complement of the space spanned by Universum examples. Furthermore Zhang et al. [13] demonstrated the feasibility and effectiveness of Universum data and Chen et al. [20] proposed a trick for selecting informative Universum data, named as in-between Universum examples. Further discussions on the effect and selection of Universum data are discussed.

Motivated by the success of Universum learning, in this paper, we propose a novel supervised dimensionality reduction method for multi-view data containing Universum data. As mentioned above, DCCA only focuses on target samples with aim to find two basis vectors for each view to ensure that within-class correlation is mutually maximized and between-class correlation minimized. Due to the fact that Universum data cannot be classified to any of the target classes, it may carry contrary prior information to the target samples. Then we aim to project the Universum data into the common low-dimensional space with the minimal correlation. Consequently, the projection makes the within-class correlation maximized for the two-view target samples and the correlation of two-view Universum data minimized. In the scenario of multi-view learning, one critical issue is to effectively utilize the information stemming from different view data, not only inter-view but also intra-view information are important [21, 22]. The intra-view information depicts the relationship of the samples within a certain view. Universum linear discriminant analysis (ULDA) [17] has validated that introducing domain knowledge inferenced from Universum data into dimensionality reduction can improve the separability of the target samples. For C-class classification problem, Universum data can be treated as the $C + 1$-th class because it does not belong to any of the C classes. Different class samples should be projected far from each other in the dimension reduced space. Thus, we aim to find the projection vectors to ensure maximize the scatter of target samples and Universum data within each view, resulting a new dimensionality reduction algorithm, Universum DCCA (UDCCA). It makes within-class correlation as maximal as possible for the target samples of different views, while as minimal as possible for the Universum data of different views, and separates target samples from Universum data as far as possible within each view. UDCCA considers both inter-view and intra-view information, hence leading to more discriminant projections, and can be solved analytically through a generalized eigenvalue decomposition problem.

The rest of this paper is organized as follows. Section 2 reviews the related work including CCA and DCCA. Section 3 presents the framework of the proposed method. Section 4 reports all the experimental results and analysis on three real-world multi-view dataset. The conclusions and future work are listed in Sect. 5.

2 Related Work

2.1 Canonical Correlation Analysis (CCA)

Given a two-view dataset $\{(\mathbf{x}_1, \mathbf{y}_1), \cdots, (\mathbf{x}_n, \mathbf{y}_n)\}$, where $(\mathbf{x}_i \mathbf{y}_i)$ denotes the i-th paired example. Let $\mathbf{X} = [\mathbf{x}_1, \ldots, \mathbf{x}_n] \in R^{p \times n}$, $\mathbf{Y} = [\mathbf{y}_1, \ldots, \mathbf{y}_n] \in R^{q \times n}$ be the data matrices, n is the number of target samples, p and q are the dimensions of X-view and Y-view data, respectively. For simplicity, $\mathbf{X}$ and $\mathbf{Y}$ are centralized in advance. CCA attempts to find two projection vectors $\mathbf{w}_x$ and $\mathbf{w}_y$ such that the correlation between $\mathbf{w}_x^T \mathbf{X}$ and $\mathbf{w}_y^T \mathbf{Y}$ are maximized, shown as the following optimization problem,

$$\max_{\mathbf{w}_x, \mathbf{w}_y} \frac{\mathbf{w}_x^T \mathbf{X} \mathbf{Y}^T \mathbf{w}_y}{\sqrt{\mathbf{w}_x^T \mathbf{X} \mathbf{X}^T \mathbf{w}_x \mathbf{w}_y^T \mathbf{Y} \mathbf{Y}^T \mathbf{w}_y}} \tag{1}$$

As the scales of $\mathbf{w}_x$ and $\mathbf{w}_y$ have no effect on the correlation, (1) can be rewritten to,

$$
\begin{aligned}
\max_{\mathbf{w}_x,\mathbf{w}_y} \quad & \mathbf{w}_x^T \mathbf{X}\mathbf{Y}^T \mathbf{w}_y \\
s.t. \quad & \mathbf{w}_x^T \mathbf{X}\mathbf{X}^T \mathbf{w}_x = 1, \mathbf{w}_y^T \mathbf{Y}\mathbf{Y}^T \mathbf{w}_y = 1
\end{aligned}
\tag{2}
$$

CCA can be formulated as a generalized eigenvalue problem,

$$
\begin{pmatrix} & \mathbf{X}\mathbf{Y}^T \\ \mathbf{Y}\mathbf{X}^T & \end{pmatrix}\begin{pmatrix} \mathbf{w}_x \\ \mathbf{w}_y \end{pmatrix} = \lambda \begin{pmatrix} \mathbf{X}\mathbf{X}^T & \\ & \mathbf{Y}\mathbf{Y}^T \end{pmatrix}\begin{pmatrix} \mathbf{w}_x \\ \mathbf{w}_y \end{pmatrix}
\tag{3}
$$

where $(\mathbf{w}_x, \mathbf{w}_y)$ is the eigenvector corresponding to the eigenvalue.

2.2 Discriminant Canonical Correlation Analysis (DCCA)

For a labeled multi-view dataset, DCCA tries to guarantee that within-class correlation is maximized and between-class correlation is minimized in the common low-dimensional space. Due to the fact that the samples have been mean-normalized, DCCA can be expressed as:

$$
\begin{aligned}
\max_{\mathbf{w}_x,\mathbf{w}_y} \quad & \mathbf{w}_x^T \mathbf{C}_w \mathbf{w}_y \\
s.t. \quad & \mathbf{w}_x^T \mathbf{X}\mathbf{X}^T \mathbf{w}_x = 1, \mathbf{w}_y^T \mathbf{Y}\mathbf{Y}^T \mathbf{w}_y = 1
\end{aligned}
\tag{4}
$$

where $\mathbf{C}_w = \sum_{i=1}^{c}\sum_{k=1}^{n_i}\sum_{l=1}^{n_i} \mathbf{x}_k^{(i)}\mathbf{y}_l^{(i)T}$ is the within-class correlation matrix, $\mathbf{x}_k^{(i)}$ is the k-th sample of the i-th class for the X-view, $\mathbf{y}_l^{(i)}$ is the l-th sample of the i-th class for Y- view. Similarly DCCA can be solved by the generalized eigenvalue problem,

$$
\begin{pmatrix} & \mathbf{C}_w \\ \mathbf{C}_w^T & \end{pmatrix}\begin{pmatrix} \mathbf{w}_x \\ \mathbf{w}_y \end{pmatrix} = \lambda \begin{pmatrix} \mathbf{X}\mathbf{X}^T & \\ & \mathbf{Y}\mathbf{Y}^T \end{pmatrix}\begin{pmatrix} \mathbf{w}_x \\ \mathbf{w}_y \end{pmatrix}
\tag{5}
$$

3 Universum Discriminant Canonical Correlation Analysis

Suppose we have a two-view dataset $\{(\mathbf{x}_1, \mathbf{y}_1), \cdots, (\mathbf{x}_n, \mathbf{y}_n), (\mathbf{x}_{u1}, \mathbf{y}_{u1}), \ldots, (\mathbf{x}_{um}, \mathbf{y}_{um})\}$. $\mathbf{X}_u = [\mathbf{x}_{u1}, \ldots, \mathbf{x}_{um}]$ and $\mathbf{Y}_u = [\mathbf{y}_{u1}, \ldots, \mathbf{y}_{um}]$ denote the paired Universum data in the two views, and m is the number of Universum samples. Universum data distribute in the same domain of the task but cannot be assigned to any of the classes. Then we aim to find projection vectors to ensure that the correlation between different view Universum data is minimized. If we only take inter-view correlation ($\mathbf{X} \leftrightarrow \mathbf{Y}, \mathbf{X}_u \leftrightarrow \mathbf{Y}_u$) into account, it means that the inter-view data structure is preserved, while the intra-view data structure ($\mathbf{X} \leftrightarrow \mathbf{X}_u, \mathbf{Y} \leftrightarrow \mathbf{Y}_u$) is ignored. Figure 1 shows the relationship among the four datasets. Most existing multi-view learning is often concerned

$$\mathbf{X} \leftarrow \qquad \rightarrow \mathbf{X}_u$$
$$\updownarrow \qquad \qquad \updownarrow$$
$$\mathbf{Y} \leftarrow \qquad \rightarrow \mathbf{Y}_u$$

Fig. 1. Relationship among four datasets

with only vertical relationships, while ignoring horizontal relationships which can also carry meaningful discriminant information for classification.

In the proposed Universum Discriminant Canonical Correlation Analysis (UDCCA), the intra-view information is also considered by introducing a scatter matrix for the target samples and Universum data within each view. The UDCCA aims to find projection vectors to ensure: (1) within-class correlation between two view target samples is maximized, (2) within-Universum correlation between two view Universum data is minimized, (3) the scatter of the target samples and Univesum data in each view is maximized. Concretely, it can be expressed as

$$\max_{\mathbf{w}_x, \mathbf{w}_y} \ \mathbf{w}_x^T (\mathbf{C}_w - \eta \mathbf{C}_{uxy}) \mathbf{w}_y + \tfrac{\alpha}{2} (\mathbf{w}_x^T \mathbf{S}_{bx} \mathbf{w}_x + \mathbf{w}_y^T \mathbf{S}_{by} \mathbf{w}_y)$$
$$s.t. \quad \mathbf{w}_x^T \mathbf{X}\mathbf{X}^T \mathbf{w}_x + \mathbf{w}_y^T \mathbf{Y}\mathbf{Y}^T \mathbf{w}_y = 1 \tag{6}$$

where $\mathbf{C}_{uxy} = \frac{1}{m} \mathbf{X}_u \mathbf{Y}_u^T$, $\mathbf{S}_{bx} = (\mathbf{m}_x - \mathbf{m}_{xu})(\mathbf{m}_x - \mathbf{m}_{xu})^T$ is the scatter matrix for $\mathbf{X}$ and $\mathbf{X}_u, \mathbf{m}_x, \mathbf{m}_{xu}$ are means of $\mathbf{X}$ and $\mathbf{X}_u$ respectively. $\mathbf{S}_{by} = (\mathbf{m}_y - \mathbf{m}_{yu})(\mathbf{m}_y - \mathbf{m}_{yu})^T$ is the scatter matrix for $\mathbf{Y}$ and $\mathbf{Y}_u$, $\mathbf{m}_y, \mathbf{m}_{yu}$ are means of $\mathbf{Y}$ and $\mathbf{Y}_u$ respectively. Parameters η and α control the balance among different terms.

In order to solve the problem, we define the following Lagrange function with the help of Lagrange multiplier technique,

$$L = \ \mathbf{w}_x^T (\mathbf{C}_w - \eta \mathbf{C}_{uxy}) \mathbf{w}_y + \frac{\alpha}{2} (\mathbf{w}_x^T \mathbf{S}_{bx} \mathbf{w}_x + \mathbf{w}_y^T \mathbf{S}_{by} \mathbf{w}_y)$$
$$- \frac{\lambda}{2} (\mathbf{w}_x^T \mathbf{X}\mathbf{X}^T \mathbf{w}_x + \mathbf{w}_y^T \mathbf{Y}\mathbf{Y}^T \mathbf{w}_y - 1) \tag{7}$$

Compute the partial derivative of the Eq. (7) with respect to $\mathbf{w}_x$ and $\mathbf{w}_y$,

$$\frac{\partial L}{\partial \mathbf{w}_x} = (\mathbf{C}_w - \eta \mathbf{C}_{uxy}) \mathbf{w}_y + \alpha \mathbf{S}_{bx} \mathbf{w}_x - \lambda \mathbf{X}\mathbf{X}^T \mathbf{w}_x,$$
$$\frac{\partial L}{\partial \mathbf{w}_y} = (\mathbf{C}_w^T - \eta \mathbf{C}_{uxy}^T) \mathbf{w}_x + \alpha \mathbf{S}_{by} \mathbf{w}_y - \lambda \mathbf{Y}\mathbf{Y}^T \mathbf{w}_y. \tag{8}$$

Setting the partial derivative equal to zero, the optimization (6) can be turned into the following generalized eigenvalue problem via a series of mathematical derivation,

$$\begin{pmatrix} \alpha \mathbf{S}_{bx} & \mathbf{C}_w - \eta \mathbf{C}_{uxy} \\ \mathbf{C}_w^T - \eta \mathbf{C}_{uxy}^T & \alpha \mathbf{S}_{by} \end{pmatrix} \begin{pmatrix} \mathbf{w}_x \\ \mathbf{w}_y \end{pmatrix} = \lambda \begin{pmatrix} \mathbf{X}\mathbf{X}^T & 0 \\ 0 & \mathbf{Y}\mathbf{Y}^T \end{pmatrix} \begin{pmatrix} \mathbf{w}_x \\ \mathbf{w}_y \end{pmatrix} \tag{9}$$

The set of eigenvectors $\left(\mathbf{w}_{x_i}, \mathbf{w}_{y_i}\right)$ of Eq. (9) corresponding to the d largest non-negative eigenvalues $\lambda_1 \geq \lambda_2 \geq \ldots \geq \lambda_d \geq 0$ compose the projection matrices $\mathbf{W}_x = [\mathbf{w}_{x1}, \mathbf{w}_{x2}, \cdots, \mathbf{w}_{xd}]$ and $\mathbf{W}_y = [\mathbf{w}_{y1}, \mathbf{w}_{y2}, \cdots, \mathbf{w}_{yd}]$ for X-view and Y-view respectively. $\mathbf{W}_x^T \mathbf{x}$ and $\mathbf{W}_y^T \mathbf{y}$ are low-dimensional representations for $\mathbf{x}$ and $\mathbf{y}$ respectively. In order to verify the discriminative property of $(\mathbf{W}_x, \mathbf{W}_y)$, we use the nearest neighbor classifier under one way of feature fusion used in [8], that is

$$\mathbf{z} = \mathbf{W}_x^T \mathbf{x} + \mathbf{W}_y^T \mathbf{y} = \left[\mathbf{W}_x^T, \mathbf{W}_y^T\right] \begin{bmatrix} \mathbf{x} \\ \mathbf{y} \end{bmatrix} \tag{10}$$

4   Experiments and Results

Several experiments were designed to evaluate the performance of the proposed UDCCA for two-view data. The methods used for comparison include CCA and DCCA. Comparisons were performed on three multi-view datasets including MFD, Ads and AWA. The parameters η and α used in UDCCA were grid searched from $2^k - 1$ ($k = 0, 1, \ldots, 10$) via cross validation and the parameters corresponding to the best results in the validation were selected. We repeated the experiments ten times and report their average results and variance in the following experiments.

4.1   MFD Dataset

The multiple-feature database (MFD) [25], from the UCI machine learning repository, composes of six feature sets of handwritten digits ("0" to "9") with 200 samples for each digit. The six feature sets describe different views of the digits as listed as follows: (1) Flourier coefficients of the character shapes (Fou,76); (2) Profile contour correlation characteristics (Fac,216); (3) Karhunen-Loève expansion coefficients (Kar,64); (4) pixel average in 2*3 windows (Pix,240); (5) Zernike moments (Zer,47); and (6) morphological characteristics (Mor,6). The dimension of each feature is indicated in the parentheses, after the feature abbreviation. Handwritten digits are often chosen in Universum learning to evaluate performance. The main reason is that if we select some digits as target samples, the other digits naturally act as Universum data.

We selected any two sets of features as X-view and Y-view, so that there are 15 combinations of the six features in total and each combination forms a two-view dataset. For each combination, the first two class samples ("0" and "1") were used as Universum data, while the rest digits served as target samples. Furthermore, each class target samples were split into training and test ones with the ratio 1:1. The results are shown in Table 1 where the best performances are highlighted in bold. From Table 1, we can obtain some meaningful observations as follows:

(1)   The superiority of DCCA compared to CCA demonstrates that the discriminant information expressed by within-class correlation is more meaningful to extracting discriminant low-dimensional features for multi-view data. UDCCA markedly

Table 1. Comparison of average (%) and variance (10^{-4}) on MFD

	X-view	Y-view	CCA	DCCA	UDCCA
1	Fac	Fou	81.99 ± 0.41	96.15 ± 0.22	**96.08 ± 0.12**
2	Fac	Kar	94.89 ± 0.40	97.61 ± 0.19	**98.41 ± 0.13**
3	Fac	Mor	69.19 ± 2.48	91.95 ± 15.0	**93.00 ± 12.2**
4	Fac	Pix	84.06 ± 3.57	97.15 ± 0.30	**98.08 ± 0.19**
5	Fac	Zer	81.93 ± 0.76	**95.36 ± 0.49**	94.71 ± 1.17
6	Fou	Kar	90.53 ± 0.87	94.20 ± 0.45	**96.30 ± 0.44**
7	Fou	Mor	72.96 ± 2.06	80.46 ± 1.83	**81.73 ± 0.53**
8	Fou	Pix	73.28 ± 2.29	93.75 ± 0.48	**94.15 ± 0.39**
9	Fou	Zer	78.66 ± 0.79	82.54 ± 0.77	**83.14 ± 0.51**
10	Kar	Mor	76.06 ± 3.54	**92.74 ± 8.92**	90.62 ± 7.12
11	Kar	Pix	92.85 ± 0.21	95.30 ± 0.32	**97.35 ± 0.23**
12	Kar	Zer	89.89 ± 1.64	93.31 ± 0.30	**94.86 ± 1.03**
13	Mor	Pix	68.78 ± 1.89	**92.11 ± 5.02**	89.26 ± 5.08
14	Mor	Zer	68.40 ± 1.07	76.85 ± 0.67	**77.60 ± 0.93**
15	Pix	Zer	78.34 ± 1.89	93.00 ± 1.37	**93.09 ± 0.83**

exceeds CCA to different degrees from 4% to 23%. Especially for the cases 1, 3, 4, 5, 8, 10, 13 and 15, UDCCA improves more than 13% in classification accuracies. UDCCA outperforms DCCA on 12 out of 15 combinations. The improvement in accuracy verifies that it is reasonable to say that not only the prior information encoded by Universum data, but also the strategy of using the priority knowledge is important for multi-view data dimensionality reduction.

(2) The performance of UDCCA is not always better than DCCA. The reason may be that the prior information provided by "0" and "1" is not enough for extracting discrimnant low-dimensional features for the other eight handwritten digits. The performance is affected by the combination of Universum data and target samples.

4.2   Advertisement Data (Ads)

The Ads dataset [26], also from the UCI repository, contains 3279 samples, including 458 advertisements ("ad") and 2821 non-advertisements ("non-ad"). Each sample is treated as a binary vector with large sparsity. It is a typical multi-view dataset containing five feature sets listed in Table 2. We obtained 10 feature combinations by selecting any two sets of features as X-view and Y-view. Then we randomly selected half of each class of each view for training and the remained for test.

Because there is not natural Universum data in the Ads dataset, we randomly chose 458 samples from non-advertisements class and averaged them with given 458 advertisements samples. The classification accuracies averaged over 10 independent trials are summarized in Table 3, with the best performances highlighted in bold.

The performance of UDCCA is clearly better than CCA and DCCA for all 10 feature combinations. The superiority of UDCCA compared to DCCA verifies that discriminant information obtained from the target samples and Universum data within

Table 2. Description of the five features of Ads

Abbreviation	Description of the feature	Dimension
alt	Information of the alt terms	111
cap	Information of the words occurring near the anchor text	19
url	Information of phrases occurring in the URL	457
origurl	Information of the image's URL	495
ancurl	Information of the anchor text	472

each view contributes more than the within-class correlation information of different views. It coincides with our motivation that is not only inter-view, but also intra-view information, meaningful for multi-view learning.

4.3 Animals with Attributes (AWA) Dataset

The AWA dataset [27] consists of 30475 images of 50 animal classes with six pre-extracted feature representations for each image. The six different features describing the image of animals from different views are listed as follows: (1) color histogram(cq,2688); (2) local self similarity (lss,2000); (3) hisgram of oriented gradient (phog,252); (4) rgSIFT descriptor (rgsift,2000); (5) SIFT descriptor (sift,2000); (6) SURF descriptor (surf, 2000). The dimension of each feature is given after the feature abbreviation in the parentheses. We selected 5 classes (antelope, bat, beaver, blue whale and bobcat) as target samples and buffalo for Universum data. We employed the PCA to reduce the dimensionality of different views, maintaining 95% of the original variance. Then we conducted the experiments with the result listed in Table 4.

UDCCA significantly outperforms CCA and DCCA with regard to classification accuracy in all cases. UDCCA largely exceeds CCA to varying degrees from 7% to 26%, especially for the cases 1, 2, 3, 5, 7, 9, 10 and 12, UDCCA improves more than 18% in classification accuracy. UDCCA outperforms DCCA by more than 10% on 10 out of 15 combinations. Although 'buffalo' does not belong to any target animals, the

Table 3. Comparison of average (%) and variance (10^{-4}) on Ads

	X-view	Y-view	CCA	DCCA	UDCCA
1	alt	cap	89.05 ± 0.33	88.40 ± 0.21	**89.44 ± 0.15**
2	alt	ancurl	94.91 ± 0.17	93.84 ± 0.30	**95.29 ± 0.20**
3	alt	orig	90.52 ± 0.19	89.15 ± 0.31	**92.85 ± 0.14**
4	alt	url	91.75 ± 0.50	90.68 ± 0.64	**95.00 ± 0.04**
5	cap	ancurl	95.34 ± 0.13	95.00 ± 0.80	**95.72 ± 0.05**
6	cap	orig	90.58 ± 0.18	90.62 ± 0.02	**91.83 ± 0.14**
7	cap	url	93.10 ± 0.05	93.10 ± 0.67	**94.59 ± 0.03**
8	ancurl	orig	95.21 ± 0.05	94.48 ± 0.47	**95.65 ± 0.07**
9	ancurl	url	95.61 ± 0.25	94.52 ± 0.26	**96.33 ± 0.13**
10	orig	url	93.90 ± 0.10	91.70 ± 0.22	**95.12 ± 0.13**

Table 4. Comparison of average (%) and variance (10^{-4}) on AWA

	X-view	Y-view	CCA	DCCA	UDCCA
1	cq	lss	35.88 ± 0.74	45.12 ± 3.45	**53.44 ± 0.62**
2	cq	phog	34.85 ± 2.87	44.54 ± 1.85	**54.47 ± 0.99**
3	cq	rgsift	37.75 ± 3.71	45.39 ± 2.50	**55.95 ± 0.97**
4	cq	sift	43.61 ± 1.19	45.10 ± 2.35	**54.96 ± 1.05**
5	cq	surf	34.56 ± 0.86	46.13 ± 1.53	**52.42 ± 3.77**
6	lss	phog	34.66 ± 3.26	41.68 ± 4.00	**51.41 ± 1.87**
7	lss	rgsift	34.61 ± 3.81	43.50 ± 3.13	**53.53 ± 1.11**
8	lss	sift	41.19 ± 1.47	40.42 ± 0.57	**52.67 ± 1.97**
9	lss	surf	28.61 ± 5.51	43.37 ± 2.17	**51.76 ± 4.68**
10	phog	rgsift	37.56 ± 2.32	43.55 ± 0.47	**53.54 ± 1.21**
11	phog	sift	43.91 ± 0.64	40.81 ± 2.64	**50.98 ± 1.42**
12	phog	surf	32.90 ± 1.81	42.21 ± 1.80	**52.94 ± 3.64**
13	rgsift	sift	41.06 ± 1.32	49.24 ± 2.52	**54.03 ± 2.51**
14	rgsift	surf	37.86 ± 2.75	46.84 ± 1.59	**52.79 ± 2.05**
15	sift	surf	42.35 ± 2.07	44.61 ± 3.06	**53.97 ± 3.20**

images of buffalo reflect the attribute of animal images, hence useful for extracting more discriminant low-dimensional features of multi-view data. The improvement further validates that (1) selecting approximate Universum data is important for Universum learning, (2) both inter-view and intra-view correlation are meaningful for the dimensionality reduction of multi-view data.

5   Conclusions and Future Work

Encouraged by the success of multi-view learning and Universum learning, we present a novel supervised dimensionality reduction method, Universum Discriminant Canonical Correlation Analysis (UDCCA), for multi-view data, which contains both target samples and Universum data. UDCCA focus on not only inter-view information expressed by within-class correlation and within-Universum correlation between different views, but also the intra-view information of the target samples and Universum data within each view. Concretely, UDCCA can seek the desirable projection directions to ensure the maximal within-class correlation for the target samples, minimal correlation for the Universum data of different view, and maximal separability for the target samples and Universum data within each view. To evaluate the performance of the proposed method, a series of experiments were performed on several benchmark multi-view datasets with markedly improved results.

There are several lines for future studies. (1) Universum data selection. The Universum data is easy to obtain, even easier than unlabeled target samples. However, not all Universum samples are helpful for improving the performance, thus selecting effective and informative Univsersum data will be useful. (2) Algorithm design.

The trick of incorporating Universum learning into multi-view data can be generalized to other extensions of CCA and the scenario where Universum data are not paired.

Acknowledgments. This work was supported by National Natural Science Foundations of China (NSFC) under Grant no. 61403193, 11471159, 61661136001 and NUAA Research Funding under Grant no. NP2014081 and NZ2015103.

References

1. Glenisson, P., Mathys, J., Moor, B.D.: Meta-clustering of gene expression data and literature-based information. SIGKDD Explor. Newsl. **5**(2), 101–112 (2003)
2. Chen, Q., Sun, S.: Hierarchical multi-view fisher discriminant analysis. In: Leung, C.S., Lee, M., Chan, Jonathan H. (eds.) ICONIP 2009. LNCS, vol. 5864, pp. 289–298. Springer, Heidelberg (2009). doi:10.1007/978-3-642-10684-2_32
3. Xu, C., Tao, D., Xu, C.: A survey on multi-view learning. arXiv: 1304.5634 (2013)
4. Sun, S.: A survey of multi-view machine learning. Neural Comput. Appl. **23**(7), 2031–2038 (2013)
5. Li, Y., Yang, M., Zhang, Z.: Multi-view representation learning: a survey from shallow methods to deep methods. arXiv Preprint, **14**(8), 1–27 (2016)
6. Hotelling, H.: Relations between two sets of variates. Biometrika **28**(3), 321–377 (1936)
7. Hardoon, D.R., Szedmak, S., Shawe-Taylor, J.: Canonical correlation analysis: an overview with application to learning methods. Neural Comput. **16**(12), 2639–2664 (2004)
8. Sun, T., Chen, S., Yang, J., Shi, P.: A novel method of combined feature extraction for recognition. In: 8th IEEE International Conference on Data Mining, pp. 1043–1048 (2008)
9. Vapnik, V.N.: Estimation of Dependences Based on Empirical Data, vol. 4(356). Springer, New York (2006)
10. Weston, J., Collobert, R., Sinz, F., Bottou, L., Vapnik, V.: Inference with the Universum. In: 23rd International Conference of Machine Learning, pp. 1009–1016 (2006)
11. Sinz, F.H., Chapelle, O., Agarwal, A., Schölkopf, B.: An analysis of inference with the Universum. Adv. Neural. Inf. Process. Syst. **26**, 1369–1376 (2008)
12. Dhar, S., Ramakrishnan, N., Cherkassky, V., Shah, M.: Universum learning for multiclass SVM. arXiv preprint arXiv:1609.09162 (2016)
13. Zhang, D., Wang, J., Wang, F., Zhang, C.: Semi-supervised classification with universum. In: SIAM International Conference on Data Mining, pp. 323–333 (2008)
14. Pan, S., Wu, J., Zhu, X., Long, G., Zhang, C.: Boosting for graph classification with universum. Knowl. Inf. Syst. **1–25**, 53–77 (2016)
15. Wang, Z., Zhu, Y., Liu, W., Chen, Z., Gao, D.: Multi-view learning with Universum. Knowl. Based Syst. **70**, 376–391 (2014)
16. Li, D., Zhu, Y., Wang, Z., Chong, C., Gao, D.: Regularized matrix-pattern-oriented classification machine with Universum. Neural Process. Lett. **45**(3), 1077–1098 (2017)
17. Chen, X., Chen, S., Xue, H.: Universum linear discriminant analysis. Electron. Lett. **48**(22), 1407–1409 (2012)
18. Chen, X., Ma, D.: Universum principal component analysis. In: International Conference on Information Technology and Management Engineering, pp. 236–241 (2014)
19. Qiu, J., Zhang, Y., Pan, Z., Yang, H., Ren, H., Li, X.: A novel semi-supervised approach for feature extraction. In: 2016 International Joint Conference on Neural Networks (2016)

20. Chen, S., Zhang, C.: Selecting informative Universum sample for semi-supervised learning. In: 21st International Joint Conference on Artificial Intelligence, vol. 1, pp. 1016–1021 (2009)
21. Sun, S., Xie, X., Yang, M.: Multiview uncorrelated discriminant analysis. IEEE Trans. Cybern. **46**(12), 3272–3284 (2015)
22. Kan, M., Shan, S., Zhang, H., Lao, S., Chen, X.: Multi-View discriminant analysis. IEEE Trans. Pattern Anal. Mach. Intell. **38**(1), 188–194 (2016)
23. Fisher, R.: The use of multiple measurements in taxonomic problems. Ann. Eugenics **7**(2), 179–188 (1936)
24. Maćkiewicz, A., Ratajczak, W.: Principal components analysis. Comput. Geosci. **19**(3), 303–342 (1993)
25. Breukelen, M.V., Duin, R.P.W., Tax, D.M.J., Hartog, J.E.D.: Handwritten digit recognition by combined classifiers. Kybernetika **34**(4), 381–386 (1998)
26. Kushmerick, N.: Learning to remove internet advertisements. In: 3rd Proceedings of the Annual Conference on Autonomous Agents, pp. 175–181 (1999)
27. Lampert, C.H., Nickischm, H., Harmeling, S.: Learning to detect unseen object classes by between-class attribute transfer. In: CVPR, pp. 951–958 (2009)

Color Image Segmentation by Multilevel Thresholding Based on Harmony Search Algorithm

Viktor Tuba[1], Marko Beko[2], and Milan Tuba[3(✉)]

[1] Graduate School of Computer Science, John Naisbitt University, Belgrade, Serbia
[2] Computer Engineering Department,
Universidade Lusófona de Humanidades e Tecnologias, Lisbon, Portugal
[3] Department of Technical Sciences,
State University of Novi Pazar, Novi Pazar, Serbia
tuba@ieee.org

Abstract. One of the important problems and active research topics in digital image precessing is image segmentation where thresholding is a simple and effective technique for this task. Multilevel thresholding is computationally complex task so different metaheuristics have been used to solve it. In this paper we propose harmony search algorithm for finding optimal threshold values in color images by Otsu's method. We tested our proposed algorithm on six standard benchmark images and compared the results with other approach from literature. Our proposed method outperformed other approach considering all performance metrics.

Keywords: Harmony search algorithm · Metaheuristic algorithms · Image segmentation · Otsu's method · Multilevel thresholding · Color images

1 Introduction

Digitalization and especially digital images and their processing facilitated significant progress in numerous scientific fields. Representation of digital images as matrices of integer numbers enables powerful processing by mathematical methods [9, 15].

Digital image processing can be roughly divided into three main categories. The first group is low level image processing which includes image enhancement and normalization but without any knowledge about the image content. This group can contain filters for smoothing, denoising, light level and contrast adjustment, etc. The second group is middle level processing where some elements in the image are detected such as edges, coutures, faces, using morphological operations and segmentation. The last group is high level processing where some machine learning is included. High level image processing is image understanding and recognition of previously detected objects such as face recognition, character recognition, etc.

© Springer International Publishing AG 2017
H. Yin et al. (Eds.): IDEAL 2017, LNCS 10585, pp. 571–579, 2017.
https://doi.org/10.1007/978-3-319-68935-7_62

One common task in digital image processing is image segmentation. This task belongs to the middle level image processing algorithms. The goal of image segmentation is to group pixels that belong to one object together and to separate pixels from the different objects. The simplest example is separating objects from the background where all object's pixels are converted to white (or black) while background pixels are black (or white). Often, beside separating background pixels from objects, different objects need to be differentiated, thus several classes need to be determined.

Image segmentation is used in various applications and represents an active research topic [7,14]. Numerous techniques for this task were proposed such as clustering based segmentation [6], region growing [16], graph based segmentation [12], etc. One of the simplest, but very efficient, methods for image segmentation is thresholding where pixels are grouped based on their intensity. All pixels between two threshold values are set to the same intensity level, i.e. they belong to the same class. Several methods were proposed in the past for finding optimal threshold values. Some of the widely used methods are entropy based such as Kapur's and Tsallis' methods and variance based Otsu's method. Grey scale image thresholding was intensively studied in the past while color image segmentation by thresholding methods represents recent research topic and it is more complex then the previous one. In both cases finding threshold values represents a combinatorial problem similar to traveling salesman that does not have deterministic solution except checking all possible solutions which for rather small input can last for thousands of years. In recent years, for solving these kind of problems nature inspired algorithms, especially swarm intelligence algorithms, were successfully used. Image segmentation by using different swarm intelligence algorithms and methods was researched in the past. For example, in [2] modified artificial bee colony algorithm was used for finding optimal thresholds by Kapur's, Tsallis' and Otsu's method for satellite image segmentation. Improved bat algorithm was applied for image segmentation by Kapur's and Otsu's method in [1] while in [3] cuckoo search and firefly algorithm were used.

In this paper we proposed harmony search algorithm (HS) for finding optimal threshold values by Otsu's method for color images. The proposed method is tested on six standard benchmark images and the results are compared to other approaches from literature.

The rest of the paper is organized as follows. In Sect. 2 Otsu's method for multilevel thresholding for color images is described. Our proposed harmony search algorithm for thresholding is presented in Sect. 3. Simulation results along with comparison results are presented in Sect. 4 while conclusion and suggestions for further work are given in Sect. 5.

2   Otsu's Method for Multilevel Thresholding

Otsu's method is often used thresholding technique where optimal thresholds are determined by putting pixels into the same class based on variance. The goal is to maximize between-class variance. Originally, Otsu's method was proposed for gray scale images binarization but later it was extended for multilevel thresholding. In recent years it was also used for color image multilevel thresholding. Otsu's method definition for color images is given below.

Assume that I is digital image represented by red, green and blue component where each of them takes value from the range $[0, L-1]$. For each component histogram can be made by counting the pixels with the same component intensities. Histograms can be denoted as $h^c(i)$ where $c \in \{R, G, B\}$ and $i = 0, 1, \ldots, L-1$. The probabilities of pixels of component c at level i are p_i^c. If N_i^c is the number of pixels of component c with level i, probability p_i^c is defined as $p_i^c = \frac{h_i^c}{n*m}$ where n and m represent the image dimensions.

Optimal k threshold values are determined by maximization of between-class variance:

$$f^c(t_0^c, t_1^c, \ldots, t_{k-1}^c) = \sum_{i=0}^{k} \sigma_i^c \tag{1}$$

where σ^c functions are:

$$\sigma_0^c = w_0^c \left(\sum_{i=0}^{t_0^c-1} \frac{i p_i^c}{w_0^c} - \sum_{i=0}^{L-1} i p_i^c \right)^2, \qquad w_0^c = \sum_{i=0}^{t_0^c-1} p_i^c,$$

$$\sigma_1^c = w_1^c \left(\sum_{i=t_0^c}^{t_1^c-1} \frac{i p_i^c}{w_1^c} - \sum_{i=0}^{L-1} i p_i^c \right)^2, \qquad w_1^c = \sum_{i=t_0^c}^{t_1^c-1} p_i^c, \tag{2}$$

$$\vdots$$

$$\sigma_k^c = w_k^c \left(\sum_{i=t_{k-1}^c}^{L-1} \frac{i p_i^c}{w_k^c} - \sum_{i=0}^{L-1} i p_i^c \right)^2, \qquad w_k^c = \sum_{i=t_{k-1}^c}^{L-1} p_i^c,$$

where t_l^c represent threshold values that are used to separate classes for component c, $c \in \{R, G, B\}$.

Optimal threshold values for RGB images are searched for each component by Otsu's method thus when $(k+1)$-level thresholding is needed than $3k$ thresholds need to be determined. In the case of RGB images, pixels of each component are separated into $k+1$ classes thus $(k+1)^3$ different color shades are defined. Color image thresholding is more complex than gray scale thresholding because dimension of the problem is larger and search space is $[0, L-1]^3$ instead of $[0, L-1]$.

3   Harmony Search Algorithm for Multilevel Thresholding

Harmony search algorithm (HS) was proposed by Geem et al. [5]. It was developed based on artificial phenomena, music harmony or more precise process how skilled musicians try to produce pleasant harmony. Process is simplified so that the two main parts of optimization algorithms, exploration and exploitation, are implemented without unnecessary details. Harmony search algorithm has been used in various applications but it was also applied to multilevel thresholding. For example, in [4,10,11] harmony search was used for gray scale image multi-level thresholding problem and in [8] harmony search and Kapur's method were used for mammograms image segmentation. Harmony search was not applied to regular color images which is proposed in this paper.

The main idea of harmony search is that musician has three choices when tries to produce harmony: play any tune from memory, play something similar to the tune by adjusting the pitch or composing new, random notes. The first two options are used for implementing exploitation while the third option is used for exploration. Harmony search algorithm pseudo code is shown in Algorithm 1.

Algorithm 1. Pseudo-code of the BSO algorithm

1: **Initialization**
2: Define harmony memory accepting rate r_{accept}, pitch adjusting rate r_{pa} and other parameters.
3: Generate harmony memory (HM) with random harmonies.
4: **repeat**
5:     **for** i = 0; i < n_{var}; i++ **do**
6:         **if** $rand < r_{accept}$ **then**
7:             Choose a value from HM for the variable i.
8:             **if** $rand < r_{pa}$ **then**
9:                 Adjust the value by adding certain amount.
10:            **end if**
11:        **else**
12:            Choose a random value
13:        **end if**
14:    **end for**
15:    Accept the new harmony if it is better
16: **until** Maximal iteration number is reached.
17: **return** The best solution (harmony) among all harmonies.

For the first choice, accepting rate parameter $r_{accept} \in [0, 1]$ is used for selecting new solution (harmony) based on previous solutions (which is memory of tunes). For the second choice, two parameters are introduced, pitch bandwidth b_{range} and pitch adjusting rate r_{pa}. Pitch adjusting rate is used to control frequency of pitch adjustment according to the equation:

$$x_{new} = x_{old} + b_{range} \, \epsilon \qquad (3)$$

where x_{old} is the existing pitch stored in the memory and x_{new} is a new pitch. Parameter ϵ is random number from uniform distribution in the range $[-1, 1]$. The last possible choice is randomization where completely new random solution is generated.

In this paper harmony search algorithm was adjusted for optimizing Otsu's method for color images. Into original HS algorithm two adjustments were introduced. The first one is consequence of the fact that threshold values are sorted numbers, $t_0 < t_1 < \cdots < t_k$. The second adjustment is necessary since HS algorithm searches solutions in real number space while threshold values are integers. These two conditions are satisfied in a way that each generated solution was rounded and sorted before further use. Objective function was set to be the average value of fitness functions for three components defined by Eq. 1.

4 Simulation Results

Our proposed method was implemented in Matlab R2016a and simulations were performed on the platform with Intel® Core™ i7-3770K CPU at 4 GHz, 8 GB RAM, Windows 10 Professional OS.

The proposed algorithm was tested on 6 standard benchmark images (Fig. 1). All images are free for download from https://www2.eecs.berkeley.edu/ Research/Projects/CS/vision/bsds/BSDS300/html/dataset/images.html.   All test images are RGB images of the dimension 481×321. Parameters for harmony search algorithm were set empirically. Number of harmonies was set to 100 and maximal iteration number was 2000. Parameter r_{accept} was 0.9, pitch adjustment r_{pa} was 0.3 and b_{range} was 0.2. For each test image the proposed method was run 15 times.

(a) Butterfly   (b) Starfish   (c) Rhino

(d) Horse   (e) Flower   (f) Train

Fig. 1. Test images

In order to test the quality of our proposed algorithm we compared it with other approach from literature where firefly algorithm (FA) was used [13]. In [13] three versions of firefly algorithm, BFA, LFA and CFA were proposed for optimizing Otsu's method for color image thresholding. Beside mean of objective function for 15 runs two standard image similarity measures were calculated: peak signal to noise ration (PSNR) and structural similarity index matrix. These two measures were calculated as:

$$PSNR = 20 \log_{10} \frac{255}{\sqrt{\frac{1}{NM} \sum_{i=1}^{N} \sum_{j=1}^{M} (x_{i,j}^* - x_{i,j})^2}}, \tag{4}$$

where N and M are dimensions of the image, $x^*(i,j)$ is pixel value of the original image at position (i,j) while $x(i,j)$ is value of segmented image.

Structural similarity index matrix is defined as:

$$SSIM = \frac{(2\mu_o\mu_s + c_1)(2\sigma_{os} + c_2)}{(\mu_o^2 + \mu_s^2 + c_2)(\sigma_o^2 + \sigma_s^2 + c_2)} \quad (5)$$

where μ_o and μ_s are averages of the original and the segmented image respectively, σ_{os} is covariance of the original and the segmented image, σ_o^2 and σ_s^2 are variances for the original and the segmented image, while $c_1 = (k_1 L)^2$ and $c_2 = (k_2 L)^2$ for L=256, $k_1 = 0.01$ and $k_2 = 0.03$. Larger values for PSNR are better, while SSIM has value in the range $[-1, 1]$ where SSIM is equal to 1 for two identical images.

Comparison results are presented in Table 1. Best results are in bold. Our proposed method outperformed all three version of the FA proposed in [13] for all metrics. Value of the objective function was significantly larger compared to ones obtained by FA optimization methods which finally resulted in better PSNR and SSIM metrics. The largest gap between the proposed HS and FA based methods was for test image Rhino. The maximal Otsu's fitness function for 2 threshold values was 2107.28 while HS found threshold values where fitness function was 3040.00. On the other hand, the smallest improvement was for Flower test image, 1302.61 with LFA to 1687.50 with HS. PSNR was larger for all test images for all tested thresholding levels which means that images segmented by the proposed

Table 1. Comparison of performance for RGB test images

Image	m	Objective function				PSNR				SSIM			
		BFA	LFA	CFA	HS	BFA	LFA	CFA	HS	BFA	LFA	CFA	HS
Butterfly	2	3515.92	3402.61	3617.38	**3947.37**	10.866	10.243	11.026	**12.213**	0.6399	0.6402	0.6394	**0.6512**
	3	3629.37	3638.81	3640.72	**4068.82**	14.297	15.173	14.927	**16.214**	0.7133	0.6936	0.7047	**0.7982**
	4	3691.66	3669.02	3690.81	**4135.99**	17.562	17.283	17.602	**19.210**	0.7835	0.7669	0.7639	**0.8291**
	5	3822.81	3792.55	3811.01	**4176.75**	19.554	19.328	19.715	**21.214**	0.8472	0.8317	0.8274	**0.8973**
Star fish	2	1986.97	1972.10	1985.11	**2777.55**	11.513	13.272	12.267	**14.827**	0.7320	0.7461	0.7392	**0.7919**
	3	2017.18	2081.66	2088.41	**2956.86**	14.868	14.792	14.901	**16.349**	0.7831	0.7706	0.7593	**0.8481**
	4	2107.25	2109.91	2098.77	**3040.98**	18.382	18.281	18.332	**20.193**	0.8032	0.7996	0.8106	**0.8862**
	5	2251.73	2178.24	2201.62	**3083.94**	19.191	20.037	20.097	**22.524**	0.8529	0.8274	0.8461	**0.8987**
Rhino	2	2004.99	2081.84	2107.28	**3040.00**	9.881	11.368	11.206	**12.733**	0.6837	0.7106	0.7083	**0.7685**
	3	2216.72	2205.22	2192.77	**3222.82**	13.463	14.122	13.974	**16.356**	0.7153	0.7342	0.7311	**0.7991**
	4	2251.33	2267.18	2222.9	**3300.98**	16.182	16.001	16.189	**18.377**	0.7316	0.7628	0.7528	**0.8010**
	5	2388.16	2371.97	2382.28	**3342.59**	18.068	17.926	17.874	**20.018**	0.7829	0.8152	0.7902	**0.8193**
Horse	2	2635.11	2671.03	2587.99	**3302.50**	10.517	12.015	11.739	**14.004**	0.6402	0.7261	0.7264	**0.7524**
	3	2688.04	2683.31	2660.37	**3475.32**	14.701	14.826	14.519	**16.817**	0.7026	0.7418	0.7302	**0.7781**
	4	2717.37	2716.03	2700.83	**3550.59**	16.576	16.478	17.005	**19.219**	0.7792	0.7902	0.7886	**0.8004**
	5	2782.7	2763.44	2746.67	**3597.91**	18.269	20.027	20.157	**22.557**	0.8218	0.8142	0.8213	**0.8502**
Flower	2	1159.57	1302.61	1288.92	**1687.50**	11.284	13.721	12.826	**14.123**	0.6820	0.7227	0.7301	**0.7524**
	3	1420.23	1472.71	1392.44	**1836.72**	16.168	14.916	16.026	**18.293**	0.7211	0.7529	0.7329	**0.7847**
	4	1681.16	1592.88	1562.39	**1907.58**	21.174	20.177	20.291	**22.857**	0.7938	0.8111	0.8102	**0.8259**
	5	1690.00	1623.71	1607.35	**1950.53**	20.844	21.002	20.926	**23.117**	0.8315	0.8268	0.8331	**0.8447**
Train	2	1829.01	1803.55	1831.63	**2641.22**	12.648	12.579	12.739	**14.236**	0.6826	0.6901	0.6883	**0.7016**
	3	1903.28	1873.77	1894.00	**2806.65**	14.282	14.138	14.620	**16.432**	0.6869	0.7132	0.7039	**0.7593**
	4	1937.42	1903.18	1917.22	**2885.33**	18.548	18.207	18.442	**19.982**	0.7385	0.7835	0.7893	**0.7981**
	5	1975.56	1955.28	1977.61	**2928.84**	20.031	20.379	20.715	**21.285**	0.8193	0.8352	0.8374	**0.8523**

Table 2. Threshold values for Butterfly test image

m	BFA			LFA			CFA			HS		
	R	G	B	R	G	B	R	G	B	R	G	B
2	14, 98	7, 107	4, 136	16, 102	8, 115	4, 143	16, 100	9, 105	6, 138	84, 156	98, 170	69, 141
3	13, 69, 144	6, 96, 174	3, 82, 167	15, 71, 149	7, 93, 171	3, 85, 164	15, 68, 147	8, 98, 177	4, 80, 168	59, 119, 184	96, 160, 210	66, 122, 178
4	12, 51, 105, 146	6, 71, 124, 178	3, 64, 108, 172	13, 49, 107, 145	5, 74, 126, 181	3, 61, 110, 175	13, 50, 104, 144	6, 72, 123, 179	3, 61, 111, 175	53, 97, 151, 198	78, 117, 167, 212	61, 100, 147, 192
5	12, 54, 96, 133, 167	5, 52, 114, 153, 192	2, 46, 107, 140, 179	11, 56, 99, 136, 171	4, 50, 115, 151, 195	2, 41, 109, 146, 184	11, 57, 95, 135, 168	4, 54, 116, 151, 191	2, 44, 109, 148, 184	47, 79, 123, 170, 206	76, 106, 145, 181, 221	46, 73, 115, 162, 201

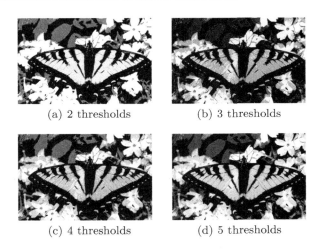

(a) 2 thresholds   (b) 3 thresholds

(c) 4 thresholds   (d) 5 thresholds

Fig. 2. Example of segmented test image Butterfly

Otsu's method optimized by harmony search algorithm were more similar to the originals than these in [13]. Obtained SSIM also indicate that our proposed method provided better segmentation. Rather different fitness function values as well as PSNR and SSIM were obtained due the fact that harmony search found completely different optimal thresholds. In Table 2 are presented threshold values obtained by our proposed method, along with threshold values found in [13], while in Fig. 2 segmentations of Butterfly test image by these threshold values are shown.

5   Conclusion

In this paper harmony search algorithm was proposed for color image segmentation by Otsu's method. Otsu's between-class variance method was used for finding optimal threshold values for each color component and harmony search algorithm was used for finding maximal value of Otsu's fitness function. We compared our proposed method with approach from literature where firefly algorithm was implemented for Otsu's optimization. Our proposed method was superior in all tests compared to firefly based segmentation. In further work different thresholding methods such as Kapur's and Tsallis' method can be tested.

Acknowledgment. This research is supported by the Ministry of Education, Science and Technological Development of Republic of Serbia, Grant no. III-44006.

References

1. Alihodzic, A., Tuba, M.: Improved bat algorithm applied to multilevel image thresholding. Sci. World J. **2014**, 1–16 (2014). Article ID 176718
2. Bhandari, A., Kumar, A., Singh, G.: Modified artificial bee colony based computationally efficient multilevel thresholding for satellite image segmentation using Kapur's, Otsu and Tsallis functions. Expert Syst. Appl. **42**(3), 1573–1601 (2015)
3. Brajevic, I., Tuba, M.: Cuckoo search and firefly algorithm applied to multilevel image thresholding. In: Yang, X.-S. (ed.) Cuckoo Search and Firefly Algorithm. SCI, vol. 516, pp. 115–139. Springer, Cham (2014). doi:10.1007/978-3-319-02141-6_6
4. Cuevas, E., Zaldívar, D., Perez-Cisneros, M.: Otsu and Kapur segmentation based on harmony search optimization. Applications of Evolutionary Computation in Image Processing and Pattern Recognition. ISRL, vol. 100, pp. 169–202. Springer, Cham (2016). doi:10.1007/978-3-319-26462-2_8
5. Geem, Z.W., Kim, J.H., Loganathan, G.: A new heuristic optimization algorithm: harmony search. Simulation **76**(2), 60–68 (2001)
6. Gong, M., Liang, Y., Shi, J., Ma, W., Ma, J.: Fuzzy c-means clustering with local information and kernel metric for image segmentation. IEEE Trans. Image Process. **22**(2), 573–584 (2013)
7. Li, Y., Jiao, L., Shang, R., Stolkin, R.: Dynamic-context cooperative quantum-behaved particle swarm optimization based on multilevel thresholding applied to medical image segmentation. Inf. Sci. **294**, 408–422 (2015). Innovative Applications of Artificial Neural Networks in Engineering
8. Maleki, F., Nooshyar, M., Fatin, G.Z.: Breast cancer segmentation in digital mammograms based on harmony search optimization. Tech. J. Eng. Appl. Sci. **4**(4), 477–484 (2014)
9. Nikolic, M., Tuba, E., Tuba, M.: Edge detection in medical ultrasound images using adjusted canny edge detection algorithm. In: 24th Telecommunications Forum (TELFOR), pp. 691–694. IEEE (2016)
10. Oliva, D., Cuevas, E., Pajares, G., Zaldivar, D., Perez-Cisneros, M.: Multilevel thresholding segmentation based on harmony search optimization. J. Appl. Math. **2013**, 1–24 (2013). Article ID 575414
11. Ouadfel, S., Taleb-Ahmed, A.: Performance study of harmony search algorithm for multilevel thresholding. J. Intell. Syst. **25**(4), 473–513 (2016)
12. Peng, B., Zhang, L., Zhang, D.: A survey of graph theoretical approaches to image segmentation. Pattern Recogn. **46**(3), 1020–1038 (2013)
13. Rajinikanth, V., Couceiro, M.: RGB histogram based color image segmentation using firefly algorithm. Procedia Comput. Sci. **46**, 1449–1457 (2015)

14. Tuba, E., Tuba, M., Jovanovic, R.: An algorithm for automated segmentation for bleeding detection in endoscopic images. In: International Joint Conference on Neural Networks (IJCNN), pp. 4579–4586 (2017)
15. Tuba, E., Tuba, M., Dolicanin, E.: Adjusted fireworks algorithm applied to retinal image registration. Stud. Inf. Control **26**(1), 33–42 (2017)
16. Zhao, Y.Q., Wang, X.H., Wang, X.F., Shih, F.Y.: Retinal vessels segmentation based on level set and region growing. Pattern Recogn. **47**(7), 2437–2446 (2014)

Finding Sentiment in Noise: Non-linear Relationships Between Sentiment and Financial Markets

Zeyan Zhao$^{(\boxtimes)}$, Stephen Kelly, and Khurshid Ahmad

Trinity College Dublin, Dublin, Ireland
{zhaoz,Stephen.Kelly,Khurshid.Ahmad}@scss.tcd.ie

Abstract. Sentiment analysis, especially in economics and finance, start-ed with a great fanfare in the late 1990's when (negative) sentiment proxies were introduced in econometric schemes used in finance for forecasting price returns. Intuitively, the relationship between returns and the proxies suggests a polar opposition - high negative sentiment low returns and vice versa. The parametric (vector autoregression) analysis has suggested a linear relationship between two variables especially in ostensibly well-regulated and stable markets like New York Stock Exchange. We have examined this return-sentiment relationship over a period of many years in stable (NYSE), quasi-stable (Copenhagen Stock Exchange and the Chicago Board of Trade's commodity markets) and the emergent markets (Shanghai Stock Exchange). A non-parametric method (locally weighted regression) with variations across the board has graphically shown the relationship between sentiment and returns in a non-linear fashion. Our study is largely based on newspaper report (Shanghai and Copenhagen), opinion pieces (New York and Chicago) and blogosphere (Chicago) - totalling 39.2 million tokens regressed with time series comprising 15,871 data points. Evidence from both models shows that negative sentiment actively influences stock and commodity markets respectively. The analytical conclusions determine that sentiment represents a meaningful performance in predicting momentary market-wide drops in valuation.

Keywords: Sentiment · Non-linear · Financial markets · Time series analysis

1 Introduction

Fisher Black worked on financial models that can be used to elegantly compute return on assets. These models have been extensively used in trading derivatives.

Z. Zhao—The research leading to these results has received funding from the EU FP7 Slandail project under grant agreement no. 607691. In this study we used the text analysis system Rocksteady, developed as part of the Faireachain project for monitoring, evaluating and predicting the behaviour of markets and communities (20092011). Support for Rocksteady's development was provided by Trinity College, University of Dublin and Enterprise Ireland (Grant IP-2009-0595).

H. Yin et al. (Eds.): IDEAL 2017, LNCS 10585, pp. 580–591, 2017.
https://doi.org/10.1007/978-3-319-68935-7_63

He was clear about the distinction between information and noise in market trading: those using information have a reasonable expectation to make a profit, but those leveraging noise will generally fail to do so. Nevertheless, Fisher regarded noise as an important ingredient in the "existence of liquid markets": "Noise in the sense of a large number of small events is often a causal factor, much more powerful than a small number of large events. Noise makes trading in financial markets possible, and thus allows us to observe prices for financial assets" [1].

In Black's rationalist framework, investor sentiment will perhaps be regarded as noise especially as such sentiment is generally quite ephemeral and the impact of sentiment on stock price returns is generally remedied in time. On the contrary for scholars like Shiller [2], investor sentiment is a real phenomenon and can be used to determine the direction in which markets move. It was noted that the unaccounted for fluctuations in financial instruments traded on New York Stock Exchange, like closed-end funds issued specifically for a country, can be attributed especially to negative news about the country in the New York Times [4]. These studies coupled with studies of the so-called contrarian traders and noise traders led to the belief that whilst the rationalist framework advocated by Black and others has some merit, one has to include the impact of investor psychology in the manner in which assets are priced [2,3].

A number of different proxies for investor sentiment have been proposed including trading volume, order flow, dividend premiums [25]. Following on from work in content analysis of newspaper texts including reports and opinion, in politics and social sciences [5], scholars in economics and finance have analysed opinion pieces to extract market sentiment. This noise/sentiment appears to improve forecast price return with negative sentiment appearing to contribute to the market downturn by as much as 5–10 basis points [28]. The impact of sentiment is generally reversed in four to five trading days; though during the recession this period could be longer and could be as large as 10–12 basis points [30] (1 basis point change in Dow Jones Industrial Average at current prices is USD 540 million).

The rest of this paper is organised as follows. We revise literature from the domain of financial and text analysis that extracts information from a financial text to create a proxy that acts as a variable to be incorporated into statistical estimations of price change. Section 3 describes the content analysis method used to extract sentiment from a text, vector autoregression and the locally weighted regression methods used to model returns and sentiment together to determine linear and non-linear relationships. Finally, we present our results and findings for the impact of sentiment across different financial markets and a brief discussion and conclusion are then presented.

2   Method

2.1   Noise and Quantitative Proxies

Price return of an asset at time, r_t has a number of interesting properties, including the auto-regressive nature of dependence of prices at one moment in time with an n past price returns; the weighted average of lagged returns, denoted symbolically as a lagged operator $L_n r_t$, and denotes the symbolic weight:

$$\alpha L_n r_t = \alpha_1 r_{t-1} + \alpha_2 r_{t-2} + ... + \alpha_n r_{t-n} \tag{1}$$

The error in the estimation is denoted as ϵ_t; the error is generally assumed to be normally distributed in time and has a zero mean and unit standard deviation $(NID(0,1))$. This model is based on a number of assumptions one of which relates to the closed nature of the model-forecast return depends only upon returns. This notion of using other correlated variables has been advocated by leading econometricians (e.g. [6]) for macroeconomic analysis in general. Sims assertion that the other variables may be used in detecting the influence of public's change of taste, a subjective argument at its best one may argue, and this change introduces what he calls 'disequilibrium economics': This disequilibrium idea is one of the fore-runners of modern sentiment analysis. Business cycle fluctuations may also be due to calendar effects, asset trading is pursued mainly in weekdays and during day-light hours, trading is impacted by public holidays, especially before and after long holidays - January trading volumes are lower due to the post end-of-Georgian calendar a month previously. These fluctuations can be added into Eq. (1) as dummy exogenous variables - the inclusion of exogenous variables in forecasting returns has a rich history of arguments of pros and cons of including these variable have been discussed since the introduction of these variables about 50 years ago [19].

Sentiment analysis focuses on how the basic market dictum, that price (return) of an asset contains all the necessary information relevant to its potential for buying/selling, breaks down and one has to rely on other sources of data. For example, a new invention related to (an individual or) a firm may lead to a mispricing of the asset; during wars and periods of economic boom and bust there are many instances of over- and under-pricing, or in anticipation of a central bank announcement or before/after a summit meeting asset prices change irrespective of other critical economic performance data. (See, for example, Engle's Nobel Lecture [23]). It has been suggested that one has to assume that the key stylised facts, especially first and second moments of return will show a time-dependence; one theoretically well-founded model for dealing with this heteroskedastic behaviour is the generalised auto-regressive model that will relate the asset price returns to volatility measures like mean and standard deviation. Alternatively, other empirical volatility measures can be added on to Eq. (1) for a well-grounded estimate of the forecast return: lagged squared returns $(L_n r_t^2)$, lagged (detrended) volume $(\rho L_n V l m_t)$, or other option-traded volatility indices are used as like VIX_t are used in an empirical fashion and added to Eq. (1) above.

2.2   Noise in Market Collateral Texts and Sentiment Proxies

The 1990's innovation, based on the earlier work of Philip Stone and his predecessor Harold Laswell in political science in the 1940's [5], led to the harvesting of sentiment from the opinion pieces published in upmarket and business oriented (daily) newspapers like the Wall Street Journal [28,30] and sometimes the New York Times. The choice of opinion pieces, comprising often unattributed remarks on the behaviour of an individual or a firm, is an interesting one as these near-gossip column factoids were expected to influence investor sentiment. Furthermore, sentiment harvesting was conducted by a bag-of-words (BoW) model [17] that uses a thesaurus of emotions created by Stone et al. (1966) [5] that is essentially a word list classified according to many emotion categories.

The other evolution on harvesting investor sentiment has come from scholars who have opted for a lower frequency harvesting of sentiment from documents generated by a firm to report on its own health and well-being. Scholars have used newer specialised thesauri [22,32] on the sentiment harvesting from news reports and opinion pieces, whilst others have used the low-frequency documents (Form 10-K in the USA filed every six months by individual firms) together with an up-to-date thesaurus [20].

Table 1. Inclusion of sentiment and volatility: E denotes exogenous variables, S denotes sentiment variables, V_1 denotes volatility-covariant measure and V_2 denotes implied volatility measure; DW, J, 1987, NBER, EP and BC shorts for the Day of the Week dummy, the January dummy, the 1987 crash index dummy, the NBER recession dates dummy, the Economic Policy News proxy and Business Conditions Index respectively.

Return	Regression scheme	Eq#	Reference	Exogenous variable
	$r_t = \alpha L_5 r_t + \epsilon_t$	(1a)		
E	$r_t = \alpha L_5 r_t + \lambda_1 Exog_t + \epsilon_t^i$	(1b)	[19]	NA
S + E	$r_t = \alpha L_5 r_t + \beta L_5 s_{t+1} Exog_t + \epsilon_t^{ii}$	(1c)	[28,29]	DW + J + 1987
S + E + V_1	$r_t = \alpha L_5 r_t + \xi L_5 r_t^2 + L_5 s_{t+1} Exog_t + \epsilon_t^{iii}$	(1d)	[30]	DW + NBER
S + E + V_2	$r_t = \alpha L_5 r_t + \xi VIX_t + s_{t+1} Exog_t + \epsilon_t^{iv}$	(1e)	[20] use [21]	EP + BC

2.3 Inclusion of Sentiment in an Auto-Regressive Model of Price Returns

We have tabulated (see in Table 1) the regression schemes used by the various authors discussed above and this show that (i) following Sims [6] these authors have added more and more correlated variables; (ii) following Balestra and Nerlove (1966) [19] have added interesting exogenous variables as new vectors to an existing auto-regression scheme; and (iii) following Engle [23] either second moment of the price return, option price index volatility, or an heteroskedastic measure of volatility has been used. The output from the VAR modes is also a fitted regression model. A VAR model implies that dependent variable Y is a linear function of the predictors (with n different lags of each predictor), plus statistical noise. It has been designed to test the influences between internal

(5 lags of the dependent variable) and external variables (5 lags of quantitative and qualitative variables) within five working days.

2.4 Graphical Relationship Between Residuals and Sentiment - Robust Locally Weighted Regression

A time-ordered plot of the residual obtained from the Eq. (1a) and the newspaper sourced sentiment shows two noisy time series superposed on another (Fig. 1a). However, if we plot sentiment against the residual and then do a simple regression on the two, one see a sorts of relationship between (Fig. 1b) possibly obscured.

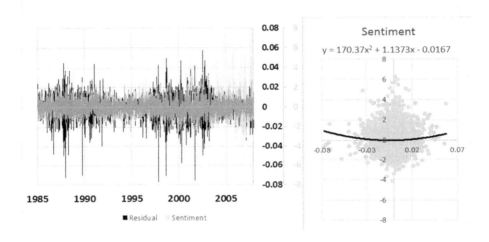

Fig. 1. a: time varying behaviour of residuals as extracted from Eq. 1b and sentiment harvested from WSJ AbOM; b: relationship between sentiment and residuals using DJIA and WSJ between 1984 and 2007

It has been argued that "the visual information on a plot can be greatly enhanced [...] by computing and plotting smoothed points"[7] under the rubric of robust locally weighted regression that essentially does piece-wise regression.

We will use a non-linear function - locally weighted scatterplot smoothing (also known as LOWESS) [7] to explore the relationship between residuals and sentiment values. Similar to a moving average method, the smoothed value in LOWESS process is decided by neighbouring data points defined within a span[1] and is weighted as the regression weighting function and is also determined for the data within the span:

$$w_i = (1 - |(x - x_i)/d(x)|^3)^3 \qquad (2)$$

where x is the predictor value associated with the response value to be smoothed, x_i are the nearest neighbours of x as defined by the span, and $d(x)$ is

---

[1] A percentage of the total number of data points in the data set.

the distance along the abscissa from x to the most distant predictor value within the span.

The three components of our algorithm are:

1. Use VAR for regressing residuals from market returns.
2. Calculate the weights then smoothed values between residuals and negative variable for each data point in a span using Eq. (2).
3. Output graph of the smoothed values.

3   Quantitative and Qualitative Data

3.1   Data Description

We had looked at the impact of sentiment in three markets (New York, Copenhagen and Shanghai) in a period of time where there were documented instances of market downturns sparked off by the mortgage crisis, currency crisis, or stock market crashes.

To begin with, we looked at the rather well established and ostensibly well-regulated New York Stock Exchange and studied the impact of sentiment on Dow Jones Industrial Average over a 23 year period that has included three major business downturns (1990–91; 2001; and 2007–2009). There had been major market downturns in this period, including the 1987 Black Monday, Denver loans scandal, Asian and Russian currency crisis (1997–99), the Dotcom boom (2000) and 2007 sub-prime loan crisis. Stock market manipulations in many of these crises [8,9] and the international impact of such crises have been documented [10]. Moving on from the well established markets we looked at the impact of sentiment on price returns in a relatively smaller market, the Copenhagen Stock Exchange over a period of 13 years i.e. 2002–2015: Denmark (with some serious reservations) was becoming part of a federated economic and political system - the European Union - with the introduction of the Euro (2002) in other EU countries, and the establishment of a European Central Bank and EU-wide regulatory authorities [11,12]. The well-documented sub-prime mortgage crisis (2007–2009) and the volatility spill overs across national boundaries [13] did hit Copenhagen.

We then looked at what was once called an 'emerging' market economy, China and its key Shanghai Stock Exchange over a 15 year period, 2000–2015. The market, though has its roots in 19th century stock markets (much like NYSE and Copenhagen), it has its own share of volatility ever since it was re-established in 1997 [14] and more volatility and reputation repair in the following 15 years [15,16]. These sentiment inducing volatilities, either motivated by business cycles or through market manipulations, provide a basis to see the robustness of price return forecast (with respect to sentiment). In addition to the three market indices, we also look at the commodities market using West Texas Intermediate (WTI) crude oil index. WTI, also known as Texas light sweet, is a grade of crude oil used as a benchmark in oil pricing.

US textual data are available from Wall Street Journal's (WSJ) Abreast of the Market column. We also collect and analyse Danish and Chinese economic news from newswire and newspapers in order to inspect whether one corpus is representative of another. Moreover, we also test editorial news about commodity markets from the Financial Times to see whether the results hold. Table 2 shows a brief description of the data sources.

Table 2. Data description

Financial data			
Case study	Exchange/Index	Est.d/Mkt cap	Period/Data point
I. US	NYSE/DJIA	1885/$19.6T	1984–2007/5997
II. Denmark	CSE/OMXC	1808/€0.2T	2002–2015/3143
III. China	SSE/SHCOMP	1990*/$3.5T	2000–2015/3474
IV. Oil	NYME/WTI		2000–2014/3553
News articles			
Case study	Name	Type of text	Period/No. of news/tokens
I. US	WSJ AofM	Opinion	1984–2007/6160/6.8m
II. Denmark	Local Newspapers	News	2002–2015/27194/2.7m
III. China	Xinhua	News	2000–2015/18953/17.4m
IV. Oil	FT	Editorial	2000–2014/23157/12.3m

*International Settlement in 1842.

It is inescapable to have missing data in such analysis as the data comes from varying types of sources. We collect textual corpus and financial information from online news aggregators and data providers. Normally, news articles are published at a different frequency than financial data is published. We deal with this mismatch of information by aligning the news data to financial data and the latter is the dependent series.

3.2 Measuring Affect

A dictionary of sentiment words is provided to Rocksteady (a sentiment analysis system developed at Trinity College Dublin) to perform text analysis. These dictionaries include a list of words that are classified according to some unifying subject and usually are organised by scholars with domain knowledge. In constructing our negative sentiment proxy, we used the General Inquirer Harvard IV dictionary [5] with a list of 2,291 negative terms. Rocksteady stores the collection of the captured news which, in turn, can be used to conduct a historical analysis over a selected period. Sentiment terms can be obtained from news reports (facts), editorials and commentaries (opinions). We use the BoW model, together with the Harvard GI dictionary, to extract sentiment term frequencies from the corpus.

The sentiment time series are used to construct our investor sentiment proxies. Computing the frequency of these words as they appear in a text can indicate the tone of the document, reviewing the overall fashion of sentiment. We standardise the sentiment proxy by dividing the total frequency of negative sentiment by the total number of words in the text and order this number according to time. This enables us to examine the level of sentiment in each text. This allows us to align the sentiment variable with other time series data such as stock price returns and compute these variables using the multivariate regression models such as vector autoregression evaluating the inter-relationship between the sentiment proxy and financial data.

4   Results

4.1   Linear Relationship - VAR

We first test our hypothesis that sentiment has a statistically significant impact on returns by incorporating 'sentiment' in the regressive model of returns using vector autoregression (VAR). In a model similar to Tetlock (2007) [28] and carried out previously by the authors [26,31,32], we will use the following regression equation to test the hypothesis:

$$r_t = +L_5 r_t + L_5 log V_t + L_5 S_t + Exog_{(t-1)} + \epsilon_t \tag{3}$$

$L_5 S_t$ will be replaced by $L_5 S_t^{DJIA}$, $L_5 S_t^{OMXC}$, $L_5 S_t^{SHCOMP}$ or $L_5 S_t^{WTI}$ accordingly when we analyse a text corpus from different sources. Running this model allows a linear relationship to be created between the sentiment measure extracted from news and financial returns.

Table 3. The negative effects of news sentiment on financial instruments

Case study	DJIA				OMXC [31]	SHCOMP [32]	WTI [27]
Period	1984–1999	1984–1991	1992–1999	2000–2007	2002–2015	2000–2015	2000–2014
Observations	3709	1714	1994	1992	3143	3474	3553
$L_n S_t$							
L_1	−4.8	−4.3	−7.4	0.1	−6.0	−9.5	−2.1
L_2	2.2	1	3.9	−3	0.9	7.0	−8.5
L_3	−1	−2.7	0.2	1.3	0.6	−0.5	−8.5
L_4	4.9	7.1	4.1	−2.7	−0.7	1.8	−6
L_5	2.5	0.6	4.3	4.4	4.4	−0.6	1
$\chi^2(5)$[joint]	20.20	7.07	15.60	8.1	5.13	15.5	14.9

We use vector autoregression to examine the impact of sentiment measures. The assumption of our model is that the expectation of regression residuals is independently and identically distributed (i.i.d). The endogenous variables include five lags of market returns, relevant market trading volumes, and sentiment measures. The exogenous variables include dummy variables of day-of-the-week, month-of-the-year, local holidays, 1987 financial crash effects, and five lags of the volatility proxy (we use GARCH(1,1) conditional variance of returns and VIX index as our volatility measures in the research). The trading volumes have been detrended from log volume. Robust standard errors [18] are used to reduce the Heteroskedasticity of residuals.

The first model we compute uses DJIA returns as the dependent variable for the period from 1984 to 1999 (results are presented in Table 3). The negative impact of 1st day lag negative sentiment on DJIA returns is 4.8 basis points (1 basis point is a 0.01% change in return) and statistically significant at the 0.01 level; this is fully recovered within the 2nd to 5th days lag (with fourth-day lag statistically significant at 0.01 level). We then split this period into two: 1984–1991 and 1992–1999 to test the difference between normal and bull periods. The normal period (1984–1999) includes the famous Black Monday crash but the bull period doesn't have any business downturns (according to NBER business cycles). The results show that the impact of negative sentiment is more significant during the 1990s (negative 7.4 basis points 1st day lag significance at 0.01 level) than it is during the 1980s (positive 7.1 basis points. 4th day lag significance at 0.05 level). We further extended the data period to the volatile 2000s (there are two recessions during the period 2000 to 2007). The results are similar to those in the 1980s; the 4.4 basis points lag five of negative sentiment is significant at 0.05 level. Following DJIA we look at a cross-section of assets, aggregated stock market indices and crude oil futures and investigate the movement of the aggregates in three different markets (OMXC, SHCOMP and WTI). We extract negative sentiment in news reports from local news wires and newspapers. The 1st day lag negative sentiment in both OMXC and SHCOMP studies show significant coefficients (-9.5 and -6.0 respectively and significant at 0.01 level). In the WTI case, the effect is delayed to the 2nd day lag, however, is still statistically significant (-8.5 at 0.01 level). The result confirms that there exists a linear relationship between negative sentiment and market returns in different markets.

4.2 Non-linear Confirmation - Robust Locally Weighted Regression

Most researchers analysed the effect of the investor sentiment extracted from media using parametric estimates. In order to visualise the non-linear relationship between the media factor and stock returns, we examine a semi-parametric approach. This approach comprises two steps: first, we conduct a parametric method that estimates the unexplained stock return residuals from Eq. (3) excluding the lags of the sentiment measures; second, we obtain the non-parametric estimates using a locally weighted regression method. Following the confirmation of linearity, we use one of the locally weighted regressions - LOWESS method to determine the non-linear relationships between sentiment and market returns.

First, we omit the lags of the negative sentiment measures ($L_5 S_t$) from the linear regression in Eq. (3) and obtain the residual from market returns. Second, we form locally weighted regressions (Lowess) for the measures of negative sentiment. The x-axis denotes the standardised negative sentiment, and the y-axis denotes the market returns residuals.

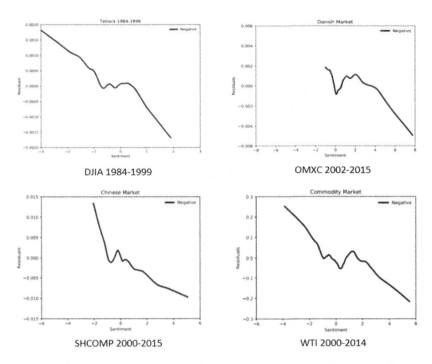

Fig. 2. Locally weighted regressions across markets: The x-axis denotes the standardised negative sentiment, and the y-axis denotes the market returns residuals.

The trend line that LOWESS produces, the smoothing parameter f has been set to 0.4 and number of iterations has been set to 1, similar to some researchers did in their previous studies [24,28,30]. Results are shown for four locally weighted regressions for measures of negative sentiment in four different markets Fig. 2. All the measures (OMXC data only shows the relationship when the value of the negative sentiment is positive) demonstrate an adverse impact on market return residuals - the residuals monotonically decreases as negative sentiment increases with the exception of a short interval near the vertical axis. The graphs in Fig. 2 show the visualised non-linear relationships between negative sentiments and VAR model residuals. Apart from a small exception near the vertical axis, the DJIA returns decrease as negative sentiment increases over different periods. This remains true for the other three markets we explore (with the exception that the relation in the left hand-side of the graph for OMXC market doesn't hold when sentiment is lower than -1 standard deviation). These effects are more significant when the negative sentiment values are near the far positive or negative side. A robustness test was carried out by eliminating the outliers through winsorising the data. The results remain consistent throughout demonstrating that the estimations weren't being influenced by the presence of outliers.

We found that both parametric and non-parametric models indicate that sentiment proxy as a qualitative measure of investor sentiment has an impact on market returns (quantitatively); the computational results show the opposition of relationship between proxies and returns - high negative sentiment occurring with low returns and vice versa.

5   Conclusion

In this paper, we conducted two stage analysis - parametric and non-parametric regressions to explore the relationship between media sentiment and market returns across DJIA, SHCOMP, OXMC and WTI. In particular, linear regressions (vector autoregression) results indicate that negative sentiment generally has a pessimism impact on the market the first two days after the news release and a reversal conclusion within five days. Correspondingly, using the non-linear method (locally weighted regression), we observed that the estimates of the effect of negative sentiment on returns are broadly consistent with the parametric coefficients. Sentiment forecasts remarkably large and tenacious slumps in the returns of small stocks, indicating sentiment estimates investors' opinions. Overall, the tests here recognise return patterns consistent with the hypothesis that the sentiment estimates of words in the text corpus are valid sentiment indicators.

References

1. Black, F.: Noise. J. Finan. **41**(3), 529–543 (1986)
2. Robert, J.S.: Irrational Exuberance. Princeon University, Princeton (2015)
3. Hirshleifer, D.: Investor psychology and asset pricing. J. Finan. **56**(4), 1533–1597 (2001)
4. Klibanoff, P., Lamont, O., Wizman, T.A.: Investor reaction to salient news in closed-end country funds. J. Finan. **53**(2), 673–699 (1998)
5. Stone, P.J., Dunphy, D.C., Smith, M.S., Olgilvie, D.M., with associates: The General Inquirer: A Computer Approach to Content Analysis. MIT Press, Cambridge (1966)
6. Sims, C.A.: Macroeconomics and reality. Econometrica: J. Econometric Soc. 1–48 (1980)
7. Cleveland, W.S.: Lowess: a program for smoothing scatterplots by robust locally weighted regression. Am. Stat. **35**(1), 54–54 (1981)
8. Hardie, I., MacKenzie, D.: Assembling an economic actor: the agencement of a hedge fund. Sociol. Rev. **55**(1), 57–80 (2007)
9. MacKenzie, D.: Aftermath of a typhoon. New Sci. **220**(2943), 6–7 (2013)
10. Aitken, M.J., Harris, F.H.D.B., Ji, S.: A worldwide examination of exchange market quality: greater integrity increases market efficiency. J. Bus. Ethics **132**(1), 147–170 (2015)
11. Hansen, J.L.: The trinity of market regulation: disclosure, insider trading and market manipulation. Int. J. Discl. Gov. **1**(1), 82–96 (2003)
12. Hansen, J.L.: Mad in a hurry: the swift and promising adoption of the eu market abuse directive. Eur. Bus. L. Rev. **15**, 183 (2004)

13. Kanas, A.: Volatility spillovers between stock returns and exchange rate changes: international evidence. J. Bus. Finan. Account. **27**(3–4), 447–467 (2000)
14. Xu, J.: Modeling shanghai stock market volatility. Ann. Oper. Res. **87**, 141–152 (1999)
15. Lee, C.F., Chen, G.-M., Rui, O.M.: Stock returns and volatility on china's stock markets. J. Financ. Res. **24**(4), 523–543 (2001)
16. Wang, P.: Restructuring to repair legitimacy-a contingency perspective. Corp. Governance Int. Rev. **18**(1), 64–82 (2010)
17. Harris, Z.S.: Distributional structure. Word **10**(2–3), 146–162 (1954)
18. Newey, W.K., West, K.D.: A simple, positive semi-definite, heteroscedastic and autocorrelation consistent covariance matrix. Econometrica **55**, 703–709 (1987)
19. Balestra, P., Nerlove, M.: Pooling cross section and time series data in the estimation of a dynamic model: the demand for natural gas. Econometrica: J. Econometric Soc. 585–612 (1966)
20. Da, Z., Engelberg, J., Gao, P.: The sum of all fears investor sentiment and asset prices. Rev. Financ. Stud. **28**(1), 1–32 (2014)
21. Whaley, R.E.: The investor fear gauge. J. Portfolio Mgmt. **26**(3), 12–17 (2000)
22. Ahmad, K., Hutson, E., Kearney, C., Liu, S.: Media pessimism and stock returns: a time-varying analysis at the firm level (2014)
23. Engle, R.: Risk and Volatility. New York University, New York (2003)
24. Yu, J.: Disagreement and return predictability of stock portfolios. J. Financ. Econ. **99**(1), 162–183 (2011)
25. Baker, M., Wurgler, J.: Investor sentiment and the cross-section of stock returns. J. Finan. **61**(4), 1645–1680 (2006)
26. Kelly, S., Ahmad, K.: The impact of news media and affect in financial markets. In: Jackowski, K., Burduk, R., Walkowiak, K., Woźniak, M., Yin, H. (eds.) IDEAL 2015. LNCS, vol. 9375, pp. 535–540. Springer, Cham (2015). doi:10.1007/978-3-319-24834-9_62
27. Kelly, S., News, Sentiment, and Financial Markets: A computational system to evaluate the influence of text sentiment on financial assets. Ph.D. thesis, Trinity College Dublin (2016)
28. Tetlock, P.C.: Giving content to investor sentiment: the role of media in the stock market. J. Finan. **62**(3), 1139–1168 (2007)
29. Tetlock, P.C., Saar-Tsechansky, M., Macskassy, S.: More than words: quantifying language to measure firms' fundamentals. J. Finan. **63**(3), 1437–1467 (2008)
30. Garcia, D.: Sentiment during recessions. J. Finan. **68**(3), 1267–1300 (2013)
31. Zhao, Z., Ahmad, K.: Qualitative and quantitative sentiment proxies: interaction between markets. In: Jackowski, K., Burduk, R., Walkowiak, K., Woźniak, M., Yin, H. (eds.) IDEAL 2015. LNCS, vol. 9375, pp. 466–474. Springer, Cham (2015). doi:10.1007/978-3-319-24834-9_54
32. Zhao, Z., Ahmad, K.: A computational account of investor behaviour in chinese and us market. Int. J. Econ. Behav. Organ. **3**(6), 78–84 (2015)

Stochastic and Non-Stochastic Feature Selection

Antonio J. Tallón-Ballesteros[1]([✉]), Luís Correia[2], and Sung-Bae Cho[3]

[1] Department of Languages and Computer Systems,
University of Seville, Seville, Spain
`atallon@us.es`
[2] Department of Computer Science, University of Lisbon, Lisbon, Portugal
[3] Department of Computer Science, Yonsei University, Seoul, Korea

Abstract. Feature selection has been applied in several areas of science and engineering for a long time. This kind of pre-processing is almost mandatory in problems with huge amounts of features which requires a very high computational cost and also may be handicapped very frequently with more than two classes and lot of instances. The general taxonomy clearly divides the approaches into two groups such as filters and wrappers. This paper introduces a methodology to refine the feature subset with an additional feature selection approach. It reviews the possibilities and deepens into a new class of algorithms based on a refinement of an initial search with another method. We apply sequentially an approximate procedure and an exact procedure. The research is supported by empirical results and some guidelines are drawn as conclusions of this paper.

Keywords: Feature selection · Ant search · Refinement · Classification · Stochastic feature selection · Filters

1 Introduction

Machine learning algorithms which construct classifiers from sample data, such as rule-based systems, radial basis functions and decision trees, have received growing attention for their wide applicability [14]. A review of them can be found in [7]. The possible inputs to a classifier can be enormous for many applications. There may be some redundancy among different inputs. A large number of inputs to any kind of classifier increase the solution and thus requires more training data and longer training times in order to achieve reasonable generalization ability. Pre-processing is often needed to reduce the number of inputs to a classifier. The application of feature selection (FS) approaches has become a real initial step for model building due to the multi-dimensional nature of many modeling tasks in some fields.

Our objective is to improve the accuracy and to reduce the complexity (measured by means of the number of inputs) of the classification models. The training of databases for classification, which have different numbers of patterns, features and classes, is dealt with by means different types of supervised machine learning

© Springer International Publishing AG 2017
H. Yin et al. (Eds.): IDEAL 2017, LNCS 10585, pp. 592–598, 2017.
https://doi.org/10.1007/978-3-319-68935-7_64

approaches. In this paper we use as a baseline feature set the resulting one with an ant colony optimisation meta-heuristic [16] that is only applied in the training set. This initial solution is going to be enhanced with a no-stochastic approach. This paper is organised as follows: Sect. 2 describes some concepts about FS; Sect. 3 reviews the roots of swarm intelligence (SI) and more concretely the optimisation based on ant colonies (AC); Sect. 4 presents the description of our proposal; Sect. 5 details the experimentation process; then Sect. 6 shows and analyzes the results obtained; finally, Sect. 7 states the concluding remarks.

2   Feature Selection

FS is the problem of choosing a small subset of features that ideally is necessary and sufficient to describe the target concept [11]. There are various ways in which FS algorithms can be grouped according to the attribute evaluation measure: depending on the type (filter or wrapper technique) or on the way that features are evaluated (individual or subset evaluation). The filter model [13] relies on general characteristics of the data (such as distance, consistency, and correlation) to evaluate and select feature subsets without involving any mining algorithm. The wrapper model [12] requires a predetermined mining algorithm and uses its performance as evaluation criterion.

The subset evaluator CFS (Correlation-based Feature Selection) [8] is the most widespread. This paper works with the aforementioned ant search method within the CFS accompanied with the heuristic Information Gain (InfoGain as abbreviation, [3]) which is considered as the baseline method due to its strong performance.

3   Swarm Intelligence

Swarm intelligence refers to a kind of problem-solving ability that emerges in the interactions of simple information-processing units [10]. Social insects (ants, bees, termites, and wasps) can be viewed as powerful problem-solving systems with sophisticated collective intelligence [2].

The meta-heuristic optimization based on ants (MOBA) was proposed by Dorigo et al. [6] and is inspired in the behaviour of real ant colonies. Depending on the amount of pheromone deposited by the ant in their route there would be some points that are more likely to be visited by the next ants [4]. The (artificial) ants define a randomised construction heuristic which makes probabilistic decisions depending on the strength of artificial pheromone trails and available heuristic information. As such, MOBA can be interpreted as an extension of traditional construction heuristics, which are readily available for many combinatorial optimization problems. Yet, an important difference with construction heuristics is the adaptation of the pheromone trails during algorithm execution to take into account the cumulated search experience. As construction algorithms work on partial solutions trying to extend these in the best possible way to complete problem solutions.

4   The Proposed Method

In the current work, MOBA, implemented following the Ant System (AS) model, is considered as search strategy in the context of CFS method after the attribute evaluation phase. MOBA guides the search by means of a heuristic evaluator. As heuristic evaluator, we have considered InfoGain for CFS as evaluator resulting an hybrid approach. Moreover, CFS, and InfoGain compute different kinds of measure to evaluate the relevance, such as correlation and information, respectively.

The probabilistic transition rule is defined in the same way as in [9] and is the most widely used in AS [5]:

$$p_{ij} = \frac{\tau_{ij}^{\alpha} \cdot [\eta_{ij}]^{\beta}}{\sum_{il \in \mathcal{N}(x)} \tau_{il}^{\alpha} \cdot [\eta_{il}]^{\beta}}, \quad \forall ij \in \mathcal{N}(x) \tag{1}$$

where p_{ij} represents the probability that current ant at feature i would travel to feature j, τ_{ij} is the amount of pheromone on the ij edge, η_{ij} is the heuristic desirability of the ij transition and $\mathcal{N}(x)$ the set of current feasible components. Lastly, α and β are parameters that may take real positive values –according to the recommendations on parameter setting in [5]– and are associated with heuristic information and pheromone trails, respectively. All ants update pheromone level with an increase of small quantities, depending directly on the heuristic desirability of the ij transition given by the measure (merit) of the subset attribute evaluator used as heuristic evaluator and inversely proportional to the subset size.

First of all, some feature selectors are applied independently to the training set of all data sets in order to obtain a list of attributes, for each of them, considered for training and test phases. In this way, two reduced sets (reduced training and test sets) are generated, where only most relevant features are included. It is important to point out that the FS is performed only with training data; the reduced test set has the same features as the reduced training set. These reduced sets are taken as input to the classifier.

The baseline method, MOBA based on AS, gets an initial set of features. On the other hand, approaches considering the individual performance of every feature may have some flaws because a group containing only some potentially good individual attributes may be more effective than a group consisting of only good attributes. A newly method called Leave-k-out ReliefF has been proposed very recently [17].

This paper tries to simplify the initial subset after the baseline approach removing the worst ranked attribute according to Leave-k-out ReliefF. The current paper presents SnS-FS (Stochastic and no-Stochastic Feature Selection) methodology, a refinement of an initial stochastic feature selection task with a no-stochastic method to reduce a bit more the subset of features to be retained.

Recently, we proposed an approach to combine different subsets that are obtained independently [18]. The current proposal introduces a very different approach because one procedure is run after the other whereas in the previous reference both methods are or at least could be run in a parallel mode because there is not data dependency.

5   Experimentation

Table 1 describes the data sets employed. All of them are publicly available at the UCI repository [1]. The following five have been used: Hepatitis, Ionos., Lymph., Satimage and SPECTF. Missing values have been replaced in the case of nominal variables by the mode or, when concerning continuous variables, by the mean, taking into account the full data set. Table 2 depicts the feature subset selection methods applied in the experimental process.

Table 3 summarises the main parameters along with their symbols and numerical or conceptual values for all the feature subset selection methods used for the experiments. On one hand, in relation to F1 (MOBA based on AS), the n_a and gen parameters have been set to fix values to our choice. For τ_0 parameter, in [5] there is a suggestion to assign a small positive constant and hence we have defined a value of 0.5. The trade-off between α and β parameters may influence in the behaviour of the algorithm thus for their determination a preliminary experimental design by means of a five-fold cross validation on the training set has been carried out with a couple of values for each one parameter (1 and 2).

The experimental design uses the cross validation technique called stratified hold-out that consists of splitting the data into two sets: training and test set, maintaining the class distribution of the samples in each set approximately equal as in the original data set. Their sizes are approximately 3n/4 and n/4, where n is the number of patterns in the problem. For some problems a prearranged distribution was applied by the donors of the data and we do not alter it.

Table 4 shows the number of features with two data pre-processing approaches in the context of the framework SnS-FS. Ten runs with different seeds were conducted for F1 and the most frequent solution was incorporated for the evaluation and hence as input to F2. The computational cost for the data preparation in an Intel i7, 2.5 GHz, 8 GB RAM is around one second for the execution of F1 with one seed. Once the solution is obtained with F1, F2 takes 4 seconds for the slowest problem (Satimage) and only one for the remaining ones.

The new proposal has been assessed with classifiers C4.5 [15], SVM [19] with a polynomial kernel and PART. The values of parameters were fix with the

Table 1. Summary of the data sets used

Data set	Size	Train	Test	Feat.	Cl.
Hepatitis	155	117	38	19	2
Ionos.	351	263	88	33	2
Lymph.	148	111	37	38	4
Satimage	6435	4435	2000	36	6
SPECTF	267	80	187	44	2

Table 2. List of feature selectors for the experimentation

Attribute evaluator	Search method	Heuristic evaluator	Additional FS	Reference	Abb. Name
CFS	*AntSearch*	*InfoGain*	−	[16]	*F*1
CFS	*AntSearch*	*InfoGain*	*Leave − k − outReliefF*	[17]	*F*2

Table 3. Parameter values for F1 and F2 feature selection approaches

FS method	Parameter	Symbol	Value
F1	*Number of ants*	n_a	10
	Number of generations	gen	10
	Pheromone trail influence	α	1
	Heuristic informacion value	β	2
	Pheromone initial value	τ_0	0.5
F2	*Number of features dropped*	k	1

Table 4. Number of selected features for each feature selector

Data set	Selected features	
	F1	F2
Hepatitis	9	8
Ionos.	10	9
Lymph.	8	7
Satimage	21	20
SPECTF	8	7

default ones because this is the recommendation of the own researchers that introduced the algorithms. These methods are implemented in Weka tool [20].

6   Results

This section details the results obtained, measured in Correct Classification Ratio (CCR) or accuracy in the test subset.

We have reported in Table 5 the results with the baseline approach (F1) and the new methodology (F2) that we have coined as SnS-FS. Improvements in favour of F2 are represented in boldface. None special font-type has been used for the cases of ties or losses. From an analysis of the results, we can assert the following. Taking into account the new proposal compared to the baseline approach, as aforesaid, in all cases the number of features is lower. Besides, depending on the classifier some progress happen. PART gets more accurate

Table 5. Accuracy test results for the new proposal compared to the baseline method

Data set	C4.5		SVM		PART	
	F1	F2	F1	F2	F1	F2
Hepatitis	89.47	89.47	89.47	89.47	86.84	**92.10**
Ionos.	87.50	**88.63**	89.77	88.63	93.18	93.18
Lymph.	81.08	81.08	83.78	83.78	70.27	70.27
Satimage	86.25	85.65	84.50	84.40	83.55	**84.85**
SPECTF	69.52	69.52	63.64	**64.17**	76.47	76.47

results in two out of five problems what is very remarkable. C4.5 increases the accuracy in one problem and in three data sets a tie is get. SVM improves the result in one problem and two ties. According to Wilcoxon signed-ranks test, since there are 5 data sets, the T value at $\alpha = 0.10$ should be less or equal than 1 (the critical value) to reject the null hypothesis. It should be mentioned that for classifiers C4.5 and PART the F2 performance is significantly better than F1.

7 Conclusion

This paper presented a methodology called SnS-FS that in a first step applies a feature selection procedure based on swarm intelligence and later in a second step runs other feature selection procedure to diminish the number of features retained to be prompted to the supervised machine learning algorithm.

PART and C4.5 take advantage of the new methodology enhancing the accuracy in the best cases and keeping the same results with a lower number of attributes in a good number of problems. Losses sometimes happen but are not the rule of thumb. On the other hand, SVM is able to improve the results but in two out of the five problems worst accuracies are get what mean that SVM is robust and minor changes in the number of features may not very convenient with the reported test-bed.

Acknowledgments. This work has been partially subsidised by TIN2014-55894-C2-R project of the Spanish Inter-Ministerial Commission of Science and Technology (MICYT), FEDER funds and P11-TIC-7528 project of the "Junta de Andalucía" (Spain).

References

1. Bache, K., Lichman, M.: UCI machine learning repository (2013)
2. Bonabeau, E., Dorigo, M., Theraulaz, G.: Swarm Intelligence: From Natural to Artificial Systems, vol. 1. Oxford University Press, Oxford (1999)
3. Cover, T.M., Thomas, J.A.: Elements of Information Theory. Wiley, Hoboken (1991)

4. Dorigo, M., Di Caro, G., Gambardella, L.M.: Ant algorithms for discrete optimization. Artif. Life **5**(2), 137–172 (1999)
5. Dorigo, M., Maniezzo, V., Colorni, A.: Ant system: optimization by a colony of cooperating agents. IEEE Trans. Syst. Man Cybern. Part B Cybern. **26**(1), 29–41 (1996)
6. Dorigo, M., Stützle, T.: Ant colony optimization: overview and recent advances. In: Gendreau, M., Potvin, J.Y. (eds.) Handbook of Metaheuristics, pp. 227–263. Springer, Boston (2010)
7. Duda, R.O., Hart, P.E., Stork, D.G.: Pattern Classification. Wiley, New York (2012)
8. Hall, M.A.: Correlation-based feature selection for machine learning. Ph.D. thesis, The University of Waikato (1999)
9. Jensen, R., Shen, Q.: Finding rough set reducts with ant colony optimization. In: Proceedings of the 2003 UK Workshop on Computational Intelligence, vol. 1 (2003)
10. Kennedy, J.: Swarm intelligence. In: Zomaya, A.Y. (ed.) Handbook of Nature-inspired and Innovative, pp. 187–219. Springer, Heidelberg (2006)
11. Kira, K., Rendell, L.A.: A practical approach to feature selection. In: Proceedings of the Ninth International Conference on Machine Learning (ICML 1992), pp. 249–256, San Francisco, CA. Morgan Kaufmann (1992)
12. Kohavi, R., John, G.H.: Wrappers for feature subset selection. Artif. Intell. **97**(1–2), 273–324 (1997)
13. Liu, H., Setiono, R.: A probabilistic approach to feature selection - a filter solution. In: Proceedings of the Thirteenth International Conference on Machine Learning (ICML 1996), pp. 319–327, Italy. Morgan Kaufmann (1996)
14. Michie, D., Spiegelhalter, D.J., Taylor, C.C.: Machine learning, neural and statistical classification (1994)
15. Quinlan, J.R.: C4. 5: Programming for Machine Learning. Morgan Kauffmann, Burlington (1993)
16. Tallón-Ballesteros, A.J., Riquelme, J.C.: Tackling ant colony optimization metaheuristic as search method in feature subset selection based on correlation or consistency measures. In: Corchado, E., Lozano, J.A., Quintián, H., Yin, H. (eds.) IDEAL 2014. LNCS, vol. 8669, pp. 386–393. Springer, Cham (2014). doi:10.1007/978-3-319-10840-7_47
17. Tallón-Ballesteros, A.J., Riquelme, J.C.: Low dimensionality or same subsets as a result of feature selection: an in-depth roadmap. In: Ferrández Vicente, J.M., Álvarez-Sánchez, J.R., de la Paz López, F., Toledo Moreo, J., Adeli, H. (eds.) IWINAC 2017. LNCS, vol. 10338, pp. 531–539. Springer, Cham (2017). doi:10.1007/978-3-319-59773-7_54
18. Tallón-Ballesteros, A.J., Riquelme, J.C., Ruiz, R.: Merging subsets of attributes to improve a hybrid consistency-based filter: a case of study in product unit neural networks. Connection Sci. **28**(3), 242–257 (2016)
19. Vapnik, V.N.: The nature of Statistical Learning Theory. Springer, Heidelberg (1995)
20. Witten, I.H., Frank, E., Trigg, L.E., Hall, M.A., Holmes, G., Cunningham, S.J.: Weka: practical machine learning tools and techniques with java implementations (1999)

Understanding Matching Data Through Their Partial Components

Pablo Álvarez de Toledo[1], Fernando Núñez[1], Carlos Usabiaga[2],
and Antonio J. Tallón-Ballesteros[3]([✉])

[1] Department of Industrial Organization and Business Management I,
University of Seville, Seville, Spain
[2] Department of Economics, Quantitative Methods and Economic History,
Pablo de Olavide University, Seville, Spain
[3] Department of Languages and Computer Systems, University of Seville,
Seville, Spain
atallon@us.es

Abstract. In this paper we develop a previous work on matching data
[2], inserting their contents in the more general framework of contingency tables and dealing with the dimensions problem generated by the combination of the multiple characteristics that define each row and column category. Two concepts related to the matching process are defined: propensity to match and similarity in the matching. Both measures can be divided into partial components which allow a better understanding of the underlying structure of the data. We illustrate our methodology taking as an example a labor market where each worker category and each job category is defined by the combination of two attributes: location and occupational level.

Keywords: Contingency tables · Propensity to match · Factor decomposition · Clustering · Labor matching data

1 Introduction

In this paper, we analyze interdependences between variables on matching data, expanding a previous contribution [2]. First, we insert their contents in the more general framework of contingency tables (see, for example [1, 3, 5, 8]) in multivariate statistics. Second, we deal with the "dimensions problem" generated by the combination of the multiple characteristics that define each row and column categories and make their number soar. We propose a factor approach to disentangle the mix of the effects of the multiple interactions between the different variables.

To explain our methodology, we can imagine that we are analyzing the matching process between worker categories and job categories in the labor market. The notion of *propensity to match* between each row (worker) category and each column (job) category in two-way contingency tables plays a central part in our

© Springer International Publishing AG 2017
H. Yin et al. (Eds.): IDEAL 2017, LNCS 10585, pp. 599–606, 2017.
https://doi.org/10.1007/978-3-319-68935-7_65

paper. The second major point that this paper addresses, deals with the dimensions problem generated when we define the row and column (worker and job) categories combining categories of several variables (for example sex, age, location, occupation, etc.). If more variables are combined, the number of combined categories may be very high and the large number of rows, columns and cells in the contingency table makes it difficult to get an overall picture. In addition, the number of cells in the table can be so high that, even with a large number of observations, in many cells the observed frequency may be zero, although, actually, its probability would not be zero ("zero frequency problem"). To solve these problems we propose a factor decomposition of the multiple interactions among the different variables that are combined and, consequently, of the clusters/biclusters based on those variables. This factor decomposition of matching data (structured in terms of contingency tables) constitutes a novelty in this field and the main extension respect to our previous works[1]. Using a Cobb-Douglas (or Log-Log) functional form we estimate the total propensity to match between each row combined category and each column combined category as a multiplicative function of the partial propensities to match between row individual categories and column individual categories. In this way we disentangle the mix of the effects of the multiple interactions between the different variables. We make an analogous analysis for the similarities used in the clustering process. The similarity measure is that proposed, in a matching context, by [2]: two row (column) categories are more similar, the more they resemble in the way they match with column (row) categories. The bicluster analysis can be clarified relating the clusters and biclusters obtained with combined variables with the clusters and biclusters obtained for each variable separately.

Let X and Y be two categorical variables with the same R categories each one, that are cross-classified in a R x R square contingency table, and n_{ij} be the observed frequency in a random sample from a given population, for cell (i, j), i, $j=1$, ..., R. As usual in this methodology, the final row and column of the contingency table represent the marginal frequency distributions and n is the total sample size. Let us suppose that the table also has the following characteristics: (a) The variables considered are mainly categorical. (b) The order in which the categories are displayed is, in principle, arbitrary, although it is the same for rows and columns. (c) Categories of X (in rows) and Y (in columns) are obtained through the combination of categories of different variables. Each cell (i, j) corresponds to the match of combined categories from both sides of the table: $i = \{i_1, i_2, ..., i_k, ..., i_m\}$ in rows side, matches with $j = \{j_1, j_2, ..., j_k, ..., j_m\}$ in columns side, with an observed frequency n_{ij} and marginal totals n_{i+} and n_{+j}. The total sample size must be equal to $n = \sum_{\forall i, \forall j} n_{ij} = \sum_{\forall i} n_{i+} = \sum_{\forall j} n_{+j}$. (d) As a consequence of c), the number of cells in the table, R^2, can be so high that, even with a large number of observations, in many cells the observed frequency may be zero, although, actually, its probability is not zero ("zero frequency

---

[1] Among the general references in the voluminous clustering literature are [9] and [4]. References about biclustering are generally more specific. [6] and [7] offer an overview about the subject.

problem"). Very low observed frequencies may also be problematic in order to estimate the corresponding probabilities. (e) The large number of rows, columns and cells makes it difficult to get an overall picture (dimensions problem). (f) On the main diagonal each category in rows matches with its corresponding category in columns ($i_1 = j_1, i_2 = j_2, ..., i_k = j_k, ..., i_m = j_m$).

In addition to the observed frequencies in each cell and marginal totals, we will use the totals corresponding to each of the categories of each variable separately. That is, the total frequency of each category i_k of v_k in rows $\eta_{ik+} = \sum_{\forall i/i_k \in i} \eta_{i+}$ (η_{ik+} will be obtained by adding the marginal totals η_{i+} for all the combined categories i in which appears i_k), and the total frequency of each category j_k of v'_k in columns $\eta_{+jk} = \sum_{\forall j/j_k \in j} \eta_{+j}$ and therefore $n = \sum_{\forall i_k} \eta_{i_k} = \sum_{\forall j_k} \eta_{j_k}$. Finally, we will use also the frequency with which each category i_k of each variable v_k on the side of the rows matches with each category j_k of the corresponding variable v'_k on the side of the columns $\eta_{i_k j_k} = \sum_{\forall ij/i_k \in i \wedge j_k \in j} \eta_{ij}$; and therefore $n = \sum_{\forall i_k j_k} \eta_{i_k j_k}$. The relative frequencies give us estimates of four kinds of probabilities: the joint probability mass function (empirical probabilities) $p_{ij} = \frac{\eta_{ij}}{\eta}$; the marginal probability mass functions $p_{i+} = \frac{\eta_{i+}}{\eta}$ and $p_{+j} = \frac{\eta_{+j}}{\eta}$; the separate probability of each of the categories of each variable v_k in rows and each variable v'_k in columns, or individual probabilities $p_{i_k+} = \frac{\eta_{i_k+}}{\eta}$ and $p_{+j_k} = \frac{\eta_{+j_k}}{\eta}$; and the probability that each category of each variable on the side of the rows matches with each category of the corresponding variable on the side of the columns, or individual matching probabilities $p_{i_k j_k} = \frac{\eta_{i_k j_k}}{\eta}$.

2 Proposal: Matching and Clustering

Consider the hypothesis of independence between X and Y. In this case, the joint probability would be the product of the corresponding marginal probabilities $p_{ij} = p_{i+}p_{+j}$ and also $p_{i_k j_k} = p_{i_k+}p_{+j_k}$. Additionally, consider the hypothesis of independence between the m variables combined in rows and in columns. Then, the probability of each combination would be the product of the probabilities of their components $p_{i+} = \prod_{k=1}^{m} p_{i_k+}$ and $p_{+j} = \prod_{k=1}^{m} p_{+j_k}$. In what follows, we will denote these probabilities obtained under both hypotheses of independence, *random* probabilities (denoted as pr), because we would get them if we randomly combine and match each category of each variable, given their probabilities separately: $pr_{i+} = \prod_{k=1}^{m} p_{i_k+}$; $pr_{+j} = \prod_{k=1}^{m} p_{+j_k}$; $pr_{ij} = pr_{i+}pr_{+j}$; and $pr_{i_k j_k} = p_{i_k+}p_{+j_k}$.

We can measure the *total propensity to match* between the row and column combinations in cell (i, j) as the ratio of the probability estimated from the observed frequencies to the random probability $pm_{ij} = \frac{p_{ij}}{pr_{ij}}$. Values higher than one mean that the propensity is greater than in the random case, and vice versa. Moreover, it is straightforward that, by using pr_{ij} as weights, the weighted mean of the pm_{ij} values in all the cells is one. If we take as given the row and column combinations of categories and focus exclusively in the row-column matching, we can define alternatively the *row-column propensity to match* as $pm'_{ij} = \frac{p_{ij}}{p_{i+}p_{+j}}$.

Due to the dimensions problem exposed in Sect. 1, the analysis of the total propensity to match in each cell may require the consideration of a very large number of cases. Furthermore, in each case there is a mix of effects from the multiple interactions between the different variables on both sides. We will quantify separately these effects using a much smaller number of *partial propensities* to match or combine: propensity to match between each category of each variable on the side of the rows, and each category of the corresponding variable on the side of the columns[2], or *partial propensities to match* $pm_{i_k j_k} = \frac{p_{i_k j_k}}{pr_{i_k j_k}}$; *propensity* of the categories of the different variables on the side of the rows *to combine* between them $pm_{i+} = \frac{p_{i+}}{pr_{i+}}$; and propensity of the categories of the different variables on the side of the columns to combine between them $pm_{+j} = \frac{p_{+j}}{pr_{+j}}$.

From the above equations, we can relate the observed frequencies with the propensities to match or combine, the individual probabilities and the total sample size

$$n_{ij} = np_{ij} = npr_{ij}pm_{ij} = n(\prod_{k=1}^{m} p_{i_k+})(\prod_{k=1}^{m} p_{+j_k})pm_{ij} \tag{1}$$

$$n_{i+} = np_{i+} = npr_{i+}pm_{i+} = n(\prod_{k=1}^{m} p_{i_k+})pm_{i+} \tag{2}$$

$$n_{+j} = np_{+j} = npr_{+j}pm_{+j} = n(\prod_{k=1}^{m} p_{+j_k})pm_{+j} \tag{3}$$

Also, we obtain the relation between the total propensity to match, the row-column propensity to match and the propensities of the categories of the different variables to combine between them on the side of the rows and on the side of the columns: $pm_{ij} = pm'_{ij}pm_{i+}pm_{+j}$. As can be seen, the total propensity to match includes the effects of the other three propensities.

The set of total propensities in each cell which corresponds to a given set of partial propensities is not univocally determined, but, on the contrary, there are infinite solutions. Bearing this in mind, for a given sample data, we can try to specify a regression equation that explains as much as possible the total propensities in terms of the partial propensities. From Eq. 1, we can regress the equation

$$n_{ij} = n(\prod_{k=1}^{m} p_{i_k+})(\prod_{k=1}^{m} p_{+j_k})\widehat{pm}_{ij} + \epsilon \tag{4}$$

where $\widehat{pm}_{ij}$ is the estimated value of pm_{ij}: $\widehat{pm}_{ij} = (\prod_{k=1}^{m} pm_{i_k j_k}^{\alpha_k})pm_{i+}^{\beta}pm_{+j}^{\gamma}$; $\widehat{n}_{ij}$ is the estimated value of n_{ij}: $\widehat{n}_{ij} = n(\prod_{k=1}^{m} p_{i_k+})(\prod_{k=1}^{m} p_{+j_k})\widehat{pm}_{ij}$; α_k, β, and

---

[2] We have not considered the propensity to match of each category of each variable on the side of the rows, with each category of other variables on the side of the columns because the analysis would be more complex and the methodological gain marginal.

γ are parameters to be estimated; and ϵ is an error term. The use of the Cobb-Douglas functional form for $\widehat{pm}_{ij}$ reflects the idea that the effects of the partial propensities multiply each other (rather than add to each other).

The equations of $\widehat{pm}_{ij}$ and $\widehat{n}_{ij}$ can be completed with a residual factor rf, calculated as the ratio of the observed value to the estimated value. From Eq. 1 and the above equation on $\widehat{n}_{ij}$: $rf = \frac{n_{ij}}{\widehat{n}_{ij}} = \frac{pm_{ij}}{\widehat{pm}_{ij}}$, and then

$$pm_{ij} = \widehat{pm}_{ij} rf = (\prod_{k=1}^{m} pm_{i_k j_k}^{\alpha_k}) pm_{i+}^{\beta} pm_{+j}^{\gamma} rf \tag{5}$$

$$n_{ij} = \widehat{n}_{ij} rf = n(\prod_{k=1}^{m} p_{i_k+})(\prod_{k=1}^{m} p_{+j_k}) \widehat{pm}_{ij} rf \tag{6}$$

From 5 and the equality $pm_{ij} = pm'_{ij} pm_{i+} pm_{+j}$, we also obtain the factor decomposition of the row-column propensity to match

$$pm'_{ij} = (\prod_{k=1}^{m} pm_{i_k j_k}^{\alpha_k}) pm_{i+}^{\beta-1} pm_{+j}^{\gamma-1} rf \tag{7}$$

The factorial Eq. 5 allows us to quantify separately the contributions of each of the partial propensities to the total propensity to match in each cell, as well as the contribution of the residual factor (remember that values higher than one mean that propensities are greater than in the random case, and vice versa). The residual factor measures the part of the total propensity not explained by the partial propensities. The residual factor increases with lower values of observed frequencies, which makes the estimation of Eq. 5 less accurate, but it may reflect also more complex interactions than the corresponding to the partial propensities. Furthermore, each of the factors can be analyzed separately, examining which cases correspond to the highest and lowest values of the partial propensities (for instance, which are the locations of workers and jobs with greater propensity to match). Finally, when, as we have seen in Sect. 1, empirical probabilities are not reliable because they are based on zero or very low observed frequencies, we can use alternatively the estimated frequencies $\widehat{n}_{ij}$.

The clustering methodology is usually based on a previously defined similarity (or dissimilarity) measure between the elements that are clustered. In a matching context, we consider that row (column) categories are more similar, the more they resemble in the way they match with column (row) categories. For instance, in our scenario of the labor market, we consider that two categories of workers are more similar, the more they resemble in the way they match with the different categories of jobs and vice versa. Then, we measure the similarity between each pair of categories $i_A - i_B$ (rows of the contingency table) as the overlapping or percentage of coincidence of their row profiles (distribution of their conditional probabilities p_{ij}/p_{i+} of matching with each of the different column categories j): $sim_{i_A-i_B} = \sum_j min(\frac{p_{i_A j}}{p_{i_A +}}, \frac{p_{i_B j}}{p_{i_B +}})$; or also, taking into account that $pm'_{ij} = \frac{p_{ij}}{p_{i+}p_{+j}}$: $sim_{i_A-i_B} = \sum_j p_{+j} min(pm'_{i_A j}, pm'_{i_B j})$.

Its value will be between one (if the row profiles are identical) and zero (if their intersection is null). This measure of similarity can be related to the Manhattan or City Block distance metric: $dist_{i_A - i_B} = \sum_j \left| \frac{p_{i_A j}}{p_{i_A +}} - \frac{p_{i_B j}}{p_{i_B +}} \right|$ with $sim_{i_A - i_B} = 1 - 0.5 dist_{i_A - i_B}$. We can measure the similarity between each pair of categories $j_A - j_B$ (columns of the contingency table) in an analogous way: $sim_{j_A - j_B} = \sum_i min(\frac{p_{ij_A}}{p_{+j_A}}, \frac{p_{ij_B}}{p_{+j_B}}) = \sum_i p_{i+} min(pm'_{ij_A}, pm'_{ij_B})$.

Again, due to the dimensions problem, the detailed analysis of similarities between each pair of rows or columns may require the consideration of a very large number of cases and, in each case, there is a mix of effects from the multiple interactions between the different variables on both sides. We will quantify separately these effects using a much smaller number of *partial similarities* for each of the m variables ($k = 1, 2, ..., m$), namely: (a) partial similarities between each pair of categories $i_{kA} - i_{kB}$ of each variable v_k (in the rows of the contingency table) measured as the percentage of coincidence of their row profiles (considering only the variable v_k): $sim_{i_{kA} - i_{kB}} = \sum_{jk} min(\frac{p_{i_{kA} jk}}{p_{i_{kA} +}}, \frac{p_{i_{kB} jk}}{p_{i_{kB} +}}) = \sum_{jk} p_{+jk} min(pm_{i_{kA} jk}, pm_{i_{kB} jk})$. (b) partial similarities between each pair of categories $j_{kA} - j_{kB}$ of each variable v_k (in the columns of the contingency table) measured as the percentage of coincidence of their column profiles (considering only the variable v_k):

$$sim_{j_{kA} - j_{kB}} = \sum_{ik} min(\frac{p_{i_k j_{kA}}}{p_{+j_{kA}}}, \frac{p_{i_k j_{kB}}}{p_{+j_{kB}}}) = \sum_{ik} p_{i_k +} min(pm_{i_k j_{kA}}, pm_{i_k j_{kB}}).$$

The values of partial similarities are also between one (if the row profiles are identical) and zero (if their intersection is null). They will be high if the partial propensities to match in the equation of $sim_{i_{kA} - i_{kB}}$, or in that of $sim_{j_{kA} - j_{kB}}$, are similar for each of the r_k categories of j_k and i_k, respectively. As with the propensities, the set of similarities $sim_{i_A - i_B}$ (or $sim_{j_A - j_B}$) which corresponds to a given set of partial similarities $sim_{i_{kA} - i_{kB}}$ (or $sim_{j_{kA} - j_{kB}}$), is not univocally determined, but, on the contrary, there are infinite solutions. Using some of the equations described above, we could relate them but, however, this relation is too complicated. Instead, just as we make with propensities, for a given sample data, we could try to specify a regression equation that explains as much as possible the similarities $sim_{i_A - i_B}$ in terms of the partial similarities $sim_{i_{kA} - i_{kB}}$ (or the similarities $sim_{j_A - j_B}$ in terms of the partial similarities $sim_{j_{kA} - j_{kB}}$). Assuming a Cobb-Douglas functional form: $sim_{i_A - i_B} = \prod_{k=1}^m sim_{i_{kA} - i_{kB}}^{\alpha_k} + \epsilon$ and $sim_{j_A - j_B} = \prod_{k=1}^m sim_{j_{kA} - j_{kB}}^{\alpha_k} + \epsilon$; where α_k are parameters to be estimated and ϵ is an error term (the two equations have the same functional form but they would be estimated with different data, so the estimates of α_k would be different). As [2] shows, in a square contingency table, with corresponding categories in rows and columns, the similarities $sim_{i_A - i_B}$ between each pair of categories $i_A - i_B$ (in rows) are highly correlated (positively) with the similarities $sim_{j_A - j_B}$ between the corresponding pair of categories $j_A - j_B$ (in columns). The same applies for $sim_{i_{kA} - i_{kB}}$ and $sim_{j_{kA} - j_{kB}}$. In this case, it may be convenient to use the same measure of similarity in the clustering process of rows and columns in order to obtain the same grouping on both sides, so we will use the arithmetic means of the above expressions for rows and columns and for the Cobb-Douglas decomposition. Based on those former similarity measures, we use a hierarchical method of clustering which allows getting the corresponding contingency tables with a decreasing number of clusters (rows and columns) and biclusters (combinations of a row cluster and a column cluster). Also, the order in which the categories are displayed (order that was, in principle, arbitrary) is changed so that the categories with greater similarity are closer. Using the same measure of similarity in the clustering process of rows and columns, we get the same grouping and ordering on both sides and we obtain a main diagonal of biclusters with (usually) high propensity to match -for example, in our scenario of labor matching, groups of categories location-occupation of workers and jobs with high similarity and propensity to match between them. It is also interesting the analysis of high propensities to match out of the main diagonal possibly corresponding to isolated specific cases (for instance a particular combination location-occupation of workers with high propensity to match with a particular combination location-occupation of jobs). Instead of the propensity to match for each of the cells of the original contingency table, now we can calculate the row-column propensity to match for each bicluster.

As an illustrative example, which is not explained due to lack of space, we represent in Fig. 1 the contingency table of the matching process in the Andalusian labor market expressed as a "matching map", showing in each cell or bicluster the corresponding propensity to match. We can use different levels of "zoom" depending on whether you want to have an overview of the whole map or a more detailed view of a particular area. Once more, the detailed analysis of clusters and biclusters may require the consideration of a very large number of

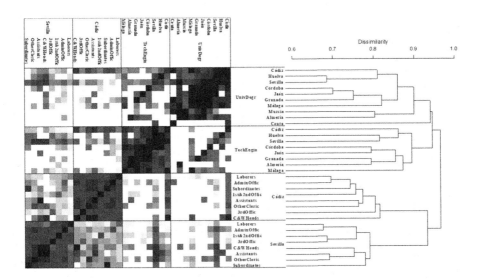

Fig. 1. Contingency table, represented as a matching map (A darker cell represents a higher propensity)

cases and, in each case, there is a mix of effects from the multiple interactions between the different variables. The analysis can be clarified linking it to the clusters obtained for each variable separately. Given the correspondence between clusters and similarities, and the relation between the similarities of combined categories of different variables and the partial similarities of each variable, an analogous relation can be obtained between the clusters for combined variables and for each variable separately.

3   Conclusions

We develop the concept of propensity to match between each row category (worker category in our example) and each column category (job category in our example) in two-way contingency tables. Using a Cobb-Douglas functional form we estimate the total propensity to match between each row combined category and each column combined category as a multiplicative function of the partial propensities to match between row individual categories and column individual categories. In this way, we disentangle the mix of the effects from the multiple interactions between the different variables.

We apply the biclustering procedures to our contingency table using a measure of similarity that can also be analyzed from its partial components. Based on this similarity measure, we can carry out a clustering process which sorts the categories (in rows and columns), and groups them in a lower number of clusters and the cells in a lower number of biclusters. Biclustering contributes to solve the dimensions problem reducing the number of rows, columns and cells in the contingency table and increasing cell frequencies. Finally, we represent the contingency table as a "matching map" which admits different "zoom" levels.

References

1. Agresti, A.: Categorical Data Analysis. Probability and Statistics. Wiley, Somerset (2013)
2. Álvarez de Toledo, P., Núñez, F., Usabiaga, C.: An empirical approach on labour segmentation. Applications with individual duration data. Econ. Model. **36**, 252–267 (2014)
3. Bishop, Y.M.M., Fienberg, S.E., Holland, P.W.: Discrete Multivariate Analysis: Theory and Practice. MIT Press, Cambridge (1975)
4. Everitt, B.S., Landau, S., Leese, M., Stahl, D.: Cluster Analysis. Probability and Statistics. Wiley, Chichester (2011)
5. Fienberg, S.E., Rinaldo, A.: Three centuries of categorical data analysis: log-linear models and maximum likelihood estimation. J. Stat. Plann. Infer. **137**(11), 3430–3445 (2007)
6. Govaert, G., Nadif, M.: Co-clustering. Wiley, New York (2013)
7. Padilha, V.A., Campello, R.J.G.B.: A systematic comparative evaluation of biclustering techniques. BMC Bioinform. **18**(1), 55 (2017)
8. Stigler, S.: The missing early history of contingency tables. Annales de la Faculté des Sciences de Toulouse. **11**(4), 563–573 (2002)
9. Xu, R., Wunsch, D.: Survey of clustering algorithms. IEEE Trans. Neural Netw. **16**(3), 645–678 (2005)

Author Index

Printed in the United States
By Bookmasters